Dmitry Yu. Murzin
Engineering Catalysis

Also of interest

Chemical Product Technology
Murzin, 2018
ISBN 978-3-11-047531-9, e-ISBN 978-3-11-047552-4

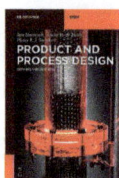

Product and Process Design
Driving Innovation
Harmsen, de Haan, Swinkels 2018
ISBN 978-3-11-046772-7, e-ISBN 978-3-11-046774-1

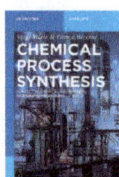

Chemical Process Synthesis.
Connecting Chemical with Systems Engineering Procedures
Bezerra, 2018
ISBN 978-3-11-046825-0, e-ISBN 978-3-11-046826-7

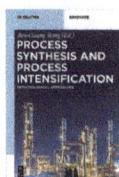

Process Synthesis and Process Intensification.
Methodological Approaches
Rong (Ed.), 2017
ISBN 978-3-11-046505-1, e-ISBN 978-3-11-046506-8

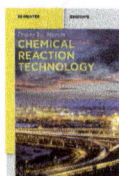

Chemical Reaction Technology
Murzin, 2015
ISBN 978-3-11-033643-6, e-ISBN 978-3-11-033644-3

Dmitry Yu. Murzin

Engineering Catalysis

2nd Edition

DE GRUYTER

Authors
Prof. Dmitry Yu. Murzin
Åbo Akademi University
Process Chemistry Centre
Biskopsgatan 8
20500 Turku/Åbo
Finland
dmurzin@abo.fi

ISBN 978-3-11-061442-8
e-ISBN (PDF) 978-3-11-061443-5
e-ISBN (EPUB) 978-3-11-061469-5

Library of Congress Control Number: 2019955320

Bibliographic information published by the Deutsche Nationalbibliothek
The Deutsche Nationalbibliothek lists this publication in the Deutsche Nationalbibliografie;
detailed bibliographic data are available on the Internet at http://dnb.dnb.de.

© 2020 Walter de Gruyter GmbH, Berlin/Boston
Cover image: zorazhuang / E+ / Getty Images
Typesetting: Integra Software Services Pvt. Ltd.
Printing and binding: CPI books GmbH, Leck

www.degruyter.com

To the memory of Elena Murzina

Preface to the first edition

> If you do not stay the road will lead you
> Andrei Platonov

I wrote this book – devoted to the engineering of heterogeneous catalysts and reactions – in the reverse order of chapters. The last chapter on the technology of industrial catalytic reactions was written first, with the intention that the needs of technological implementations will define the reactions and reactors, the reactors setting the requirements for catalytic materials, their preparation and characterization, and so on. It is up to the readers to decide if this approach was successful. In any case, the book is based on many literature sources, reflecting my experience as a researcher, chemical engineer, manager and educator, first in a governmental research institute, then in industry and finally in academia, while always maintaining very close ties to industrial catalysis and catalytic engineering.

https://doi.org/10.1515/9783110614435-202

Preface to the second edition

Apparently, this book devoted to the engineering aspects of heterogeneous catalysis found its readership, as the publisher suggested to prepare the second edition. One of the reasons was that the author, with experience in both industry and academia, tried to combine both perspectives on catalysis, which are sometimes antagonistic as illustrated in a somewhat provocative slide

Industrial perspective: should work in practice	Academic perspective: publish in high-impact factor journals
Real feedstock	Model feedstock
Continuous reactors	Often batch reactors
Simple and reliable catalysts	Fancy catalysts
Detailed mass balance and adequate anlaytics	Do not spent too much time on analytics
Deactivation: an issue to consider	Deactivation better not to discuss
Kinetic and reactor modeling	Kinetic and reactor modeling: What is this?

presented by the author at the International Congress on Catalysis in Beijing, 2016. The first edition of the book was corrected and expanded in scope with more coverage of industrial preparation of catalysts and their technological implementations.

https://doi.org/10.1515/9783110614435-203

About the author

Dmitry Yu. Murzin studied Chemical Technology at the Mendeleev University of Chemical Technology in Moscow, Russia (1980–1986) and graduated with honors. He obtained his PhD (advisor Prof. M.I. Temkin) and DrSc degrees at Karpov Institute of Physical Chemistry, Moscow in 1989 and 1999, respectively. He worked at Universite Louis Pasteur, Strasbourg, France and Åbo Akademi University, Turku, Finland as a post-doc (1992–1994). In 1995–2000 he was associated with BASF, being involved in research, technical marketing and management. Since 2000 Prof. Murzin holds the Chair of Chemical Technology at Åbo Akademi University. He serves on the editorial boards of several journals in catalysis and chemical engineering field. He is an elected member of Academia Europaea and the Finnish Academy of Science and Letters.

Prof. Murzin is the coauthor with Prof. T. Salmi of a monograph *Catalytic Kinetics* (Elsevier, 2005, second edition in 2016), and author of *Chemical Reaction Technology* (de Gruyter, 2015) and *Chemical Product Technology* (de Gruyter, 2018). He holds several patents and is an author or coauthor of approximately 750 journal articles and book chapters.

https://doi.org/10.1515/9783110614435-204

Contents

1 The basics

1.1 Catalytic concepts

1.1.1 Definitions

Catalysis is a phenomenon related to the acceleration of rates of chemical reactions.

Such acceleration was recognized long time ago and put into practice even before the concept of catalysts was formulated. For example, a German chemist Döbereiner discovered that oxidation of hydrogen can occur in the presence of Pt already in 1823. Such invention resulted in substantial utilization of Döbereiner lamps just within few years after the discovery.

The available data on the conversion of starch to sugars in the presence of acids, combustion of hydrogen over platinum, decomposition of hydrogen peroxide in alkaline and water solutions in the presence of metals, were summarized by a Swedish scientist J. J. Berzelius in 1836 [1], who proposed the existence of a certain body, which "effecting the (chemical) changes does not take part in the reaction and remains unaltered through the reaction".

Catalytic power or force, according to Berzelius, meant that "substances (catalysts) are able to awaken affinities which are asleep at this temperature by their mere presence and not by their own affinity".

This concept was immediately criticized by Liebig, as this theory was placing catalysis somewhat outside of other chemical disciplines [2]. A catalyst was later defined by Ostwald as "a compound that increases the rate of a chemical reaction, but which is not consumed by the reaction" [3]. This definition allows for the possibility that small amounts of the catalyst are lost in the reaction or that the catalytic activity slowly declines.

From these definitions, a direct link between chemical kinetics and catalysis is apparently clear as, accordingly, catalysis is a kinetic process. There are, however, many issues in catalysis that are not directly related to kinetics, such as mechanisms of catalytic reactions, elementary reactions, surface reactivity, adsorption of reactants on solid surfaces, synthesis and structure of solid materials and catalytic engineering.

An important issue in catalysis is selectivity towards a particular reaction. For example, transformation of synthesis gas (a mixture of CO and hydrogen) can lead either to methanol (on copper) or high alkanes (on cobalt). For consecutive reactions it could be desirable to obtain an intermediate product. In the oxidation of ethylene such target is ethylene oxide, but not CO_2 and water.

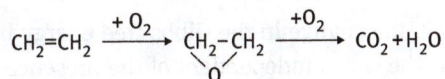

$$CH_2=CH_2 \xrightarrow{+\ O_2} CH_2-CH_2 \xrightarrow{+O_2} CO_2 + H_2O$$

https://doi.org/10.1515/9783110614435-001

Although formaly catalysts remained unchanged during a reaction, they are involved in chemical bonding with the reactants during the catalytic process in a cyclic process: the reactants are bound to one form of the catalyst, and the products are released from another, regenerating the initial state (Fig. 1.1).

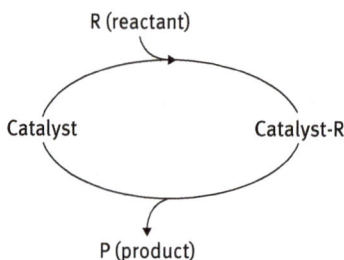

Fig. 1.1: Catalytic cycle.

It should be also noted that although it is usually assumed that a catalyst participates in the process but remains unchanged at the end, there could be major changes in its structure and composition.

Potential energy diagrams for the catalytic and non-catalytic reactions presented in Fig. 1.2 indicate that in both cases the reactions should overcome a certain barrier, which is lower in the presence of a catalyst.

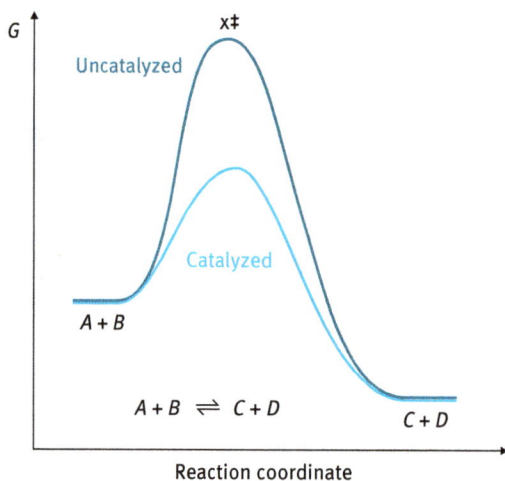

Fig. 1.2: Potential energy diagram for catalytic and non-catalytic reactions. Changes of Gibbs energy along the reaction coordinate.

As follows from Fig. 1.2, the change in the Gibbs free energy between the reactants and the products ΔG is the same, independent of the presence or absence of a catalyst, but providing, however, an alternative reaction path.

A lower value of activation energy implies higher reaction rates, which could be expressed through the rate constant k dependence on temperature:

$$k = AT^m e^{\frac{-E_a}{RT}} \qquad (1.1)$$

where A is the pre-exponential factor, E_a is the activation energy related to the potential energy barrier and m is a constant. Equation (1.1) was proposed by Kooij and van't Hoff [4] to explain the temperature dependence of reaction rates and could be derived from the transition state theory of Eyring and Polanyi [5]. Arrhenius [6] applied a slightly simplified form:

$$k = k_o e^{\frac{-E_a}{RT}} \qquad (1.2)$$

A decrease in activation energy and an increase in temperature lead to an increase of the rate constant and thus of the reaction rate. Catalysts substantially accelerate chemical reactions, enabling them to be carried out under the most favorable thermodynamic regime, and at much lower temperatures and pressures.

Equation (1.2) corresponds to an elementary reaction. In the case of complex reactions, it is appropriate to discuss the so-called apparent activation energy, which can be expressed through the reaction rate r according to

$$E_{act,\ apparent} = -R \frac{\partial \ln r}{\partial 1/T} = RT^2 \frac{\partial \ln r}{\partial T} \qquad (1.3)$$

Figure 1.3 also shows that if a catalyst is active in enhancing the rate of the forward reaction it will do the same with a reverse reaction. As mentioned above, the catalyst

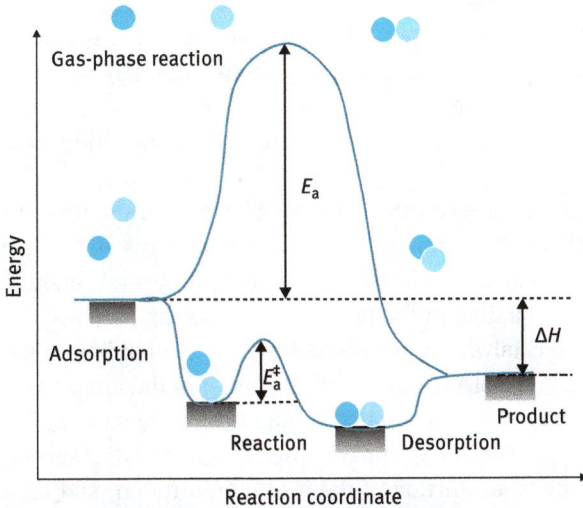

Fig. 1.3: Potential energy diagram with catalyst binding.

affects only the rate of the reaction, but not the thermodynamics (for example, the Gibbs energy of the reaction is the same), the approach to equilibrium or the equilibrium composition. Thus, in the case of a thermodynamically unfavorable process, there is no hope of finding an active catalyst that will beat thermodynamics.

Instead, as a first step the reaction conditions (temperature, pressure and reactant composition) must be optimized to maximize the equilibrium concentration of the desired product. Once suitable reaction conditions are identified the catalyst screening is performed with the aim of finding a suitably active and selective material.

The statement – that thermodynamics frequently limits the concentration of a desired product and that a catalyst search will not help – sounds very straightforward and obvious. The author knows, however, a case where modern methods of high throughput experimentation were used in industry to find an active catalyst for a reaction with a maximum conversion allowed by thermodynamics of only 1%.

Several years ago I started to be involved in a project related to the synthesis of dimethyl carbonate from methanol and CO_2. The literature was full of papers describing various catalytic systems and the mechanisms behind their catalytic action. Unfortunately, explicit data on conversions and thermodynamics were absent from the literature. The PhD student doing the experiments was initially very frustrated when, after a couple of months of catalyst screening, very low conversions in the range of 0.1% were obtained. After repeated advises finally thermodynamics was calculated and it became clear that the reaction is heavily thermodynamically limited. This switched the strategy from catalyst screening to finding a means of shifting reaction equilibrium, in that case by in situ water removal.

Figure 1.2 represents a simplified situation, not taking into account the binding of the reactant to the catalyst (called adsorption in heterogeneous catalysis), which is illustrated in Fig. 1.3. The energy barrier between the catalyst-substrate and the transition state should be lower than between the substrate and the transition state in an uncatalyzed reaction (Fig. 1.3).

If the energy is lowered too much when the substrate is bound to the reactant during adsorption, and the activation energy in the catalytic reaction is still rather high compared to non-catalytic reactions, then the reaction rate is slow and catalysis is not effective. In addition, if the binding of the products is too strong they will not be desorbed (released) from the catalyst. At another extreme, if the binding is too weak then the catalytic cycle could not proceed as effectively. It is thus intuitively clear that bonding between a catalyst and a reactant should not be too strong. This principle was first put forward by Sabatier [7], who proposed that a catalytic reaction has an optimum (maximum) rate as a function of the heat of adsorption, and then further developed by Balandin [8], who introduced volcano plots, relating catalytic activity with adsorption energy (Fig. 1.4).

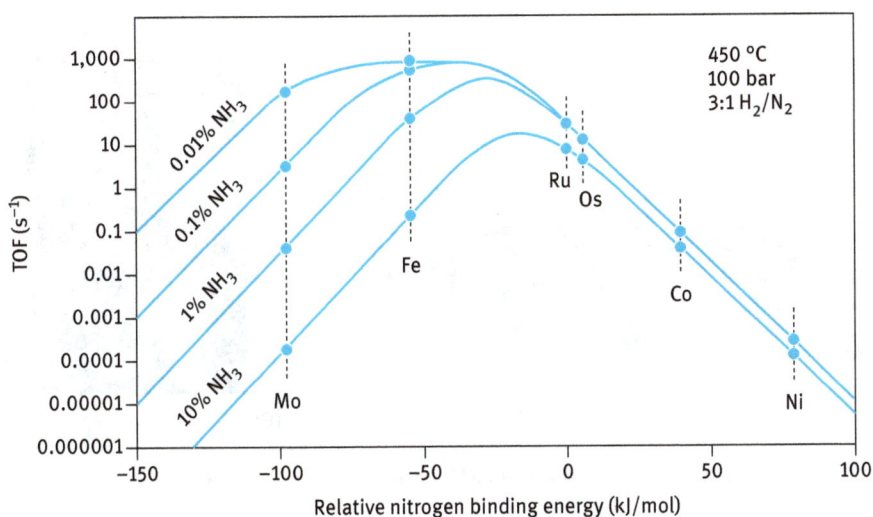

Fig. 1.4: Activity (turnover frequency) in ammonia synthesis with different catalysts as a function of the relative nitrogen binding energy, 450 °C, 100 bar, H_2/N_2 =3. (Reproduced with permission from [9]).

Sabatier is known for his experimental work on hydrogenation of organic compounds, for which he was awarded the Nobel Prize in 1912. Together with Senderen, he discovered that traces of nickel catalyze methanation of CO_2:

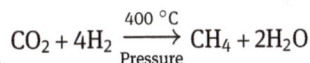

$$CO_2 + 4H_2 \xrightarrow[\text{Pressure}]{400\ °C} CH_4 + 2H_2O$$

This reaction is witnessing now a renaissance associated with a strong desire to diminish CO_2 emissions. This reaction has also a special application, as it can be applied to produce water from exhaled CO_2 in confined spaces (e.g., submarines or the International Space Station). An important issue related to the latter applications is efficient and rather small-scale generation of hydrogen.

1.1.2 Length and time scales in catalysis

Heterogeneous catalysis covers a wide scale, from a molecular scale (nm) of an active site (Fig. 1.5A) to a catalytic reactor (m) scale (Fig. 1.5B). Catalysis occurs on the surface of solid materials, representing the chemistry in two dimensions. Active sites of heterogeneous catalysts are related to a molecular or atomic level arrangement of atoms, responsible for catalytic properties (Fig. 1.5A). At this sub-nanometer level, chemical reactions proceed that involve the rupture and

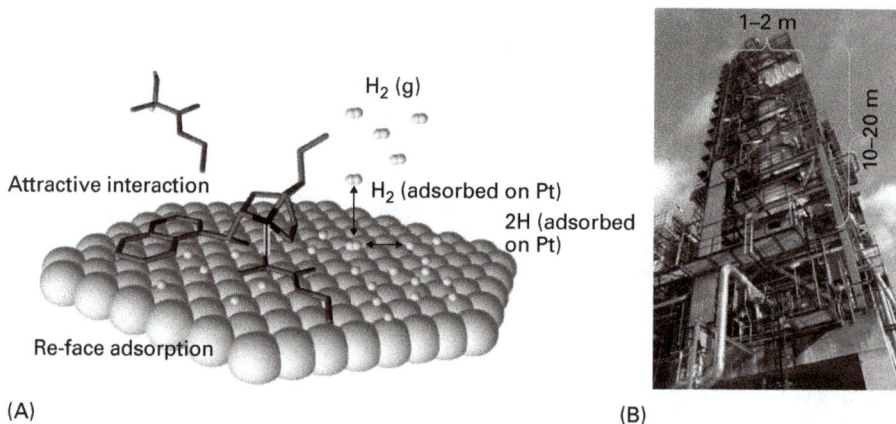

Fig. 1.5: Length scales in catalysis ranging from (A) an active site to (B) a reactor.

formation of chemical bonds. The reactivity of solids and mechanisms of catalytic reactions will be briefly reviewed here in Chapter 1.

Catalytically active particles are typically between 1 and 10 nm in size (Fig. 1.6A) and are located inside the pores of a support material (μm level, Fig. 1.6B). Chemical composition, texture of materials and pore structure are very important issues, related not only to catalysis, but also to the transport of molecules through the pores to the active sites. Of particular interest are the size, shape, structure and composition of the active particles.

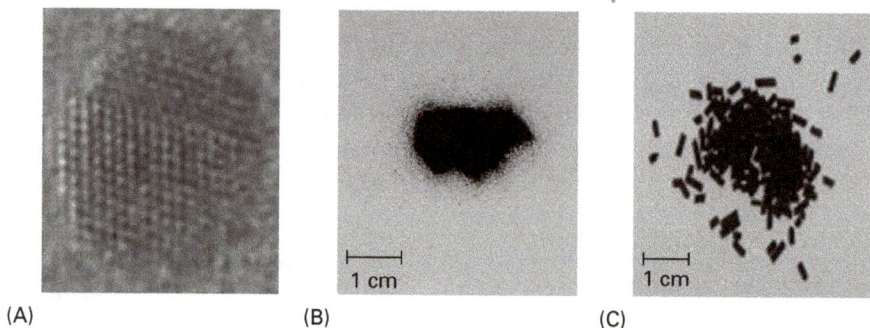

Fig. 1.6: Different levels in catalysis. (A) active site level (subnanometer scale), (B) catalyst particle powder (μm) level, (C) shaped catalyst (mm) level. Photos courtesy of Dr. E. Toukoniitty.

Catalysts in the form of powders can be used in industrial processes only in limited cases. Shaped catalysts, in the form of extrudates, pellets and tablets on a mm scale (Fig. 1.6C), are introduced into industrial reactors. Engineering of such materials requires addressing porosity, mechanical strength, and attrition resistance in addition to activity, kinetics and mass transfer.

The engineering of catalytic materials through tailor-made preparation and the understanding of kinetic and transport phenomenon will be discussed in Chapters 2 and 3. Chapter 3 also covers the modeling of catalytic reactors.

It should be noted that in industrial reality the mass and heat transport through the catalyst bed can be as important as intrinsic kinetics.

Finally, the performance of a catalyst should be viewed as part of an overall process that includes feed treatment, separation, etc., in addition to chemical reactions. For example, issues such as catalyst mechanical stability, sensitivity to trace impurities in the reactant feed (as industrial feeds are rarely pure) and leaching of the active phase can be very crucial to industrial operation. Basic principles of designing catalytic processes (plant level, Fig. 1.5B) will be described in Chapter 4, with examples for several catalytic processes, representing different type of materials (supported metals, oxides, zeolites), catalyst shapes (powder, extrudates, tablets, irregular grains) and different type of reactors (with fixed and moving catalysts).

The time scale of catalytic events varies from the picosecond level of bond breaking and formation to the seconds/minutes level for completion of a catalytic cycle. Pore diffusion may take seconds and minutes, while the residence times of molecules in reactors may vary from seconds (fluidized catalytic cracking) upwards (coke can stay for months or years on the catalyst before eventual regeneration or disposal).

1.1.3 Catalytic trinity: activity, selectivity, stability

The main requirements of a catalyst for an industrial process depend on the trinity of catalysis: activity, selectivity and stability. Activity refers to the ability to conduct a process within a reasonable contact time, which influences the reactor dimensions and process capacity. Insufficient activity in principle could be compensated by higher catalyst amounts or some other means, such as higher temperature. Catalyst selectivity is probably the most important characteristics of a catalyst, which should be also sufficiently stable under operation conditions.

The reaction rate is calculated as the converted amount of the substance with time relative to the amount of catalyst (mass or volume), specific surface area or to the amount of active sites. In cases when the rate is defined per catalyst mass, the unit of measurement is mol $g^{-1} h^{-1}$.

The rate per amount of sites available for the reaction (exposed sites) is denoted in heterogeneous catalysis as the turnover frequency (TOF) [or turnover number (TON)], which is defined as the number of molecules reacted per site per unit time:

$$\text{TOF} = \text{Amount of reacted substrate}/(\text{active site x time}) = \text{mol mol}^{-1}\text{s}^{-1} = \text{s}^{-1} \quad (1.3)$$

Even if the TOF concept is widely applied, its application is not straightforward, because TOF depends on kinetics, i.e., the concentration of reagents and temperature, and thus strictly speaking should be referred to particular conditions. Typical values of TOF for industrially relevant reactions are in the range $10^{-2} \div 10^2 \text{ s}^{-1}$.

Exercise 1.1: Fig. E1.1 displays a fixed-bed reactor where there is a certain molar flow in the reactor $\dot{n}_{A_{in}}$ and out of the reactor $\dot{n}_{A_{out}}$. In such reactors, during catalyst screening conversions at constant space velocity [volume flow rate (m^3 s^{-1}) relative to catalyst (kg)] are compared.

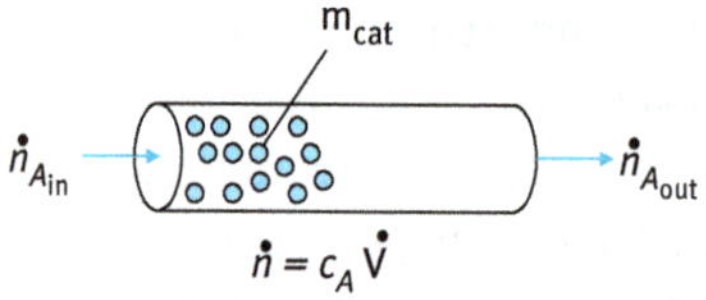

$$\dot{n} = c_A \dot{V}$$

Fig. E1.1. Fixed-bed reactor.

Calculate:

1. The conversion rates per catalyst volume
2. Weight
3. Surface
4. TOF (metal density 1,015 sites cm^{-2})

for a continuous reaction of benzene hydrogenation over 5 cm^3 5 wt% Pd on a support with the catalyst density of 0.9 g cm^{-3} and the catalyst surface area of 250 m^3 g^{-1} (5% of which is covered by Pd) when the molar flow of benzene in the reactor is 4 mol h^{-1} and the outlet flow is 3.7 mol g^{-1}.

Hints: The rates are defined in the following way:

- per volume $r_A = -dn_A/(V_{cat}dt) = -d\dot{n}_A/V_{cat}$
- per weight $r_A = -dn_A/(m_{cat}dt) = -dn_A/V_{cat}\,p_{cat}dt)$
- per surface $r_A = -dn_A/(S_{cat}dt) = -dn_A/(S_{rel}\,m_{cat}dt)$
- per atom of Pd $r_A = -dn_A/(n_{pd}dt)$

The number of mols of Pd should be calculated using the following strategy. From the overall surface, the surface of Pd atoms could be computed. This in turn gives the number of atoms of Pd if the surface density is known. Finally the number of moles of the metal can be easily calculated.

Exercise 1.2 Lilja et al. [10] investigated esterification of propanoic acid with methanol over a hetero-geneous fiber catalyst. The temperature dependence is presented in Fig. E1.2.

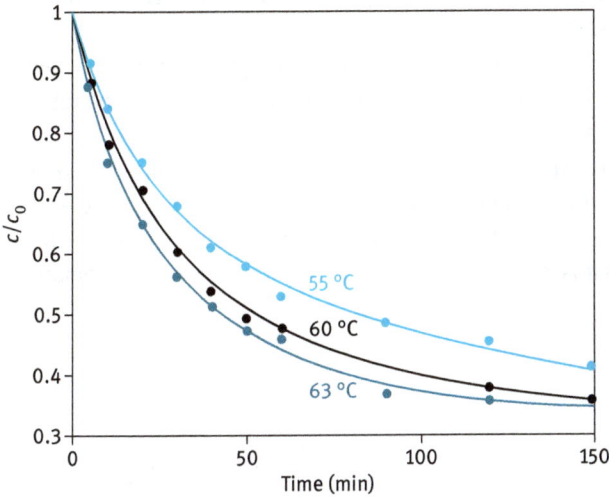

Fig. E1.2: Esterification kinetics of propanoic acid with methanol (the initial molar ratio 1:1) as a function of temperature.

Calculate:

1. The activation energy, based on initial rates (10 min).
2. Gibbs energy of the reaction at 60 °C based on experimentally recorded equilibrium data.
3. Reaction enthalpy at 60 °C, assuming that the entropy change during the reaction is negligible.

Hints:

- Initial rates ($\Delta C/dt$) should be calculated from the slopes for the initial period when the rates are not changing with time.
- Composition at equilibrium defines equilibrium constant ($K_{eq} = C_{products}/C_{reactants}$) and subsequently Gibbs energy ($\Delta G = -RT\ln K$).
- Enthalpy of the reaction is defined through Gibbs energy: $\Delta G = \Delta H - T\Delta S$.

Exercise 1.3: Steam reforming of heptane was performed over a 6 wt% cobalt catalyst supported on γ-alumina. The amount of catalyst was 15 cm^3. Knowing conversion and specific activity calculates the molar feed rate, the rate per catalyst volume and catalyst weight, and apparent activation energy.

°C	Conversion of heptane (%)	Specific catalytic activity (mol/(mol(Co) s))
350	17	0.00051
400	67	0.0199
450	97	0.00288

The turnover number is a useful concept, but is limited by the difficulty of determining the true number of active sites. The situation is somewhat easier for metals, as chemisorption, which will be discussed further (Chapter 2, Section 2.2.4), could be used to measure the exposed surface area.

For some reactions (structure insensitive) the rate is independent on size, shape and other physical characteristics, while for structure-sensitive reactions the rate depends on the detailed surface structure.

In order to increase turnover number it is obvious that the number of exposed sites should be increased, which is possible with small sizes of active particles providing high surface area per unit volume (Fig. 1.7). This can be achieved with the catalyst particles in the nanoscale range, thus placing catalysts in the domain of nanomaterials.

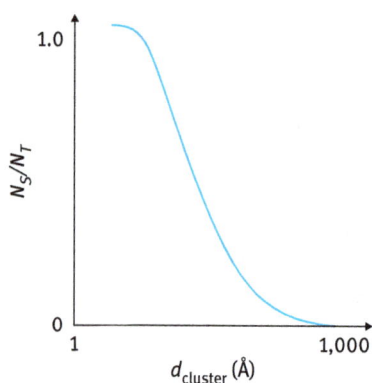

Fig. 1.7: Ratio of surface (N_S) to total (N_T) atoms as a function of cluster size $d_{cluster}$.

It should be noted that many granting agencies are somewhat reluctant to consider heterogeneous catalysis as a part of nanoscience and nanotechnology, especially in comparison with some emerging applications. This is probably because catalysis is a more established concept, with many dozens of decades of experience in synthesis of such materials as supported metals and zeolites.

Small particles of the active catalyst alone cannot provide thermostable highly active catalysts because of sintering of them at conditions of catalyst preparation and catalysis. Moreover, the separation of reaction products from nanosized catalysts is far from being trivial, in many cases even impossible. Therefore an active phase (responsible for activity and selectivity) is usually deposited on a thermostable support (Fig. 1.8), which also provides the required shape, mechanical strength and pore structure.

Expensive active compounds (such as precious metals) are typically prepared with the objective of maximizing the active surface area per unit weight of the active compound, which is achieved at low loadings and rather small metal clusters.

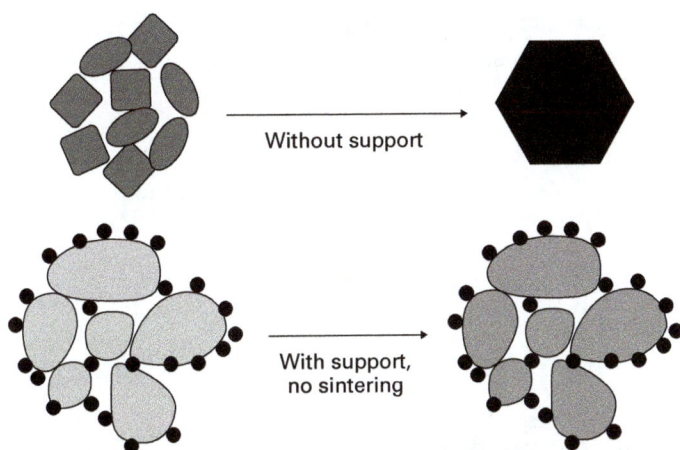

Fig. 1.8: Supported active phase. (Reproduced with permission from [11]).

For the less expensive active phase (supported base metal catalysts) the target is maximum active surface area per unit volume, thus allowing higher loading of the support with the active phase.

Fig. 1.8 depicts a simplified view on the catalyst nanocluster. In fact, the cluster can contain different types of sites (Fig. 1.9A), with different binding to neighbor atoms and thus differences in the ability to bind (adsorb) reacting molecules. As a consequence, we can observe notable differences between the reactivity of different surface atoms and active sites (edges, corners and faces as shown on Fig. 1.9B).

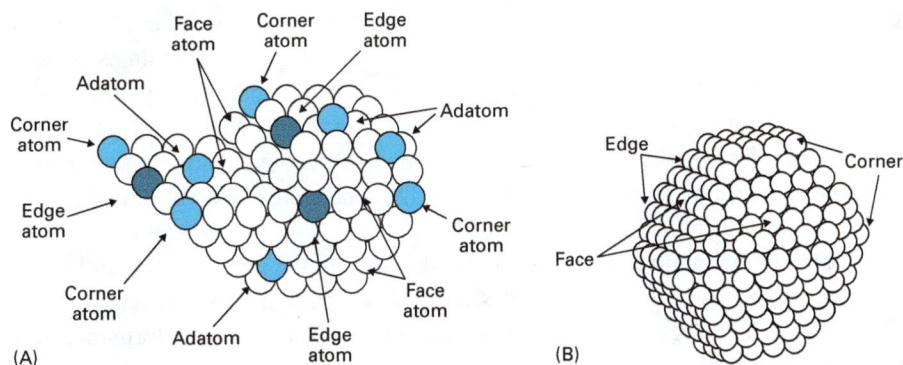

Fig. 1.9: Different types of sites in a nanosized catalyst.

By controlling the amount of different surface sites (atom) we can, in principle, control catalyst reactivity and selectivity. There are however practical challenges in the preparation of a cluster with a desired size and a narrow cluster size distribution.

Relative amounts of surface atoms with different reactivity vary as a function of metal particle size, thus leading to structure sensitivity (TOF dependence on the cluster size, Fig. 1.10).

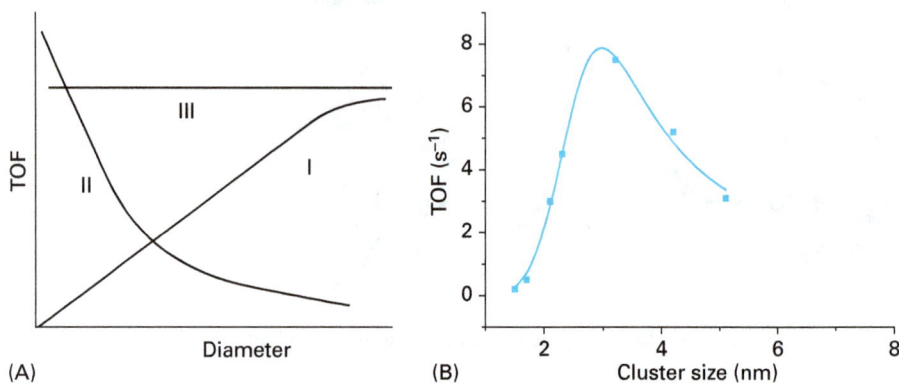

Fig. 1.10: (A) I, II correspond respectively to structure sensitivity, while III represents a structure insensitive reaction and (B) ethene hydrogenation on Pd/TiO$_2$ (data from [12]).

The temperature dependence of heterogeneous catalytic reactions was partially addressed above with the point that, in general, activity increases with temperature. Increasing the temperature, however, shortens the life span of the catalyst and typically increases the rate of undesirable reactions. Moreover, a catalyst with an extremely high reaction rate will certainly exhibit large mass- and heat-transfer limitations.

Although the catalyst must possess high activity and long-term stability, the most important characteristic is selectivity, the ability to direct transformation of reactants to the desired product along one specific pathway. For many reacting systems, various reaction paths are possible, and the catalyst type often determines the path that will be followed. A catalyst can increase the rate of one reaction without increasing the rate of others. In general, selectivity depends on pressure, temperature, reactant composition, conversion and the nature of the catalyst. Therefore, selectivity should be referred to specific conditions.

There are two definitions of selectivity that are commonly used. Integral selectivity is defined as the ratio of the desired product per consumed reactant, while differential selectivity corresponds to the ratio of the desired product formation rate to the rate of reactant consumption.

Although there is always a desire to have a stable catalyst for a particular reaction, strong catalyst deactivation does not necessarily mean that in such a case a particular catalyst cannot be applied at all. In fact, it is only in theory that the catalyst remains unaltered during the reaction. Actual practice is far from this ideal, as the progressive loss of activity could be associated with coke formation, attack of poisons, loss of volatile agents and changes of crystalline structure, which cause a loss of mechanical strength. There are industrial examples, discussed in Chapter 4, that show successful industrial implementation of catalytic reactors in combination with continuous regeneration, when the catalyst life is merely few seconds. Because of the extreme importance of catalyst deactivation, the kinetic aspects of this phenomenon will be treated in a separate section (Section 3.4.5).

In summary, it can be concluded that the target priorities in catalyst development and applications are typically: selectivity > stability > activity.

1.1.4 Composition of catalysts

The catalytic properties of activity, selectivity and stability are closely related to the composition of the catalyst. Typical catalytic materials are shown in Fig. 1.11. Most catalysts are multicomponent and have a complex composition. Components of the catalyst include the active agent itself and may also include a support, a promoter, and an inhibitor.

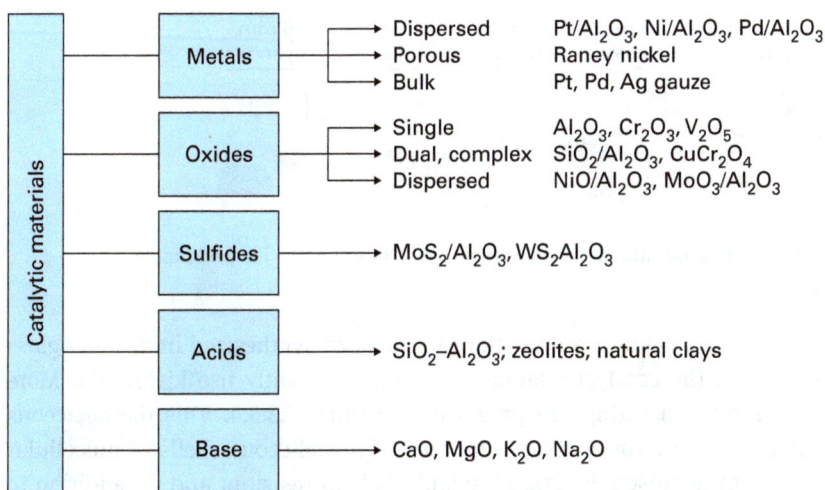

Fig. 1.11: Catalytic materials.

For example, metal catalysts are not in their bulk form but they are generally dispersed on a high surface area insulator (support) such as Al_2O_3 or SiO_2. Many heterogeneous catalysts need supports, which, from an economic viewpoint, are a means of spreading expensive materials and providing the necessary mechanical strength, heat sink/source, help in optimization of bulk density and in dilution of an overactive phase. The supports also provide geometric functions (an increase in surface area, optimization of porosity, crystal and particle size) and chemical functions (improvement of activity, minimization of sintering, and poisoning, as well as the beneficial effect of spillover, which will be discussed later).

Sintering of supports can start at a temperature substantially lower than the melting point, thus supports with high melting points are applied, such as alumina or silica. In particular, alumina is mainly preferred because of favorable bulk density, thermal stability and price. It is applied in the form of y-alumina in many reactions, such as hydrotreating, hydrocracking, hydrogenation, reforming, oxidation, and so on. Another phase of alumina (α-alumina) that stable at very high temperatures is used for steam reforming of natural gas.

Silica has a lower bulk density, which means that more reactor volume is needed for the same active phase loading on the support. Silica is more liable to sintering above 900 K than alumina and is volatile in the presence of steam and elevated pressures. There are still several applications (polymerization, oxidation, hydrogenation) where silica is used, and a very specific example is the oxidation of sulfur dioxide to trioxide (Fig. 1.12).

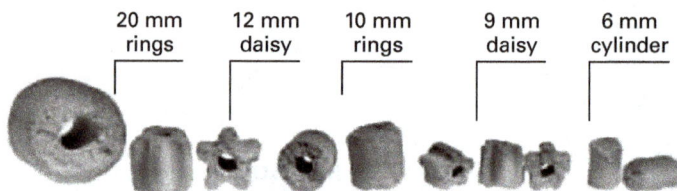

| 20 mm rings | 12 mm daisy | 10 mm rings | 9 mm daisy | 6 mm cylinder |

Fig. 1.12: Vanadium-based catalysts supported on silica for sulfur dioxide oxidation.

Although there were catalyst formulations using silica synthesized from water glass (sodium silicate), the catalyst manufacturers predominantly use kieselguhr. More details of the manufacturing are provided in Section 2.3.5.4. This diatomaceous earth (diatomite) is a form of silica composed of the siliceous shells of unicellular aquatic plants of microscopic size. Kieselguhr is heat resistant and, in addition to catalytic applications, has been used as an insulator, a component in toothpaste and an abrasive in metal polishes. The specific features of the diatomaceous earth (surface area 20–40 m^2) are related to small amounts of alumina and iron as part of the skeletal structure and a broad range of pore sizes. The main components of the sulfur dioxide oxidation catalyst include SiO_2 as a support, vanadium, potassium and/or caesium and various other additives. The reaction actually occurs within a molten salt consisting of potassium/caesium sulfates and vanadium sulfates, coated on the solid silica support. Vanadium is present as a complex sulfated salt mixture and not as vanadium pentoxide (V_2O_5).

Aluminosilicates, either amorphous or crystalline (zeolites), can be used as supports and catalysts (as zeolites) because of their acidic properties: Active carbon is another support that is often applied, mainly for the synthesis of fine chemicals in various hydrogenation reactions. Other supports include clays, ceramic (cordierite), titania (for selective oxidations and selective catalytic reduction of NOx) as well as magnesium aluminate (steam reforming of natural gas).

While selecting a support for a particular reaction, various physical and chemical factors should be carefully considered. The physical factors are the desired surface area and porosity, thermal conductivity, mechanical stability, resistance to coking, and the possibility of applying a desired geometrical shape; while the chemical factors are the activity of the support, chemical interactions between the active catalyst phase or the products/reactants and the support, and the resistance to poisoning. Details on the various supports used in catalysts and their preparation will be given in Chapter 2.

Sometimes both the metal and the support act as catalysts, which is called bifunctional catalysis. An example of this is platinum dispersed on alumina used in the gasoline reforming. Platinum is active in dehydrogenation reactions of linear alkanes to linear alkenes which further undergo skeletal isomerization on alumina. Formed branched olefins are hydrogenated on platinum. The active agent, discussed above is the component, which causes the main catalytic action, while a promoter (being inactive, when added into the catalyst in small amounts) improves activity, selectivity or stability, prolonging the catalysts lifetime.

There are various types of promoters that improve the catalyst in different ways. Textural promoters are inert substances, which inhibit sintering of the active catalyst by being present in the form of very fine particles. Usually they have a smaller particle size than that of the active species, are well-dispersed, do not react or form a solid solution with the active catalyst and have a relatively high melting point. Oxides (with melting points in parentheses) such as Al_2O_3 (2,027 °C), SiO_2 (1,700 °C), ZrO_2 (2,687 °C), Cr_2O_3 (2,435 °C), CeO_2 (2,600 °C), MgO (2,802 °C) and TiO_2 (1,855 °C) can be used as textural promoters.

Structural promoters change the chemical composition, produce lattice defects, change the electronic structure and the strength of chemisorption. For example, Pb, Ag, or Au can be used as structural promoters for palladium catalysts applied in selective hydrogenation of acetylene to ethylene.

An inhibitor is the opposite of a promoter. When added in small amounts, it diminishes activity, selectivity or stability. Inhibitors are useful for reducing the activity of a catalyst for an undesirable side reaction. For example, silver supported on alumina is an excellent oxidation catalyst, widely used in the production of ethylene oxide from ethylene. However, under the same conditions complete oxidation to carbon dioxide and water also occurs, resulting in mediocre selectivity to C_2H_4O. It was found that the addition of halogen compounds to the catalyst inhibits this complete oxidation and results in satisfactory selectivity [13].

Metal oxides usually consist of bulk oxides. As semi-conductors, metal oxides catalyze the same type of reactions as metals but are used in processes that require higher temperatures. Often a multicomponent mixture of various oxides is used to increase the catalytic activity.

For example: transition metals such as MoO_3 and Cr_2O_3 are good catalysts for the polymerization of olefins; a mixture of copper on chromium oxides, called copper chromite, is used for hydrogenation; and a mixture of iron and molybdenum oxide, called iron molybdate, is used for formaldehyde formation from methanol.

1.2 Reactivity of solids

Heterogeneous catalysts in industrial applications have a variety of different geometrical shapes and porosity. Therefore, understanding of mass and heat transfer in heterogeneous catalysis is essential for the description of the processes involved in transporting the reactants to the catalyst surface and the transfer of energy to and from catalyst particles.

The following steps are usually followed:
1. External diffusion: transfer of the reactants from the bulk fluid phase to the fluid-solid interface (external surface of the catalyst particle).
2. Internal diffusion (if the particle is porous): transfer of the reactants into the pores of a catalyst particle.
3. Adsorption: physisorption and chemisorption of the reactants at the surface sites of the catalyst particle. During adsorption the reactants become attached to the surface either by physical (van der Waals) or chemical forces, thus significantly decreasing the entropy compared to the gas- or liquid-phase. This leads to negative ΔS, thus for the Gibbs energy ($\Delta G = \Delta H - T\Delta S$) to be negative, the heat of adsorption (ΔH) should be negative and adsorption should be exothermic.
4. Surface reaction: the chemical reaction of adsorbed species to produce adsorbed products. It can proceed by formation of precursors on the catalyst surface, and surface diffusion of these as well as of other adsorbed species to the reaction sites.

Langmuir advanced the first kinetic approach in catalysis based on physicochemical understanding of the underlying processes [14], by applying the mass action law to reactions on solid surfaces. This was done by supposing that the rate of an elementary reaction is proportional to the surface concentration (coverage) of reactive species adsorbed on the surface. The Langmuir model of uniform surfaces assumes that all the surface sites are identical, binding energies of the reactants are the same independent of the surface coverage, and interactions between adsorbed species may be neglected. This ideal adsorbed layer is considered to be similar to ideal solutions with fast surface diffusion, thus allowing an application of the mass action law.

Typically two main types of mechanisms are considered for a bimolecular reaction. Either one reactant (A) is adsorbed and the other (B) approaches from the fluid phase (the Eley-Rideal mechanism, see Fig. 1.13A) or a reaction occurs because of the interaction of two adsorbed reactants A and B (the Langmuir-Hinshelwood mechanism, see Fig. 1.13B). The products are then released either directly into the fluid phase or first through adsorbed product species followed by desorption.

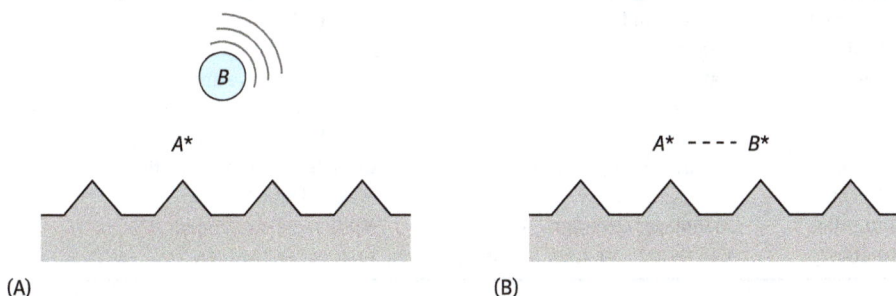

(A) (B)

Fig. 1.13: (A) the Eley-Rideal and (B) the Langmuir-Hinshelwood mechanisms of catalytic reactions.

Detailed analysis of catalytic kinetics for various reaction mechanisms, including Eley-Rideal and Langmuir-Hinshelwood ones, will be provided in Chapter 3.

Further steps in heterogeneous catalysis are:
5. Desorption: the release of adsorbed products by the catalyst.
6. Internal diffusion: the transfer of products to the outer surface of the catalyst particle.
7. External diffusion: the transfer of products from the fluid-solid interface into the bulk fluid stream.

1.2.1 Physisorption and chemisorption

Adsorption in heterogeneous catalysis takes place before surface reactions and thus deserves special consideration.

In the case of chemisorption of a molecule (adsorbate), a chemical bond is formed between the molecule and the surface (adsorbent). For chemisorption, the adsorption energy is comparable to the energy of a chemical bond. The molecule may chemisorb intact or it may dissociate with breaking chemical bonds. The chemisorption energy is 30–70 kJ mol^{-1} for molecules and 100–400 kJ mol^{-1} for atoms.

In physisorption, the bond is a van der Waals interaction and the adsorption energy is typically 5–10 kJ mol^{-1}, which is much weaker than a typical chemical bond. As a consequence the chemical bonds in the adsorbing molecules remain intact. As the van der Waals interactions between the adsorbed molecules are similar to the van der Waals interaction between the molecules and the surface, many layers of adsorbed molecules may be formed. Some other features of physisorption and chemisorptions are presented in Tab. 1.1.

Tab. 1.1: Comparison between chemisorption and physisorption.

Features	Chemisorption	Physisorption
Adsorbent specificity	Sorbent and crystallographic plane dependent	Independent on the surface structure
Type	Can be dissociative and irreversible	Non-dissociative and reversible
Temperature	A wide range	Near or below boiling point of adsorbate
Heat of adsorption	A wide range (40–800 kJ mol^{-1})	Close to heat of liquefaction (5–40 kJ mol^{-1})
Saturation	Monolayer coverage	Multilayers are formed
Kinetics	Can be activated, for example, slow	Fast, non-activated

Fig. 1.14 shows potential energy diagrams of chemisorption and physisorption. When approaching the surface a molecule encounters the weak van der Waals interactions. If the molecule loses energy upon interactions with the surface, it may be trapped in the weak attractive potential at a distance corresponding to the sum of the van der Waals radii of the molecule and the surface atom, and thus become physisorbed on the surface. Moving even closer to the surface will result in repulsion.

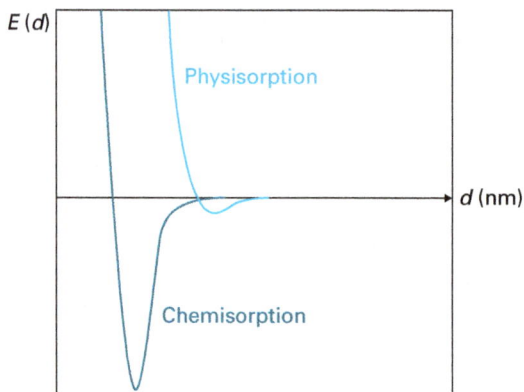

Fig. 1.14: Potential energy curves for chemisorption and physisorption.

After rearranging its electronic configuration and interacting with the electron clouds of the metal, the molecule, however, may become chemisorbed.

The equilibrium position of this new chemisorbed state is at a shorter distance than for physisorption, Moreover, for chemisorption the potential energy curve is dominated by a much deeper chemisorption minimum. The chemisorption curve intersects with the physical adsorption curve at a value close to zero in the energy axes. For a clean surface this barrier between physisorption and chemisorption is below zero, thus chemisorption on clear surfaces is usually fast and non-activated. Dissociative chemisorption can occur for a diatomic molecule.

The structure of solid surfaces is fairly complicated (Fig. 1.15) as they are usually complex and rough, and consisting of high Miller index (to be discussed later) planes with different energetics.

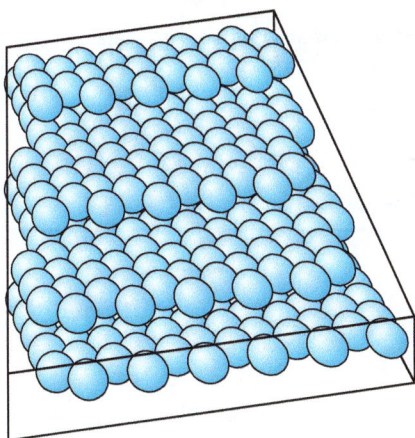

Fig. 1.15: Surface structure.

Moreover, adsorbed species can form different types of complexes on the surfaces (some examples are given in Fig. 1.16), which might interact with each other (lateral interactions).

As a result, the heat of adsorption might decrease with coverage, with the minimum of adsorption energy decreasing in magnitude. This leads to a situation when such a decrease of the adsorption heat will lead to an activation energy of adsorption (Fig. 1.17).

Physisorption at low temperatures is a standard method for determination of the surface area of a catalyst, pore size and volume, and pore size distributions; this will be discussed in detail in Chapter 2 along with the application of chemisorption measurements for active surface area, surface site energetics and determination of catalytic sites.

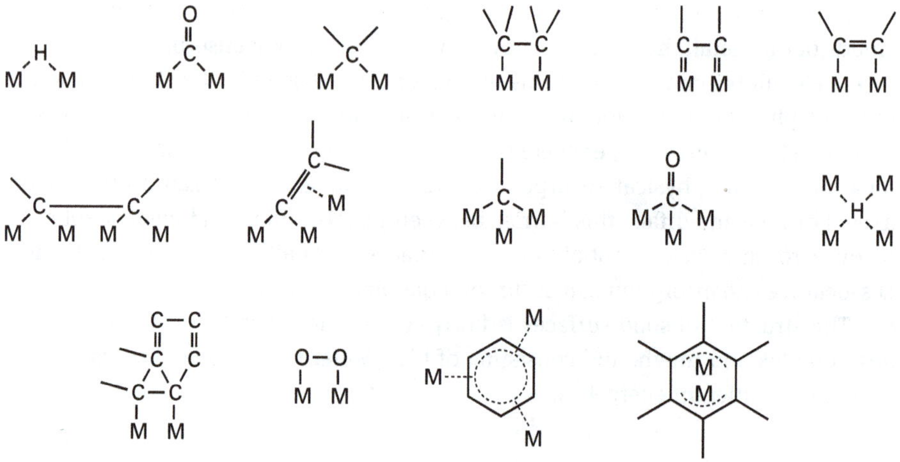

Fig. 1.16: Examples of the complexes that could be formed on the solid surfaces. M, metal site.

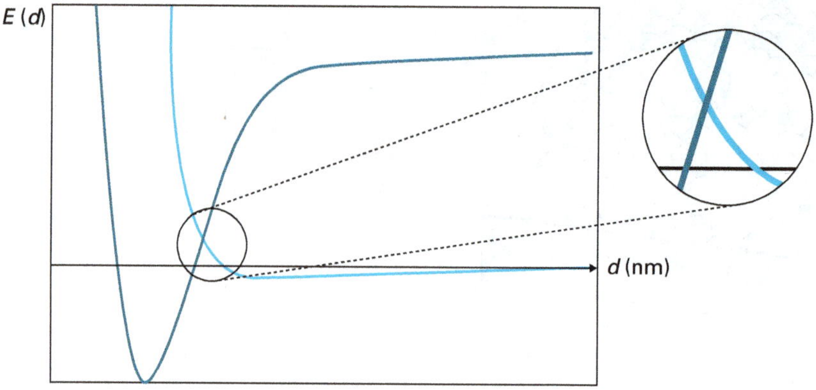

Fig. 1.17: Chemisorption proceeding with activation.

1.2.2 Basics of chemisorption theory

Chemisorption of catalytically active metals that have 6–10 d electrons is usually considered for analyzing the behavior of d electrons. There is an overlapping of orbitals, when, if large, gives a broad band and strong interactions. Such band is a collection of a large number of molecular orbitals that form a continuum when they are close to each other. In metals the bands are filled with valence electrons up to the Fermi level or to the highest occupied state (E_f) analogously to the highest occupied molecular orbital (HOMO). If an empty level exists below E_f it will be filled instantaneously. A band behaves like a molecular orbital, in a sense that lower levels are bonding, upper levels are antibonding and middle levels are non-bonding.

Behavior of *d* orbitals is important in understanding catalytic activity of the transition metals. Details on the reactivity of transition metals on the basis of d-band analysis are provided in specialized textbooks [15].

When an atom is chemisorbed, bonding and antibonding chemisorption orbitals should be considered. The strength of chemisorption depends on the position of the bonding orbital. If interactions of an atomic orbital and a d-band are strong then there is substantial splitting between the bonding and antibonding orbitals. The antibonding orbitals are unoccupied, leading to strong chemisorption (Fig. 1.18A). In the opposite case of weak interactions, small splitting can result in an antibonding orbital falling below E. Such an orbital is occupied, leading to repulsive interactions and thus no chemisorption (Fig. 1.18B).

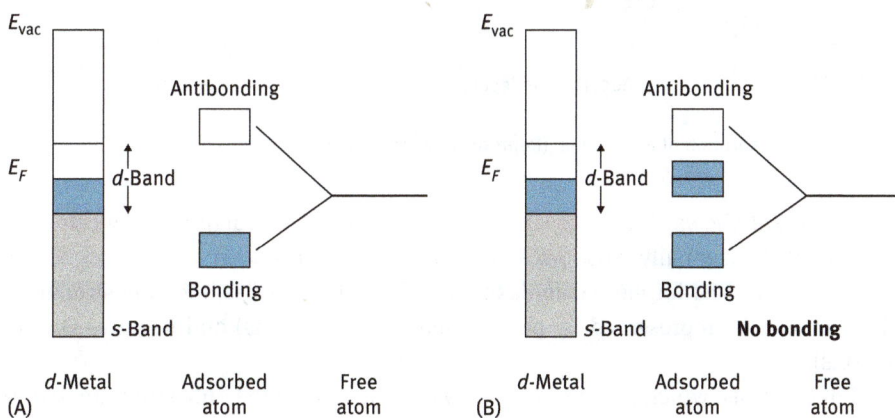

Fig. 1.18: Chemisorption of an atom. (A) strong chemisorption, (B) no bonding. (From [15]).

In the case when a molecule is chemisorbed, chemisorption orbitals are formed from both bonding and antibonding molecular orbitals of a molecule (Fig. 1.19).

Filling of the orbital of the molecule, which was originally antibonding, makes interactions with the surface stronger, while at the same time weakening the intramolecular bond of the adsorbed molecule and facilitating, for example, dissociation of a molecule.

Filling of the bonding orbital is called donation, favoring "on-top" adsorption (discussed in Section 1.2.3). Filling of the antibonding orbital, which binds a molecule additionally to the surface, is called "back donation".

1.2.3 Surface crystallography

The crystallographic structure of the metals that are most important for catalysis is presented in Fig. 1.20(A).

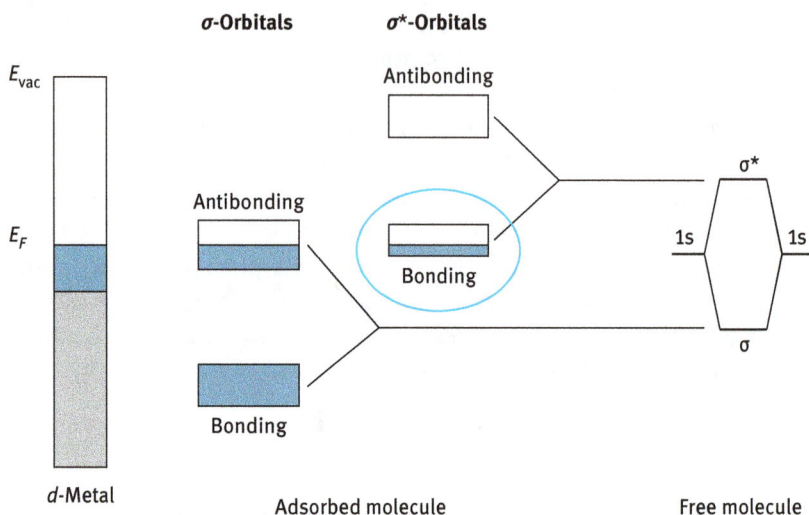

Fig. 1.19: Chemisorption of a molecule. (Reproduced with permission from [15]).

Fig. 1.20B shows three most important metal crystal structures: face-centered cubic (fcc), hexagonally close-packed (hcp) and body-centered cubic (bcc). Other structures can exist for more complex materials such as oxides, the bulk structures of which can be represented by positive metal ions (cations) and negative O ions (anions).

For catalysis, which is a surface phenomenon, it is not the bulk structure that is important, but the exposed surface. As energy is required to cleave the surface, the total energy is higher and the surface free energy is always positive.

Reactivity of the surface depends on the number of surrounding atoms, or in other words, on packing. Dense regular packing (Fig. 1.21) has lower energy.

In order to discuss surface crystallography – the two-dimensional analogue of bulk crystallography – the nomenclature of various metal planes (Fig. 1.20) should be explained. The assignment of indices is done using the following approach. A particular surface plane of a crystal is considered and is shown in Fig. 1.22.

Firstly the intercepts on the x-, y- and z-axes are identified. In Fig. 1.22 the intercept on the x-axis is at $x = a$ [at $(a,0,0)$], corresponding to the unit cell distance. The surface is parallel to the y- and z-axes, thus there are no intercepts on these two axes and they are considered to be at infinity (∞). The intercept coordinates on the x-, y- and z-axes (a, ∞,∞) are converted to fractional coordinates by dividing by the respective cell-dimension (a) to give the fractional intercepts: $(1, \infty,\infty)$. The reciprocal values of the fractional intercepts are 1, 0 and 0, yielding the Miller indices (100), which is specified without being separated by any commas or other symbols (Fig. 1.23).

The surfaces in Fig. 1.23 are different in terms of their coordination or number of neighbors. The most stable solid surfaces are those with a high surface atom density

bcc   hcp   fcc

*Lanthanide series

**Actinide series

(A)

bcc   hcp   fcc

(100) (110) (111)

(B)

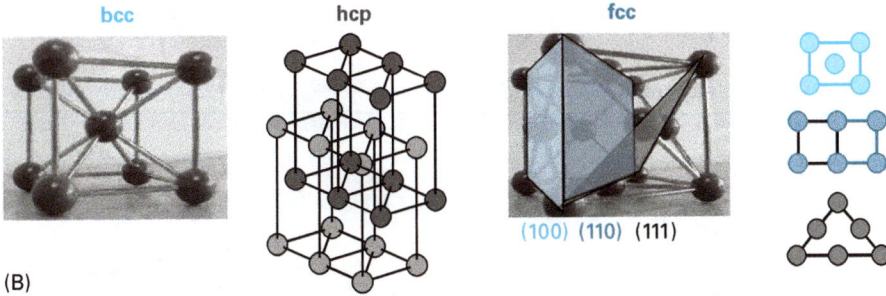

Fig. 1.20: (A) Crystallographic structure of some catalytically active metals. (B) Metal crystal structures.

that contains surface atoms with a high coordination number. For face-centered cubic structures (representative of Ni, Cu, Pt, Pd, Ir, Au, Ag, etc.) the following order of stability is valid: fcc (111) > fcc (100) > fcc (110). The most stable surface is then the most closely packed. At the same time, the more open the surface is the more reactive it is. It should be mentioned that surfaces of real catalysts are usually rough and consist of high Miller index planes.

Figure 1.24 shows the most common adsorption sites: on-top, bridge (long or short) and hollow (three-fold or four-fold).

From the discussion above it is apparently clear that the classical treatment of adsorption and kinetics, which assumes the same adsorption strength for all sites, is an oversimplification. In addition to different energetics of adsorption an adsorbed molecule can bind to several metal sites and there could be interactions between

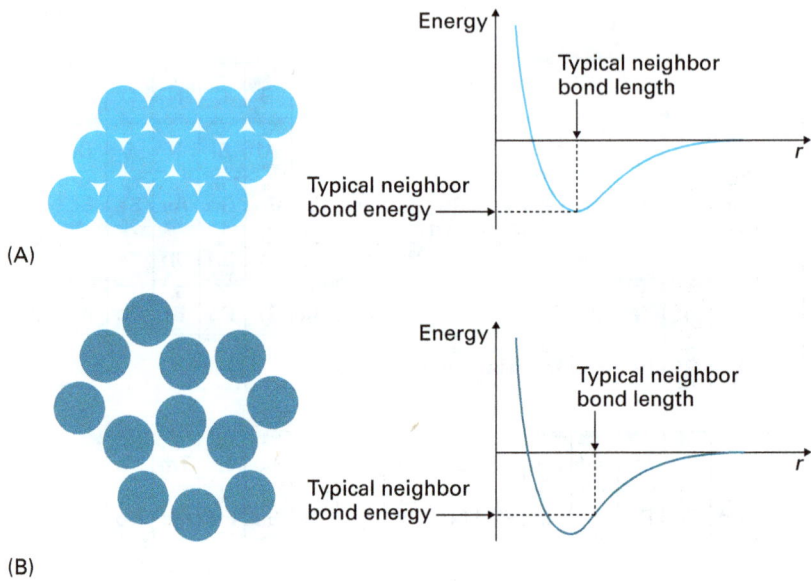

(A)

(B)

Fig. 1.21: (A) Dense, regular packing. (B) Non-dense, random packing.

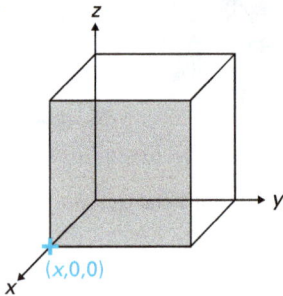

Fig. 1.22: Assignment of indices.

adsorbed species. This complicates the mathematical treatment of adsorption and kinetics, which will be addressed in detail in Chapter 3.

1.2.4 Mechanisms of some catalytic reactions

There are countless heterogeneous catalytic reactions with very many different types of catalysts. It is thus impossible in a textbook focused on engineering rather than on mechanisms of catalysis to describe even a small fraction of these, therefore only a few generic examples for some typical reactions will be mentioned here.

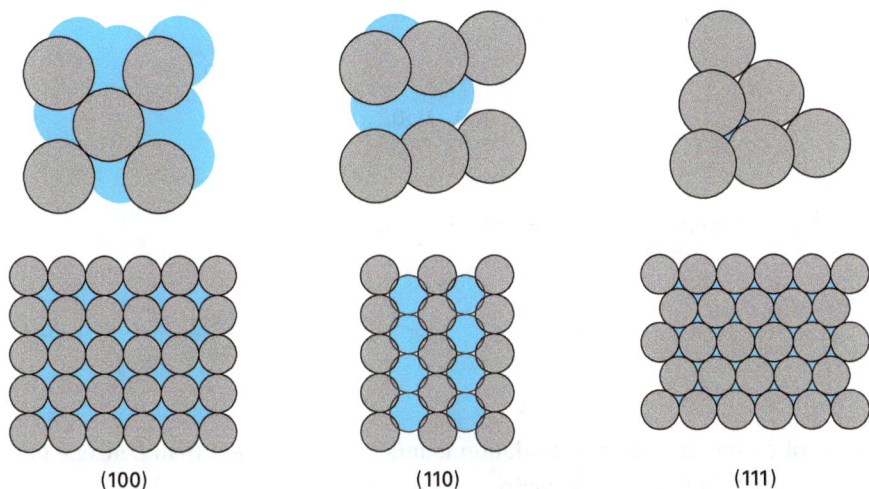

Fig. 1.23: Various metal surface planes.

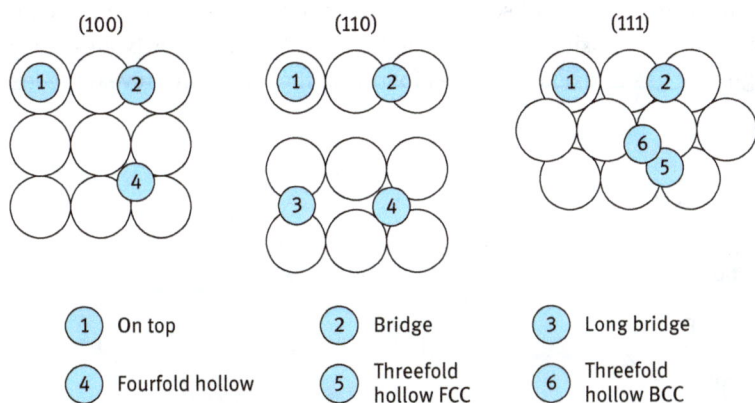

1 On top	2 Bridge	3 Long bridge
4 Fourfold hollow	5 Threefold hollow FCC	6 Threefold hollow BCC

Fig. 1.24: Adsorption sites. (From [15]).

1.2.4.1 Oxidations

In industry, typical oxidation reactions over heterogeneous catalysts are usually partial or selective oxidations, such as:

– ethylene epoxidation

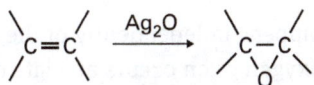

- oxidation of propene to acrolein

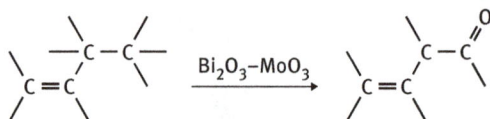

- oxidation of benzene or butane to maleic anhydride

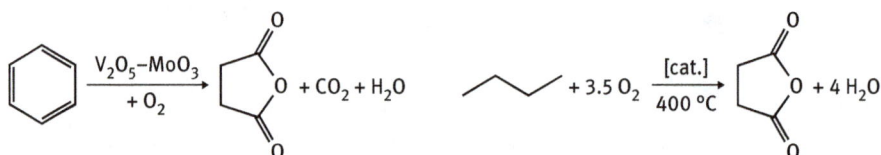

In terms of carbon use, butane oxidation is much more efficient than benzene oxidation, as according to stoichiometry in benzene oxidation two moles of CO_2 are formed per one mole of reacting substrate.

In the selective oxidation reactions, CO_2, CO and even water are undesirable products.

Partial oxidation reactions (and selective catalytic reduction, Fig. 1.25) are typically carried out using such oxides as, for example, V_2O_5, MoO_3, or molybdates. They proceed through changes in oxidation states following the Mars-van Krevelen redox mechanism.

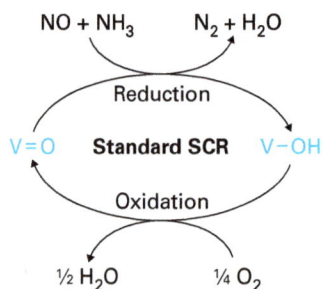

Fig. 1.25: Mechanism of selective catalytic reduction.

In this mechanism (kinetics will be presented in Chapter 3) reduction of the oxide is dependent on the organic reactant (NO and ammonia in the case of selective catalytic reduction (SCR), or CO in Fig. 1.26) and takes place in the absence of gas-phase oxidants (O_2, H_2O). Isotope labeling studies were used to confirm that oxygen in the product comes from the lattice oxygen (Fig. 1.26A), but not directly from the gas-phase species.

Re-oxidation of oxide happens independently of the reduction step (Fig. 1.26B). Dissociative adsorption of oxygen often occurs at a site different from an oxidation site, followed by diffusion through the bulk in order to refill the surface vacancy.

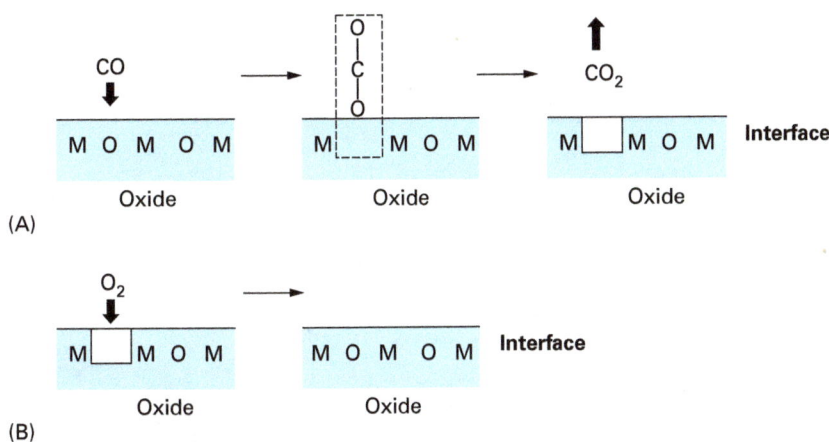

Fig. 1.26: The Mars-van Krevelen redox mechanism. (A) oxide reduction and (B) re-oxidation.

For oxidations not involving changes in oxidation state, such as oxidation of CO on metals, a different mechanism could operate. For example, the substrates can co-adsorb on the surface, which is followed by a reaction between the adsorbed species according to the Langmuir-Hinshelwood mechanism (Fig. 1.27).

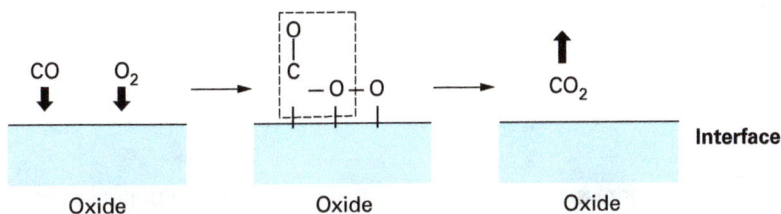

Fig. 1.27: Oxidation following a Langmuir-Hinshelwood mechanism.

Alternatively one reactant is adsorbed and the other reacts from the fluid phase.

1.2.4.2 Hydrogenation
From the fundamental as well as an industrial viewpoint, hydrogenations can be considered to be the most advanced field of catalysis. Various hydrogenation reactions are employed in industry, such as hydrogenation of double, triple and aromatic bonds, and C=O bonds to name a few. Different mechanisms were proposed for the hydrogenation of olefinic double bonds [16]. The Horiuti-Polanyi mechanism assumes that both the alkene and hydrogen must be activated by being adsorbed on the surface (see Fig. 1.28).

1. $\quad H_2 + 2\ * \longrightarrow$ 2 H | *

2. $\quad$ >C=C< + 2 * $\longrightarrow$ >C−C< (with * below each C)

3. $\quad$ >C−C< (with * * below) + H | * $\longrightarrow$ >C−C< + 2 * (with * H below)

4. $\quad$ >C−C< (with * H below) + H | * $\longrightarrow$ >C−C< + 2 * (with H H below)

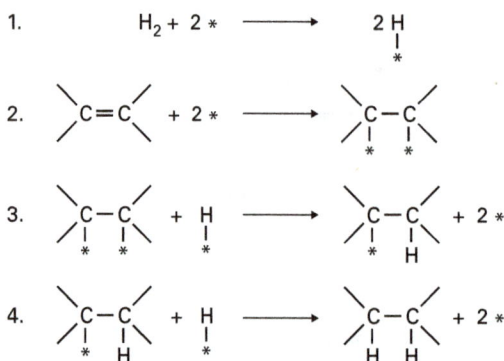

Fig. 1.28: The Horiuti-Polanyi mechanism.

It should be noted that double-bond hydrogenation is stereo selective (Fig. 1.29), corresponding to addition to the less crowded face of the double-bond (*syn* addition of both H atoms).

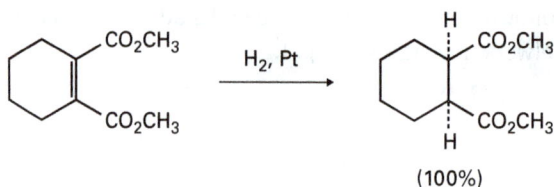

$$H_2,\ Pt$$

(100%)

Fig. 1.29: Syn addition to the double-bond.

The mechanism, which addresses stereo selectivity and is consistent with liquid-phase hydrogenation kinetics, assumes the addition of hydrogen to the adsorbed organic molecule, followed by isomerization in the adsorbed state of the formed olefin-hydrogen complex (Fig. 1.30). The adsorbed product is displaced from the surface by the incoming substrate. See [17] for a detailed discussion on the mechanism.

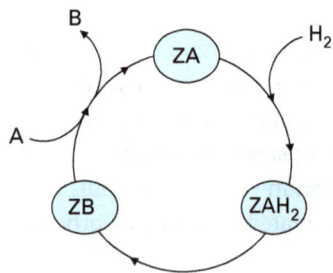

Fig. 1.30: Mechanism of catalytic hydrogenation. (From [17]).

1.2.4.3 Catalysis by solid acids

Crystalline aluminosilicates-zeolites can be used in catalysis because of the presence of acidic sites, which are generated by the substitution of silicon by aluminum in the framework of silica. A detailed discussed on the properties of zeolites and their preparation will be given in Chapter 2. Here we can mention that differences in charges between Si^{+4} and Al^{+3} are balanced by protons, which is the reason for strong Brønsted acidity of zeolites, caused by the presence of these protons.

The mechanism of acid-catalyzed reactions is thus connected to the presence of protons and Lewis acid sites; the Lewis acid sites can act as solid acids by accepting electrons. For example, double-bond isomerization in olefins such as butene starts with the proton addition to the double-bond, forming a secondary butyl cation:

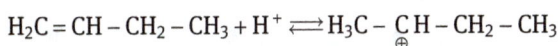

$$H_2C = CH - CH_2 - CH_3 + H^+ \rightleftharpoons H_3C - \underset{\oplus}{C}H - CH_2 - CH_3$$

This is followed by an internal hydride shift:

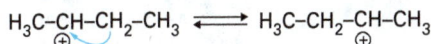

$$H_3C-\underset{\oplus}{C}H-CH_2-CH_3 \rightleftharpoons H_3C-CH_2-\underset{\oplus}{C}H-CH_3$$

and the subsequent proton elimination:

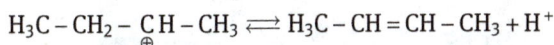

$$H_3C - CH_2 - \underset{\oplus}{C}H - CH_3 \rightleftharpoons H_3C - CH = CH - CH_3 + H^+$$

Skeletal isomerization of olefins also proceeds with the addition of proton and further rearrangement of carbenium ions because of their different stabilities:

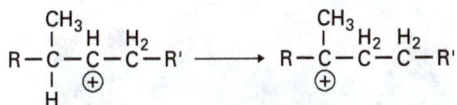

Thus, tertiary carbenium ions are more stable than secondary ones, with primary ions the least stable. Because of the different mechanisms of proton addition, the skeletal isomerization of alkanes is more difficult than that of olefins, requiring stronger acid sites.

Catalytic cracking of heavy organic compounds is an industrially important reaction in oil refining that will be covered in detail in Chapter 4. During cracking, secondary carbenium ions are formed at random followed by cracking at the bond at the beta position to the charged carbon atom, giving an olefin and a primary carbenium ion:

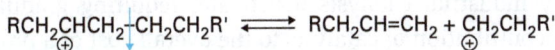

$$RCH_2\underset{\oplus}{C}HCH_2{-}CH_2CH_2R' \rightleftharpoons RCH_2CH=CH_2 + \underset{\oplus}{C}H_2CH_2R'$$

The primary ion undergoes a hydride shift to secondary ions, as discussed above. A detailed account of the mechanism of catalytic cracking is provided in Section 4.3.

1.3 Catalysis in industry and for environmental protection

The immense importance of catalysis in the chemical industry is demonstrated by the fact that roughly 85–90% of all chemical products have seen a catalyst during the course of production. The number of catalytic processes in industry is very large, with the catalysts coming in many different forms, such as heterogeneous catalysts in the form of porous solids (Fig. 1.31), homogeneous catalysts dissolved in the liquid reaction mixtures and biological catalysts in the form of enzymes. In fact, enzymes could either be in the same phase as the liquid reactants or in their solid form (either poor soluble enzymes or immobilized ones).

Fig. 1.31: Shapes of heterogeneous catalysts. (Reproduced with permission from [18]).

In theory, catalysts are not consumed in a chemical reaction, thus in an ideal scenario for cost calculations only the cost of the initial catalyst should be considered. In practice many industrial catalysts deactivate, requiring gradual replacement. Nonetheless, the contribution of catalysts to the overall cost of a product is on average not very high (~3%) and subsequently a part of catalyst sales in relation to the gross domestic product (GDP) is marginal (~0.1%). However, the share in GDP of products made by catalysts could be as high as 25%. The global catalyst market was estimated to be worth 27.7 billion dollars in 2015 and is projected to increase to

33.5 billion USD in 2019. The main contribution to the global demand for catalysts is associated with environmental applications (Fig. 1.32), while the second most important contribution is related to crude oil refining (Fig. 1.33).

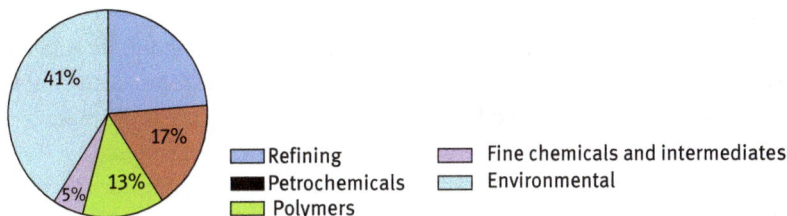

Fig. 1.32: Global catalyst value. (From [19]).

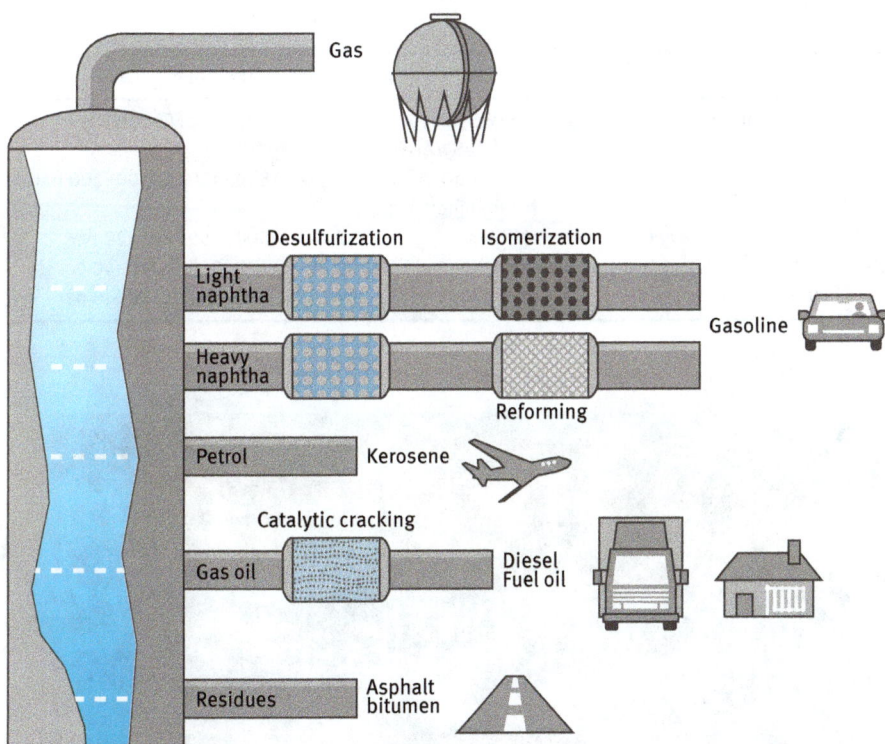

Fig. 1.33: Future perspectives in catalysis. (From [20]).

Most growth in catalyst demand is outside the Organization for Economic Cooperation and Development and is driven to a substantial extent by environmental air quality and fuel specifications. The annual growth rate of emission control catalysts (~12%) is higher than the average for the catalyst manufacturing industry (~8%).

In addition to tightening the regulations for mobile emissions and for fuel quality, other trends driving future demand for catalysts are: improving process efficiency by more efficient use of feedstock and higher yields; and increasing energy efficiency, expanding simultaneously the feedstock base to include coal, natural gas and biomass.

The contribution of catalysts to improving the environment is not only related to end-of-pipe technologies (for example, the cleaning of wastes), but also to the prevention of pollution by avoiding the formation of waste and unwanted by-products through improving processes and selectivity in the existing ones.

Tab. 1.2 gives some examples of catalytic processes in oil refining for the production of gasoline, diesel, kerosene, heating oils, while a photo of catalysts used in refining as well as in the synthesis of petrochemicals is presented in Fig. 1.34.

Tab. 1.2: Major catalytic processes in oil refining.

Process	Catalyst	Conditions
Hydrodesulfurization	$NiS/WS_2/Al_2O_3$ $CoS/MoS_2/Al_2O_3$	300–450 °C, 100 bar H_2
Hydrocracking	$MoO_3/CoO/Al_2O_3$ Ni/aluminosilicate	320–420 °C, 100–200 bar H_2
Catalytic cracking of heavy fractions	zeolites	500–550 °C, 1–20 bar
Catalytic reforming of naphtha	Pt(+other metal)/alumina	470–530 °C, 15–40 bar H_2
Isomerization of light gasoline	Pt/alumina	400–500 °C, 20–40 bar

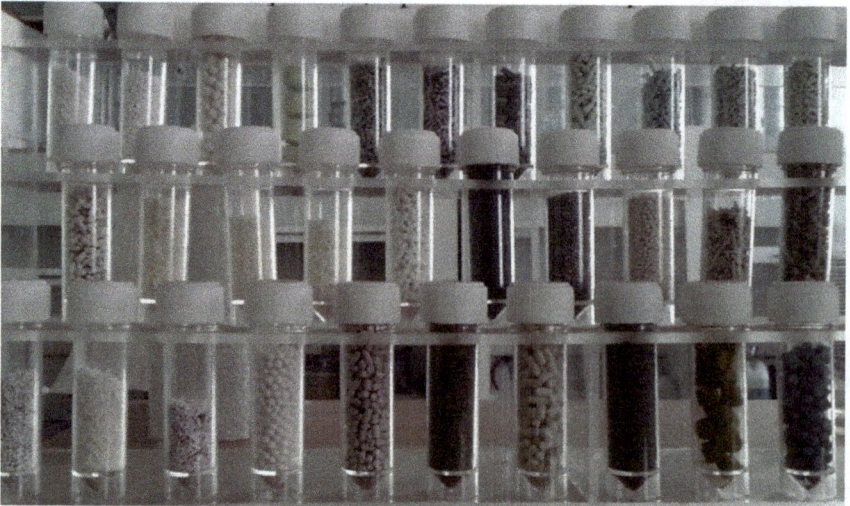

Fig. 1.34: Refining and petrochemical catalysts (photo of the author).

For various reasons there is a current trend to consider other feedstocks – rather than oil – as a source of fuel, olefins and aromatics. Using feedstocks such as natural gas, coal, and especially biomass, would require fundamental changes in the chemical industry and refining.

It should be noted that the transport sector uses nearly 28% of all energy generated, and oil and gas are mainly used for the production of fuels, while the chemical industry accounts for only 5–10% of the oil consumption. Because the first generation of fuels and chemicals obtained from biomass (for example, bioethanol) has many shortcomings, lignocellulosic biomass, not competing with the food chain, has attracted a lot of attention. Lignocellulose contains cellulose ($\approx$40 wt%), hemicellulose ($\approx$25 wt%), extractives (~5%) and lignin (the balance). With 700 billion metric tons available on Earth, cellulose is the most abundant organic material.

Cellulose is a linear polysaccharide, built from glycoside units (the number of monomers can be as high as 15,000), which are linked by β-1,4-glycoside bonds, resulting in the formation of a crystal structure of cellulose with intra- and intermolecular hydrogen bonds. As there is not much space around the glycoside (bridging) oxygen in cellulose, there is limited access to this bond for homogeneous or heterogeneous catalysts. Hemicellulose (a polysaccharide) is composed of several types of monomers, which imposes steric hindrance because of the presence of side chains and prevents the formation of a crystalline structure. Lignin is a three-dimensional polymer composed of propylphenol units. Since structure of biomass is different from petroleum, technologies of biomass processing in general and in particular involving catalysis differ from those widely used in petroleum refining. The oxygen-to-carbon ratio in lignocellulosic biomass is close to unity, which is unsuitable for the production of transportation motor fuels, i.e. hydrocarbons. At the same time, a minor amount of oxygen is needed in gasoline, and thus the strategy for using biomass for fuels should include substantial, but not complete, removal of oxygen. In addition, the reactions in petroleum refining are often conducted in the gas or liquid-phase in the presence of organic media, while catalytic reactions of sugars and polyalcohols occur in an aqueous environment. Acidic catalysts based on aluminosilicates and zeolites might be unstable and prone to deactivation because of the impurities and quality of the lignocellulose, which depends substantially on the type of biomass and geographical location.

The same feedstock shift towards renewable resources is currently being investigated for the production of chemicals. Many intermediates used in the chemical industry contain oxygen, which means that the carbon-to-oxygen ratio is close to that for biomass in at least some chemicals such as methanol, acetic acid, and ethylene glycol, and in the range 0.3–0.6 for other important chemicals. These substances do not require a very high degree of oxygen removal. Some of the typical reactions used in biomass processing (oxidation, hydrogenation, cracking, etc.) are similar to processes used in the chemical industry. Conversely, because of a large number of hydroxyl and other functional groups, dehydration, aldol condensation,

ketonization, and decarboxylation, which are less frequent in the chemical industry, would be required in biomass processing, when and if the concepts of biorefi-
nery are realized in industry.

In large-scale production of chemicals with oil and gas as the main feedstock,
there are many important reactions such as various hydrogenations (Ni, Co, Cu, Pd,
or Pt catalysts), dehydrogenations (on oxides, Ni, Fe, Cu), oxidations (over metals,
reducible oxides), and acid-catalyzed (zeolites, aluminosilicates) reactions, such as
alkylation, hydration, dehydration, condensation, etc. Some of these reactions are
presented in Tab. 1.3, but this is very far from being conclusive.

Tab. 1.3: Some catalytic processes for the synthesis of basic chemicals (From [18]).

Process or product	Catalyst	Conditions
Hydrogenation		
Methanol synthesis	$ZnO-Cr_2O_3$	250–400 °C, 200–300 bar
$CO + 2H_2 \rightarrow CH_3OH$	$CuO-ZnO-Cr_2O_3$	230–280 °C, 60 bar
Fat hardening	Ni/Cu	150–200 °C, 5–15 bar
Benzene to cyclohexane	Raney Ni	liquid-phase 200–225 °C, 50 bar
	noble metals	gas-phase 400 °C, 25–30 bar
Aldehydes and ketones to alcohols	Ni, Cu, Pt	100–150 °C, 30 bar
Esters to alcohols	$CuCr_2O_4$	250–300 °C, 250–500 bar
Nitriles to amines	Co or Ni on Al_2O_3	100–200 °C, 200–400 bar
Dehydrogenation		
Ethylbenzene to styrene	Fe_3O_4 (Cr, K oxide)	500–600 °C, 1.4 bar
Butane to butadiene	Cr_2O_3/Al_2O_3	500–600 °C, 1 bar
Oxidation		
Ethylene to ethylene oxide	Ag/support	200–250 °C, 10–22 bar
Methanol to formaldehyde	Ag cryst.	~600 °C
Benzene or butane to maleic anhydride	V_2O_5/support	400–500 °C, 1–2 bar
o-Xylene or naphthalene to phthalic anhydride	V_2O_5/TiO_2 $V_2O_5-K_2S_2O_7/SiO_2$	400–450 °C, 1.2 bar
Propene to acrolein	Bi/Mo oxides	350–450 °C, 1.5 bar
Ammoxidation		
Propene to acrylonitrile	Bi molybdate (U, Sb oxides)	400–450 °C, 10–30 bar
Methane to HCN	Pt/Rh nets	800–1,700 °C, 1 bar
Oxychlorination		
Vinyl chloride from ethylene +HCl/O_2	$CuCl_2/Al_2O_3$	200–240 °C, 2–5 bar
Alkylation		
Cumene from benzene and propene	H_3PO_4/SiO_2	300 °C, 40–60 bar
Ethylbenzene from benzene and ethylene	Al_2O_3/SiO_2 or H_3PO_4/SiO_2	300 °C, 40–60 bar

As mentioned above, environmental catalysis constitutes a substantial part of all catalyst demand.

Photos presented in Figs. 1.35 and 1.36 are not directly related to catalysis, reflecting, however, different problems associated with emissions in the environment. Pictures in Fig. 1.35 were taken by the author on two successive days (Saturday and Sunday) from the same hotel room in Beijing. Apparently, a lower demand for coal utilization during Saturday resulted in better visibility on the next day. The other photo (Fig. 1.36) is somewhat older and not directly related to the personal experience of the author. During a drive along the Cuyahoga River in downtown Cleveland, Ohio, my brother, who is residing in one of the suburbs nearby, showed the place where the river caught fire because it became so polluted with chemicals.

Fig. 1.35: Different levels of smog in Beijing (photo of the author).

Historically, supported Pt/Al_2O_3 catalysts were first used in the 1940s for the catalytic purification of off-gases by oxidation. The catalytic control of NOx emissions from nitric acid and power plants, as well as various automobile exhaust emissions, became driven by legislature. As non-catalytic methods require high temperatures and non-selective methods use an excess of reducing agents, the SCR (Selective Catalytic Reduction) of NOx had been developed to operate at moderate temperatures and with low amounts of reductants. For stationary sources, ammonia SCR is applied, with the ammonia dosed in stoichiometric quantities:

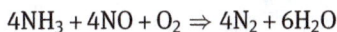

$$4NH_3 + 4NO + O_2 \Rightarrow 4N_2 + 6H_2O$$

In the presence of sulfur, undesired side reactions

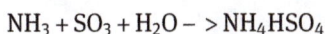

$$SO_2 + 1/2\,O_2 - > SO_3$$

$$NH_3 + SO_3 + H_2O - > NH_4HSO_4$$

Fig. 1.36: A fire tug fights flames on the Cuyahoga River near downtown Cleveland, Ohio, where oil and other industrial wastes caught fire on June 25, 1952 [21].

can also occur. Apart from the parallel side reactions, selectivity is almost 100% at the same conversion level. The potential drawback in NH_3 SCR is ammonia slippage that requires careful control of ammonia dosage.

Although different catalysts with different operation ranges can be used for ammonia SCR (Fig. 1.37A), vanadium pentoxide, supported on titania and wash-coated on a ceramic monoliths, is often applied in the form of blocks, which are mounted in a larger structure (Fig. 1.37B).

The catalytic converters for automotive exhaust emissions, whose level is driven by local legislature and is thus different in various parts of the world, are also monolithic supports (ceramic or metallic) coated with platinum and other noble metals. The monolith structures are demonstrated in Fig. 1.38.

The converters are placed behind the engine and the conventional silence module (Fig. 1.38C). For conventional three-way catalysts (Fig. 1.39) the operation of a catalytic exhaust control system is critical, particularly the monitoring of the air-fuel ratio (Fig. 1.40). In an excess of fuel the concentration of unburned hydrocarbons is high, while with an excess of air the conversion of NOx is diminished, caused by the lack of the reducing agent. The following reaction might occur (where HC are hydrocarbons):

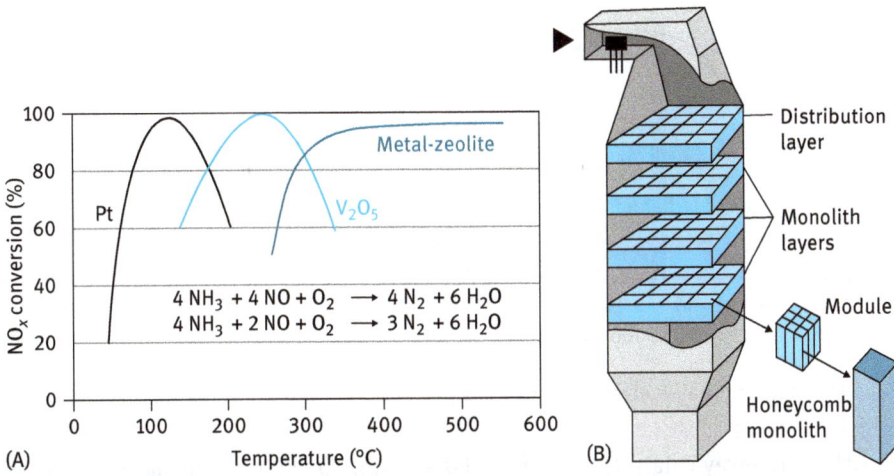

Fig. 1.37: Catalysts for ammonia SCR: (A) temperature dependence of NO_x conversion, (B) monolithic blocks.

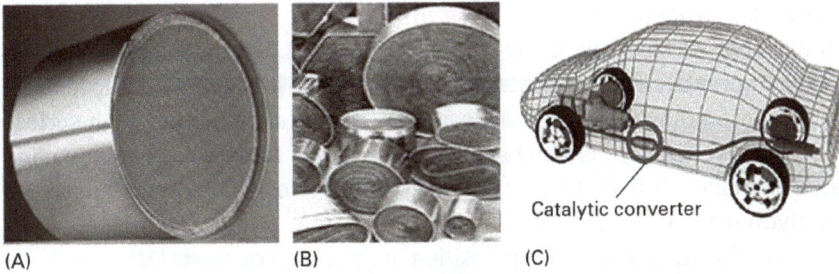

Fig. 1.38: Typical converters: (A) ceramic, (B) metallic, (C) placement.

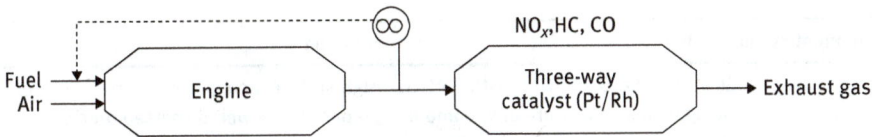

Fig. 1.39: Three-way catalyst.

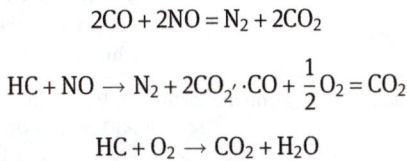

$$2CO + 2NO = N_2 + 2CO_2$$

$$HC + NO \rightarrow N_2 + 2CO_2, \cdot CO + \frac{1}{2}O_2 = CO_2$$

$$HC + O_2 \rightarrow CO_2 + H_2O$$

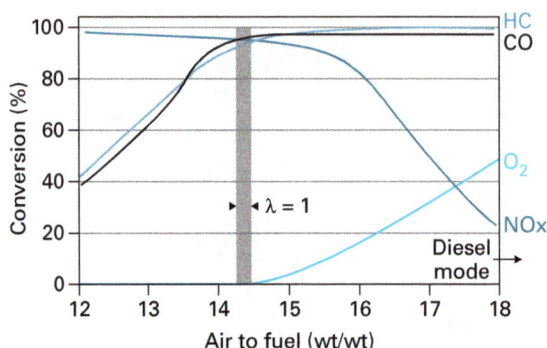

Fig. 1.40: Effect of the air-fuel ratio on emissions.

The three-way catalysts contain Pt, Pd and Rh. The latter is the most expensive component in three-way catalysts, being, however, needed because of its activity and high selectivity in reduction of NO to dinitrogen with low formation of ammonia. Effective utilization and stabilization of Rh is thus important. Palladium oxide displays moderate activity and selectivity in NO reduction, while Pt is the least selective resulting in ammonia.

Multilayered washcoating strategy is applied in the design of three-way catalysts to separate catalyst components, which can otherwise interact leading to deactivation. Such multilayered washcoats comprise one or more porous, high surface area supports, which are thermally stable, for example, La-stabilized alumina and stabilized zirconia in different layers. Ceria–zirconia mixed oxide acts as an oxygen storage. Hydrated alumina is used as a binder.

Table 1.4 illustrates a typical composition of three-way catalysts [22].

Tab. 1.4: Typical composition of three-way catalysts. (Reproduced with permission from [22]).

Component	Composition	Function
Supports	– 10–12% La_2O_3/Al_2O_3 with Al_2O_3 being usually a mixture of γ, δ and θ – ZrO_2	High surface area, porous carrier; enables preparation of well-dispersed precious metals, especially Pt and PdO, prevent their sintering Lanthana stabilizes alumina against loss of surface area above 700 °C ZrO_2 is a noninteracting support for Rh, which is used in a separate layer
Catalytic phase(s)	PdO, Pt, Rh (usually 0.1–2 g/L on monolith)	Pt, PdO oxidize CO and hydrocarbons; Rh, PdO reduce NO, using CO and hydrocarbons as reducing agents.

Tab. 1.4 (continued)

Component	Composition	Function
Oxygen storage material	Solid solution of 40 at% of ceria and 60% of zirconia; about 30–50% of washcoat	ZrO_2/CeO_2 stores oxygen during oxidizing cycle releasing it during reduction cycle; zirconia and rare-earth oxides (Pr_2O_3 and Nd_2O_3 stabilize CeO_2 against sintering)
Additives and promoters	Less than 1 wt% of total catalyst	Different types are used (e.g., nickel oxide diminishes the formation of H_2S)

More stringent regulations regarding fuel usage resulted in the introduction of lean burn and diesel engines with approximately 25–30% less fuel consumption. Engine manufactures in Europe mainly tuned engines to minimize particulates, which increases NOx emissions. In the USA engine out emissions are mainly tuned to minimize NOx leading to a high concentration of particulates (Fig. 1.41).

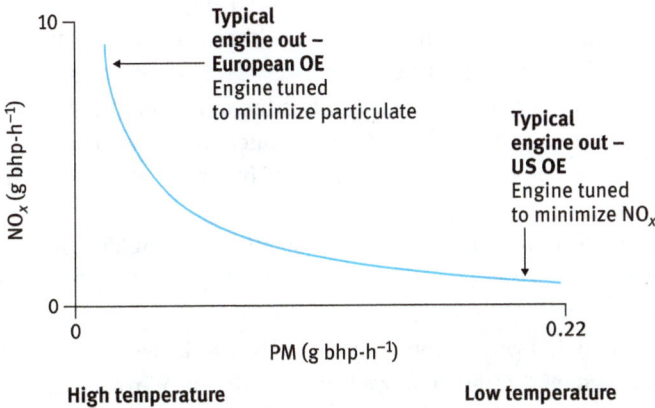

Fig. 1.41: Engine emissions map.

Conventional three-way catalysts do not work under an excess of air, so other technologies have been introduced for diesel and lean burn small, medium and large (off-road) engines.

Figure 1.42 illustrates the NOx storage concept developed by the Toyota Motor Corporation, which involves periodic operations and low sulfur fuels. In the lean period (excess of air) NOx is oxidized into NO_2 and stored as nitrates, while during the rich (in fuel) period metal oxide is regenerated. The sulfur level in fuel has to be minimized, as the storage material (Ba) has a higher affinity to sulfates than nitrates.

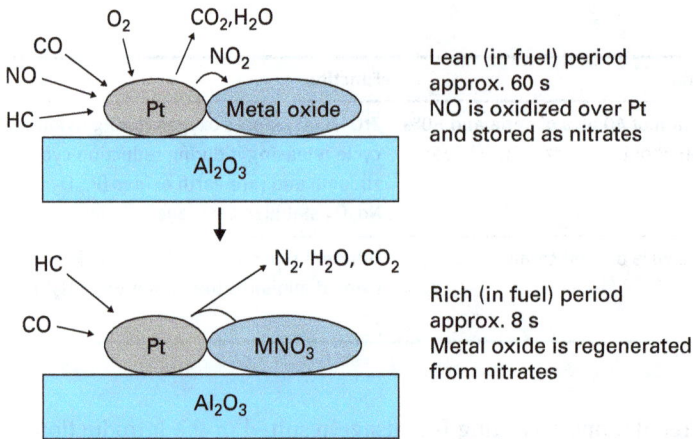

Fig. 1.42: NOx storage concept. (From [23]).

Desulfurization, if done to regain activity, requires rich conditions and elevated temperatures (> 700 °C) provoking thermal aging. NOx sensors are needed to control the periodic operation and desulfurization, calling for an advanced engine control.

SCR of NOx emissions from diesel engines might use either ammonia (or in fact urea for easier dosage, Fig. 1.43) or hydrocarbons.

Hydrocarbons SCR cannot rely on passive control as the amounts of unburned hydrocarbons from the engine are not sufficient for the reduction of NOx over a catalyst. Instead it requires an active control (the addition of fuel after diesel particulate filter upstream SCR catalyst, Fig. 1.44).

Catalysts in exhaust gas cleaning should operate constantly under highly transient and very demanding conditions (extremely high flow rate: GHSV < 200,000 h^{-1}, almost zero pressure drop, simultaneous oxidation and reduction, presence up to ~12 vol% H_2O in the feed). Silver on alumina was recently reported to be an efficient catalyst for hydrocarbons SCR with a potential to be employed industrially for off-road engines. High fuel penalty (Fig. 1.45) and negligible low temperature activity could be among the drawbacks of the catalyst, restricting its application.

Nevertheless, hydrocarbon lean NO_x catalyst system using a silver-based catalyst was developed by General Electric for large locomotive applications, and is provided by Tenneco for on- and off-road emission control markets.

A special case worth mentioning is development of SCR systems for different types of ships, including large cruise ships (Fig. 1.46), as they use marine fuels with a high sulfur content.

For example, already from 2009 the sulfur content in on-road diesel was limited to 10 ppm in some regions of the world, while for ships operating outside designated emission control areas the current limit for sulfur content of marine fuel oil is 3.50% mass/mass. Obviously, SO_2 emissions for fuels containing large amounts of

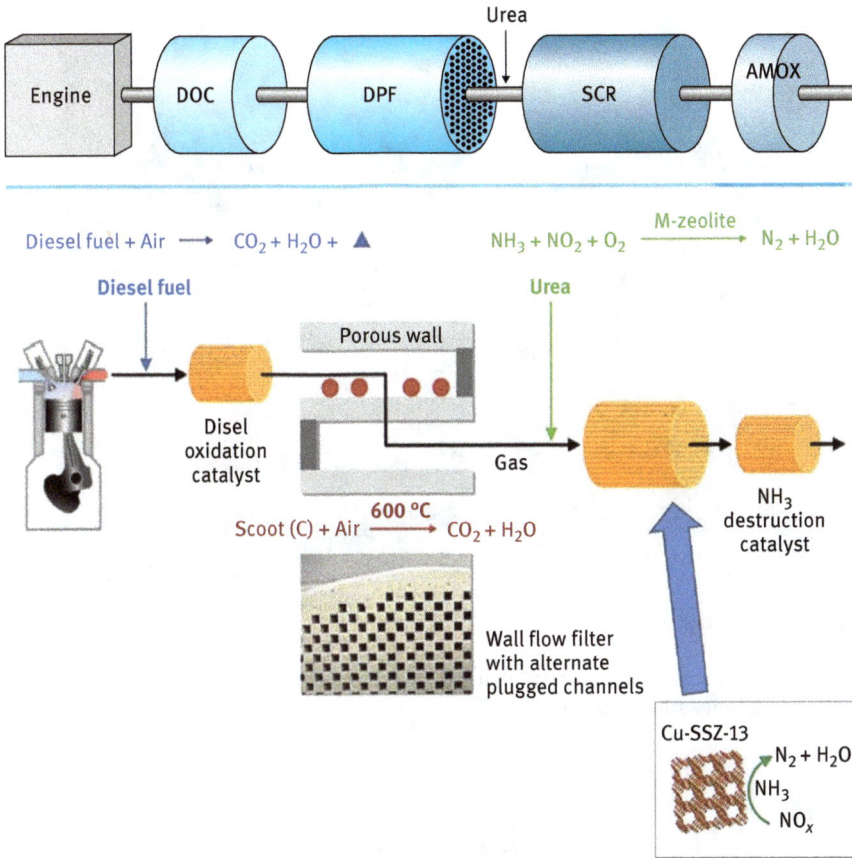

Fig. 1.43: SCR with ammonia in diesel engines. (From [24].) DOC, diesel oxidation catalyst; DPF, diesel particulate filter; AMOX, ammonia oxidation catalyst.

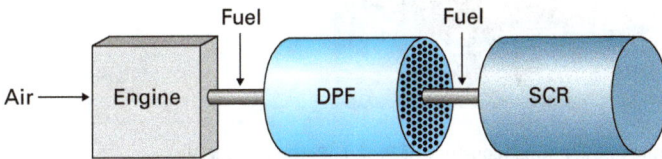

Fig. 1.44: Hydrocarbons SCR.

sulfur are quite high being orders of magnitude higher than from on-road diesel engines. A substantial cut to 0.50% m/m will apply after 1 January 2020. An even stricter limit of 0.10% m/m is already applied in emission control areas such as Baltic Sea area, the North Sea area, the North American area, covering designated

Fig. 1.45: Laboratory-scale installation at Turku University of Applied Sciences for HC SCR using a monolithic block with a silver catalyst.

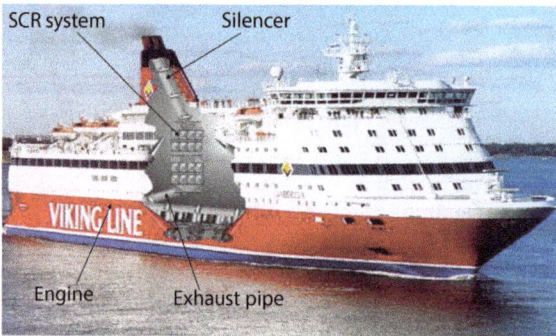

Fig. 1.46: Urea SCR for marine applications (courtesy of Prof. L. Pettersson, KTH).

coastal areas of the United States and Canada, the United States Caribbean Sea area around Puerto Rico and the United States Virgin Islands.

Heterogeneous catalysts, such as platinum group metals, also have numerous applications in the removal of volatile organic compounds and hazardous air pollutants from the chemical and pharmaceutical manufacturing processes. Monoliths are applied in addition to pellets, which require more volume than monoliths and have higher pressure drop. Pt and Pd can be used for CO oxidation, and combinations of Pt, Pd, Rh and Ir are applied while in the presence of CO and hydrocarbons various. In special cases of certain pollutants, such as halogenated organics, chromium containing catalysts can be used.

An interesting example of hazardous air pollutant removal is ozone abatement in high-flying commercial aircrafts [25]. Intake air fed to the passenger cabin above 12,000 m altitudes can contain up to ca. 1–4 ppm of ozone, which can cause headache, chest pain as well as irritations of the eye, nose and throat. This altitude is typically used because air resistance is decreased upon elevation of altitude diminishing the fuel consumption. Removal of ozone to the level of 0.1 vppm is done using 1% Pd on alumina supported in a monolith at 100–160 °C (Fig. 1.47).

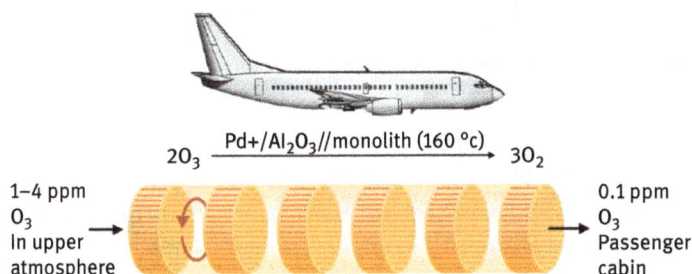

Fig. 1.47: Pd catalyst for ozone removal.

High space velocities of 200,000–500,000 h and low concentration of ozone require efficient contact of ozone with the catalyst. Because of variable flows, pressure drop, lightweight requirements to decrease fuel penalty, vibration and mechanic shock, ceramic and more lately metallic monolith were applied. A monolith washcoat similar to the automobile converter maximizes geometric surface area and mass transfer at the same time allowing minimum pressure drop.

Catalyst deactivation is possible by contaminations of sulfur, phosphorus and hydrocarbon compounds present in compressor lubricating oils, hydraulic fluids and ground-level air. To eliminate odor from hydrocarbons entering the cabin because of long waits behind other aircrafts on the runaway, a small amount of platinum is added to the catalyst. Another source of contamination is deicing fluids, which can enter the air intake system. Catalyst regeneration can be done by applying washing solutions after 10,000–20,000 flight hours.

In addition to the application of catalysts in various environmental and biomass-related reactions and the use of CO_2 as a feedstock, it can be predicted that more selective catalysts will be developed for the production of basic and specialty chemicals (for example, in oxidation processes). Moreover, catalysts will contribute to the production of new functional polymers with unique properties as well as production of fine chemicals, pharmaceuticals and consumer electronics.

In particular fine chemicals are usually produced in several reaction steps using stoichiometric methods and result in high waste-to-desired product ratio, the E-factor, proposed by Sheldon (Tab. 1.5) [26].

Tab. 1.5: The annual production of different chemicals and the amount of formed by-products per amount of desired product.

Industry	Production (t year^{-1})	By-products (kg kg^{-1})
Oil refining	10^6–10^8	0.01–0.1
Bulk chemicals	10^4–10^6	> 1–5
Fine chemicals	10^2–10^4	5–50
Pharmaceuticals	10–10^3	25–100

One of the reasons for high E factors and the use of classical synthetic methods in the synthesis of fine chemicals and pharmaceuticals – besides the fact that synthetic chemists are unfamiliar with heterogeneous catalysis – is that commercially available general purpose heterogeneous catalysts do not possess the required selectivity. Although selectivity of these catalysts, bearing metals of nanoparticle size, could be enhanced by carefully adjusting the size and environment of metal nanoparticles, it obviously requires dedicated time-consuming work. When the priority is in getting the product to the market as fast as possible, such process development might not be seen as an important factor.

Atom efficiency is another metric used to evaluate how sustainable or "green" the process is. It is defined as the molecular weight of the desired product divided by the total molecular weight of all products. For stoichiometric oxidation of a secondary alcohol:

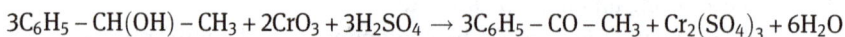

$$3C_6H_5 - CH(OH) - CH_3 + 2CrO_3 + 3H_2SO_4 \rightarrow 3C_6H_5 - CO - CH_3 + Cr_2(SO_4)_3 + 6H_2O$$

an atom efficiency is 42%, while the theoretical E-factor is 1.5.

If the same reaction is conducted catalytically:

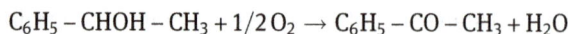

$$C_6H_5 - CHOH - CH_3 + 1/2 O_2 \rightarrow C_6H_5 - CO - CH_3 + H_2O$$

an atom efficiency is 87%, with the E-factor equal to zero if water is not considered as a waste.

Various hydrogenations, carbonylation and selective oxidation reactions are of note among 100% atom efficient processes.

1.4 Fuel cells and electrocatalysis

A fuel cell is an electrochemical cell, where typically the chemical energy of the fuel (hydrogen) and an oxidant (oxygen) is converted to electricity in a continuous mode, provided there is a supply of reactants. In hydrogen–oxygen fuel cell (Fig. 1.48), only water as the by-product in the overall chemical reaction is generated.

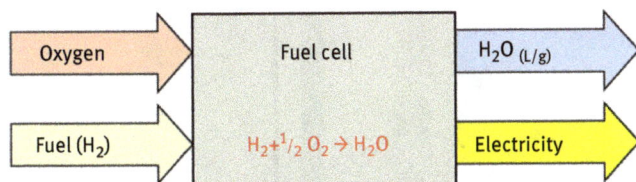

Fig. 1.48: Principle of a fuel cell as an electrochemical energy conversion device. (Reproduced with permission from [27]).

Fuel cells are used for primary and backup power for different buildings in remote areas, as well as for vehicles operating with fuel cells (e.g., automobiles, buses, boats, motorcycles or submarines).

The Gibbs free energy, ΔG°, of the overall fuel cell reaction

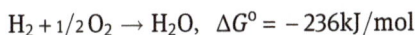

$$H_2 + 1/2\,O_2 \rightarrow H_2O, \quad \Delta G^{\circ} = -236 kJ/mol$$

is related with the equivalent electric cell potential ΔE° by

$$\Delta G^{\circ} = -nF\Delta E^{\circ} \tag{1.4}$$

where F is the Faraday constant and n denotes the number of electrons exchanged in the overall chemical process.

Fuel cells consist of an anode, a cathode and an electrolyte allowing hydrogen ions to move from anode to cathode. Two coupled, however, spatially separated half-cell redox catalytic reactions at the anode and cathode sides are presented in Fig. 1.49.

On the anode (often platinum) surface, dihydrogen is split into protons and electrons in the hydrogen–oxygen reaction (HOR), while water is generated from oxygen protons and electrons on the cathode side (often nickel) in the oxygen-reduction reaction (ORR). Different types of ion conductors can be used, including KOH, solid polymers, solid oxides (yttria, zirconia), phosphoric acid, lithium and potassium carbonates. A typical fuel cell produces a voltage from 0.6 to 0.7 V at full load. A schematic view on the processes in a fuel cell is given in Fig. 1.50.

Besides HOR and ORR relevant for fuel cells, Fig. 1.50 also shows the oxygen evolution reaction and hydrogen evolution reaction.

Requirements for a certain cell current (j_{cell}) imply that the anode potential shifts more positive by $\eta_{act,HOR}$. A negative shift of the cathode potential by $\eta_{act,ORR}$ even further diminishes the observed cell potential V, which is thus smaller than

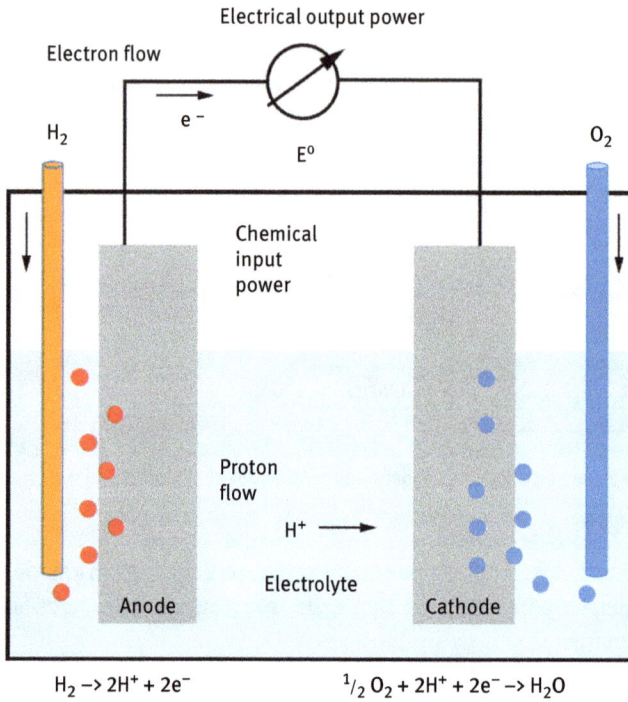

$$H_2 \rightarrow 2H^+ + 2e^-$$ $$\tfrac{1}{2}O_2 + 2H^+ + 2e^- \rightarrow H_2O$$

Fig. 1.49: Principle of a hydrogen and oxygen fuel cell galvanic element. (Reproduced with permission from [27]).

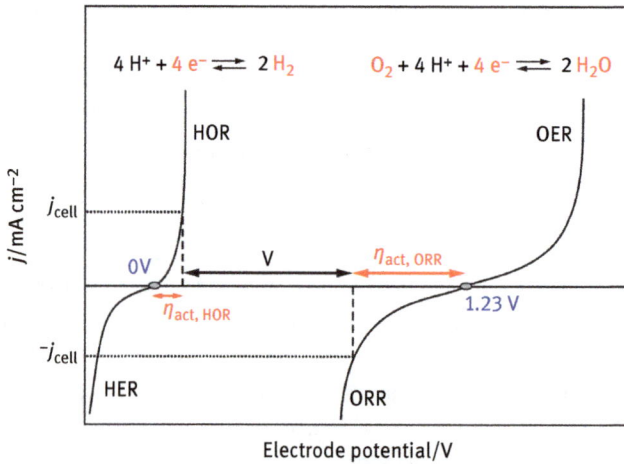

Fig. 1.50: Schematic origin-of-activation overpotentials in a hydrogen-oxygen fuel cell. (Reproduced with permission from [27]).

the difference between the standard potentials of the hydrogen electrode (0 V) and the oxygen electrode (1.23 V).

While Pt-based metals or alloys are often considered as efficient catalysts for ORR, there are several challenges in applying them, including agglomeration of nanoparticles, Ostwald ripening, oxidation of noble metals and leaching of the nonnoble component in alloy catalysts. Designing of alternatives to Pt catalysts, a preferably cost-efficient, more stable and not containing precious metals, is an important research task.

References

[1]   Berzelius, J.J. (1836) Quelques idees sur une nouvelle force agissant dans les combinaisons des corps organiques, Ann. Chim. Phys. 61: 146

[2]   Liebig, J. (1840) Die Organische Chemie in ihre Anwendung auf Agricultur und Physiologie. Braunschweig.

[3]   http://www.nobelprize.org/nobel_prizes/chemistry/laureates/1909/ostwaldlecture.html

[4]   Arnaut, L.G., Formosinho, S.J., Burrows, H. (2006) Chemical Kinetics: From Molecular Structure to Chemical Reactivity. Amsterdam, Elsevier.

[5]   Laidler, K.J. (1987) Chemical Kinetics. New York: HarperCollins.

[6]   Robson Wright M. (2004) Introduction to Chemical Kinetics. Wiley.

[7]   Rothenberg G. (2008) Catalysis: Concepts and Green Applications. Chichester, Wiley-VCH, Weinheim.

[8]   Balandin, A.A. (1969) Modern state of the multiplet theory of heterogeneous catalysis. Adv. Catal. Rel. Subj. 19: 1.

[9]   Rostrup-Nielsen, J.R. (2012) Perspective of Industry on modeling catalysis. In: Deutschmann, O., Ed. Modeling and Simulation of Heterogeneous Catalytic Reactions: From the Molecular Process to the Technical System. Weinheim, Germany: Wiley-VCH, pp. 283–301.

[10] Lilja, J., Aumo, J., Salmi, T., Murzin, D. Yu., Mäki-Arvela, et al. (2002) Kinetics of esterification of propanoic acids with methanol over a fibrous polymer-supported sulphonic acid catalyst. Appl. Catal. 228: 253.

[11] Geus, J.W., van Dillen, A.J. (2008) Preparation of supported catalysts by deposition-precipitation. In: Knözinger, H., Schueth, F., Weitkamp, J. Handbook of Heterogeneous Catalysis. Weinheim, Germany: Wiley-VCH.

[12] Binder, A., Seipenbusch, M., Muhler, M., Kasper, G.J. (2009) Kinetics and particle size effects in ethene hydrogenation over supported palladium catalysts at atmospheric pressure. J. Catal. 268: 150.

[13] Moulijn, J.A., Makkee, Mi., van Diepen, A.E. (2011) Chemical Process Technology. Chichester, Wiley.

[14] Thomas, J.M., Thomas, W.J. (1997) Principles and Practice of Heterogeneous Catalysis. Weinheim, Wiley.

[15] Chorkendorff, I., Niemanstverdriet, J.W. (2003) Concepts of Modern Catalysis and Kinetics. Weinheim, Germany: Wiley-VCH.

[16] Bond, G.C. (1962) Catalysis by Metals. London: Academic Press.

[17] Murzin, D.Yu., Kul'kova, N.V. (1995) Kinetics and mechanism of the liquid-phase hydrogenation. Catal. Today 24: 35.

[18] Hagen, J. (2006) Industrial Catalysis: A Practical Approach, Weinheim, Germany: Wiley-VCH.

[19] TCGR's Intelligence Report: Business shifts in the global catalytic process industries.

[20] Dutch National Research School, Future Perspectives in Catalysis, (2009) Available at: http://www.vermeer.net/pub/communication/downloads/future-perspectives-in-cata.pdf

[21] http://image.cleveland.com/home/cleve-media/width600/img/opinion_impact/photo/cuyahoga-river-a55214b544fc6cac.jpg

[22] Bartholomew, C.H., Farrauto, R.J. (2006) Fundamentals of Industrial Catalytic Processes, Second Edition, Wiley.

[23] Klingstedt, F., Arve, K., Eränen, K., Murzin, D.Yu. (2006) Towards improved catalytic low-temperature NOx removal in diesel powered vehicles. Accounts Chem. Res. 39:273–282.

[24] www.catalysts.basf.com/mobilesources.

[25] Farrauto, R.J., Armor, J.N. (2016) Moving from discovery to real applications for your catalyst. Appl. Catal. B. Gen. 527: 182.

[26] Sheldon, R.A. (2007) The E factor: fifteen years on. Green Chem. 9:1273.

[27] Strasser, P. (2012) Fuel cells, in Chemical Energy Storage. Ed. R. Schlögl. Berlin: De Gruyter.

2 Engineering catalysts

2.1 Catalyst design

Heterogeneous catalysts, as discussed in Chapter 1, are nanomaterials and thus their active sites should be designed on the nanometer level, which requires understanding of catalytic phenomena on the molecular level.

While at the reactor level bulk density (mass per unit volume of bulk catalyst, packed density) is important, at the mesoscale level particle density (mass per unit volume of pellet, apparent density) and true solid density (mass per unit volume of solid) play a crucial role.

These latter parameters are related to catalyst porosity as catalytic materials usually possess a well-developed pore structure (Fig. 2.1).

Fig. 2.1 illustrates different types of pores. Open pores, which could be interconnected, transport (passing) or dead-end, are accessible while reactants cannot access closed pores. Porosity here could be defined as the volume of pores per total volume of the solid, including the pores. The following parameters are typically used to characterize porosity of catalytic materials: specific surface area (total surface area per mass of solid, $m^2\ g^{-1}$), specific pore volume (total pore volume per mass of solid, $cm^3\ g^{-1}$), pore size (nm) and its distribution (which can be broad or narrow, mono- or multi-modal).

For pore size the following classification, proposed by International Union of Pure and Applied Chemistry (IUPAC) is commonly accepted (Fig. 2.2), even if it is a bit awkward to refer to materials with a pore size of less than one nanometer (for example, zeolites) as microporous materials.

Shaped catalysts particles (mm range) are applied industrially on a macroscopic level. Thus, catalysis engineering requires knowledge of catalytic materials, their activity, stability, mechanical strength, porosity, preparation and characterization, understanding of reaction mechanisms and kinetics, as well as catalyst deactivation. Preparation and characterization of catalytic materials is described in Chapter 2, while the principles of kinetic modeling, including a description of deactivation, will be presented in Chapter 3.

In addition to intrinsic kinetics, transport phenomena and reactor engineering constitute an essential part of catalytic engineering. This is related to the fact, that under industrial conditions, the rates of chemical transformations are very often affected by the rates of other processes, and particularly by mass transfer processes. The main mass transfer effects in heterogeneously catalyzed processes are: internal diffusion inside the porous catalyst particles/layers, external diffusion in the laminar film surrounding the catalyst particles, and – in multiphase reactors – the gas-liquid or liquid-liquid mass transfer resistance at the phase boundary. The description of mass transfer effects and diffusion phenomena will be addressed in Chapter 3.

https://doi.org/10.1515/9783110614435-002

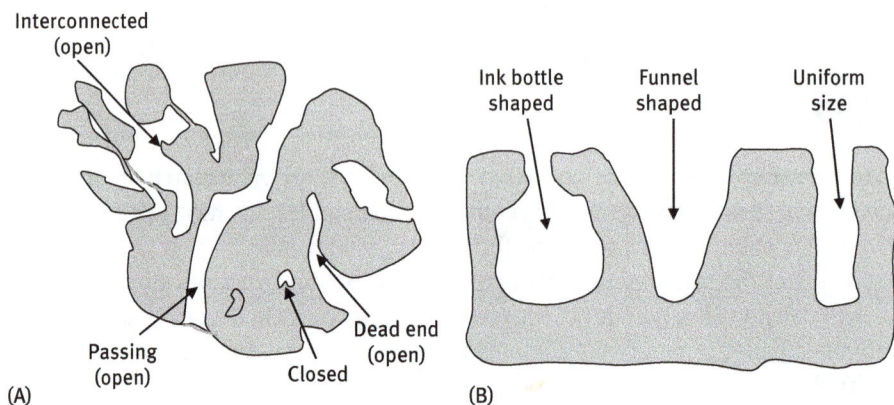

Fig. 2.1: Pore structure with different types of pores. A ([1]) and B ([2]).

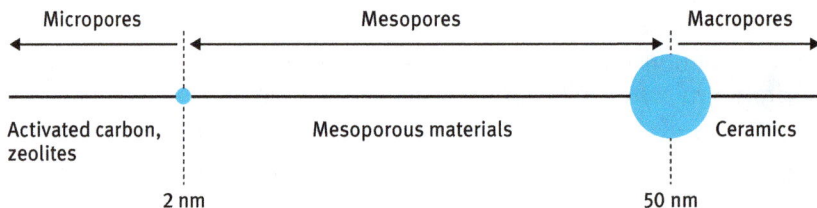

Fig. 2.2: Classification of pores.

Finally, for the understanding of the events occurring at macroscopic or reactor level (m scale) not only microscopic understanding (nm scale – active site) is required, which is mainly the goal of chemists working in the field of heterogeneous catalysis, but also description at the mm (shaped catalyst pellet) and micron scale.

2.1.1 Being in shape

The shaping of catalysts and supports is a key step in the catalyst preparation procedure. The shape and size of the catalyst particles should promote catalytic activity, strengthen the particle resistance to crushing and abrasion, minimize the bed pressure drop, lessen fabrication costs and distribute dust build-up uniformly. While small particle size increases activity by minimizing the influence of internal and external mass transfer, bed pressure drop (Fig. 2.3) increases. Thus, there is an apparent contradiction between the desire to have small catalyst particles (less diffusional length, higher activity) and to utilize large particles displaying lower pressure drop.

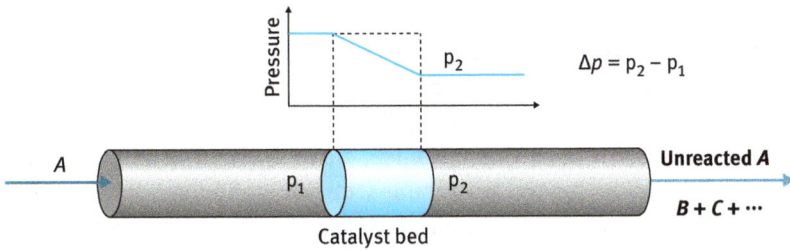

Fig. 2.3: Pressure drop in fixed-bed reactors. Courtesy of Dr. E. Toukoniitty.

There are no precise guidelines about what should be the exact value of pressure drop. It is decided on a case-by-case basis depending on a particular process technology. As a rule of thumb, the size of catalyst particles in fixed beds is exceeding 1–2 mm to avoid high pressure drop, even if larger particles of 10–12 mm are also applied. In addition to the size, the shape is important affecting the bed porosity ε. Bed porosity for spheres, Raschig rings and cylindrical particles vary between 0.35 and 0.4, 0.5 and 0.8 depending on the wall thickness and 0.3 and 0.35, respectively.

Thus, the best operational catalysts have a shape and size that represents an optimum economic trade-off. The requirements of the shape (Fig. 2.4) and size are

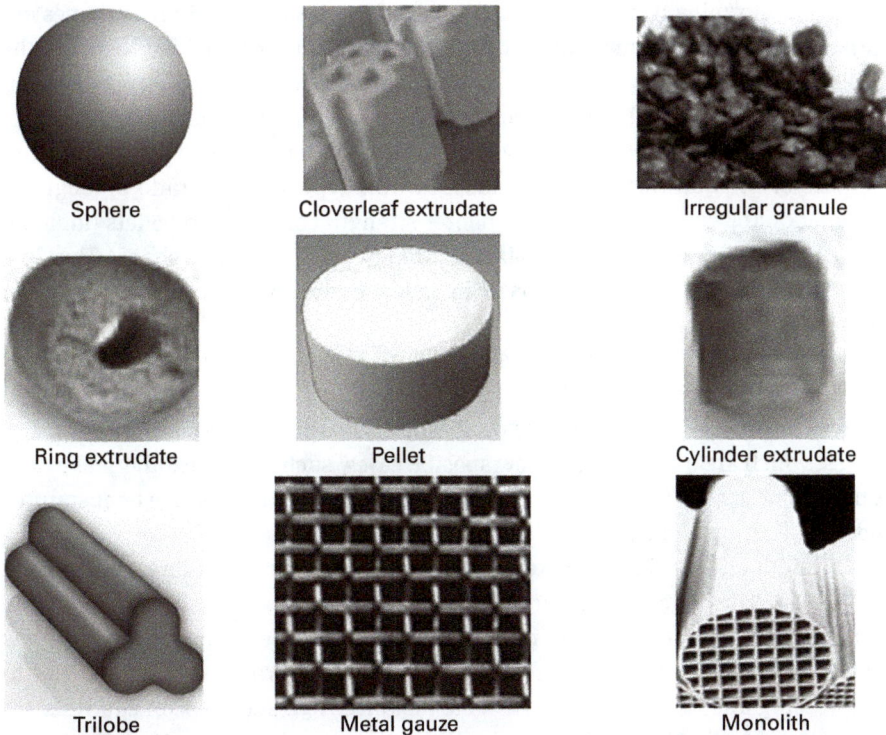

Fig. 2.4: Various catalyst shapes.

mainly driven by the type of reactor. For reactors with fixed beds (see Chapter 3), relatively large particles are applied (several mm) to avoid pressure drop. For moving-bed reactors, spherical particles are preferred because they allow a smooth flow. Catalyst powders of various sizes are utilized in slurry three-phase reactors and in fluidized bed reactors, where mechanical stress is found because of collisions between catalyst particles and with the reactor walls and formation of shear force due to cavitations at high velocities. In the case of slurry reactors, resistance to attrition is important, the size of the particles should allow easy filtration, while the bulk density is defined by settling requirements when easy settling is required. For fluidized bed reactors, attrition resistance is important, as well as particle size distribution.

Compared to other shapes, spherical particles have low manufacturing costs, but possess relatively high-pressure drop and large diffusional length, and are not that common. Similarly, irregular granules, which are low surface area materials, have applications limited to just a few reactions, for example, ammonia synthesis. Extrudates of various profiling are very common (details of extrusion will be provided in Section 2.3.7.3).

Pressure drop can be regulated by making special types of extrudates, ranging from cylindrical to rings to cloverleaf extrudates. In the particular case of natural gas steam reforming (Chapter 4), many types of catalysts with different shapes are available from catalyst manufacturers. Catalysts in the forms of rings, monoliths with several holes, wagon wheels as well as some other complex geometrical shapes (Fig. 4.61) afford a low diffusional path in addition to low pressure drop. The mechanical stability typically deteriorates with an increase of complexity and decrease of the wall thickness. Extrudates with a typical aspect ratio of length to diameter of approximately 3–6 have poorer strength compared to pellets (tablets), as the pellets possess good mechanical strength and a regular shape. This is a very common type of catalysts used in many hydrogenation, dehydrogenation and oxidation reactions.

Monoliths are mainly applied when high fluid flow rates are required (off-gas cleaning for example), as they have low pressure drop.

Low residence time and low pressure drop is a feature of low surface area metal gauzes, which are utilized in very few specific cases, such as very exothermal oxidation of ammonia to NO, when longer residence times lead to excessive temperatures and volatilization of the active catalytic phase.

Different shapes can be ranked in the following order according to their relative pressure drop: monolith<rings<pellets<extrudates<powder. It is also notable that for fixed-bed applications pressure drop should be high enough to allow even flow distribution across the bed, but not too high which would then lead to compressing and recycling costs. In some instances egg-shell type of catalysts are industrially applied to suppress internal mass transfer limitations (Fig. 2.5).

Uniform   Egg-shell   Egg-white   Egg-yolk

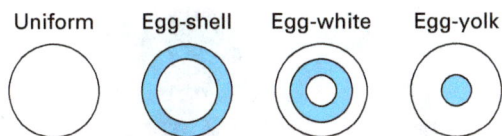

Fig. 2.5: Various concentration profiles in catalyst particles: active phase.

Such egg-shell catalysts with larger grain sizes have the same or similar intrinsic activity as powders of a smaller size with a uniform distribution of the active phase.

The choice of catalyst shapes is thus not straightforward and involves careful considerations of hydrodynamics, heat and mass transfer limitations, potential pressure drops, mechanical strength, thermal resistance to sintering and phase transition, more efficient heat conductivity for strongly exo- and endothermic reactions, as well as manufacturing methods and associated costs.

Moreover, negative effects of a noncatalytic phase (carriers, binders, rheology improvers, lubricants, etc.) and solvents on catalytic behavior should be avoided.

As a rule of thumb for the same surface-to-volume ratio the manufacturing costs are higher for complicated shapes (monoliths, wagon wheels, etc.) compared to more simple ones (i.e., extruded cylinders).

Catalysts development performed in academia in the laboratory scale is mainly limited to powder catalysts with a size that ensures an absence of mass transfer limitations. Catalyst testing is typically conducted in batch and flow reactors, although the use of structured catalysts, such as monoliths, solid foams and fibers as well as microstructured devices, is growing.

Cordierite is widely used as a support material for monoliths ($2MgO \times 2SiO_2 \times 5Al_2O_3$), especially where the operating temperature is below 1,200 °C. The surface area of cordierite is low and the active metal cannot be deposited directly on the monolith. Therefore it is first wash-coated, for example, with alumina, to increase the surface area (Fig. 2.6A) resulting in a wash-coat thickness of 15–100 μm. Fig. 2.6B shows a monolith block for a heavy duty vehicle that does not have any open passages between the channels.

As the thickness of the coating for squared channels is much higher than in other parts (Fig. 2.6A), especially at the corners, other geometries are possible, allowing more uniform wash-coating (Fig. 2.7).

The manufacturing of monolithic catalysts is inherently more costly than of other shapes such as pellets or powders. The economic benefits of using monoliths should be thus clearly demonstrated by exceeding higher catalyst costs and investments in research and development. In particular when the annual volume of each catalyst is small, it is difficult to justify the dedicated research. In addition, cordierite has some limitations in terms of durability when in contact with alkali and alkaline-earth above 700 °C.

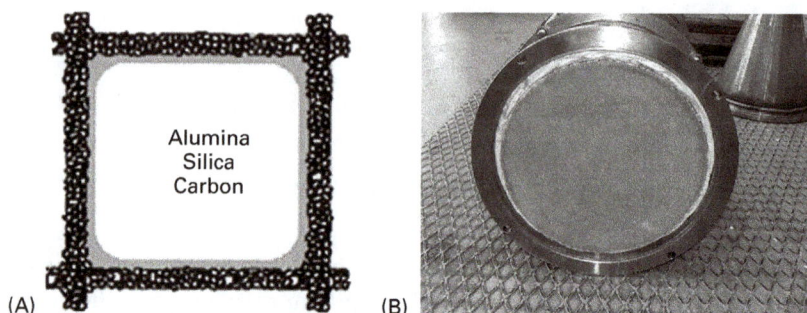

Fig. 2.6: Monoliths: (A) wash-coating principle and (B) a monolithic block for exhaust gas abatement from heavy duty vehicles.

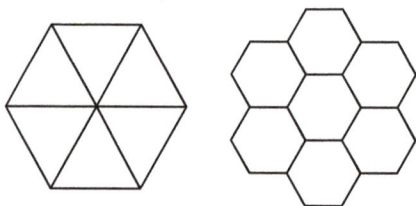

Fig. 2.7: Geometry of channels in monoliths.

Fig. 2.8 shows one of the operation modes of the liquid-phase applications, the use of cordierite is limited to a pH range of roughly pH 4–9.

Other types of structured materials have been recently suggested for use as catalysts for mainly academic purposes. Some of them are also produced commercially and are used in few specific applications related to catalysts.

Similar to monoliths, catalytic foams (Fig. 2.9) require wash-coating, as the surface area of the foams made of α-alumina with several additives is just few m^2 g^{-1}.

Foams, although less mechanically stable than metal or cordierite monoliths, have at the same time several advantages. The pressure drop can be compared to packed beds with a similar flow pattern. Because of the template preparation method (which will be covered in Section 2.3) a desired shape can be produced to fit the reactor size.

Wash-coating is also required for spinning disk reactors, where the two disks are co- or counter-rotating, with each at a different rotational speed (Fig. 2.10). Although efficient mixing and heat transfer can be achieved in such reactors, advantages of using them for heterogeneous catalytic reactions are yet to be demonstrated.

As a final example of different catalyst shapes, it is worth mentioning fiber catalysts with polyethylene as a base material further modified by grafting different functional groups (pyridine, amines, carboxylic and sulfonic acid, or combination

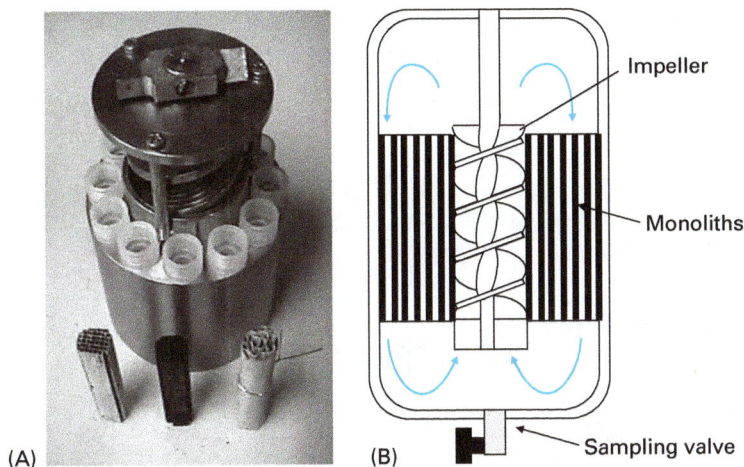

Fig. 2.8: A screw impeller reactor housing 12 pieces of structures catalysts. (A) Monolithic blocks (left) and a corrugated metallic monolith (right). (B) General principle of operation.

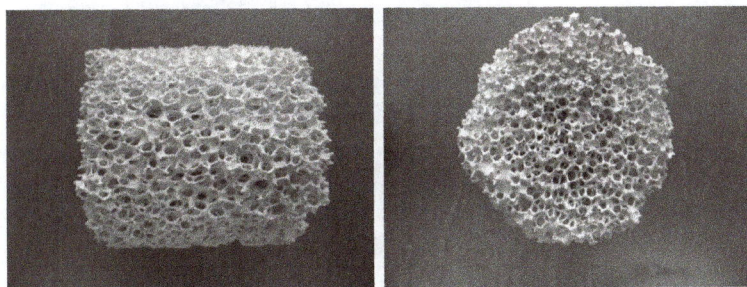

Fig. 2.9: Foam blocks, made of α-alumina, require wash-coating and a subsequent introduction of the active catalytic phase.

of them). Such materials are produced commercially and could be used as acid catalysts similar to ion exchange resins or can be utilized in metal scavenging. For catalytic reactions requiring a metal function, such metals can be immobilized on the ion exchange fiber (Fig. 2.11) with a subsequent reduction with hydrogen or a chemical reducing agent.

The fiber based catalysts can be produced in several shapes, with the diameter of the fiber ranging from 10 to 50 μm, ensuring a short diffusional path. The swelling of such fibers in many solvents and low mechanical strength are a disadvantage. This is significantly improved in some activated carbon cloth materials, which have very high specific surface area (above 1,500 $m^2 g^{-1}$), are mechanically rather strong and can be used (after deposition of the catalytic phase) in slurry and fixed-bed reactors (Fig. 2.12).

Spinning disks reactor

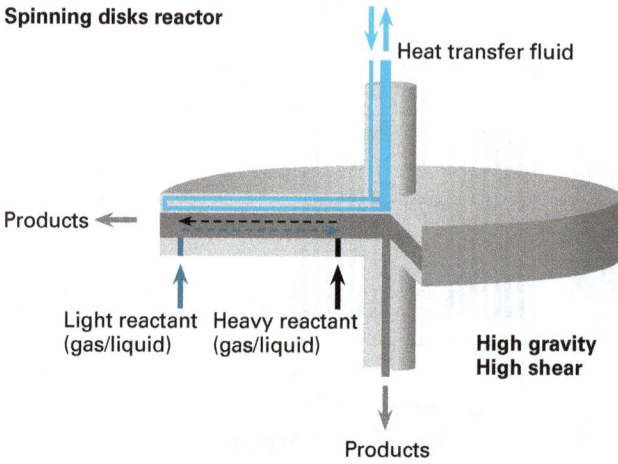

Fig. 2.10: Spinning disk reactor. (From [3]).

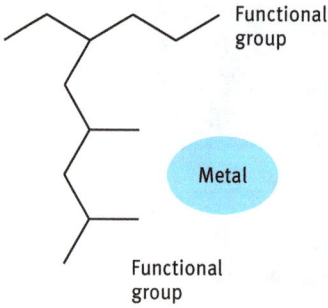

SEM micrograph of the catalyst fibers
fiber diameter = 10 μm, fiber length = 4,000 μm

Fig. 2.11: Polyethylene fiber catalyst.

(A)

(B)

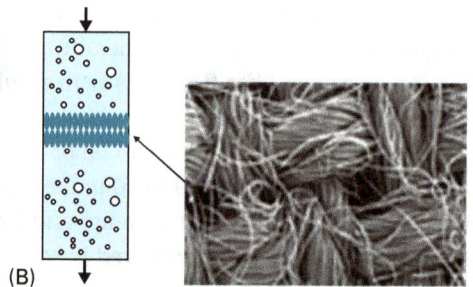

Fig. 2.12: Active carbon cloth. (A) General appearance. (B) Application in a fixed-bed reactor with several layers of carbon cloths.

2.1.2 Catalysis informatics and high-throughput experimentation

In the recent years, catalysis informatics and high-throughput experimentation became an essential part of catalyst development [4], enabling optimization of operating conditions and discovery of new catalyst materials. Data mining in catalysis informatics should help in predicting the required catalyst composition and structure, allowing to achieve the required catalytic performance. To this end analysis of available data on the catalytic reactions (activity, selectivity, stability, reaction parameters) as well as surface and bulk characterization should be done with appropriate statistical tools and physical models to integrate this information.

Catalytic informatics is indispensable in high-throughput combinatorial studies that generate enormous amount of data. As an example, it can be mentioned that a catalyst with five transition-metal components can give over 150 million possible compositions.

On top of that not only the elemental composition should be considered but other parameters as well, such as the precursor type, preparation (impregnation, deposition, ion exchange, etc.) and post-treatment methods (calcination, reduction), catalyst shape and reaction parameters to name a few [5].

High-throughput experimentations in heterogeneous catalysis involve high-throughput synthesis, characterization and screening techniques using robotics and special software.

High-throughput qualitative or semiquantitative primary screening focused on discovery is typically performed on small samples and can be often done in unconventional reactors. During primary screening, both false positives and especially false negatives should be minimized. Evolutionary strategies in high-throughput development of heterogeneous catalysts are based on creating new catalysts from the best catalysts of the previous generation by qualitative (replacing one or more chemical elements by others) and quantitative mutations (changing concentration of one or more compounds), as well as crossover (exchanging components of two materials of the previous generation). Selection of the best catalysts is based on the position of catalysts according to ranking.

For primary screening, one of the applied options is preparation of solid catalysts as thin-film spots on the surface of wafer substrates allowing deposition of hundreds to thousands of small spots of differently composed materials on a support. Such approach involves a limited number of preparation steps without, however, a guarantee that the catalysts would have the same structure as bulk catalysts because of small amounts and potential interactions with the support. Obviously, hydrodynamics as well as mass and heat transfer are different from lab-scale reactors.

An interesting option in the high-throughput catalysis is to use single pellets that might, but not necessarily, behave like pellets in an industrial catalyst bed.

The next level is secondary screening when the so-called primary "hits" are confirmed and optimized using reactors, catalyst shapes and analytics resembling real-life applications.

Table 2.1 gives a comparison between these stages of screening.

Tab. 2.1: Comparison between stages 1 and 2 in combinatorial screening (modified from [6]).

	Stage 1	Stage 2
Throughput	Very high (up to 10,000 experiments per day)	Moderate (up to 100 experiments per day)
Synthesis	Methods from inorganic chemistry, small amounts (micrograms)	Methods similar to industrial preparation, large amounts (0.5–20 g)
Testing	Fast testing, conditions different from lab reactors	Lab reactors, considerations of hydrodynamics, heat and mass transfer
Analytics	Dedicated, qualitative and semiquantitative	Quantitative (conversion, selectivity), generic
Application	Discovery of novel materials	Improvement of known systems, investigation of the influence of promoters, kinetics and deactivation

Conventional fixed-bed microreactors with complete analysis of reactants and products and evaluation of the mass balance closure can be used in the tertiary screening phase. Alternatively, piloting can proceed directly from secondary screening.

Preparation of industrial catalysts involves shaping to obtain granulates, beads or pellets as will be discussed later. Such parallel shaping of catalyst powders would be desirable on commercial synthesis platforms not yet being available. Industrial preparation should be also robust, easy to operate and control, cost-efficient in terms of precursors and ecologically friendly avoiding toxic and harmful waste.

Machine learning can assist high-throughput screening efforts by finding correlations between structure and performance [7]. For analysis of macroscopic catalysis data, partial least squares, principal component analysis and artificial neural networks (ANN) are used [8, 9]. These and other statistical classification and regression models help in extracting information from the experimental data, being limited in their ability to generalize beyond the input data. For ANN, the training set must be representative of the test set. After a proper training, in the subsequent execution phase the ANN is fed with new cases providing an output, which should be properly validated.

A particular challenge in heterogeneous catalysis is finding adequate descriptors because many parameters in catalyst preparation (cluster size, phases, synthesis conditions, pre- and post-treatment, etc.) can influence activity, selectivity and stability beyond just chemical composition.

Quantum chemical modeling can be utilized to generate larger datasets than in experiments and complement experimental data, followed by training machine learning models.

Each catalyst comprising a dataset of different catalysts is described based on the important physicochemical properties (electronic structures, physical and atomic properties) defining each material. Thereafter, machine learning tools can be used to find proper descriptors for stability, activity and selectivity.

More details on the theoretical background and quantum chemical calculations in catalysis will be given in Section 2.2.17. Here, it can be mentioned that the energy of the d-band center with respect to the Fermi level is an often used descriptor, as it is connected to the interactions between adsorbate valence states and the d-states of a transition metal surface. Thus, after correlating the band center with adsorption energies, the next step is to relate the adsorption energy of the relevant reactants with catalyst activity through linear scaling relations. Computation of the d-band center typically requires quantum mechanical (QM) calculations at the DFT (density functional theory) level.

QM calculations are computationally expensive, thus being limited to small catalytic systems. Development of interatomic potentials (i.e., mathematical functions for computing the potential energy of a system of atoms) with machine learning after training with data generated by QM can substantially accelerate simulations, keeping accuracy comparable with QM methods. Figure 2.13 describes a workflow in catalyst design involving machine learning.

As a conclusion, machine learning combined with computational modeling and experiments is an emerging path in knowledge-based development of heterogeneous catalysts.

2.2 Toolbox in catalysis

2.2.1 General overview of the characterization methods

Catalyst characterization is required for understanding relationships among physical, chemical and catalytic properties, i.e., relating catalyst structure and function, as well as elucidating catalysts deactivation, designing procedures for regeneration and choosing catalyst properties to minimize such deactivation. Characterization includes the determination of physical properties such as pore size, surface area, particle strength and morphology of the carrier and location of active species within the carrier, elemental and molecular surface composition, elucidation of the nature of adsorbed species, active phase – support and adsorbate – adsorbate interactions, to name a few. Monitoring changes in the physical and chemical properties of the

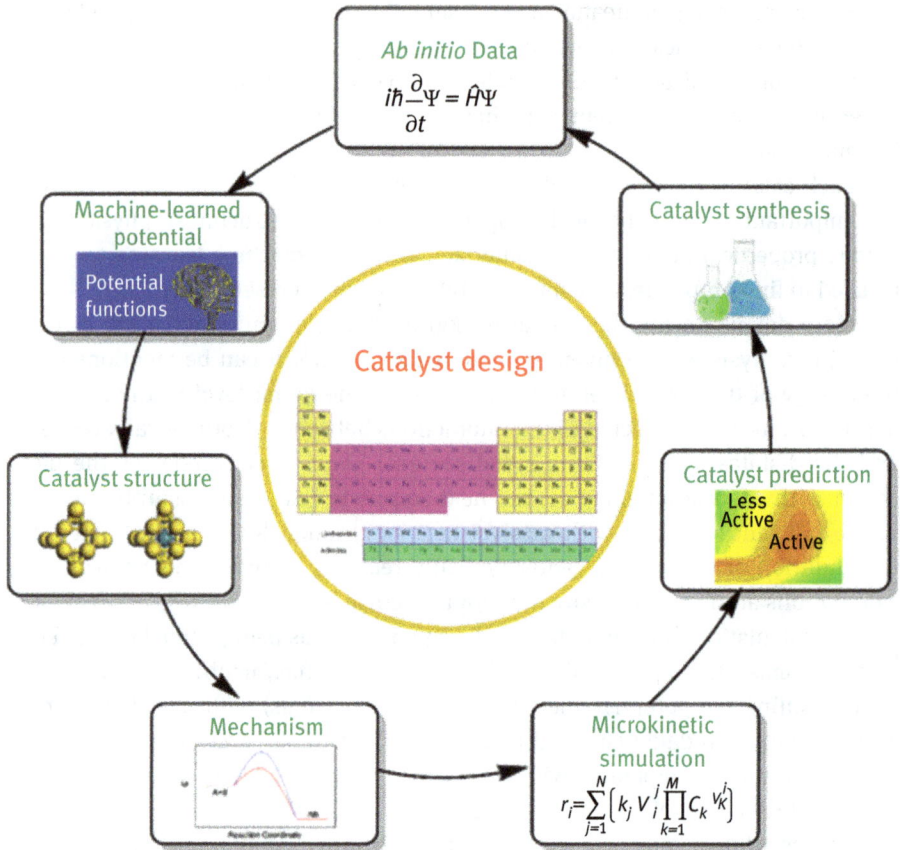

Fig. 2.13: Approaches to catalyst design involving machine learning (Reproduced with permission from [10]).

catalyst during preparation, activation and reaction stages is important for properties control. Each tool has advantages and limitations and several tools are typically needed to provide complimentary data.

In order to get information about the structure surface species various vibrational spectroscopies, such as infrared (IR) spectroscopy, can be applied, which although being of low sensitivity, can be arranged in situ. High-resolution electron energy loss spectroscopy (HREELS), Raman and sum frequency generation (SFG) are among the other methods used. Electron spectroscopy of chemical analysis or X-ray photoelectron spectroscopy (XPS) and Auger spectroscopy is mainly applied to the study of the surfaces, but can also give information about the adsorbed species.

IR spectroscopy is used also to study the nature of surfaces, either measuring the spectra of solid compounds directly (for example, investigating acidic OH groups) or by utilizing probe molecules, for which important information about the number of acid sites and their type (Lewis, Brønsted) can be obtained. Electron microscopy in various forms, XPS, and EXAFS (extended X-ray fine structure), are among methods widely employed to characterize surface structures (Fig. 2.14), while X-ray diffraction is a part of the characterization toolbox aimed at studying the bulk of the material.

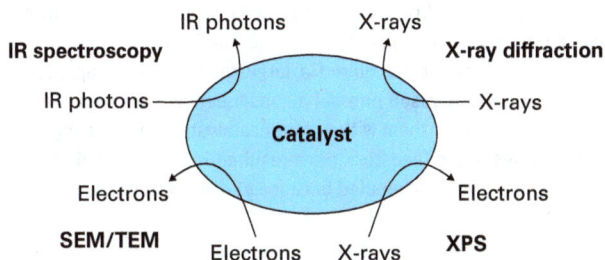

Fig. 2.14: Input–output structures of some analytical methods.

In electron microscopy the overall contrast is caused by the different adsorption of photons or diffraction phenomena. Transmission electron microscopy (TEM, Fig. 2.14) is suitable for the examination of supported particles in the range down to few nanometers. Scanning electron microscopy (SEM) provides topographical images formed from backscattered primary or low energy secondary electrons. An energy dispersive X-ray spectroscopy is often attached to an SEM instrument. Electron spectroscopy of chemical analysis (ESCA) or photoelectron spectroscopy (XPS) is a surface sensitive technique, with the depth of the sample being in the range of – two to 10 atom layers.

Surface science methods like XPS require an ultra-high vacuum, creating what is often called a pressure gap. Such conditions of high vacuum are necessary, as they provide a clear path for the emitted electrons (or other particles) and prevent collisions between the electrons and gas molecules. The average distance molecules travel between collisions is inversely proportional to pressure.

More detailed information about EXAFS, nuclear magnetic resonance (NMR), electron spin resonance (ESR) and Mössbauer spectroscopy will be provided in the corresponding sections.

As this textbook is written for chemical engineers, the description of the methods is uneven, in some instances being very basic and focusing more on the application and interpretation side of the methods, rather than on the detailed physics behind them. The limitations of the methods are specifically addressed, as this information is of real importance for chemical engineers, who are not typically experts in physical methods of catalyst characterization.

Finally, a few words about classical elemental analysis. A catalyst is dissolved in strong acids under (for example) microwave heating, followed by analysis with inductive coupled plasma-MS (ICP-MS, when a sample solution is introduced into the core of inductively coupled argon plasma, which generates temperature of approximately 8,000 °C) or atomic absorption spectroscopy (AAS). In this way, essential information about the catalyst bulk composition can be obtained as it is not related directly to surface properties of the catalyst.

One example from my industrial experience is worth mentioning in the context of elemental analysis. Once, when making such analysis of a dehydrogenation catalyst for an industrial client, a high sulfur content was unexpectedly observed. It turned out that the purchase department of that company (without informing the plant operator) acquired a large quantity of cheap coal-derived benzene, instead of the usual petroleum-based one. As a consequence, sulfur was able to penetrate through two upstream reactors (one of them with another catalyst), finally resting on the dehydrogenation catalyst. Small savings in the feedstock price resulted in large spending for the catalysts, which had to be replaced earlier than scheduled because of deactivation.

2.2.2 Adsorption methods

Adsorption methods provide information about the total surface area of the catalyst and porosity (by physisorption measurements) and the number, type and nature of adsorbed species (by chemisorption). Temperature-programmed methods (such as desorption) are also extremely useful, for example, in investigating heterogeneity of surfaces. Chemisorption and thermodesorption methods can be combined with microcalorimetric measurements, providing information about heat of adsorption, acidity distribution, surface heterogeneity and lateral interactions between adsorbed species.

Either static or dynamic methods could be applied in studying adsorption. In static methods adsorption is performed gravimetrically or volumetrically. The volumetric method uses two chambers, one a dosing section, while a sample is located in the other one. The precision of the analysis depends on the accurate determination of the volumes, including the dead space in the sample chamber, careful temperature and pressure measurements. In gravimetric measurements there is no need for such precise calibration of volumes, as the quantity of adsorbed gas is measured directly using a microbalance.

Dynamic methods can be arranged in single flow or pulse flow modes. The first option relies on a flow of a gas containing the molecules to be adsorbed through a catalytic bed. Although no vacuum system is required, making the method easy to use, the single flow method is used much less than the pulse technique because of the possibility of slow adsorption and a need to employ pure gases. In the pulse technique there is a sequence of pulses going through the sample until no further adsorption is seen. The amount of adsorbed molecules, which should be rather strongly bound to the surface, is determined by summation of the gases adsorbed in all pulses.

2.2.3 Physisorption methods

Low-temperature adsorption at a gas condensation temperature (i.e., for nitrogen this temperature is 77 K) is generally accepted as a standard procedure for the determination of surface areas and pore size distributions of a wide range of porous materials. Physisorption measurements performed at a constant temperature allow the determination of an adsorption isotherm, whose features provide important information about the porous solid. The theory of nitrogen adsorption that was developed in the late 1930s by Brunauer, Emmett and Teller (thus the well-known name of the BET method and BET theory) has substantial shortcoming because of the assumptions involved. Moreover, it does not include capillary condensation in the pores in the mathematical treatment, leading to large deviations between experimental data and model predictions at higher values of relative pressures. With clear understanding of all the flaws of the BET theory, the BET method is still very widely used throughout the world to calculate the monolayer capacity, although some other methods are also applied. Pore size distribution can be calculated from the desorption branch of the isotherm. Evaporation from the mesopores is not as easy as condensation, resulting in the differences between adsorption and desorption branches, or in other words, hysteresis. By examining the isotherm step-by-step in the domain of hysteresis, the mesopore volume and the mesopore size distribution can be obtained.

The monograph by J.B. Condon [11] is an excellent source of information on various methods for measuring and calculating surface area and porosity.

2.2.3.1 Surface area determination
Surface area measurements can be performed with commercially available instruments (see Fig. 2.15).

The following procedure could be applied for taking measurements (Fig. 2.15B). The lines (V_1) are filled with a known pressure of nitrogen (starting from as low as 10–20 Torr up to 700–800 Torr). Thereafter, the valve to the flask with the known quantity of solid material (V_2) is open and the pressure (p') is stabilized. As $p_{line}V_1 = p'_{line} (V_1 + V_2) + RTn_{adsorbed\ moles}$ the amount of adsorbed nitrogen can be calculated.

Different types of isotherms are obtained, depending on the type of the support material (Fig. 2.16).

In a type I isotherm there is a monolayer adsorption on an ideal surface with saturation and there are no interactions between adsorbed species. In type I, as well as in types II and III, there is no hysteresis, showing that these materials are free from mesopores. For a type II isotherm, multilayers starting at point B are formed. For a type III isotherm, the incoming molecules are bound stronger to the surface when it is gradually occupied, which is a sign of lateral interactions. Isotherm VI represents a

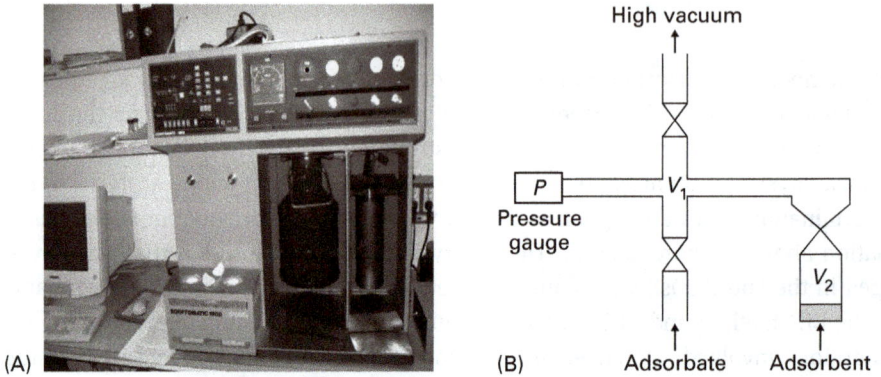

Fig. 2.15: (A) Instrument for sorption measurements (Carlo Erba Sorptometer 1900). (B) Principle scheme of measurements.

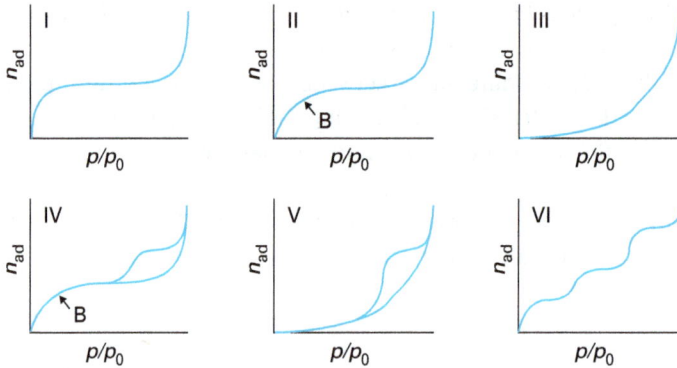

Fig. 2.16: IUPAC classification of isotherms.

layer-by-layer adsorption. Types IV (the most typical for catalytic materials) and V isotherms exhibit hysteresis in mesoporous. The formation of poly-layers and then capillary condensation also occur during adsorption, but there are different equations governing evaporation and condensation during adsorption and desorption (as will be explained later), thus for the same amount adsorbed there is a difference in relative pressures observed in the gas-phase.

The hysteresis behavior could be even more complicated, as shown in Fig. 2.17A, where the H1 type of isotherm corresponds to a narrow distribution of uniform pores, while the H2 type reflects more complex (non-uniform) pore structures. The shape of pores in H1 and H2 isotherms are nearly cylindrical, while slit-shaped pores with uniform and non-uniform sizes and shapes result in H4 and H3 isotherms, respectively.

The specific surface area can be calculated when knowing the monolayer capacity, using the following expression:

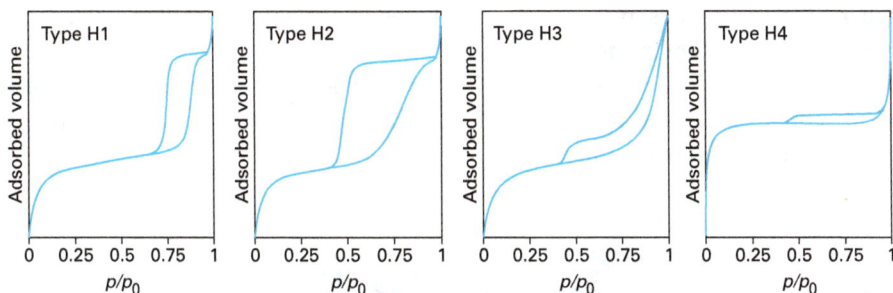

Fig. 2.17: Hysteresis behavior for different types of pores: adsorbed amount as a function of relative pressure.

$$s = n_m A_m N_A \qquad (2.1)$$

where s is the specific surface area (m^2 kg^{-1} or typically m^2 g^{-1}), m_m is the monolayer capacity (mol g^{-1}), A_m is the area occupied by one molecule (m^2 molecule^{-1}) and N_A is the Avogadro number equal to 6.023×10^{23} molecules mol^{-1}. Tab. 2.2 contains values of the areas occupied by common adsorbates. Typically, nitrogen is used, however, krypton and argon adsorptions are also used, especially for low surface area materials (< 1 m^2 g^{-1}).

Tab. 2.2: Values of the areas occupied by typical adsorbates.

Adsorbate	Boiling point (K)	A_m (nm^2/molecule)
N_2	77.3	0.162
Ar	87.4	0.142
CO_2	194.5	0.17
Kr	120.8	0.152

In practice, the adsorbed volume is measured and the volumetric monolayer capacity (recalculated for standard pressure and temperature) is reported. Thus, instead of eq. (2.1) we get:

$$s = \frac{V_m A_m N_A}{V} \qquad (2.2)$$

where V_m is the volumetric monolayer capacity, V is the volume of one mole at standard pressure and temperature (22.4×10^{-3} m^3 mol^{-1}).

There is, therefore, a need to calculate the value of the monolayer capacity from experimental data for physisorption when multilayers are formed. The monolayer capacity can be calculated using the method developed by Brunauer, Emmett and

Teller for multilayer adsorption, which was based on the approach of Langmuir, who considered adsorption in a monolayer.

Layers of adsorbed molecules (Fig. 2.18) are divided in the first layer by the heat of adsorption $\Delta H_{ad,1}$ and in the second and subsequent layers by $\Delta H_{ad,2} = \Delta H_{cond}$, equal to the heat of condensation.

Fig. 2.18: Schematics of desorbed layers.

Adsorption in the first layer can be described as $a_0 P S_0 = b_1 S_1$ where a_0 and b_1 are adsorption coefficient on the zero layer (bare surface) and desorption coefficient from the first layer, respectively, S_0 denotes the number of empty sites available for adsorption and S_1 is the number of adsorbed molecules in the first layers. Consequently $S_1 = (a_0 P/b_1)S_0 = y S_0$ with $y = (a_0 P/b_1)$. Analogously for the second layer $a_1 P S_1 = b_2 S_2$, $S_2 = (a_1 P/b_2)S_1 = xy S_0$ with $x = a_1 P/b_2$. As only adsorption in the first layer is different: $a_1 = a_2 = a_3 = ... = a_n$; $b_2 = b_3 = ... = b_n$ and subsequently $a_2 P S_2 = b_3 S_3$. $S_3 = (a_2 P/b_3)S_2 = x^2 y S_0$. For adsorption in the nth layer: $S_n = x^{n-1} y S_0 = x^n (y/x)S_0 = x^n c S_0$ with $c = y/x$. The total amount of adsorbed molecules is defined as $\sum_{n=1}^{\infty} n S_n = \sum_{n=1}^{\infty} n c x^n S_0 = c S_0 \sum_{n=1}^{\infty} n x^n$.

The monolayer capacity should also include the empty sites $\sum_{n=0}^{\infty} S_n = S_0 + \sum_{n=1}^{\infty} c x^n S_0 = S_0 + c S_0 \sum_{n=1}^{\infty} x^n$. Making use of infinite geometrical progression $\sum_{n=1}^{\infty} n x^n \Rightarrow \frac{x}{(1-x)^2} \sum_{n=1}^{\infty} x^n \Rightarrow \frac{x}{1-x}$ the equation for the monolayer capacity can be easily obtained:

$$\frac{N_{ads}}{N_{mono}} = \frac{V_{ads}}{V_{mono}} = \frac{cS_0 \dfrac{x}{(1-x)^2}}{S_0 + cS_0 \dfrac{x}{1-x}} = \frac{cx}{(1-x)(1-x+cx)} \qquad (2.3)$$

As adsorption-desorption in the nth layers can be considered as liquid-vapor equilibrium, condensation-evaporation equilibrium gives $a P_0 = b$ where P_0 is the saturation pressure. For adsorption of nitrogen $P_0 = 1$ bar at 77 K. Further transformations give $x = (a/b)P = P/P_0$.

Linearization of eq. (2.3) results in:

$$\frac{1}{V}\frac{x}{1-x} = \frac{1}{cV_m} + \frac{c-1}{cV_m}x \qquad (2.4)$$

which can be presented in the graphical form (see Fig. 2.19), allowing for the calculation of the monolayer volume:

$$V_m = \frac{1}{i + s} \tag{2.5}$$

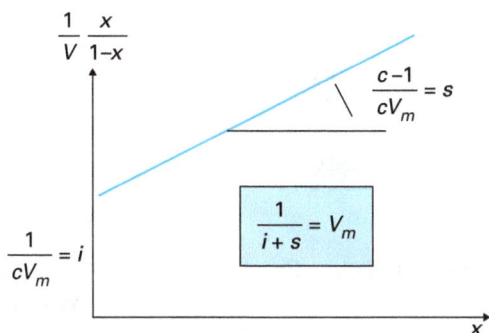

Fig. 2.19: Linearization of BET equation.

In eq. (2.5) the slope is $s = (c - 1)/cV_m$ and the intercept is given by $1/cV_m$, allowing for the calculation of the monolayer volume as the reciprocal value of the sum of the slope and the intercept.

The linearization should be performed for the relative pressures between 0.05 and 0.3, as at higher values the deviations are becoming substantial because of various shortcomings in the model; for example, not accounting for capillary condensation in the pores.

By taking the volume of the monolayer from eq. (2.2) it is possible to calculate the specific surface area. Typically the plots for catalytic materials are presented with the adsorbed volume calculated in cm^3 g^{-1} at standard temperature and pressure (STP). For N_2 adsorption at 77 K the specific surface area (m^2 g^{-1}) is then the monolayer volume (cm^3 g^{-1} at STP) multiplied by coefficient 4.3532. Some typical values for catalytic materials are given in Tab. 2.3. While reporting the surface areas of catalytic materials one should be cautious, keeping in mind the precisions of volumetric flow measurements. Reporting the values that are directly available from a physisorption instrument, for example, 124.564 m^2 g^{-1} does not make much sense even if similar values appear on a regular basis in the literature. In addition, errors in the estimation of surface areas for materials with low surface areas can be very high, thus it is recommended that there is a sufficient amount of the sample in a measurement unit (~40 m^2).

Tab. 2.3 also contains values of the mean diameter of pores. For physisorption of nitrogen the monolayer thickness is 0.354 nm, which means that in the case of zeolites or activated carbon applications of BET theory, relying on an infinite

Tab. 2.3: Typical values of surface areas for catalyst supports.

Catalyst supports	Mean d_p (nm)	S_{BET} (m² g⁻¹)
Silica gel	10	200
	6	400
	4	800
γ-Al₂O₃	10	150
	5	500
Zeolite	0.6–2	400–800
Activated carbon	2	700–1,200
TiO₂	400–800	2–50
Aerosil SiO₂	–	50–200

number of adsorbed layers is more than questionable. Even for supports like alumina, a pore diameter of ~5 nm is translated in around seven layers. As the BET method is only valid in a small pressure interval, and the physical interpretation of it is not very easy, other less popular methods are also used for the calculation of the monolayer capacity.

In the t-plot method, introduced by de Boer, it is assumed that in a certain isotherm region the micropores are already filled-up, whereas adsorption in larger pores occurs according to some simple equation, characteristic for a large class of solids. This equation should approximate adsorption in mesopores, macropores and on a flat surface in a narrow pressure range just above complete filling in micropores, but below vapor condensation in mesopores. Such an assumption may be valid, if micropores are small and super-micropores are not present. Thus the thickness of the adsorbed layer depends only on the relative pressure but not on the material type and could be expressed, for example, by the equations in Fig. 2.20.

Adsorption within this pressure region, when the thickness is independent on the material type, can be described by a simple linear dependence (Fig. 2.21).

Such linear dependence can be easily understood by expressing the thickness of the adsorbed layer through the number of adsorbed monolayers (the ratio of the number of adsorbed moles n_{ad} to the monolayer capacity n_m) and the monolayer thickness (0.354 nm or 3.5410^{-10} m for nitrogen):

$$t = \frac{n_{ad}}{n_m} 0.354 \text{ nm} \tag{2.6}$$

And subsequently:

$$S_t = n_m A_m N_A = \frac{n_{ad}}{t} 0.354 \times 10^{-9} (m) A_m N_A \tag{2.7}$$

From eq. (2.7) the number of adsorbed moles depends linearly on the thickness of the adsorbed layer with the slope inversely proportional to the surface area of mesopores.

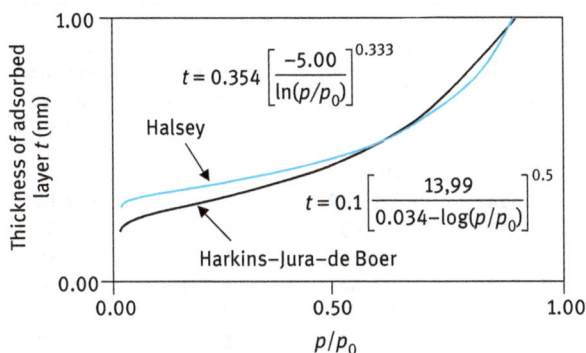

Fig. 2.20: Adsorbed layer thickness as a function of relative pressure.

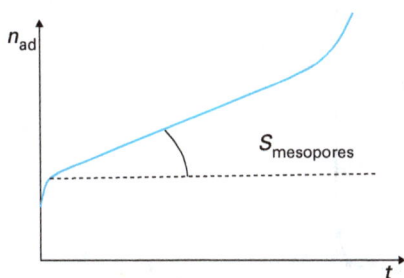

Fig. 2.21: Adsorbed amount as a function of adsorbed layer thickness for a material with micro- and mesopores.

A dependence like the one presented in Fig. 2.21 is made through plotting an experimentally obtained value of n_{ad} or V_{ad} for a particular value of the relative pressure and relating it to the thickness of the adsorbed layer for the same pressure using correlations of t vs relative pressure.

Exercise 2.1: When nitrogen was adsorbed on a catalyst at −196 °C the following data were obtained. Numerical values are given below. What is the specific surface area of the catalyst? Calculate the pore size distribution.

Adsorption			Desorption		
Rel. Press, (p/p_0)	Volume, adsorbed $(cm^3\,g^{-1})$	Rel. Press, (p/p_0)	Volume, adsorbed, $(cm^3\,g^{-1})$	Rel. Press, (p/p_0)	Volume, desorbed, $(cm^3\,g^{-1})$
0.0024	38.22	0.4477	131.796	0.9568	252.462
0.0217	55.65	0.4739	138.369	0.8321	243.642
0.0481	64.176	0.4982	147.084	0.7222	238.497
0.0764	70.35	0.5224	155.904	0.6287	231.609
0.1051	75.747	0.567	174.405	0.5508	220.941

Adsorption			Desorption		
Rel. Press, (p/p_0)	Volume, adsorbed $(cm^3\,g^{-1})$	Rel. Press, (p/p_0)	Volume, adsorbed, $(cm^3\,g^{-1})$	Rel. Press, (p/p_0)	Volume, desorbed, $(cm^3\,g^{-1})$
0.1342	80.493	0.6092	195.447	0.4943	195.321
0.1635	84.945	0.6514	216.342	0.4509	162.141
0.1929	89.124	0.6963	233.835	0.4129	129.675
0.2222	93.24	0.7484	242.382	0.3661	116.844
0.2514	97.293	0.8039	246.645	0.3211	110.502
0.2806	101.283	0.8602	249.606	0.2807	106.281
0.3094	105.504	0.9161	252.882	0.2451	103.005
0.338	109.893	0.9713	256.704	0.214	100.275
0.3662	114.681	0.9984	294.21		
0.394	119.742	0.9996	432.81		
0.4211	125.433	1	572.523		

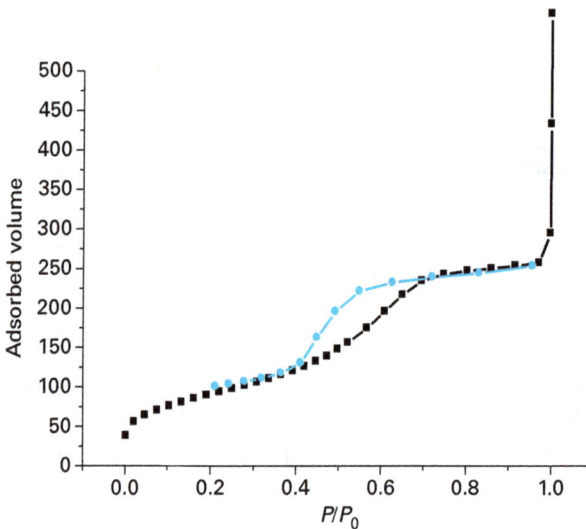

Fig. E2.1. Adsorbed volume as a function of relative pressure.

For microporous materials like zeolites and active carbon with the pore size ~0.5–2 nm there are no theoretical grounds to use the BET method with the assumption of multilayer adsorption. An ideal Langmuir equation usually overestimates the surface area. A t-plot method could be used instead for the determination of the micropore volume and the surface area.

For microporous materials a plot of the adsorbed volume vs the thickness of the adsorbed layer, defined by Harkins-Jura-de Boer equation (Fig. 2.20), gives a straight line (Fig. 2.22A). Therefore even a one point analysis at a certain value of

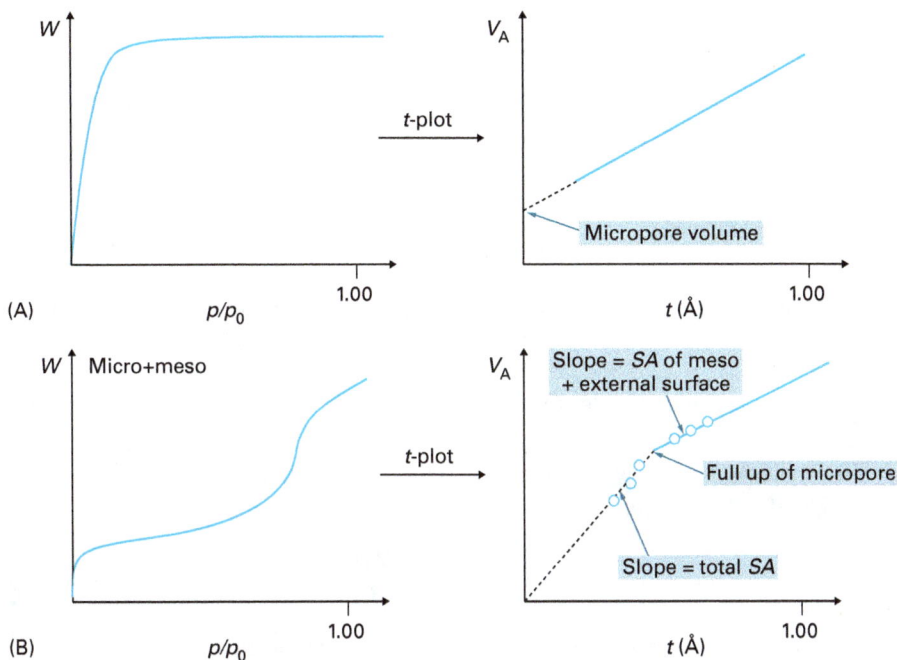

Fig. 2.22: Application of the t-plot method for determination of micropore volume.

relative pressure is sufficient, as at this point the surface is determined by the ratio between the volume adsorbed and the layer thickness at this particular pressure. Fig. 2.22B provides an illustration of the t-plot method for materials containing both micro- and mesopores.

Instead of the t-plot method for the determination of the micropore volume, a method developed by Dubinin and his co-workers [12] could be applied. The Dubinin–Radushkevich equation relates the volume of adsorbate and the total volume of the micropores V_m:

$$V = V_m \exp\left[-B\left(\frac{RT}{\beta}\right)^2 \ln^2\left(\frac{p_0}{p}\right)\right] \tag{2.8}$$

where B and β are constants and T is temperature. A linear relationship should be found between $\ln V$ and $\ln^2(p_0/p)$. Although the curve is bending at low values of $\log^2(p_0/p)$ extrapolation to zero gives the value of V_m.

2.2.3.2 Determination of pore size and pore size distribution

In the capillary condensation region ($p/p_0 > 0.4$) there is capillary condensation in the pores during both adsorption and desorption (Fig. 2.23). Evaporation from the

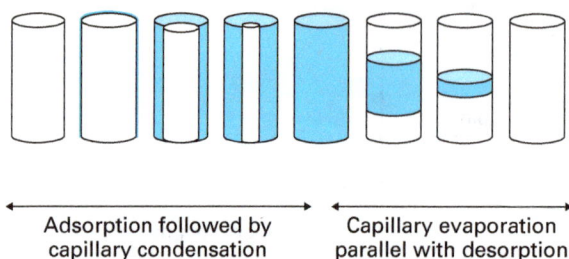

Adsorption followed by
capillary condensation

Capillary evaporation
parallel with desorption

Fig. 2.23: Representation of pore filling during adsorption and desorption. Adsorption proceeds through multilayer formation, while desorption is meniscus controlled.

pores is not as easy as condensation, resulting in differences between adsorption and desorption branches (hysteresis).

In a cylindrical pore with two radii of curvature, the saturation vapor pressure is given by the Kelvin equation:

$$\ln \frac{p_0}{p} = \left(\frac{1}{r_1} + \frac{1}{r_2} \right) \frac{\sigma V \cos \theta}{RT} \tag{2.9}$$

where p_0 is saturation pressure, σ is the surface tension of the liquid (8.72 mN m^{-1} for N_2), V is the molar volume of liquid ($N_2 = 34.6$ cm^3 mol^{-1}), θ is the contact angle of the liquid with the gas (zero for liquid nitrogen). Equation (2.9) shows the result of the interactions with the pore walls making condensation occur at a lower relative pressure than the saturation one. As the relative pressure is increased, condensation occurs first in the pores of smaller radii and then progresses. The Kelvin equation gives the possibility of estimating pore radii from a gas adsorption isotherm. At any equilibrium pressure p, pores of radii less than r will be filled with condensed vapor. Application of the Kelvin equation to all points of the isotherm at relative pressures greater than that corresponding to the monolayer volume (where capillary condensation begins to occur) yields information concerning the volume of gas adsorbed in pores of different radii.

Desorption occurs from hemispherical surfaces with both radii of curvature equal to each other, thus eq. (2.9) is rearranged:

$$r = \frac{2\sigma V \cos \theta}{RT \ln \frac{p_0}{p}} + \delta \tag{2.10}$$

The modified Kelvin equation (eq. 2.10) is the basis of Barrett, Joyner, Halenda and Dollimore-Heal methods and contains δ which is the thickness of the adsorbed layer and can be calculated from the Wheeler equation:

$$\delta = 0.734 \left[\ln \frac{p_0}{p} \right]^{-\frac{1}{3}} (\text{nm}) \tag{2.11}$$

Finally an expression is obtained for the estimation of the pore radii from the nitrogen desorption branch of isotherm:

$$r = \frac{0.93}{\ln\frac{p_0}{p}} + 0.734\left[\ln\frac{p_0}{p}\right]^{-\frac{1}{3}} (nm) \qquad (2.12)$$

The following procedure is used to calculate the pore size and pore volume distribution. The desorption branch of the experimentally measured nitrogen physisorption is applied. Each point gives the amount adsorbed and the relative pressure. By multiplying this amount by the molar volume of liquid nitrogen (34.68 cm³ mol⁻¹) the pore volume is obtained. The modified Kelvin equation gives the radius at a particular value of relative pressure. The next step is to plot the volume of pores vs radius and then calculate dV/dr at different values of r. This can be done for example, by dividing the incremental difference in volumes $(V_2 - V_1)$ by the incremental difference in corresponding radii $(r_2 - r_1)$ and relating $(V_2 - V_1)/(r_2 - r_1)$ to the average between these two radii $[(r_1 + r_1)/2]$. A typical result is presented in Fig. 2.24.

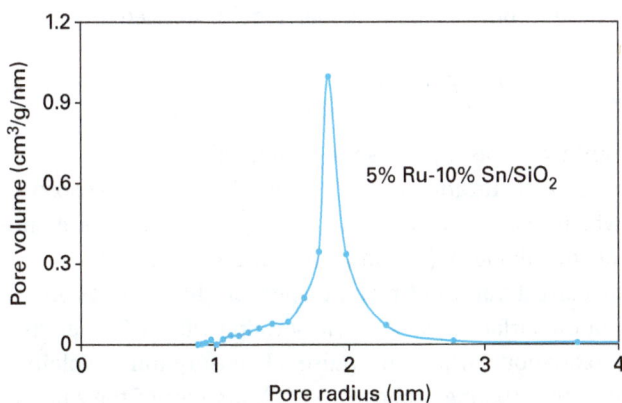

Fig. 2.24: Pore volume distribution as a function of pore radii.

Besides the suggestion that desorption proceeds from a hemispherical meniscus while adsorption occurs on a cylindrical meniscus and thus for adsorption in eq. (2.9) $r_1 = r$ and $r_2 = \infty$ there are other explanations in the specialized literature, such as changes of the contact angle caused by surface cleaning, different radii of meniscus in adsorption and the desorption presence of interconnected pores behaving differently during these processes.

For pore size distribution analysis it is not only hysteresis in nitrogen physisorption isotherm that can be used, but also mercury porosimetry. Because of the non-wetting nature of mercury (the angle of contact of mercury with solids, θ, is above

90° and ~140°) its intrusion into the pores is met with resistance and the pore radius can be related to the applied pressure following the Washburn equation:

$$r = \frac{2\sigma \cos \theta}{\Delta p} \tag{2.13}$$

where σ is surface tension normally taken as 480 mN m^{-1}. A comparison between surface area determination by mercury penetration and nitrogen physisorption typically shows a discrepancy for microporous materials, which is associated with Hg inability to penetrate small (micro)pores, uncertainty of the contact angle and surface tension values and the probability of the cracking or deforming of samples. However, mercury porosimetry remains as the most reliable method for investigation of macroporous materials.

2.2.4 Chemisorption

In addition to physisorption, which is used for the determination of specific area of supports, the exposed surface area of the active (metal) phase and metal dispersion can be evaluated using chemisorption. The metal dispersion D is defined as:

$$D = \frac{N_S}{N_T} 100\% \tag{2.14}$$

where n_s is the number of surface atoms and n_T is the total number of atoms.

In essence, chemisorption is the titration of surface sites by a certain adsorbate (hydrogen, oxygen, CO), which reacts only with the active phase to form a monolayer. The understanding of stoichiometry (as shown in Fig. 2.25 for CO adsorption) is important for getting the correct values of metal dispersion. In some instances, such as the determination of Cu surface area by chemisorption with N_2O, there is a possibility of subsurface oxidation, which in pulse chemisorption (explained below) depends on temperature (prominent above 100 °C), amount of the sample, metal loading and the catalyst bed geometry.

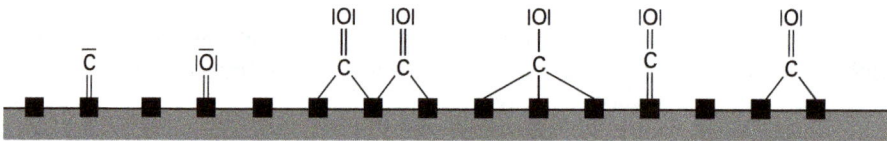

Fig. 2.25: Different stoichiometry for CO chemisorption.

From the experimentally measured adsorbed volume the number of adsorbed moles is calculated. Through knowing stoichiometry, the number of surface atoms is determined. The total number of metal atoms and subsequent metal dispersion can be calculated by knowing metal loading.

There is a relationship between the metal dispersion and the diameter of a metal cluster on a support that can be calculated using the following approach. Imagine a spherical cluster of the size d. The surface of this cluster (πd^2) is composed of metal atoms, each with a surface corresponding to the cross section of a particular atom S_{Me}. Then the number of surface atoms is defined as $N_S = \pi d^2/S_{Me}$. The total number of atoms in the cluster is proportional to the mass of the cluster and the Avogadro number and inversely proportional to metal molecular mass:

$$N_T = \frac{m_{cluster}N_A}{M} = \frac{V_{cluster}\rho N_A}{M} = \frac{\pi d^3 \rho}{6} \frac{N_A}{M} \tag{2.15}$$

where ρ is the metal density. The metal dispersion is computed from (2.14) and (2.15):

$$D = \frac{m_{cluster}N_A}{M} = \frac{V_{cluster}\rho N_A}{M} = \frac{6M}{d\rho S_{Me}N_A}100\% \tag{2.16}$$

which is finally related to the diameter of metal clusters:

$$d = \frac{6M}{D\rho S_{Me}N_A}100\% \tag{2.17}$$

Some equations, which can be used for calculating the metal particle size (in nm) are given here for several metals: Pt: 108/D(%); Pd: 107/D(%); Ru: 101/D(%); Ni: 97/D(%). An assumption in these calculations is the metal surface area. If only dense metal surfaces are considered (like 111) with smaller S_{Me}, somewhat larger values of metal clusters will be obtained. Because of uncertainties in the stoichiometry, the presence of spillover, the possible formation of subsurface species and not necessarily spherical cluster shape, verifying the estimation of metal particle size with chemisorption by some other physical methods is recommended. For example, diameter d can be measured or calculated from several techniques, such as high-resolution electron microscopy, X-ray diffraction or XPS. The overall number of adsorbed molecules in a monolayer could be determined by temperature-programmed desorption, while infrared spectroscopy using a probe molecule is widely applied for determination of the number of acid and basic sites in zeolites and mesoporous materials. These methods will be described later in detail.

In static chemisorption methods, after a high temperature evacuation the isotherm is measured, followed by a subsequent low-temperature evacuation, where physisorbed molecules are removed. In the second phase of adsorption measurements only these weakly adsorbed molecules are chemisorbed back on the surface. Subtraction of the weakly adsorbed amounts from totally adsorbed ones gives the amount of strongly and irreversible adsorbed gas, which is then used for

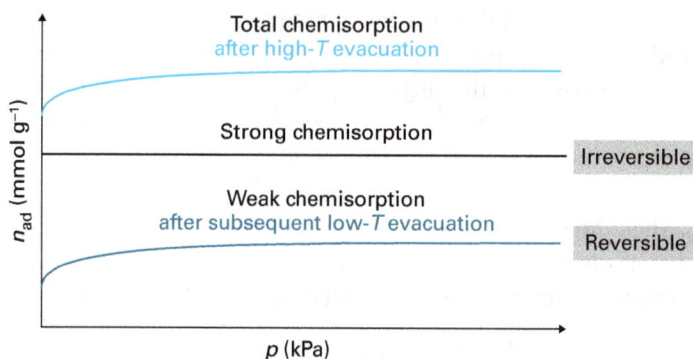

Fig. 2.26: Principles of the back sorption method.

calculating the surface area and metal dispersion. The principles of the back sorption method are illustrated in Fig. 2.26.

Although historically the static methods for chemisorption were applied, dynamic methods are also often used for measuring chemisorption (Fig. 2.27A).

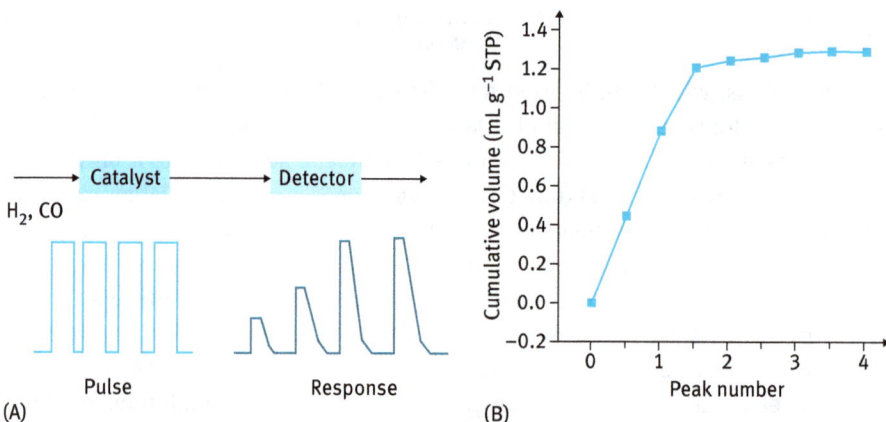

Fig. 2.27: Dynamic method for measuring chemisorption. (A) pulse-response sequence. (B) adsorbed amounts as a function of pulse number.

Determination of the metal dispersion of a supported metal catalyst (for example, palladium on a mesoporous support, Fig. 2.27B) could be performed by measuring CO chemisorption. The sample should be placed into a U-tube and reduced by hydrogen, while increasing the temperature linearly, 5 °C min^{-1} to 200 °C, then maintaining it for 2 h. After the reduction, the catalyst sample should be cooled down to 40 °C under helium flow. Chemisorption of CO is then achieved by pulsing CO through the system and completely saturating the catalyst (Fig. 2.27B). The stoichiometric ratio between metal (Pd) and CO is assumed to be unity (in fact it could depend on the cluster size, as the stoichiometry might change). The mean particle size d_s is calculated knowing the metal dispersion.

2.2.5 Temperature-programmed methods

There are several temperature-programmed techniques available for catalyst characterization. A certain process (catalyst reduction, oxidation of coke, desorption of pre-adsorbed species or their chemical reactions) is followed while the temperature is increased linearly. This non-spectroscopic technique is inexpensive and applicable to real and model catalysts.

2.2.5.1 Temperature-programmed reduction

This method is based on the reduction of a catalyst by hydrogen during temperature ramping, while monitoring hydrogen consumption with a thermal conductivity detector. Hydrogen consumption profiles are plotted as a function of temperature (or time) as shown in Fig. 2.28.

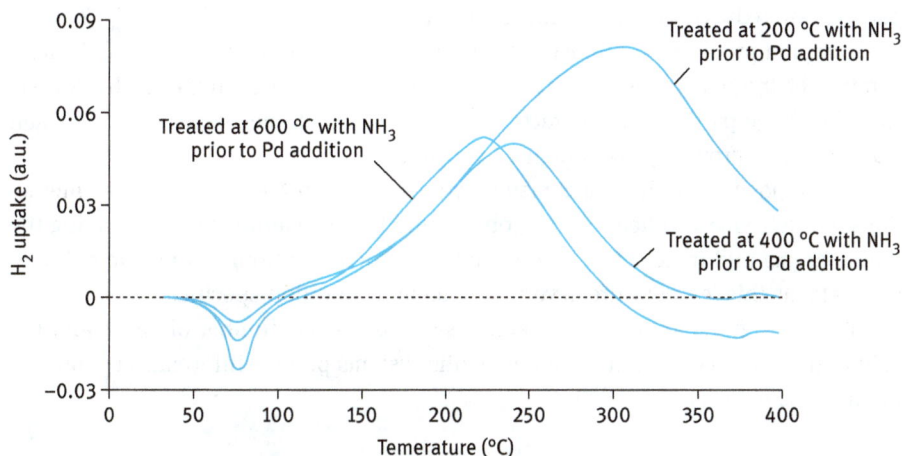

Fig. 2.28: TPR of palladium supported on *N*-functionalized carbon nanotubes, treated with ammonia at different temperatures prior to addition of palladium.

The number of peaks in the thermogram can be interpreted either by the presence of different chemical species, which are reduced at different temperatures, or by the successive reductions of one type of species. The degree of reduction or the initial oxidation state (if reduction is complete) can be elucidated from the consumed amount of hydrogen. Finally, the positions of the peaks in temperature point the active phase interactions with the support. Strong interactions with the support will result in a shift in the peak maximum to higher temperatures, giving a broader peak (Fig. 2.29A).

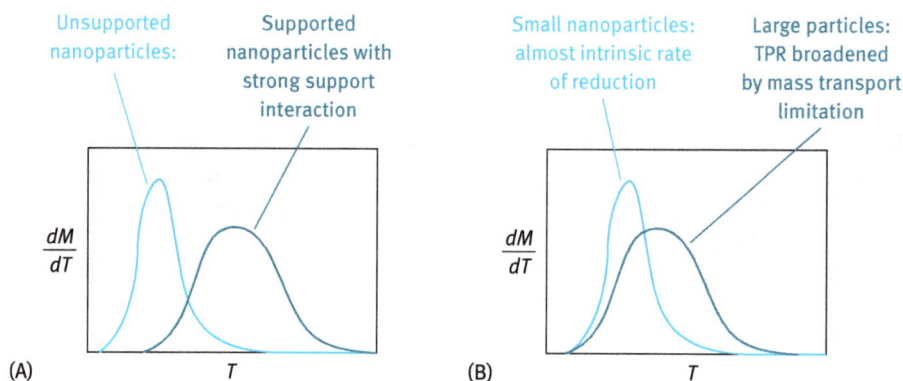

Fig. 2.29: TPR curves. (A) Influence of support. (B) Influence of the cluster size.

In the reduction processes, where mass transport limitations are prominent, large particles are reduced at higher temperatures and the peak is broadened (Fig. 2.29). In the reduction processes controlled by nucleation (reduction of nickel oxides) the temperature of the maximum is, conversely, increasing with a decrease in the particle size, thus large particles are characterized by narrow and intense peaks, while small clusters have a broader peak at higher temperatures.

As the position of the peaks may vary depending on the metal support interactions, particle size and heating rate, one should be very cautious in interpreting the thermograms and in selecting the reduction temperature from, for example, literature data, as this temperature is very sample- and condition-specific.

TPR is useful is analyzing bimetallic systems, as the number of peaks and the shift of the peaks compared to monometallic systems provide information about the interactions between the metals.

2.2.5.2 Temperature-programmed desorption

Temperature-programed desorption (TPD) involves heating a sample with a rising temperature and simultaneously detecting the residual gas in the vacuum using a mass analyzer. As the temperature rises, certain absorbed species will have enough energy to escape and will be detected as a rise in pressure for a certain mass. A vacuum is needed to prevent re-adsorption, which is thus neglected in the mathematical treatment. The desorption rate goes through a maximum. At low T the coverage is high, but the rate constant is low and overall the desorption rate is low. As the temperature increases the rate constant and thus desorption rate increases. At intermediate temperatures the desorption rate is high as both the coverage and the rate constant are high. At higher T the coverage is virtually zero and although the rate constant is enormous, the desorption rate is virtually zero. The temperature of the peak maximum provides information on the binding energy of the bound species (Fig. 2.30A) as well as decomposition path for complex organic molecules desorption (Fig. 2.30B).

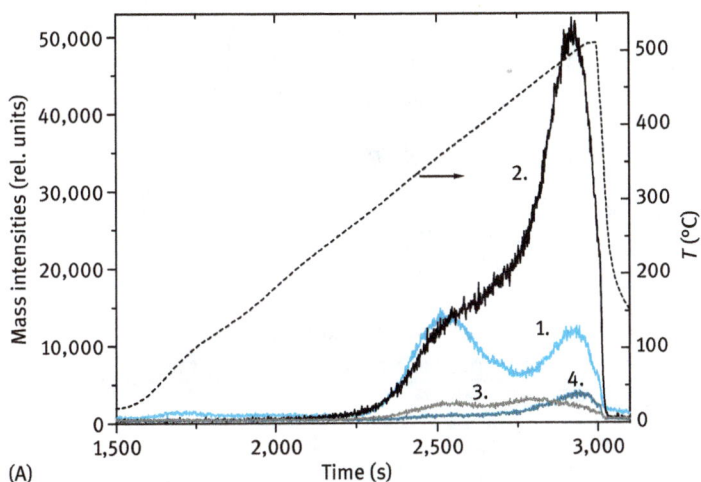

(A)

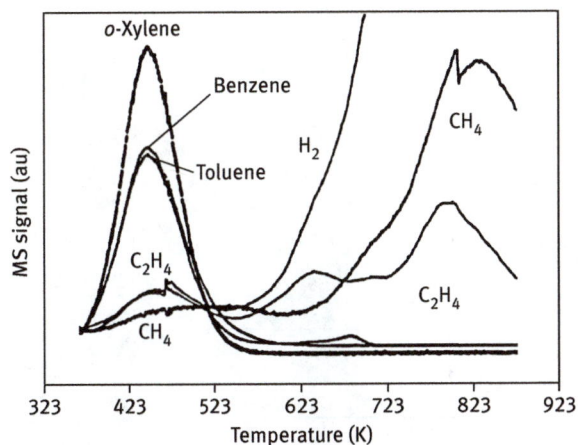

(B)

Fig. 2.30: (A) TPD of CO from Pd/C catalysts after stearic acid decarboxylation. 1, CO; 2, hydrogen; 3, CO_2; 4, methane. (B) TPD of o-xylene. (From [13].)

Activation energy of desorption can be determined from TPD experiments at different heating rates $\beta = dT/dt$. For the first-order desorption kinetics $-d\theta/dt = \theta k e^{-\Delta E_{des}/RT}$ with a linear heating rate $T = T_0 + \beta t$ the temperature dependence of the coverage is:

$$-\frac{d\theta}{dT} = \left(\frac{\theta k}{\beta}\right)e^{-\frac{\Delta E_{des}}{RT}} \tag{2.18}$$

which after differentiating gives:

$$-\frac{d^2\theta}{dT^2} = \frac{k}{\beta}\left(\theta\Delta\frac{E_{des}}{RT^2}e^{-\frac{\Delta E_{des}}{RT}} + \frac{d\theta}{dT}e^{-\frac{\Delta E_{des}}{RT}}\right) \tag{2.19}$$

At the peak maximum $(T = T_p)$ it holds that $-d^2\theta/dT^2 = 0$, therefore:

$$-\frac{d\theta}{dT} = \theta\frac{\Delta E_{des}}{RT_p^2} \tag{2.20}$$

Elimination of the term $d\theta/dT$ and further simple manipulations give a possible determination of E_{des} from TPD experiments with different heating rates β (Fig. 2.31). The slope in Fig. 2.31 is equal to $R/\Delta E_{des}$:

$$\frac{1}{T} = \frac{R}{\Delta E_{des}}\ln\frac{T^2}{\beta} + \frac{R}{\Delta E_{des}}\ln\frac{kR}{\Delta E_{des}} \tag{2.21}$$

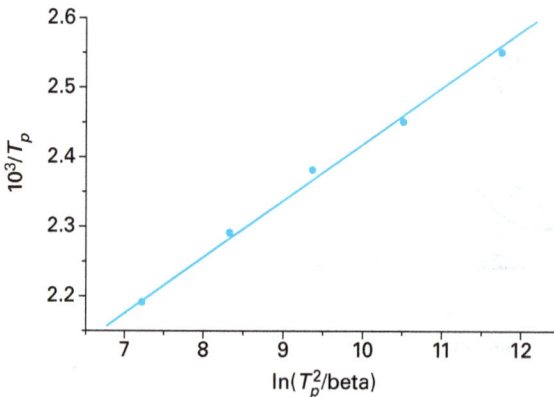

Fig. 2.31: Determination of activation energy of first-order desorption.

Exercise 2.2: The following TPD data were obtained with different heating rates (Fig. E2.2). Calculate the activation energy of desorption

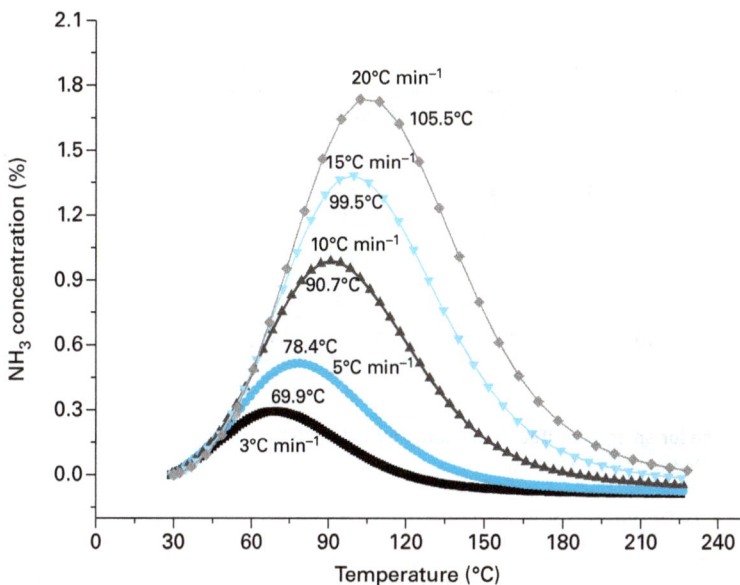

Fig. E2.2: TPD at different heating rates.

2.2.5.3 Temperature programmed oxidation (TPO)

In temperature-programmed oxidation (TPO) the nature of solids deposited on deactivated catalysts is analyzed, giving valuable information about the regeneration conditions. Oxygen consumption is measured as a function of time. The method is often used to investigate coke deposited on supported metal catalysts and zeolites (Fig. 2.32).

In many cases, such as the one shown in Fig. 2.32 for metal (nickel and vanadium) containing zeolitic catalysts used in fluid catalytic cracking (discussed in detail in Chapter 4), TPO profiles usually contain several peaks that reflect the form of carbon structure on the spent catalysts. When these peaks are not well-resolved as in Fig. 2.32, deconvolution is needed to make proper assignments.

A higher temperature peak (peak N in Fig. 2.32) is associated with more graphitic- like or highly aromatic coke with high H/C ratio. Peak M is assigned to catalytic coke produced by acid-catalyzed cracking, which is more reactive to oxygen than the graphitic coke, but less reactive than the coke associated with metals (peaks K and L).

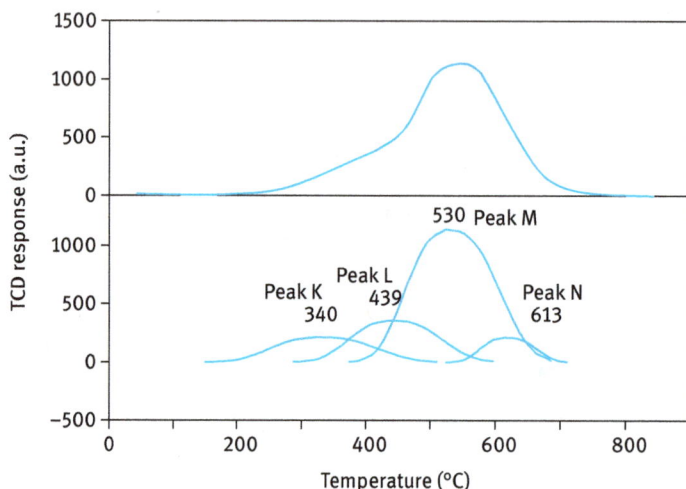

Fig. 2.32: TPO of coke for an equilibrated fluid catalytic cracking (FCC) catalyst after cracking of sour heavy gas oil. [Redrawn from [14]].

2.2.6 Calorimetry

Microcalorimetry is a very useful technique – conceptually easy but difficult to implement. Typically, during the measurements adsorption isotherm is obtained simultaneously with calorimetric data. In a differential scanning calorimeter there are two sample holders. Thermal behavior caused by the temperature difference between the sample under investigation and the reference sample is recorded (Fig. 2.33).

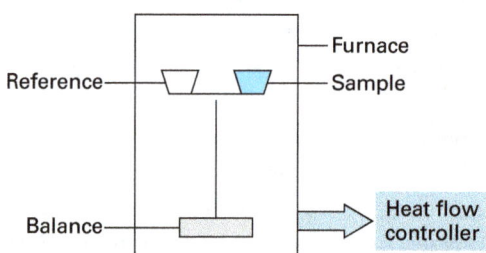

Fig. 2.33: Sketch of a differential scanning calorimeter.

In adsorption calorimetry there is a stepwise introduction of the adsorbate at a constant temperature. For each adsorption step with a slow increase of pressure the adsorbed amount and the evolved heat must be determined. The differential heat can then be calculated from the evolved heat in relation to the number of molecules adsorbed in a particular step $q_{diff} = [dQ/dn]_{T,A}$ as illustrated in Fig. 2.34A, which demonstrates that the heat of adsorption remains approximately constant up to nearly one-third of the monolayer, but decreasing thereafter. There is a wealth body of experimental data for different systems (supported metals, zeolites, single crystals) that confirms the decrease of the heat of adsorption with coverage. An example is shown in Fig. 2.34B. A linear decrease in the heat of adsorption with coverage corresponds to logarithmic (Temkin) adsorption isotherm $\theta = \ln(a_0 p)/f$

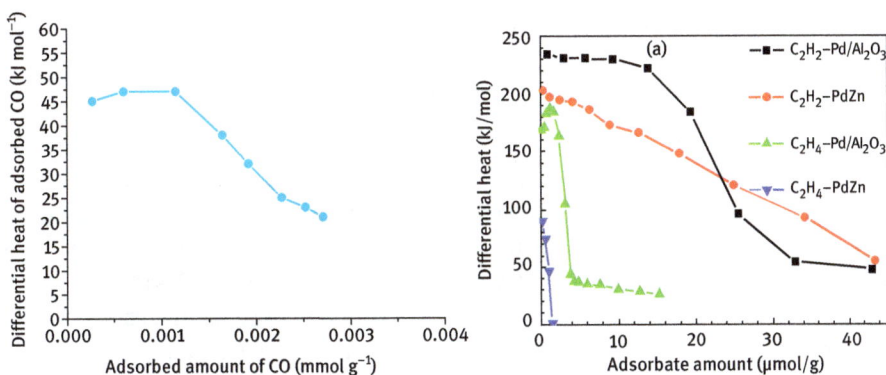

Fig. 2.34: Differential heats of adsorption as a function of coverage for (A) CO at 40 °C on Pd/N-CNT [15] and (B) ethylene on palladium catalysts [16]. Reproduced with permission.

2.2.7 X-ray diffraction

X-ray diffraction (XRD) is a versatile, non-destructive technique that reveals detailed information about the chemical composition and crystallographic structure of natural and manufactured materials.

Diffraction patterns are obtained from diffracted X-rays of a certain wave length λ (for example, CuK α radiation, $\lambda = 1.5418$ Å) by rotating the ionization chamber 2θ degrees in the scan mode within a range of 2θ values while the crystal is rotating θ degrees. The crystal planes are then reflected. They are separated by spacing d (Fig. 2.35A) when the Bragg law $n\lambda = 2d\sin\theta$ is fulfilled, with reflection order n being an integer. Diffraction patterns are compared with available databases to identify crystalline phases present in the sample.

Because of imperfection of the crystal, noise and the interference of the background, the obtained diffractograms (Fig. 2.35B) are not perfect. Width of the peak provides information about the crystallite size, which is inversely proportional to the full width at half maximum (FWHM) of the diffraction peak according to a widely used empirical Scherrer equation. Instead of the line width at half height, a more accurate parameter is the ratio between the area of the peak and its intensity.

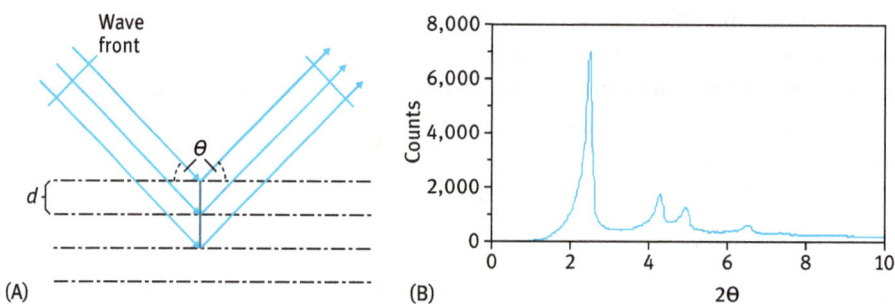

Fig. 2.35: XRD: (A) principle and (B) an XRD pattern for Na-MCM-41.

Commonly used catalytic supports such as silica and alumina typically have high surface area and low crystallinity. For example, silica has broad diffraction peaks (2θ = 20 and 25°), and also few broad diffraction peaks are obtained even for the most crystalline aluminas. Conversely, mesoporous silicas such as MCM-41 (Fig. 2.35B) have a high degree of organization and reflections found at small angles (below 8°). Zeolites are crystalline materials and have very distinct diffraction patterns.

When reflections are not seen in the pattern, there could be several reasons why. In addition to a complete absence of a certain phase, low concentrations of the active phase could be an explanation because catalytic material should be present in an amount larger than 1%. There could be not enough crystallinity in the active phase, as an amorphous phase will not be distinguished from the base line. Interference between the diffraction lines of the active phase and the support might mask the presence of an otherwise crystalline phase. Small metal clusters (below 3 nm) cannot be detected (Fig. 2.36A), thus EXAFS is most frequently used for compounds exhibiting only a short range order. Because of broadening of the peaks, the size of metal clusters is usually overestimated from XRD analysis.

XRD can be used to study if metal sintering is happening during catalysis. Fig. 2.36B illustrates a particular case of fatty acid hydrogenation for which there was no sintering of palladium. It should be mentioned that not all reflections are seen in an XRD pattern. Thus, for face-centered structures only those reflections are present, where the values of indices (Fig. 2.36B) are unmixed (all even or all odd).

Figure 2.37 illustrates information that can be obtained from an idealized diffraction pattern.

2.2.8 X-ray photoelectron spectroscopy and X-ray fluorescence

Analytical techniques, such as XPS based on the detection of ejected electrons, are the most suitable methods for the analysis of surfaces because they probe a limited depth of the sample (~5–10 first monolayers).

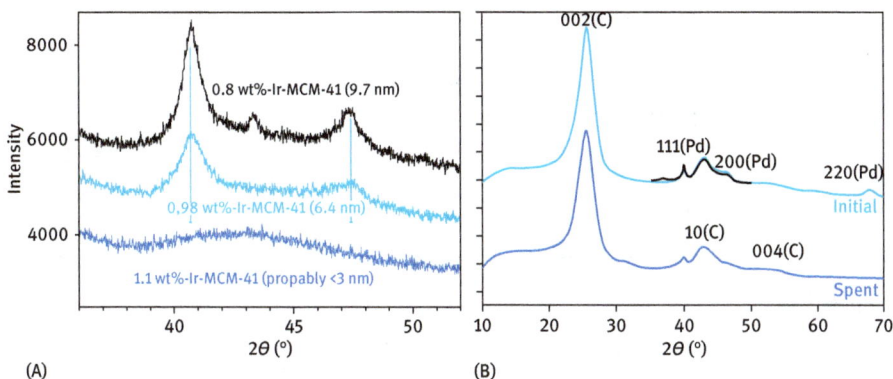

Fig. 2.36: XRD of supported metals. (A) Ir/MCM-41 catalysts prepared by different methods. (B) Fresh and spent (after hydrogenation of fatty acids) Pd/C catalysts. Reproduced with permission from [17].

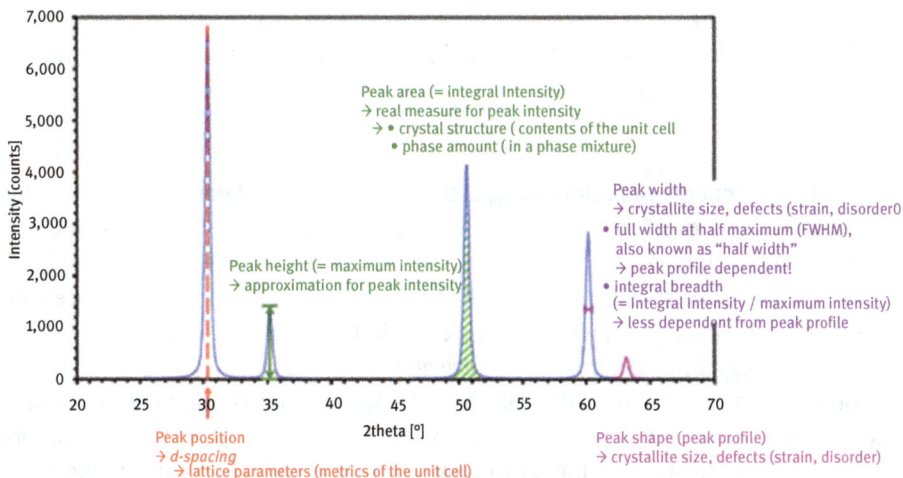

Fig. 2.37: Information content of an idealized diffraction pattern. Reproduced from [18]. Copyright © 2015 Recent Research in Science and Technology.

A limit of these techniques for applications in heterogeneous catalysis is represented by the high vacuum environment required rather than real catalytic working conditions such as atmospheric or even high-pressure environments. Nevertheless, XPS can give valuable information regarding: catalyst composition, i.e., the elements present; chemical nature of the elements; chemical nature of neighboring (coordinating) atoms; dispersion of active phase and support; and location of active phase in the particle.

In XPS measurements, ionization of the inner electron shells of atoms occurs because of exposure to monochromatic X-rays (Fig. 2.38).

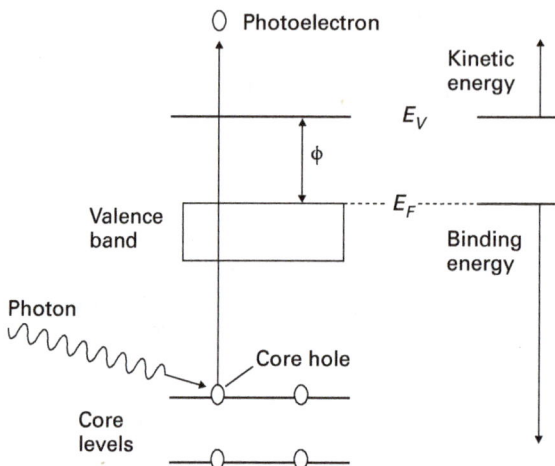

Fig. 2.38: Principles of XPS.

The difference between the kinetic energy of the emitted electrons and the X-rays energy is measured, which is then related to the binding energy:

$$h\nu = E_B + E_{Kin} + E_{corr} \tag{2.22}$$

where $h\nu$ is the energy of photons, E_{kin} is the kinetic energy of emitted electrons, E_B is the binding energy of emitted electrons, E_{corr} is the correction factor, also including the work function ϕ.

XPS is a surface sensitive technique as only electrons close to the surface can escape without energy loss, while those deeper in the bulk lose part of their kinetic energy while travelling towards the surface and cannot escape.

Several peaks are expected because of the spin-orbital coupling. The binding energy or position of a peak is specific for an element and depends on the oxidation state (electrons are stronger bound to atoms in higher oxidation states), chemical environment (for example, neighboring atoms, types of ligands, their number, etc.). Position of the main peak and satellite peaks, caused by energy loss through plasmons, help to identify the elements, and thus XPS is often called ESCA (electron spectroscopy of chemical analysis).

Quantitative analysis of XPS spectra takes into account the charging of the sample during measurements (and thus shift in the binding energy), which can be performed by relating the peaks to a reference peak, for example, C 1s peak at 284.6 eV. In addition, when the measurements take a significant time there is a possibility of surface atoms reduction altering the ratio between the peaks corresponding to different oxidation states. As an example, XPS measurements of supported Pt catalysts that contain chlorine are presented in Fig. 2.39, demonstrating that the ratio between Pt^{+2} and Pt^{+4} is changing during measurements.

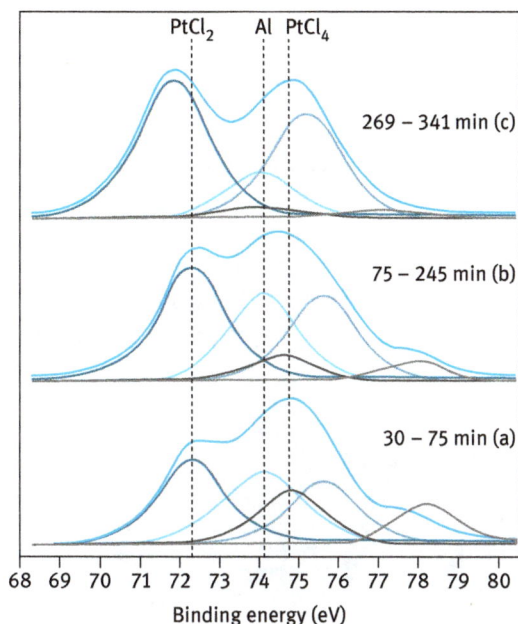

Fig. 2.39: Pt 4f photoelectron spectra of 5 wt% Pt/silica fibers catalyst during prolonged X-ray bombardment.

As already apparent from Fig. 2.39, deconvolution of XPS spectra with the contribution of several oxidation states can be challenging.

XPS data could be also used for determination of the metal particles size estimations based on the dispersed phase/support phase XPS intensity ratio, as elaborated by Davis [19].

X-ray fluorescence (XRF), widely used for elemental analysis, is the emission of characteristic "secondary" (or fluorescent) X-rays from a material that has been excited by bombarding with high-energy X-rays or gamma rays (Fig. 2.40). The term "fluorescence" is applied to phenomena in which the absorption of higher-energy radiation results in the re-emission of lower-energy radiation.

2.2.9 Infrared and Raman spectroscopies

Vibrational spectroscopies are applied in order to study chemical bonds in chemisorbed species as well as the solid substances. In the IR domain (200–4,000 cm^{-1}) the absorption of energy excites bonds to a higher vibrational state with only excitations modifying the dipolar moment taking place. Different chemical bonds absorb IR-light at certain wave lengths and the absorbed intensity can be analyzed, allowing

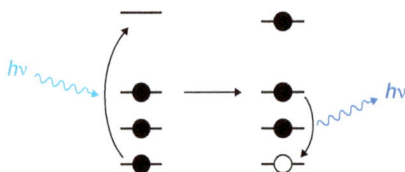

Fig. 2.40: Principles of X-ray fluorescence.

characterization of the bonds. Typically emissions of a fast pulse of light are analyzed by Fourier transform (FTIR). Transmission mode is applied for transparent samples (Fig. 2.41), which are either self-supported or KBr diluted wafers prepared under mechanical pressure.

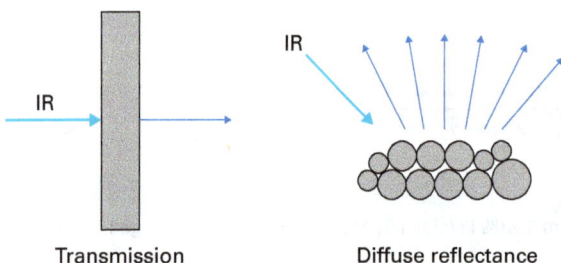

Fig. 2.41: IR spectroscopy arrangements.

Transmission is widely applied for solids and the spectrum is a result of the transmission of catalyst vs transmission of a reference. In diffuse reflectance mode (DRIFTS) catalyst grains are dispersed in a diffusing matrix, whose spectrum should be subtracted.

Infrared spectroscopy can be used for catalyst characterization by, for example, direct measurement of catalyst IR spectrum and analyzing positions of OH groups, as these positions and shapes provide information about the coordination (Fig. 2.42).

Measurements of interactions with probe molecules, such as ammonia, acetonitrile, tert-butylnitrile or pyridine, allow the determination of catalysts acidity, while basic sites could be probed with, for example, CO_2. There is a long tradition dating back to the 1950s and 1960s of using IR studies of adsorbed ammonia and pyridine molecules for the characterization of solids. The advantages of ammonia as a probe molecule for acidity are related to the molecule size, which is able to penetrate even very narrow pores and small cages of zeolites. Moreover, ammonia does not contaminate vacuum installation. However, ammonia forms dimers and the spectra are not as well-resolved as for pyridine adsorption. Fig. 2.43 illustrates adsorption of ammonia and pyridine on proton (Brønsted, H^+) and Lewis (L) sites. The band

Vicinal   Isolated   Geminal

3,720, 3,520–3,720 cm^{-1} 3,750 cm^{-1} 3,742 cm^{-1}

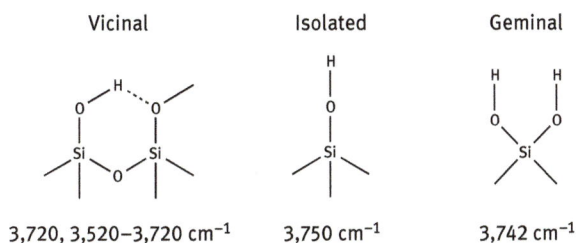

Fig. 2.42: Silanol groups on amorphous silica.

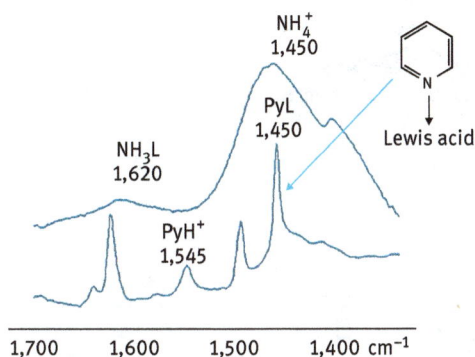

NH$_4^+$
1,450

PyL
1,450

Lewis acid

NH$_3$L
1,620

PyH$^+$
1,545

1,700   1,600   1,500   1,400 cm^{-1}

Fig. 2.43: IR spectra of ammonia and pyridine adsorption on Brønsted and Lewis acid sites.

intensity should be related to the concentration of acid sites by applying extinction coefficients.

Although pyridine is most frequently used for qualitative and quantitative acidity studies its disadvantage is related to the size, disallowing entrance to pores smaller than 0.5 nm. Acetonitrile, another often-used probe molecule, is not basic enough although it is small.

IR can be also used to elucidate metal dispersion. Fig. 2.44 shows adsorption of Pt/Si-SBA–15 catalyst, the data were obtained at University of Pierre and Marie Curie. A band at 2,079 cm^{-1} is characteristic for CO linearly adsorbed on Pt nanoparticles, while bridging CO bands adsorbed on two Pt atoms (1,860 cm^{-1}) are favored on larger faceted Pt particles. Thus, judging by low intensity of the latter signal, it can be stated that large particles of Pt are absent and the catalyst is highly dispersed.

Note that the investigation of catalysts with high metal loadings is problematic because of low transmission.

FTIR is widely applied in mechanistic studies, allowing the analysis of adsorbed substrates, reaction intermediates and deactivation, as well as in situ reaction monitoring. Fig. 2.45 illustrates FTIR spectra obtained by flowing a mixture of

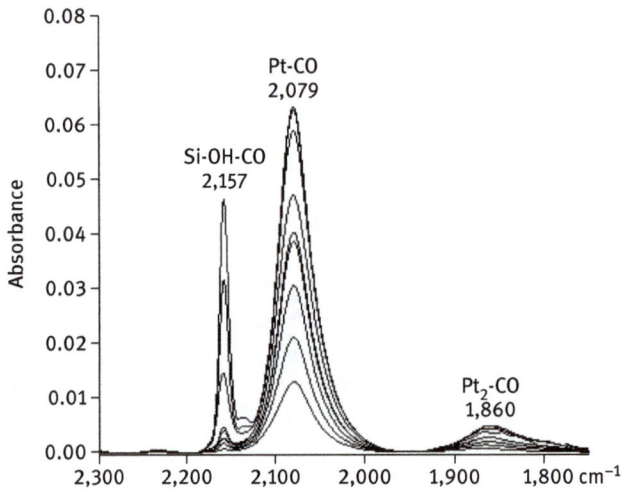

Fig. 2.44: Adsorption of CO on Pt/Si-SBA–15.

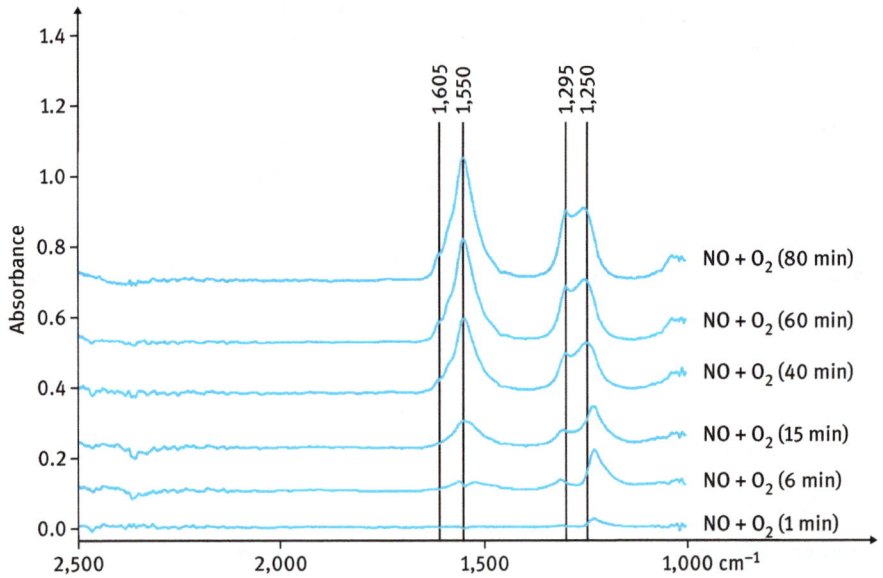

Fig. 2.45: Time-on-stream in situ FTIR on Ag/alumina at 250 °C in a flow of 1,000 ppm NO and 6 vol% O_2 in He. (Reproduced with permission from [20]).

NO and oxygen through a silver on alumina catalyst. Slow formation of the peaks at 1,250, 1,295 and 1,550 cm^{-1} and a shoulder at 1,605 cm^{-1} is attributed to generation of monodentate nitrate (1,250 and 1,550 cm^{-1}) and bidentate nitrate (1,295 and 1,605 cm^{-1}).

As a general comment on *in situ* and *operando* spectroscopies, not all observed species participate in the reaction, and some spectator species could be registered just because they do not react.

In addition to IR, ultraviolet-visible spectroscopy in the ultraviolet-visible spectral region ($50{,}000$–$2{,}500$ cm^{-1}) is also frequently applied for the analysis of solid catalysts and adsorbed species. It uses light in the visible and adjacent ranges (near-infrared, $12{,}500$–$3{,}300$ cm^{-1}). UV-vis spectra obtained *in situ* at the Fritz Haber Institute during isomerization of *n*-butane on mordenite zeolite [12] show bands at 250–370 nm (Fig. 2.46) that are ascribed to neutral conjugated double-bond oligomers, while bands at 370–500 nm correspond to carbocationic oligomers.

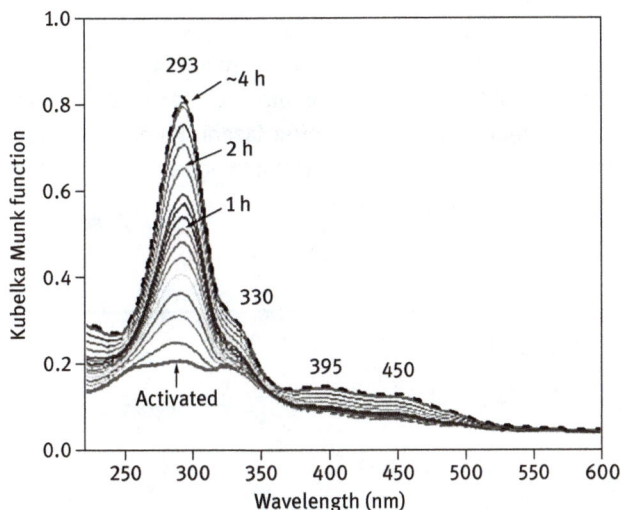

Fig. 2.46: UV-vis spectra in butane oligomerization over H-mordenite at 534 K with *n*-butane/ nitrogen ratio 5:95. Reproduced from [21].

While reflectance or transmission is transformed into adsorbance in FTIR (Fig. 2.45), in UV-vis the Kubelka Munk transform is applied in diffuse reflectance (Fig. 2.46), relating the ratio between the absorption coefficient and the scatter coefficient of the sample at a given wavelength to reflectance.

A variation of IR is attenuated total reflectance (ATR) that enables samples to be examined directly in the solid or liquid state without further preparation (Fig. 2.47).

Contrary to IR, Raman spectroscopy relies on different wavelengths of the scattered light, which is a feature only of a small fraction (10^{-6}) of emitted light. The selection rules are also different in Raman and IR spectroscopies, and some types of molecular motions appear in Raman and IR, and some only in Raman or IR. Namely, a criterion for Raman scattering is polarizability, while for IR it is a transition in the

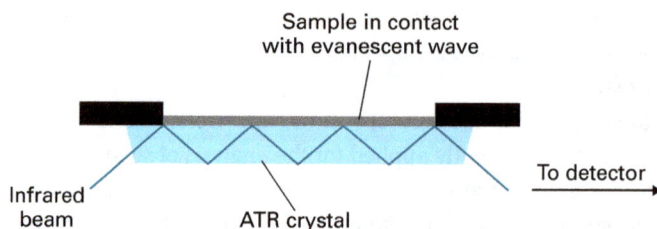

Fig. 2.47: Schematic of ATR-IR. (Reproduced from [21].)

vibrational states that induce a change in the dipole moment. A particular vibration mode in Raman is silent if polarizability is not changed by vibration.

Raman spectroscopy is more useful for studying the catalyst itself and since water does not interfere with Raman spectroscopy it could be used in aqueous systems, where IR is difficult. Conventional Raman signals have lower sensitivity that requires an enhancement, which is achieved by applying lasers. Raman spectroscopy can be combined with "operando studies", where the spectra of the catalyst are measured during a catalytic reaction (Fig. 2.48).

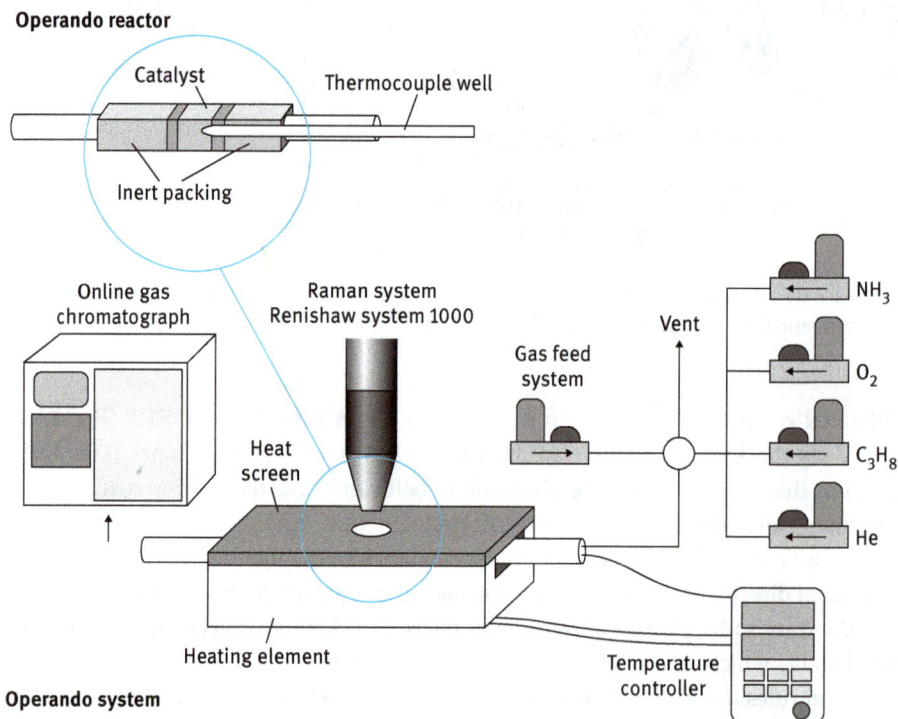

Fig. 2.48: Operando system. (Reproduced with permission from [22].)

IR-visible SFG spectroscopy has recently started to be applied for catalytic materials. It uses two input laser beams overlapping at the surface of a material, generating an output beam with a frequency that is the sum of the two input beams. One of the two input beams is a visible wavelength laser at a constant frequency, while the other is from infrared laser that is tuned to scan the system and obtain the vibrational spectrum (Fig. 2.49). SFG has the disadvantage of a relatively small frequency range $(1,600–4,000 \text{ cm}^{-1})$ and lower resolution and sensitivity, compared with IR.

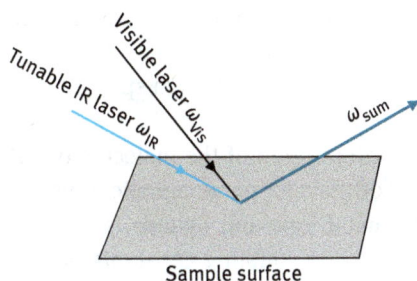

Fig. 2.49: Principle of IR-visible sum frequency generation spectroscopy.

Conversely, this method affords the possibility of obtaining vibrational spectra *in situ* at both buried and exposed interfaces, as well as under equilibrium and reaction conditions. This is in contrast to traditional vibrational spectroscopies when signals are generated from the surface.

2.2.10 Catalyst particle size measurements

The mean particle size of a catalyst can be determined by a light scattering technique using a laser diffraction and commercially available particle size analyzers. The determination is based on He-Ne laser light scattering as light is passed through the particle suspension. The scattering profile is collected by a Fourier Transforming lens and the various outputs are handled and converted to particle size distributions by an integral computer.

The particles size of powdered catalysts can be measured using a counter type apparatus that has one or more microchannels that the separate two chambers containing the electrolyte solutions. When a particle flows through one of the microchannels, it results in the electrical resistance change of the liquid-filled microchannel. This resistance change can be recorded as electric current or voltage pulses, which can be correlated to size, mobility, surface charge and concentration of the particles.

2.2.11 Electron paramagnetic/spin resonance

The technique relies on resonant absorption of microwave radiation by paramagnetic ions or molecules with at least one unpaired spin in the presence of a static external magnetic field. From an experimental perspective, the measurements are organized by keeping the wavelength constant and recoding variations in the magnetic field. As a result, complex spectra in a narrow range of energy are obtained with the superposition of different paramagnetic species and several spectra components caused by anisotropy of the environment.

EPR is applied for structural characterization of solid and liquid systems and only radicals (species with unpaired electrons) can be detected. Transition metal cations in specific oxidation states (for example, Cr^{3+} d^3, F^{3+} d^5 high spin), O^- and O_2 are most relevant to catalysis.

Signal positions depend on the electron configuration of the molecule or cation and on the energy difference between the d orbitals for paramagnetic transition metal cations. Delocalization of electrons on metal ions and ligands is reflected from the number of lines and their intensities, while coupling constants provide information on the bonding between the cation and the ligands. In addition, probe molecules can be used to study the species present in the catalyst. For example, Fig. 2.50 shows the ESR spectra recorded at 77 K for the fresh iron-modified beta

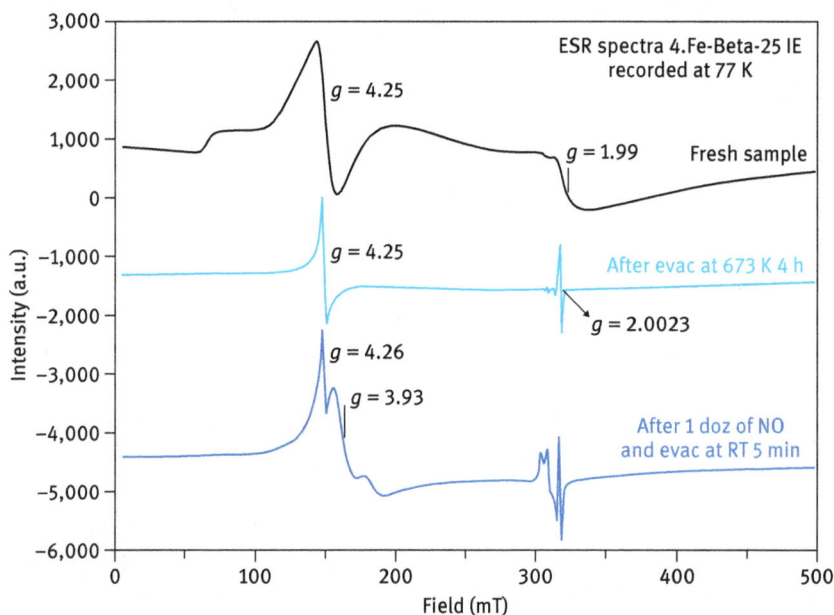

Fig. 2.50: ESR spectra of fresh, evacuated and NO adsorbed Fe-H-Beta-25 (Reproduced with permission from [23]).

zeolite, for the same material after evacuation at 300 °C for 4 h and after adsorption of NO.

The ESR spectrum for the fresh sample displays signals in two different regions of the magnetic field, a broad line that can be assigned to randomly oriented Fe^{3+} ions in oxides and/or hydroxides species. The signal at $g = 4.25$ in a low field region originates from Fe^{3+} ions in tetrahedral coordination. After evacuation at 673 K for 4 h a signal at $g = 4.25$ is more visible and a signal from iron oxide and hydroxide species is less intensive. The new narrow ESR signal ($g = 2.0023$) can originate from coke formed from residual template, but it also can be caused by Fe^{3+} in octahedral coordination, as isolated ions in cationic position attributed a sharp line at $g = 2$ to superoxide ions (O_2^-) that are associated with iron ions. Adsorption of NO results in ESR signals in the magnetic field range 300–350 mT and in another one in the low field ($g = 3.9$). After short evacuation a rhombic signal appears, with three distinct values of the diagonal terms ($g = 2.09$, $g = 2.06$, $g = 2.02$). The low field value ($g = 3.93$) originates from $(NO)_2$ biradicals. Based on this, it can be assumed that most of the iron is tetrahedral Fe^{3+} but that some Fe^{2+} is also present.

In addition to the identification of paramagnetic species on the surface or within cavities of zeolites/mesoporous materials, EPR can be used for identification of radicals, which could escape into the gas-phase. Such events could occur, for example, in partial oxidation of hydrocarbons or in hydrocarbon-assisted SCR of NOx in which the generation of radicals was detected by matrix isolation EPR [24]. Propagation of the surface reactions into the gas-phase is beneficial for the overall NOx conversion and could be used in engineering of catalytic converters. For example, a catalytic unit for reduction of NOx can contain several catalytic beds (Fig. 2.51; 3a–3e), that are arranged separately from each other in a longitudinal direction in the catalyst unit, and a number of second reaction zones (Fig. 2.51; 4a–4e) downstream of each respective catalytic bed.

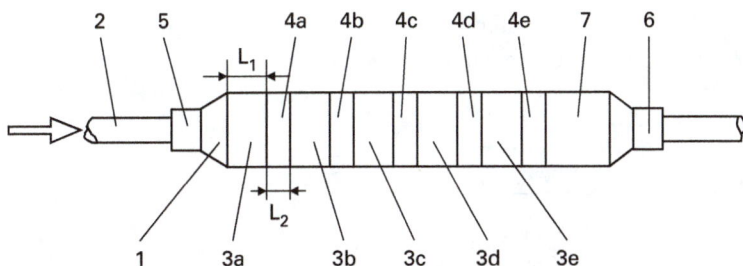

Fig. 2.51: Catalytic converter. 1, catalytic unit; 2, exhaust pipe; 3, catalytic zones; 4, gas-phase reaction zones; 5, inlet; 6, outlet; 7, oxidation catalyst. (Reproduced from [25]).

2.2.12 Mössbauer spectroscopy

Mössbauer spectroscopy provides information about the oxidation state, coordination number, ligand type, electron distribution, the strength of binding to the surrounding of the resonant atom and the electric charges of the neighbors. The technique is based on absorption by a nucleus of the resonant atom of photons emitted by a radioactive source (for iron this is ^{57}Co). The wavelength emitted has a constant value, which is slightly varied by the Doppler effect, i.e., by moving the gamma-ray source towards and away from the sample at varying speeds, typically a few mm s^{-1} (Fig. 2.52). For convenience, Mössbauer spectra allowing observations of the hyperfine interactions correspond to the intensity of the transmitted X-rays as a function of the Doppler velocity.

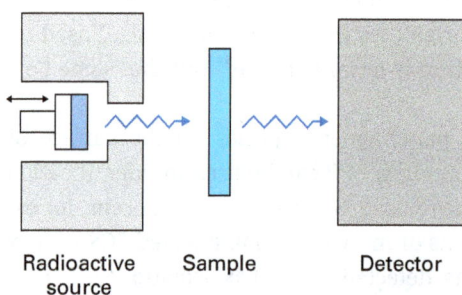

Radioactive   Sample   Detector
source

Fig. 2.52: Mössbauer apparatus.

During emission or absorption of photon a free nucleus recoils because of the conservation of momentum with a certain recoil energy. Resonance is achieved when the loss of the recoil energy is overcome by having the emitting and absorbing nuclei in a solid matrix. Identical environments of emitting and absorbing nuclei produce a single absorption line spectrum, as shown in Fig. 2.53. As the resonance occurs only in the case of exact matching the transition energies of the emitting and absorbing nucleus the effect is isotope specific.

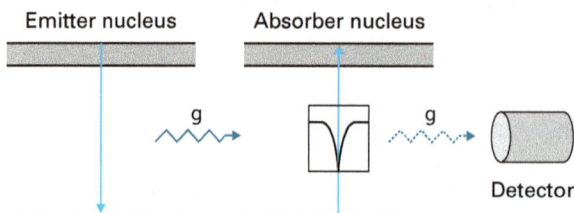

Emitter nucleus   Absorber nucleus

g   g

Detector

Fig. 2.53: Simple Mössbauer spectrum from identical source and absorber.

The strength of the signal reflects the relative number of recoil-free events and is dependent upon the y-ray energy, thus the Mössbauer effect is only detected in isotopes with very low-lying excited states. The resolution is dependent on the lifetime of the excited state. Because of these constraints the technique is limited to a few isotopes, including (relevant for catalysis) ^{57}Fe, ^{119}Sn. For isotopes such as ^{193}Ir or ^{197}Au, a very short lifetime of a radioactive source prevents their characterization by Mössbauer spectroscopy.

When the source and absorber are identical the absorption peak is seen at zero velocity (0 mm s^{-1}), while various types of changes in the environment of the resonant atoms lead to changes in the spectra. The isomer shift, which is useful for determining valence states, ligand bonding states and electron shielding, is caused by changes in the s-electron environment, allowing differentiation between Fe^{+2} and Fe^{+3} for example.

The nuclear energy levels are split (quadruple splitting) in the presence of an asymmetrical electric field caused by asymmetric electronic charge distribution or ligand arrangement. Magnetic splitting of nuclear levels with a spin of I in the presence of a magnetic field into $(2I + 1)$ substrates is caused by dipolar interactions of the nuclear spin moment with the magnetic field.

Thus ^{57}Fe Mössbauer spectroscopy is the method for the study of valency, coordination and magnetic properties of the iron species in for example iron-modified zeolites, contrary to the limited abilities of ESR or IR. In these materials, in addition to extra-framework iron species (for example, not isomorphously substituted), isolated Fe ions with a different oxidation state that are chemically anchored to the framework via −O−Fe bridges, are present in addition to dinuclear Fe-O-Fe species confined in the channels, oligomeric Fe oxi species in the large cavities and oxide-like clusters. Typically, ^{57}Fe Mössbauer spectra (see Fig. 2.54) are fitted to get the best correspondence with the data by using the line width peak intensity, isomer shifts and

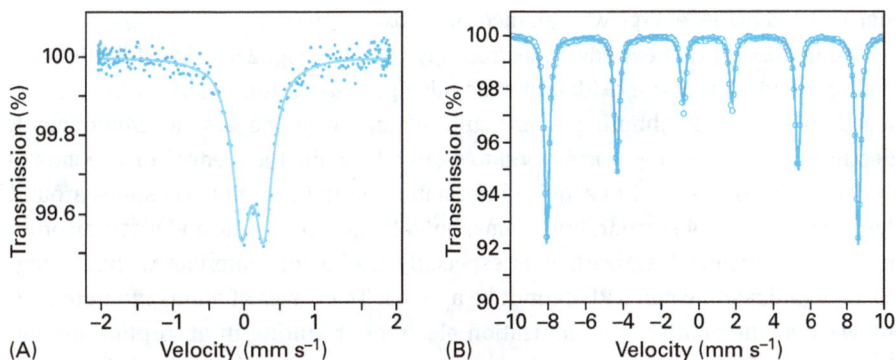

Fig. 2.54: Room temperature ^{57}Fe Mössbauer spectra of (A) Fe-ZSM-5 prepared by ion exchange and (B) spectrum with sextet: α-Fe$_2$O$_3$. (Reproduced with permission from [26]).

the quadruple coupling constant as fit parameters. The quadruple coupling constant measures the splitting in mm s^{-1} between the two lines in the doublet.

Fig. 2.54(A) displays ^{57}Fe Mössbauer spectra consisting of a single paramagnetic doublet for Fe-ZSM-5 zeolite, prepared by ion exchange. In fact, the shape of spectra depends very much on the preparation methods. In the case of ion exchange preparation methods the presence of Fe^{3+} ions in octahedral coordination gives intense central doublets (0.3 < isomeric shift < 0.4 mm s^{-1} and 0.5 < quadruple splitting < 1.1 mm s^{-1}), while the appearance of additional doublets with the isomeric shift ~1.2 mm s^{-1} and 1.5 < quadruple splitting < 2.8 mm s^{-1} (absent in Fig. 2.54A) is associated with Fe^{2+} ions in an octahedral coordination. Agglomerated iron oxide species could result in a sextet (absent in Fig. 2.54A, but present in Fig. 2.52B).

If Fe-ZSM-5 is prepared by isomorphous substitution a broad singlet with an isomer shift of 0.52 mm s^{-1} will be observed in the spectra, corresponding to tetrahedrally coordinated high spin Fe^{3+} ions and reflecting the incorporation of iron into the framework. If the same catalyst is calcined, migration of iron to an extra-framework position and increased interactions between extra-framework Fe^{3+} and the zeolite lattice will lead to an unresolved doublet in the Mössbauer spectrum. Alternative methods of Fe-ZSM-5 catalyst preparation (impregnation, reductive solid state ion exchange and chemical vapor deposition) result in spectra, which could be deconvoluted into a hyperfine magnetic sextet and two quadruple doublet lines, confirming the presence of large α-Fe$_2$O$_3$ particles and either small α-Fe$_2$O$_3$ particles or Fe^{3+} in a strong distorted environment.

2.2.13 X-ray absorption spectroscopy

This is an element-specific type of spectroscopy that requires synchrotron produced X-rays with energy close to the absorption edge of the element core shells. Information about the oxidation state of the element is retrieved from the edge (Fig. 2.55) as the edge shifts to higher energy with an increase of oxidation state.

X-ray absorption near-edge spectroscopy (XANES, Fig. 2.55), in addition to allowing elucidation of the oxidation state, also provides information on the electronegativity of the neighboring atoms and symmetry of the atomic environment. Electrons from the K edge correspond to s-orbitals, while the L edge corresponds to p-orbitals. EXAFS is used to acquire information on the coordination shells around the atom, number of surrounding atoms, bond lengths, dynamic and static disorder in the inter-nuclear distances. It is especially useful for materials without long-range organization when XRD cannot be applied. The waves of atoms interfere with the wave of the emitted after-ionization electrons, resulting in absorption oscillations seen in EXAFS spectra. Fourier transform is applied to decompose the EXAFS signal. Each peak in Fig. 2.56, presented here for illustration, corresponds to a shell of scattering atoms at a certain distance. A number of parameters is needed for the

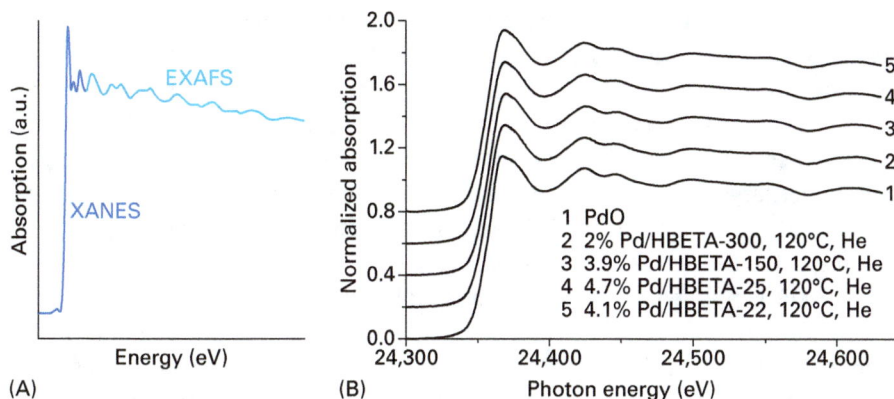

Fig. 2.55: X-Ray absorption spectroscopy. (A) XANES and EXAFS regions. (B) Pd K-XANES spectra of calcined Pd/H-beta catalysts and PdO.

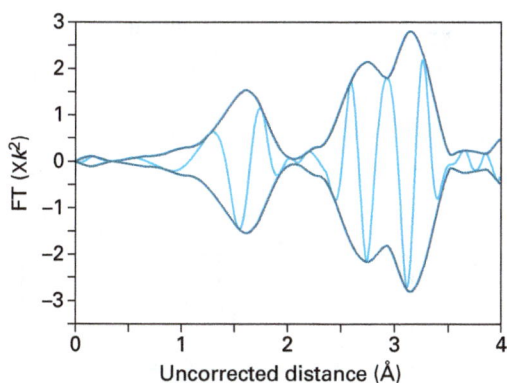

Fig. 2.56: Model fits of Pd K EXAFS spectra taken from a calcined 2% Pd/H-beta-300 catalyst.

data fitting thus precise simulation is challenging especially when too many different atoms and/or distances coexist. Blurring of a signal is possible when there is too large a discrepancy in the atom environment, as the EXAFS signal is an averaged linear combination of all sites of a particular element.

2.2.14 Nuclear magnetic resonance

Nuclear magnetic resonance (NMR), being isotope specific, concerns isotopes with a nuclear spin different from zero. It is based on splitting of the nuclear spin ground level by an external magnetic field of constant value and the subsequent absorption of energy from electromagnetic radiation of varying wavelengths. Absorption signal positions (chemical shift) depend on the nuclei electron environment shielding nucleus spin from the external magnetic field. For a nuclear spin I the number of lines is $2nI + 1$, which are separated by a coupling constant, the latter linked to the distance between the nuclei. The inherent sensitivity of NMR is low, although it is possible to get larger signals by using highly dispersed or porous materials. For the solid materials, broadening of the signals is overcome by rotating the sample 54.74° with respect to the magnetic fields (magic angle spinning, MAS).

^{1}H NMR is applied for the characterization of OH groups, while ^{13}C NMR can be used for studying adsorption of isotopically marked organic compounds. The chemical shifts in ^{29}Si MAS NMR are linked to the number of silicon atoms present as second neighbors. An example of the ^{29}Si MAS NMR spectra for a mesoporous material H-MCM-41 is given in Fig. 2.57, showing several broad overlapping lines. A computer fit to the spectrum gives the lines at −108.5 ppm (65% of the total intensity), −100.6 ppm (13%), −94.8 ppm (19%) and −80 ppm (3%). The strongest peak at −108.5 ppm can be attributed to $Si(OSi)_4$ sites, while the components at −100.6, and −94.8 ppm are caused by the sites with silanol groups $Si(OSi)_3OH$ and $Si(OSi)_2(OH)_2$, respectively. As the material has a considerable amount of Al in the framework, the contribution to the signal is expected from the sites $Si(OSi)_3(OAl)$, $Si(OSi)_2(OAl)_2$, $Si(OSi)(OAl)_3$, and $Si(OAl)_4$ in the region of −80 ppm to −100 ppm.

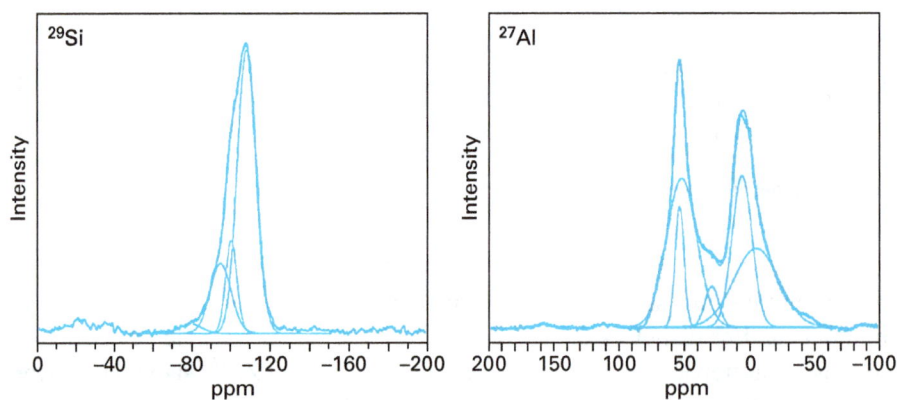

Fig. 2.57: ^{29}Si (left) and ^{27}Al (right) MAS NMR spectra of H-MCM-41 including the computer fits. (Reproduced from [27] with permission from the PCCP Owner Societies).

The ^{27}Al MAS spectrum of H-MCM-41 (Fig. 2.57) shows a signal with a maximum at 54 ppm caused by tetrahedrally coordinated framework aluminum and a broad asymmetric line around 0 ppm caused by octahedrally coordinated aluminum sites.

A small component arising from the computer fit was attributed tentatively to the distorted tetrahedral or five-coordinated Al sites. The quantitative analysis of the spectrum yields the overall absolute number of Al sites, as well as the respective values for framework tetrahedral, non-framework tetrahedral or five-coordinated and octahedral aluminum. The Si/Al ratio can be estimated from the ^{27}Al line intensity.

NMR can also be used to characterize microporosity of zeolites applying ^{129}Xe, whose chemical shifts depend on the inner void structure of the solid and charge gradients.

2.2.15 Imaging of catalysts

Several techniques are applied for the imaging of catalysts.

2.2.15.1 Electron microscopy [28, 29]

Electron microscopy uses a particle beam of electrons to illuminate a specimen and create a highly-magnified image (Fig. 2.58). Electron microscopes have a much greater resolving power and can obtain much higher magnifications than light microscopes that use electromagnetic radiation. Electron microscopy is usually combined with elemental (EDX) or structural analysis (XRD). In EDX X-rays are emitted by the atoms for which inner shell energy levels are ionized by electron excitation and then collected.

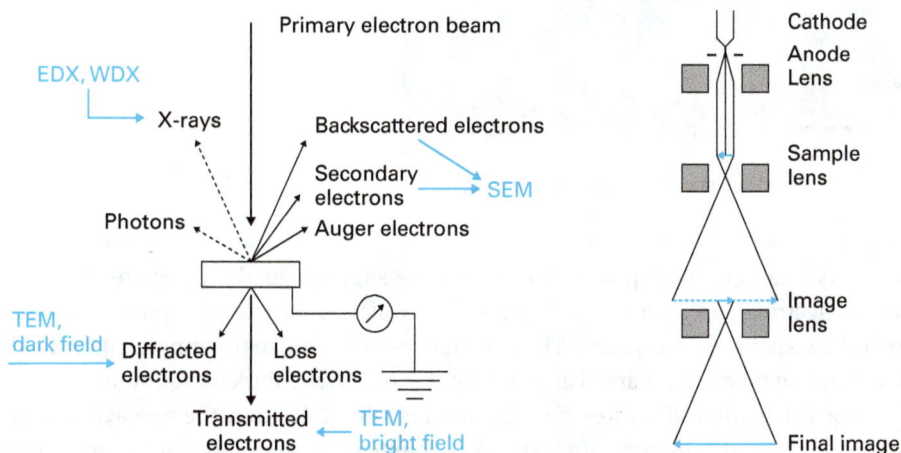

Fig. 2.58: Different electron microscopes. (Reproduced with permission from [29]).

The TEM uses a high voltage electron beam (100–300 keV) to create an image. The electrons are emitted by an electron gun and the electron beam is accelerated by an anode focused by electrostatic and electromagnetic lenses, and transmitted through the specimen that is in part transparent to electrons and in part scatters them out of the beam. The magnification range in TEM is $10^2–10^7$. The sample is put on a grid and is placed in a vacuum chamber. Images are formed because of different lateral absorptions of the beam, with heavy atoms being darker. A 10^{-7} Torr is usually used for high-resolution electron microscopy to avoid noise caused by gas molecules. The practical limit of detection is ~1 nm, although several times lower resolution is possible. Support and the supported phase are distinguished where there is good contrast between them (Fig. 2.59). Discrimination of the two phases is easier when their d-spacing and shape are different. Oxide supports are more difficult to distinguish than the metallic phase. The same difficulty applies to metals of the first transition row compared to heavier metals from the third row. Challenges in images are also present in the case of poorly organized supports, such as y-alumina, which exhibits agglomeration of small crystallites, and on well-crystallized supports with many d spacings in images.

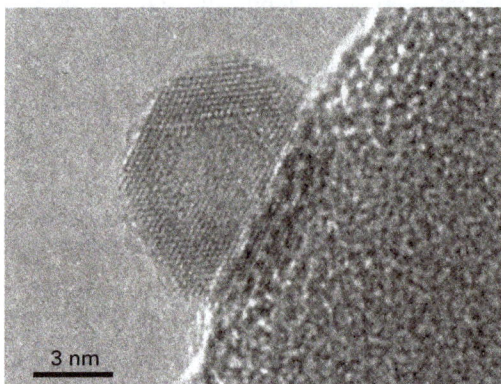

Fig. 2.59: A metal cluster on a support: TEM image.

For small particles with poor contrast with the support, good images are obtained when electrons are scattered at high angles, which is far from the position of the objective aperture in classical TEM. In the case of high-angle annular dark field (HAADF) metal particles are visible as bright sports over a dark background.

For calculation of cluster size distribution (Fig. 2.60) and the average size of particles, several hundreds of clusters are counted that counterbalance premature conclusions based on eye-catching zones (unexpectedly large particles or unusual particle shape).

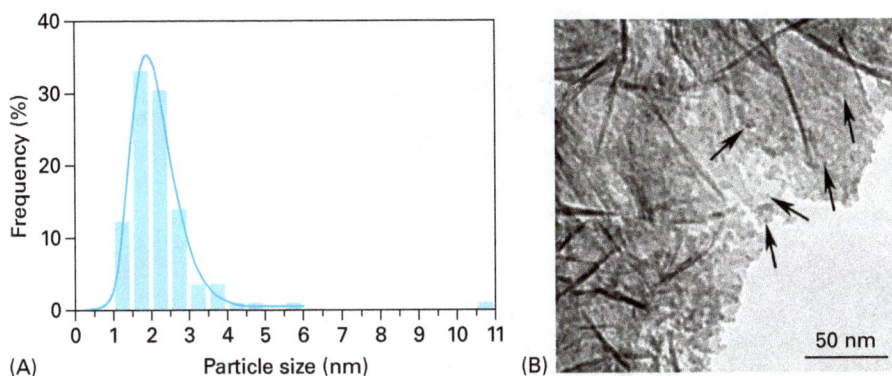

Fig. 2.60: HRTEM of Au/Al$_2$O$_3$ catalyst. (A) Image. (B) Cluster size distribution.

Another recent development in HRTEM is associated with 3D imaging, when the sample is tilted with a small (1°) increment and the images are taken after each change.

Although HRTEM analysis became very popular and is widely applied, it is important to remember the challenges and shortcomings associated with the method, such as the inability of imaging of clusters well below 1 nm, representativity of the specimen area, potential damage during preparation as well as electron beam damage. High energy of the incident electron beam may cause changes in the microstructure of crystallites or the reduction of surface ions. Applicability of TEM for in situ studies is limited by losing resolution at above ~5 mbar. In fact, all electron microscopy techniques rely on high vacuum because of the large attenuation of the electron beam in any atmosphere.

In general as technically relevant catalysts should be studied under certain gas atmospheres it is difficult to extend vacuum-based techniques, including TEM, to in situ experiments.

SEM has a magnification range 10–10^5. SEM produces images by probing the specimen with a focused electron beam (accelerating voltage 4–30 keV) that is scanned across a rectangular area of the specimen. Electrons emitted or scattered by the sample are then detected. Generally, the image resolution of SEM is about one order of magnitude poorer than that of a TEM. Fig. 2.61 shows SEM images for ZSM-5 zeolite.

Scanning transmission electron microscopy (STEM) is a merger of TEM and SEM, allowing the same resolution as TEM and the ability to perform microanalysis at a high resolution similar to SEM.

Fig. 2.61: SEM image of zeolite ZSM-5.

2.2.15.2 Scanning probe microscopy

In scanning probe microscopy the very sharp probe is brought in contact with a sample. In scanning tunneling microscopy (STM) the probe does not touch the surface (Fig. 2.62A). A constant tunneling electric current is maintained, and the method is thus limited to conducting materials. Very high resolutions (height 0.01 nm, length 0.1 nm) can be achieved.

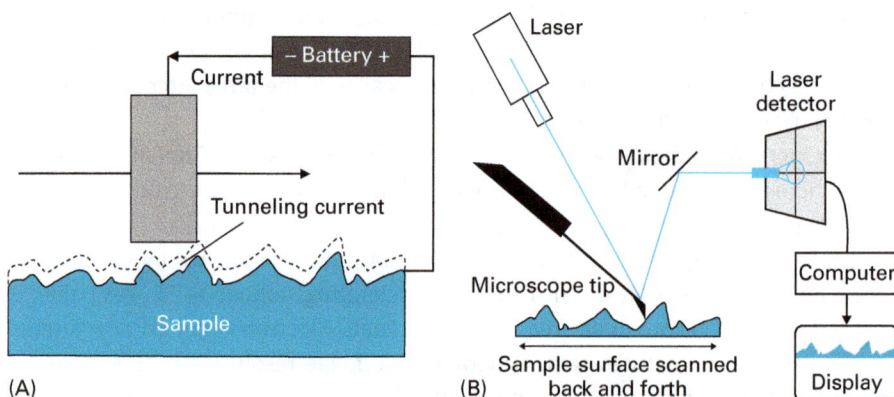

Fig. 2.62: Principles of (A) STM and (B) atomic force microscopy (AFM). (Modified from [29]).

STM studies of catalytic systems under realistic conditions (high pressure and high temperature) are possible and have been recently achieved.

In AFM the probe can touch the surface (Fig. 2.62B). As a constant small force is maintained the method is not limited to conducting materials and is suitable for all surfaces. The method allows a high resolution (length 2–10 nm, height 0.1 nm). Desulfurization reactions in oil refining are an example of applying STM to catalysis. Triangular MoS_2 nanoparticles have sulfur atoms located in the edges, which react with hydrogen, forming hydrogen sulfide. The resulting vacancy is replaced by sulfur from the sulfur-containing oil molecules. STM studies by F. Besenbacher and co-workers [30] (Fig. 2.63) demonstrated that the morphology of MoS_2 nanoparticles strongly depends on their size. When the particles contain less than 100 atoms, the relative sulfur content increases above the stoichiometric value for the bulk MoS_2. Subsequent changes occur in the edge structure to counterbalance this increase and to lower sulfur to a molybdenum ratio and improve stability of the nanoparticles that is deteriorating with sulfur increase. Refer to a review article [31] for a discussion of the progress in the application of scanning probe microscopy in catalysis.

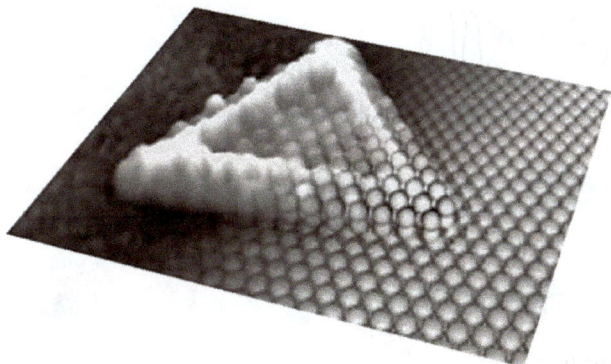

Fig. 2.63: STM of MoS_2. (Modified from [30]).

2.2.15.3 Electron microprobing

Determination of an active phase profile in catalysts grains can be performed using an electron probe microanalyzer, which is an analytical tool used to non-destructively determine the chemical composition of small volumes of solid materials. It works in a similar way to SEM by bombarding the sample with an electron beam and collecting the signals coming from the sample.

Another non-destructive method is based on applying XPS with a variation of the emission angle.

Destructive methods involve mechanical and chemical erosion by, for example, progressively removing surface atoms caused by ion bombardment. Another technique is laser ablation inductively coupled plasma mass spectrometry (LA-ICP-MS), which combines the micrometer-scale resolution of a laser probe with the speed, sensitivity and multi-element capability of ICP-MS, and rivals other microbeam techniques such as the proton microprobe and secondary ion mass spectrometry. A pulsed laser beam is used to ablate a small quantity of sample material, which is transported into the Ar plasma of the ICP-MS instrument by a stream of Ar carrier gas. Fig. 2.64 gives an example of LA-ICP, showing an LA-ICP-MS spectrum for Cu-H-MCM-41. The intensity of copper is following the Al-content, suggesting that copper species are balancing the framework charge, and thus were ion-exchanged into H-MCM-41. Some minor heterogeneity of the aluminum distribution in the samples is visible. Intensity fluctuations are also observed in the copper contents and few intensive peaks in the sodium intensity are observed, indicating that traces of sodium are still present in the samples that were synthesized first as Na-MCM-41 and then exchanged with protons.

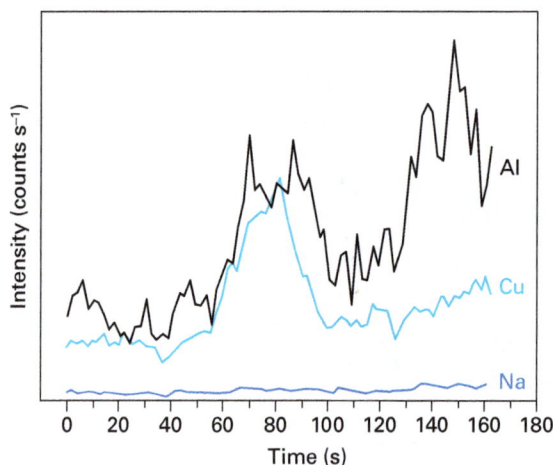

Fig. 2.64: LA-ICP of Cu-H-MCM-41. (Reproduced from [27] with permission from the PCCP Owner Societies).

2.2.15.4 Magnetic resonance imaging

In spatially unresolved conventional NMR a nucleus of non-zero nuclear spin quantum number is placed in an external magnetic field. By exposing the system to an electromagnetic energy of appropriate radio frequency a resonant absorption occurs between non-degenerate nuclear spin energy levels at a frequency proportional to the strength of the external magnetic field. The precise energy-level splitting, specific to a given isotope, is slightly modified by the electronic environment of the nuclei, making it possible to identify the presence of particular molecular species.

In order to obtain spatial resolution a spatially varying magnetic field is applied in addition to the large static field, with the resonance frequency of species within a sample becoming a function of position and strength of the applied gradient.

In pulse encoding methods, an initial and a subsequent pulsed field gradient are applied after a certain observation time, allowing the monitoring of the distance travelled during a known time and thus the measuring of diffusion, dispersion and flow processes.

Magnetic resonance imaging can be used for micro-imaging of a single catalyst pellet with a spatial resolution of ~30–50 µm, making it possible to reveal the effect of the catalytic manufacturing process on the catalytic pellet micro- and meso-scale structure, for example, tortuosity. At the same time, this spatial resolution does not make it possible to resolve individual pores in a pellet. MRI can be applied to characterizing metal ion distribution in the pellet during the catalyst preparation and in the final material, as well as to spatial distribution of coke in a spent catalyst pellet.

It is possible to investigate internal phase distribution and liquid flow fields inside fixed-bed reactors and monoliths reactors at spatial resolution of 100–500 µm, although measurement constrains do not allow for the study of large-scale industrial size reactors.

In the case of trickle bed reactors, two-phase flow investigations make it possible to measure the values of hold-up and wetting of the catalytic materials, consistent with gravimetric data and model predictions.

Fig. 2.65 shows that full pre-wetting of the bed was followed by immediate establishment of the trickle flow, yielding significantly larger catalyst wetting and hold-up compared to other methods of wetting.

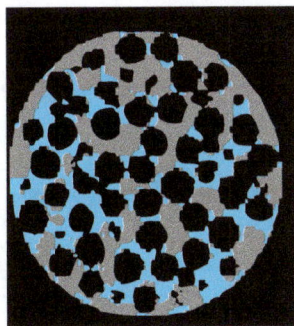

Fig. 2.65: MR images of gas and liquid distribution within a fixed-bed reactor operating at air and water velocities of 66 and 4.6 mm s^{-1}, respectively. The bed was fully pre-wetted. Red, gray, and black pixels identify water, air, and packing elements, respectively. (Reproduced with permission from [32]).

In addition to MRI related to catalytic engineering, recently there has been an interest in the imaging of reactions *in situ*.

2.2.15.5 Positron emission tomography imaging

Positron emission tomography (PET) is now a well established diagnostic technique in nuclear medicine, generating 3D images of radiolabeled molecules distributed within living human organs.

Positron-emitting short-lived radioisotopes such as [11]C (half-life 20.4 min) must be produced on-site, which is done by the irradiation of an appropriate target material with energetic beams of protons or deuterons generated by a cyclotron. Encounters between positrons and electrons lead to their mutual annihilation and resulting electromagnetic radiation. Rotating a tomograph during a scan (~10–15) allows for recoding photons emitted over 360° in the plane, making it possible to have a 3D reconstruction of the imaged object.

Data for ^{11}C methanol transformations in a catalytic bed of a zeolite are presented in Fig. 2.66. The image was acquired with a small PET scanner, developed by E. Sarkadi-Priboczki and co-workers from the Institute of Nuclear Research in Debrezen. Red and yellow colors correspond to higher radioactivity, while green and blue are related to lower radioactivity concentrations. The 3D image exhibits an inhomogeneous distribution of compounds on the catalyst, reflecting non-uniformity of the catalyst bed in terms of activity.

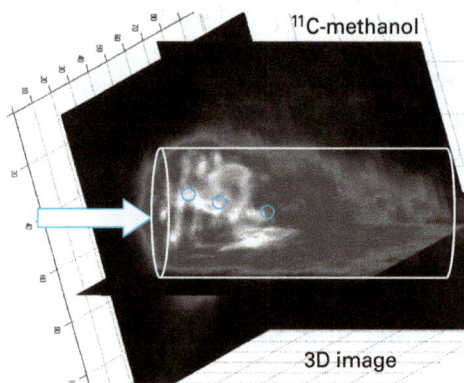

Fig. 2.66: 3D image of methanol conversion on a zeolitic material at 620 K. Courtesy of Dr. E. Sarkadi-Priboczki.

2.2.15.6 Integrated laser and electron microscopy

In fluorescence microscopy (FM) a focused laser light scans a catalyst, detecting fluorescence emission from the focal points. The technique (contrary to electron microscopy) has limited spatial resolution and only reveals fluorescent structures. The integrated laser and electron microscope (iLEMI developed by B. Weckhuysen and co-workers [33]) is an imaging tool, combining the strengths of both methods. In order to generate a fluorescence signal, fluorogenic thiophene oligomerization is used, catalyzed by the acidic sites of the zeolites. The product, oligomerized thiophene, emits green fluorescence with the intensity of the fluorescence relating to acidity. In this way a spatial resolution of ~20 nm can be achieved. The iLEM set-up shown in Fig. 2.67A consists of a custom-designed laser scanning FM mounted on a side port of a TEM with the laser beam perpendicular to the path of the electron beam. For FM imaging the grid with the catalyst sample is rotated to face the laser beam, while FM is partially retracted and the grid is tilted to enable TEM imaging.

The technique was applied to study a fluid catalytic cracking catalyst, which will be discussed in detail in Chapter 4. The active phase, zeolite Y, is embedded in a matrix consisting of clay, silica, and alumina. Using iLEM, the catalyst acidity, which is related to fluorescence intensity, could be correlated to the catalyst particle structure. Two different areas were found in the catalyst, one associated with the zeolite component giving the fluorescent products, while the other areas are mainly composed of the matrix material (Fig. 2.67B,C).

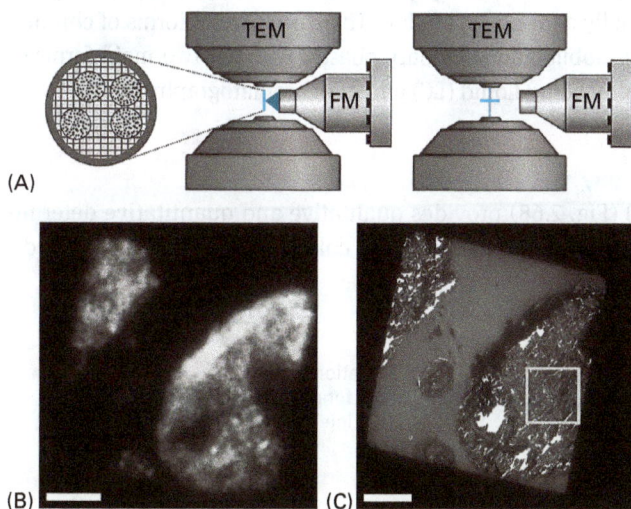

Fig. 2.67: (A) The iLEM set-up, and iLEM analysis of a sectioned FCC catalyst particle. (B) FM image. (C) TEM image, taken from the same region. (Reproduced with permission from [33]).

When the zeolitic structure is exposed to steam, Al atoms from the framework can be extracted, creating extra-framework Al species. This results in lower amounts of Brønsted acid sites and thus eventual deactivation. The iLEM images of hydrothermally deactivated catalysts show much lower fluorescence intensity compared to the fresh FCC catalyst particles, moreover a large fraction of the crystals is damaged and the large regions of clay are lost.

2.2.16 Catalytic reactions: product analysis

Various reactors for studying catalytic reactions will be discussed in Chapter 3. Investigations of catalytic reactions, in addition to providing information on kinetics, activity, selectivity and deactivation, are also important in the reaction mechanisms. Detailed product analysis is thus an essential element of each and every catalytic study. In vary many cases it can be a bottleneck, as although products might not be unknown their quantification is necessary. Moreover, rapid analysis may be needed. A brief overview of chromatographic methods is provided here. Chromatography is a physical method of separation in which the components to be separated are distributed between two phases, one of which is stationary, while the other moves in a definite direction. Chromatographic methods are widely used nowadays for analytical purposes, not only for off-line analysis but also for on-line determination of minute amounts, as well as large- scale preparative separations. In fact, not only monomers, but also polymers and oligomers, can be separated by chromatography, although in the former case it is essentially a group separation. There are several forms of chromatography that use different mobile and stationary phases, with the two main forms of instrumental chromatography being liquid (LC) and gas chromatography (GC).

2.2.16.1 Gas chromatography

Gas chromatography (GC) (Fig. 2.68) provides qualitative and quantitative determination of organic components, introduced into the column for separation either directly or in the derivatized form.

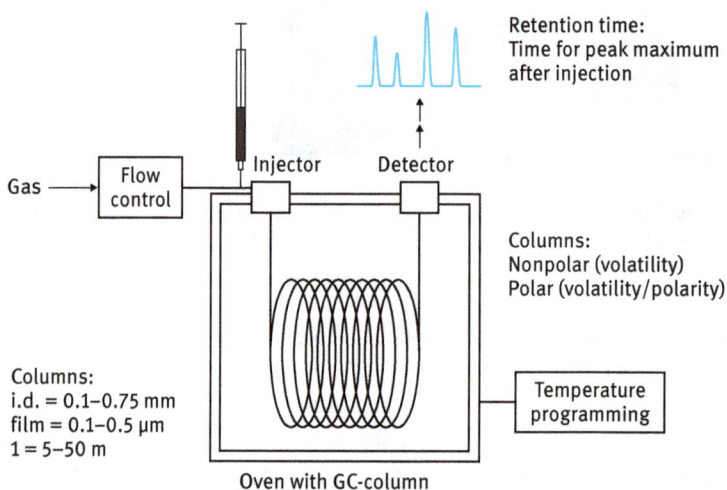

Fig. 2.68: General scheme of GC. Typically the set-up includes also a liquid injector along with an evaporator.

Vaporized products passed through the column are identified in a detector, whose response is recorded as a chromatogram. Thus, in order to be analyzed by GC, compounds in the samples must be able to get volatilized and additionally possess thermal stability.

Capillary columns are made of fused silica with a stationary phase as a thin film of liquid or gum polymer on the inside of the tube. The most common stationary phases are siloxane polymer gums with different substituents providing different polarities. The polymers are usually cross-linked in the column by photolytic or free-radical reactions, bringing strength to the polymer films. Wall-coated open-tubular columns with a liquid-phase coated directly on the inner walls, and support-coated open-tubular columns are applied. In the latter case a stationary phase is coated on fine particles, deposited on the inner walls. Among non-polar columns, those based on dimethyl polysiloxane could be mentioned. HP-5 with 95% dimethyl polysiloxane and 5% phenyl groups is slightly more polar. An even greater number of polar columns employ polyoxyethylene or polyester liquid-phases.

Capillary columns are available in a wide range of internal diameters, lengths and liquid film thicknesses. Although longer columns provide better separation, it comes at the expense of the analysis time, meaning that this time is increasing, which is usually not desitrable. In addition, longer columns lead to higher pressure and thus problems with the injection. Columns with thicker films have higher capacity, but usually require higher temperature, while thin film columns are suited for large molecules with low volatility. In principle, the analysis of components with up to 60 carbon atoms is possible.

Different types of injection systems are used in GC. Split mode, where the injected material after evaporation is split between the column and an outlet, affords rapid volatilization and homogeneous mixing with the carrier gas. Most of the sample will pass out through the split vent and only a small proportion will flow into the column. Splitless systems provide more reliable quantification, allowing analysis of rather high molecular-mass compounds as, for example, triglycerides.

Two types of detectors are mainly applied. Flame ionization detectors (FID) operate in a destructive mode with a current measuring circuit that responds to ions created by combustion in the flame. These detectors have high sensitivity for hydrocarbons, but do not detect water. Thermal conductivity detectors (TCD) are non-destructive measuring changes in conductivity caused by product in the carrier stream. They are combined with FID.

On-line coupling of capillary columns with mass spectrometers is routine nowadays enabling convenient structure identification.

An important issue, which is sometimes forgotten, is that the sensitivity for different compounds is varying for a detector; thus different peak areas are not necessarily in proportion to the weight concentrations. Knowledge of response factors is therefore necessary and calibration for components should be done properly, especially for compounds with various functional groups. Commonly, internal standard compounds are applied, for example, compounds that are not present in the sample itself are purposely added. Chemically they should be similar to the sample compounds with close retention time, however, with no peak overlapping.

What is not that often considered is a necessity to not have any leftovers in the samples from the previous injections, as heavy compounds, which are difficult to vaporize, could remain in the GC column and significantly influence subsequent analyzes. Thus regular control of retention times and response factors, as well as column cleaning or replacement in due time, should not be overlooked. For some samples related to the analysis of complex mixtures even pre-fractionation could be necessary.

In addition to such advantage of GC as accurate quantification based on internal standards, a possibility to be combined with a mass spectrometer and complete automation regarding injection and analytical runs, very high resolution should also be mentioned.

Conversely, only molecules up to about 1,000 mass units can be analyzed, as they should be stable at high temperatures. Therefore, samples should sometimes be processed before the analysis and this is important for polar compounds, for example, acids, which should be derivatized. GC and GC-MS analysis in the vapor phase require volatile derivatives that do not adsorb onto the column wall. Different derivatizations for different substances are recommended and discussed in the specialized literature on GC.

Recent advanced in gas chromatography are associated with use of multidimensional analysis (two-dimensional or 2D GC). In this version of GC a portion of the compounds separated on the first column (first dimension) is sent to a second column (second dimension) for further separation according to different physical principles. For example, if the first column is apolar with separation through different vapor pressure of components mixture, the second column is polar with polarity-based separation. An example of 2 GC is provided in Fig. 2.69.

2.2.16.2 Liquid chromatography

These chromatographic methods use liquids such as water or organic solvents as the mobile phase. Silica or organic polymers as well as anion-exchange resins are used as a stationary phase. Separation is performed either at atmospheric pressure or at a high pressure generated by pumps, which is often called high-performance liquid chromatography (HPLC) with solvent velocity controlled by high-pressure pumps, giving a constant flow rate of the solvents. Solvents are used not only as single solvents but they can also be mixed in programed proportions. In fact, even gradient elution could be applied with increasing amounts of one solvent added to another, creating a continuous gradient and allowing sufficiently rapid elution of all components.

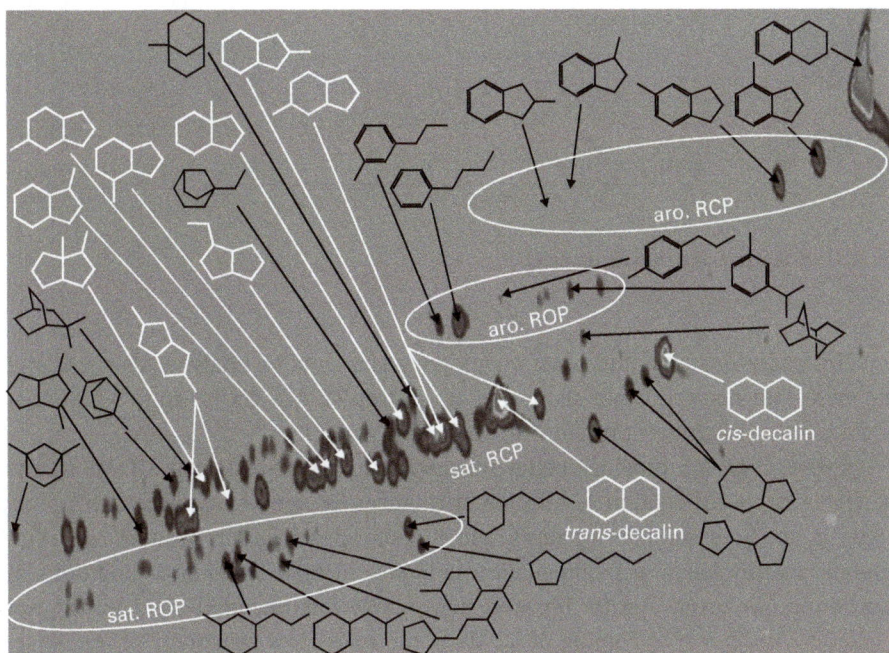

Fig. 2.69: GCxGC chromatogram of tetralin hydroconversion products. Aro- aromatics Sat. saturated; RCP-ring-contraction products; ROP- ring-opening products. (Reproduced with permission from [34]).

The most commonly used columns contain small silica particles (3–10 μm) coated with a non-polar monomolecular layer. For low-polar compounds the mobile phase is an organic solvent, while reversed phase HPLC employs mixtures of water and aceto-nitrile or water and methanol as eluents and is applied for non-ionized compounds soluble in polar solvents. UV-Vis and diode-array detectors enabling recording of UV-vis spectra, or detectors based on refractive index monitoring, are applied.

An important form of HPLC is size-exclusion chromatography (SEC), which is widely applied for the determination of molecular-mass distributions of oligomers and polymers. In SEC (Fig. 2.70), solutes in the mobile phase (for example, tetrahy-drofuran) are separated according to their molecular size. Smaller molecules pene-trate far into the porous column packing material and thus elute later than larger ones.

Among the advantages of LC are its non-destructive character and absence of derivatization. This technique can handle both small and large amounts and it can also be used for preparative isolation of compounds from mixtures. Contrary to GC there are almost no, or at least much fewer, limitations in terms of the molecular size. In addition LC can be combined with mass spectrometry, once again without derivatization.

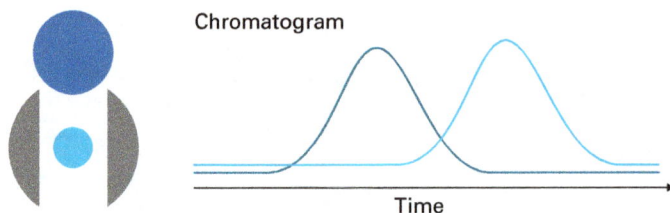

Fig. 2.70: Size-exclusion chromatography.

Thermally unstable and polar compounds can thus be analyzed as such, and the molecular mass in triple quadruple or ion-trap LC-MS can be up to 3,000 m/z, while time- of-flight versions allow up to 16,000.

Although LC-MS provides better sensitivity and selectivity than GC-MS and is excellent for quantification of selected substances in complex mixtures, this technique is not very suitable for rapid and reliable identification of unknown compounds, mainly because fragmentation is sparse at mild ionization conditions of ionization. Moreover, spectra libraries that enable identification are typically not available. Another shortcoming of LC-MS is the rather low sensitivity of the detectors for certain compounds. Finally, it may be difficult to obtain constant pressure, which in turn influences retention; clean, degassed solvents are needed and it might be challenging to find the optimum solvent mixture.

2.2.17 Theory as a part of a toolbox

It is important to emphasize that in catalyst development there is now a close coupling between theory and experiments where the theories are based upon quantum chemistry and describe single molecular situations. The current level of theory can be applied simultaneously with experiments, but these experiments cannot yet be replaced by theoretical calculations. The high sophistication of modern theoretical concepts requires that chemical engineers if not themselves proficient in theory closely collaborate with theoretical chemists. Quantum chemical calculations provide valuable information about adsorption as well as activation energy. Experimentalists, chemists and chemical engineers should, however, be careful with interpretation of the modeling results because realistic modeling of technical catalysts is too complicated and often drastic simplifications are included into calculations.

Electrons are particles with wavelike properties, therefore, they do not behave like point-like objects moving along precise trajectories. Instead their trajectory is replaced by a position-dependent wave function (Ψ). The basis of quantum chemistry is the Schrödinger equation, $\hat{H}\Psi = E\Psi$ where H is the Hamiltonian for the system (an operator containing derivatives), E is the energy of the system. The probability of finding an electron at any point in space varies with the distance from the nuclei and is proportional to $|\Psi|^2$.

Solving the Schrödinger equation is challenging because the motion of any particle is influenced by all other particles and analytic solutions can be obtained only for very simple systems, such as atoms with one electron. Thus approximations, being a trade-off between ease of computation and accuracy of the result, are needed to be able to treat molecules theoretically. For example, a Born-Oppenheimer simplification for the many-electron Schrödinger equations divides electrons into valence and inner core electrons. The inner core electrons are strongly bound and can be excluded from chemical binding calculations, while valence electrons completely determine binding properties. In such approach an atom is reduced to an ionic core interacting with the valence electrons, although the electronic wave function is still a function of the spatial coordinates of the electrons and their spin variables. Hartree-Fock approximation expresses the overall wave function as a product of single particle wave functions. This approach does not include electronic correlations, which account for interactions of electrons with each other. Additional correlations have been developed to overcome these shortcomings of the Hartree-Fock approach.

DFT instead of using many-body wave functions uses electron density as the basic variable. This theory, which became very popular to study many-body problem, is simpler than the wave function approach and is much less expensive computationally. In this theory, the properties of a many-electron system can be determined by using functionals, i.e., functions of another function, namely the spatially dependent electron density. Although DFT provides an exact path to the determination of the electronic energy of a system in a ground state, and allows computation of reaction energies and barriers of many complex catalytic systems, such calculations are computationally very demanding.

An example of DFT calculations [35] is presented in Fig. 2.71 for steam reforming of methane ($CH_4 + H_2O \leftrightarrow CO + H_2$), which will be discussed in detail in Chapter 4.

These calculations were done for the Ni (111) surface and the Ni (211) surface, representing a stepped surface. The Ni (211) step sites have lower activation barriers, are more reactive and have more strongly binding intermediates than the close-packed Ni (111) surface. In addition to pointing out these differences in activity, DFT calculations were very useful in understanding deactivation by carbon formation. It follows from Fig. 2.71 that as carbon is bound stronger at the stepped surfaces, they could thus be the potential sites of carbon nucleation. Visualization of the graphene layers growth on step sites by in situ HREM in combination with DFT resulted in a new approach of catalyst promotion by blockage of the step sites.

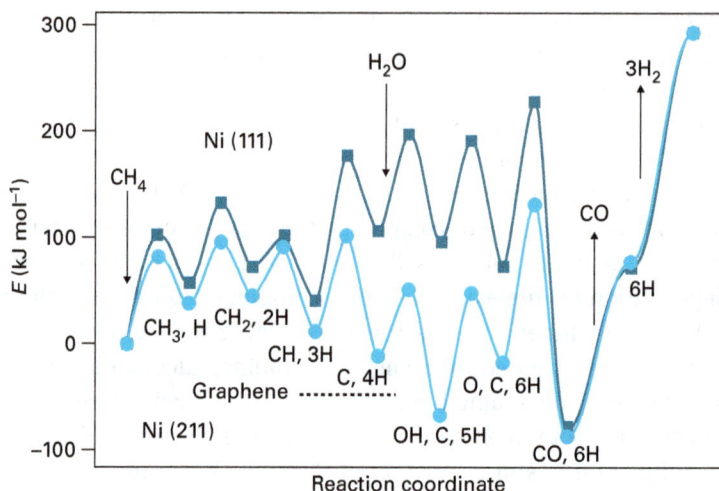

Fig. 2.71: DFT calculations for steam reforming of methane. (Reproduced with permission from [36]).

An alternative approach to complete DFT calculations is to develop descriptors of catalyst reactivity, which will enable design and engineering of more active and selective catalysts.

Thus for a methanation reaction, which is a reverse of methane steam reforming, i.e., $CO + H_2 \leftrightarrow CH_4 + H_2O$, the group of Norskov [37] demonstrated that the activity of a certain metal is given by only two descriptors: the C and O adsorption energies (Fig. 2.72).

The TOF is plotted as a function of carbon and oxygen binding energies for the stepped 211 surfaces of several transition metals at 573 K, 40 bar H_2, 40 bar CO.

The reason for volcano type behavior is that the breaking of the CO bond is the rate-determining step for relatively unreactive catalysts such as Ni. It proceeds by bond breaking the OH intermediate, yielding adsorbed C and OH on the surface. On the more reactive metals (Fe) the surface is poisoned by adsorbed carbon and oxygen because of the faster reaction. A good methanation catalyst thus should have a reasonably low CO dissociation barrier and reasonably high carbon and oxygen desorption rates as methane and water. DFT calculations identified Ni_3Fe catalysts as a cheaper alternative with a higher activity than Ni, and this was verified experimentally.

Further improvements in DFT are needed to treat, among other issues, intermolecular interactions, especially van-der-Waals forces, as well as transition states, global potential energy surfaces and some other strongly correlated systems, such as oxides.

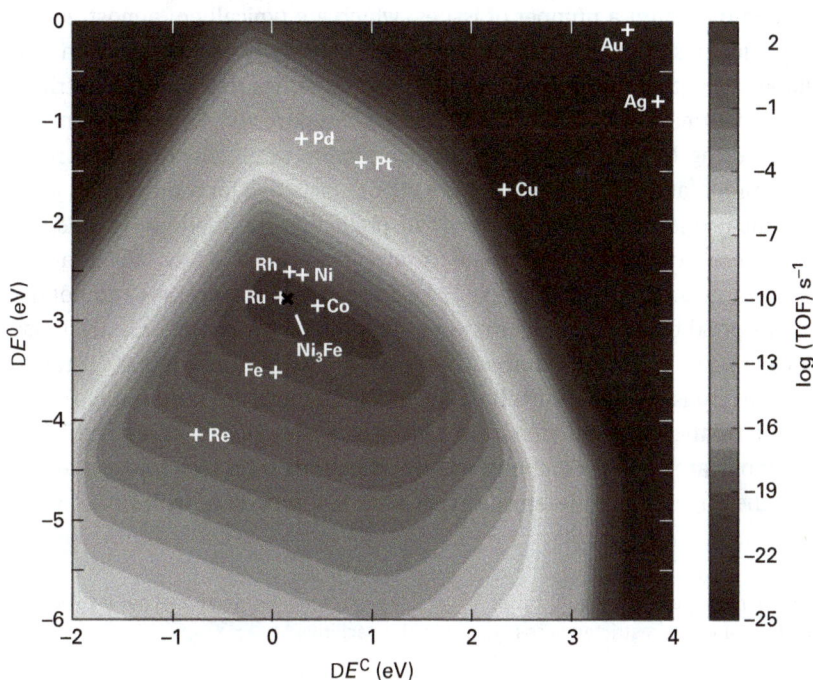

Fig. 2.72: Theoretical volcano for the production of methane from syngas, CO, and H_2. (Reproduced with permission from [37]).

2.3 Preparation of catalytic materials

2.3.1 General overview

For many years, the development and preparation of heterogeneous catalysts were considered to be more alchemy than science. A heterogeneous catalyst is a composite material, characterized by: (a) varying amounts of different components (active species, physical and/or chemical promoters, and supports); (b) shape; (c) size; (d) pore volume and distribution; (e) surface area. The optimum catalyst is the one that provides the necessary combination of properties (activity, selectivity, lifetime, ease of regeneration and toxicity) at an acceptable cost.

Although in the past the main determining factors were activity and selectivity rather than costs of catalysts, nowadays economics and ecological concerns are of significant importance. The cost/performance ratio sometimes prevents the use of new technologies for manufacturing catalysts, as they are usually associated with additional capital investment. Contrary to academic research aimed at the synthesis of well-defined structures, which could be carefully characterized with the results published in scientific journals, the industrial approach to catalyst development should take into account a number of issues, which are typically of almost no concern in academic research. In addition to activity, selectivity and preparation costs we should also mention mechanical and thermal properties (resistance to attrition, hardness and thermal conductivity), morphology and porosity, stability against deactivation assuring long-term performance, possibility of regeneration and reproducibility of preparation. As many catalysts should be active for a number of years, industry is also concerned about the ways of predicting the lifetime. To this end, acceleration tests are performed where catalysts are aged using, for example, higher temperatures. This might obviously lead to other side processes and not necessarily correspond to the actual lifetime at other process parameters. Catalyst manufacturers are usually concerned about the patent situation and report (if report) preparation methods as inventions. Judging which methods are used industrially is very difficult from the patent literature and in the majority of cases the implemented methods are kept as company secrets. The production technologies are not reported in the engineering literature except for a few processes with rather vague descriptions.

> From my personal experience even catalyst manufacturing companies personnel not involved in manufacturing can have restricted access to production facilities.

Catalyst scale-up from the laboratory to an industrial scale is not straightforward for a number of reasons. Typically, the precursors of industrial catalysts are the salts of catalytically active metals, sols, oxides and natural minerals. The selection of feedstock is determined by their constant chemical and phase composition, absence of undesired impurities, required humidity, costs, etc. Water is another source of impurities, as it is used for dissolution, dilution, washing, etc. Research in academia on the contrary uses expensive analytical grade reagents, while industry is working with technical grade substances. Some unit operations are difficult to scale-up, such as precipitation, which will be discussed below, thermal activation and catalyst formation in general. Precipitation on a large-scale could even be related with the order in which compounds are mixed. Taking all this into consideration, catalysts scaling-up still requires a lot of empirical expertise and knowledge.

Methods for catalysts preparation are very diverse and typically require several steps.

The most important of these could be identified as: (a) preparation of the primary solid by precipitation, impregnation, etc.; (b) processing of these primary solids (thermal treatment, etc.) and (c) activation (reduction, sulfidation, etc.).

Traditionally, catalysts are divided into bulk and supported catalysts.

Bulk catalysts are produced by a number of methods, such as precipitation, co-precipitation, sol-gel, flame hydrolysis, spreading or melting. Some of these methods will be discussed in the corresponding sections. For the impregnation supported catalysts the following methods are used: ion exchange, equilibrium adsorption, grafting, deposition-precipitation, immobilization of metal clusters, spreading and wetting, chemical vapor or atomic layer deposition and the anchoring of homogeneous catalysts.

2.3.1.1 Preparation of the primary solids

Precipitation and co-precipitation are mainly used for production of oxidic catalysts and support materials. Two typical examples of such catalysts are iron oxide catalysts for high temperature water-gas shift, and copper oxide-alumina catalyst for methanol synthesis. Iron oxide catalysts can be prepared by precipitation of iron carbonate from iron sulfate in the presence of sodium carbonate with subsequent washing, filtration and calcination.

The mixing and maturation procedures, pH and pH variations could influence precipitation, which involves nucleation and crystal growth. Nucleation needs conditions far from equilibrium (supersaturation), while crystal growth finally approaches equilibrium.

Low supersaturation, which could be achieved under homogeneous precipitation conditions when the precipitating agent is continuously supplied or produced in situ, leads to poor dispersion. Although high supersaturation might lead to well-dispersed catalysts they could thermodynamically be unstable, undergoing Ostwald ripening (growth of large crystals at the expense of smaller ones).

Catalyst manufacturers typically use water during precipitation, trying to avoid organic solvents for precipitation because solvent work-up and recycling are expensive. There, are, however, examples when organo-metallic precursors are used in organic solvents for the synthesis of supports by precipitation. A process for aluminum alkoxide production from aluminum metal and an alcohol, was developed by Condea Chemie. Subsequent hydrolysis also gives an alumina slurry, hydrogen and alcohol, which is recycled. Aluminosilicates could be produced by the same technology.

Water could be completely avoided when dry mixing of catalysts precursors is performed by mechanical activation. During ball milling the particles of easily decomposable salts (formates, carbonates, oxalates) are crushed and the mechanical energy is transferred to heat. When heat conductivity is poor, high temperature hot spots lead to local solid reactions, resulting in catalysts with poor dispersion and small specific surface area compared those obtained by precipitation. Iron oxide based catalysts for dehydrogenation of ethylbenzene to styrene are produced by mechanical activation when potassium hydroxide is put in contact with chromium and iron oxides.

Gelation methods are based on the continuous transformation of a solution into a hydrated solid precursor (Fig. 2.73).

The key factors are hydrolysis and condensation rates, as will be discussed later when the synthesis of silica is described. Initially, hydrophilic colloidal solutions are

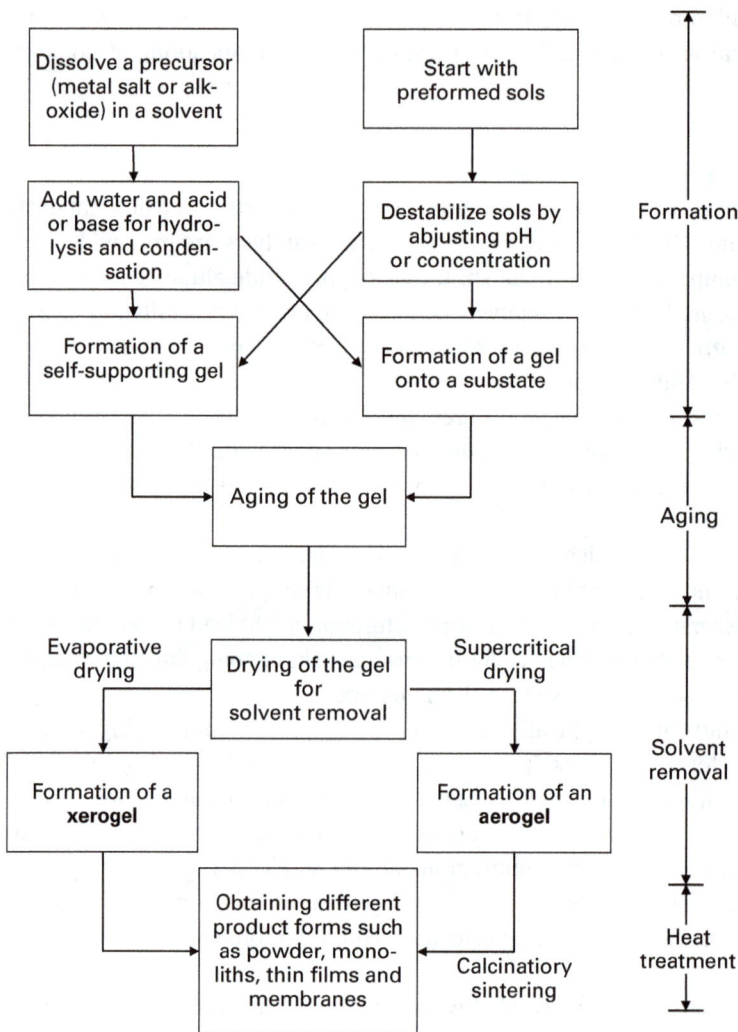

Fig. 2.73: Diagram for the different steps in the sol–gel preparation. (Redrawn from [38]).

generated with the micelles separated because of electric charges. Subsequent gelation (Fig. 2.74) depends on the pH, micelle concentration, temperature ionic strength
and temperature. The sol-gel process produces a metastable open-structure polymer
with the units bound by chemical, dipole and van der Waals forces and hydrogen
bonds. The method allows for better control of the texture, composition, structural
properties and homogeneity of the catalysts.

Removal of the solvent should be, however, carefully done, as drying can result
in either aerogels or xerogels, the xerogels being too dense to have an application

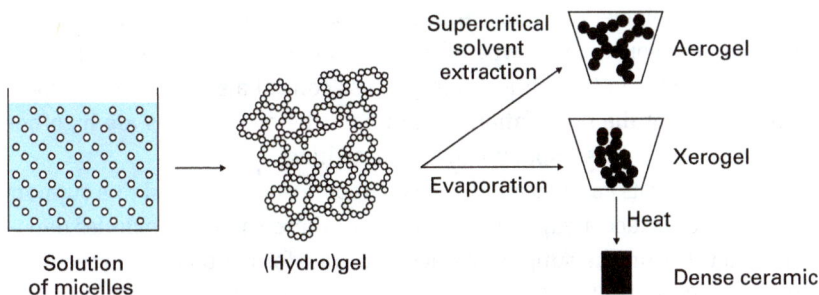

Fig. 2.74: Formation of aerogels and xerogels from micelles.

in catalysis. In addition, phase segregation and porosity changes can take place because of calcinations, thus the sol-gel method is limited to the production of carriers such as silica and alumina and to hydrothermal synthesis of zeolites, which will be discussed in a separate section.

In selective removal one of the components is removed. An example is the preparation of skeletal metals or sponge catalysts such as Raney nickel. Aluminum is removed from a relatively course powder of an alloy (NiAl$_x$) by leaching with sodium hydroxide (Fig. 2.75A). This procedure results in a highly porous nickel catalyst that possesses high activity and, also important for three-phase reactions, a high settling rate. Sponges are composed of 2–15 nm microcrystallines that have been agglomerated into macro pores. The surface area of sponge catalysts (which were and are still used commercially in various hydrogenation reactions such as hydrogenation of sugars and oils to margarine), is 50–120 m^2 g^{-1}. The catalysts prepared by selective removal have some limitations. Digestion of the less noble metal

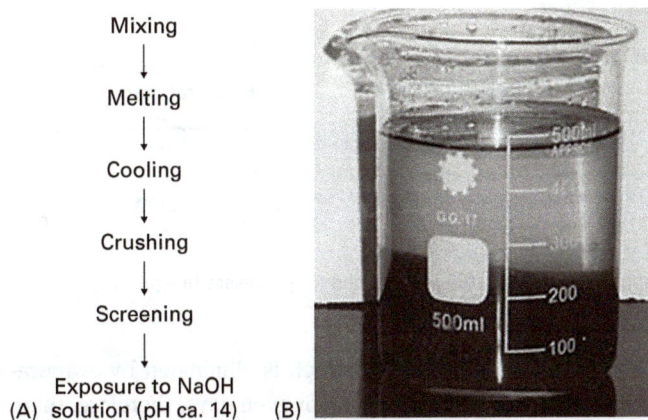

Fig. 2.75: Formation of sponge catalysts. (A) General scheme, (B) Storage of catalysts.

from a bimetallic or polymetallic alloy results in highly pyrophoric catalyst, which starts to burn spontaneously in air when dried and should be thus kept and transported under water (Fig. 2.75B). Difficulty in process control also means challenges in the reproducibility of the preparation. In addition, the overall costs are high because of the high-energy consumption needed for melting.

In supported catalysts catalytically active compounds are attached to a support with a large specific surface area. The preparation of these catalysts includes deposition of the precursor on the support surface and transformation of the precursor in the required active phase (metal, oxide, sulfide). The precursor type and concentration, nature of the support, presence of promoters, and conditions of the preparation influence chemical interactions of the precursor as well as the final active phase interactions, with the support being decisive for the performance of the catalyst in terms of activity and selectivity.

Impregnation is one of the methods of depositing the active phase on a support and could be performed by contacting a certain volume of solution with a precursor and the support (Fig. 2.76).

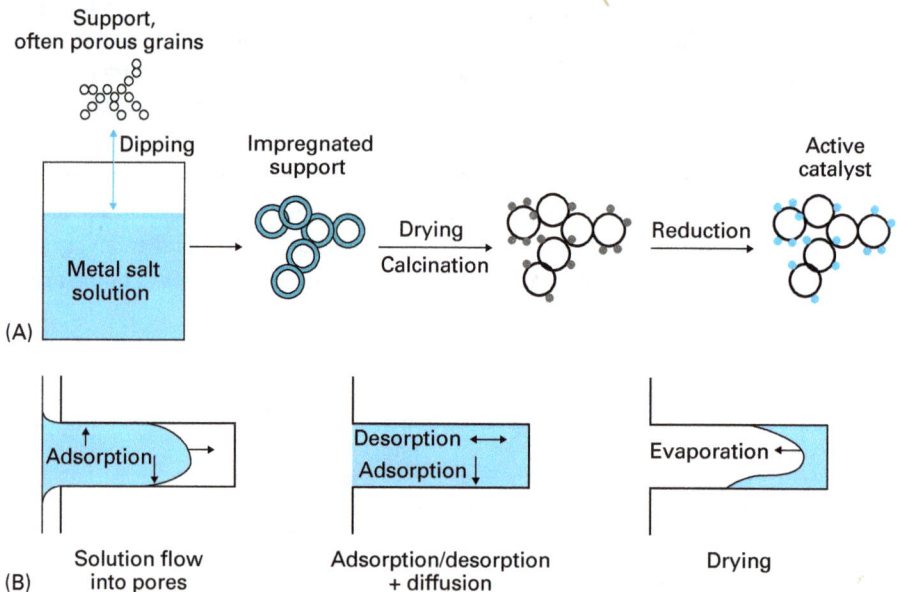

Fig. 2.76: Impregnation of supports. (A) The overall process and (B) processes in a pore.

In wet impregnation an excess of solution is used, which is eliminated by evaporation or draining. Deposition is slow and can take hours or even days, therefore, there is an apparent danger of surface restructuring. Moreover, quantitative deposition of the precursor is difficult to achieve. In pore volume impregnation the volume of the solution containing the precursor corresponds to the volume of the pores. A similar method is incipient wetness impregnation, the difference being that the volume of the solution is more empirically determined for the catalyst to start looking wet.

For both methods the operating variable is the temperature, which influences both the precursor solubility and the solution viscosity and as a consequence the wetting time. The concentration profile of the impregnated compound depends on the mass transfer conditions within the pores during impregnation and drying.

The maximum loading in these methods is limited to the solubility of the precursor in the solution. Several active components could be introduced in successive impregnations with drying and even calcinations between these impregnations.

Ion exchange is based on replacing an ion in electrostatic contact with the support by another ion. The latter ions gradually penetrate into the pore space of the support, while ions initially present in the catalyst pass into the solution until equilibrium is established, corresponding to a certain distribution of the two ions between the solid and the solution. In synthesis of zeolites for example, a sodium ion is replaced by ammonium during ion exchange, followed by the subsequent calcination and finally resulting in proton forms of zeolites.

Anchoring of a precursor from an aqueous solution can be done by adsorption (Fig. 2.77), which occurs on charged surfaces, as frequently used supports like metal oxides and active carbons develop a pH-dependent surface charge. The metal precursor should thus be selected based on the surface charge, which depends on the isoelectric point of the oxide (pH at which the surface is neutral), pH and solution ionic strength.

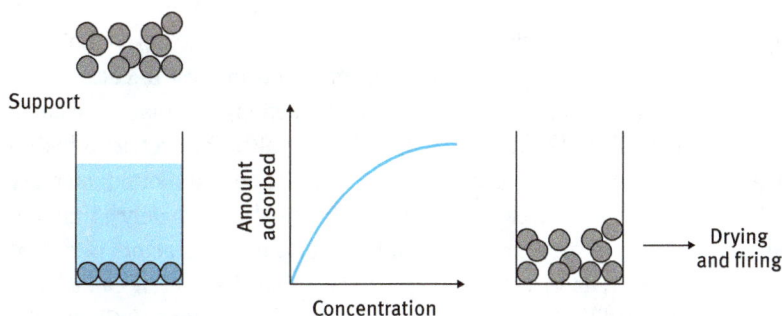

Fig. 2.77: Adsorption.

Precipitation onto the support surface (deposition-precipitation), which requires lower supersaturation than in bulk precipitation, should be carefully performed to avoid precipitation in the solution bulk (Fig. 2.78). A well-dispersed and homogeneous active phase is reached when the OH^- groups of the support interact directly with the ions present in the solution. A constant level of supersaturation is then

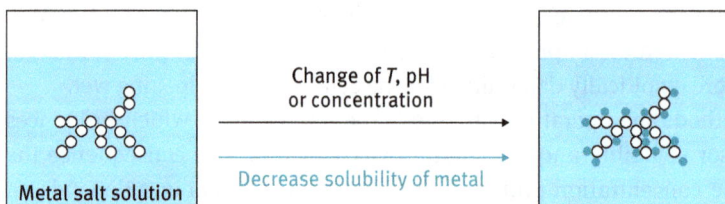

Fig. 2.78: Precipitation.

achieved, for example, by decomposing a suitable substance that releases the precipitation agent (OH⁻) continuously, as in the case of urea hydrolysis. Precipitation with urea starts with dissolving it in water and its decomposition slowly at ~90 °C, giving a uniform concentration of OH⁻ in both the bulk and pore solutions. Thus the precipitation occurs evenly over the support surface, making the use of urea the preferred method for amounts higher than 10–20%. Another option is to have a controlled and progressive addition of this precipitation agent. The use of porous supports might lead, however, to deposition-precipitation at the external surface.

A particularly interesting example in this respect is preparation of nanosized gold particles, which are very different catalysts from "bulk gold" in terms of their reactivity. The deposition precipitation method using urea was reported by Fokin already in 1913 [39] without certainly measuring the size of gold deposited on asbestos (!) and utilized in oxidative dehydrogenation of methanol to formaldehyde. In fact, gold on this support was even more active than supported silver, although utilization of supported silver catalysts became an industrial reality. This paper of Fokin written in Russian language remained in oblivion even if it was cited in a textbook of Sabatier published few years after the original discovery. Tremendous interest in catalysis by gold is mainly due to much later work of Haruta and Hutching. The latter scientist while working in industry found that gold was the best catalyst for acetylene hydrochlorination giving vinyl chloride monomer. Currently, carbon-supported mercury chloride is used commercially. For example, only in China approximately 13 million tons per year are produced in more than 90 plants with up to 100 reactors per plant having each of about 6–7 tons of catalyst per reactor. From the overall capacity of 15,000 tons, approximately 1,000 tons per year are simply lost because $HgCl_2$ is volatile. There are currently efforts to replace this catalyst with 0.1% Au/C extrudates, resulting in essentially the same productivity.

Monolithic catalysts are produced by two main methods. In the first, catalytic compounds are incorporated into the monolith before formation of the monolithic structure. A significant amount of the active compound is not accessible because of a long diffusion path. Deposition of the active phase directly on the monolith is also possible. As ceramic monoliths have low surface area wash-coating is an alternative way of active phase incorporation. This consists of depositing a layer of a high surface oxide, carbon or polymer onto the surface of a low surface area mono-lith. Introduction of the active phase is done either during wash-coating or after it by methods already presented above, such as impregnation, ion exchange, precipi-tation, deposition-precipitation. Specific methods could be also applied, such as thermal spraying, spin coating or chemical vapor deposition (CVD). CVD is an ex-ample of a gas deposition method, where deposition is done by adsorption or reac-tion from the gas-phase. Another popular method is atomic layer deposition, which will be discussed in detail later. These methods of gas deposition result in catalysts with controlled metal dispersion, which could be very high at significant loadings.

2.3.1.2 Treatment of intermediate solids or precursors

Treatment of intermediate solids includes filtration, washing, drying, thermal de-composition, calcinations and activation. Activation could also be done after form-ing the catalyst (methods briefly mentioned in Section 2.3.1.3 and described in further detail in Section 2.3.7).

In the case of cake filtration (Fig. 2.79), a relatively thin filter medium is used and separation is achieved on the upstream side of the medium. The particles should be either larger than the pores of the medium or should form bridges to cover the pores. A cake is formed from successive layers of solid deposits apparently increasing the pressure drop for the case of dead-end filtration.

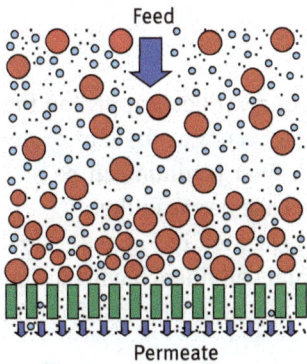

Fig. 2.79: Schematics of dead-end cake filtration [40].

An increase in the cake height during deposition of solids apparently results in an increase in the total cake resistance. In an incompressible cake with a linear in-crease of resistance with height, the filter cake porosity remains constant while the cake volume increases. Filtering at constant pressure as implemented in vacuum filtration results in decline of filtration rates as the filter cake is growing in size. Too thick filter cakes lead to prolong filter cycles because of low filtering, dewatering and washing rates. Steps in a filtering cycle include besides filtration per se also dewatering, washing, filter cake discharge, cleaning, reassembly and filling of the filter.

In continuous large-scale vacuum filters, the suspension is introduced to the filter at atmospheric pressure. Vacuum applied on the filtrate side of the medium creates the driving force for filtration. The rotary vacuum drum filter is illustrated in Fig. 2.80.

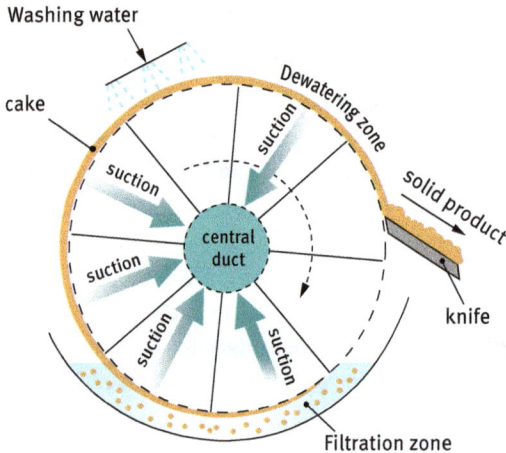

Fig. 2.80: Schematic of a rotating vacuum filter [41].

As the drum rotates, being partially submerged in the slurry, solids trapped on the drum surface are washed and dried. The cake discharge occurs at the end of the rotational cycle. The drum surface covered with a cloth filter medium can be precoated with a filter aid to improve filtration and increase cake permeability.

Horizontal filters such as the horizontal belt filter presented in Fig. 2.81 allow settling by gravity before the vacuum is applied. Horizontal belt filters having a simple design and low maintenance costs have difficulties to handle very fast filtering materials on a large scale.

The filters described earlier operate in a continuous mode. Nutsche filters, which are basically vessels divided into two compartments with vacuum applied to the lower compartment, can operate batch-wise. This might be necessary in case of stringent washing requirements or if there is a need to keep batches separated. The Nutsche (Fig. 2.82) filters can handle batches of 25 m³ and a cake volume of 10 m³ and are able thus to work with an entire charge of slurry. Sufficient holding volume is required for fast charging and emptying of the vessel. The difficulties of operation with such filters arise when cakes are slow to form, sticky and the product deteriorates during long downtime. The operational

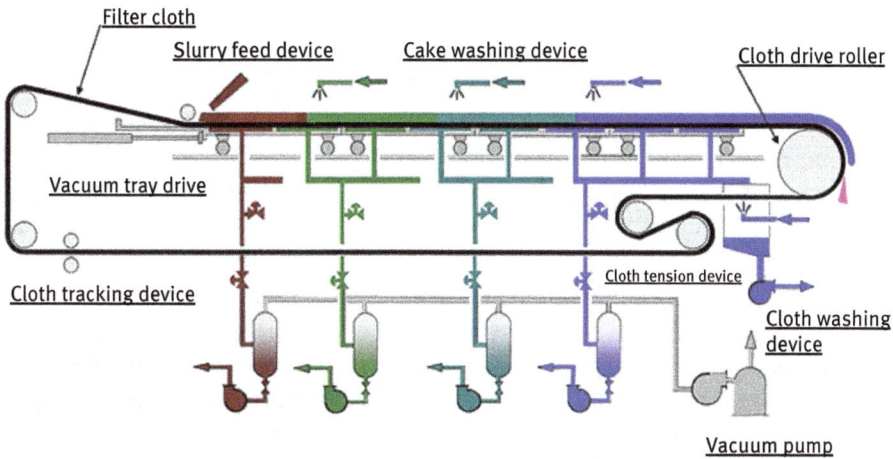

Fig. 2.81: Schematic of a horizontal belt filter [42].

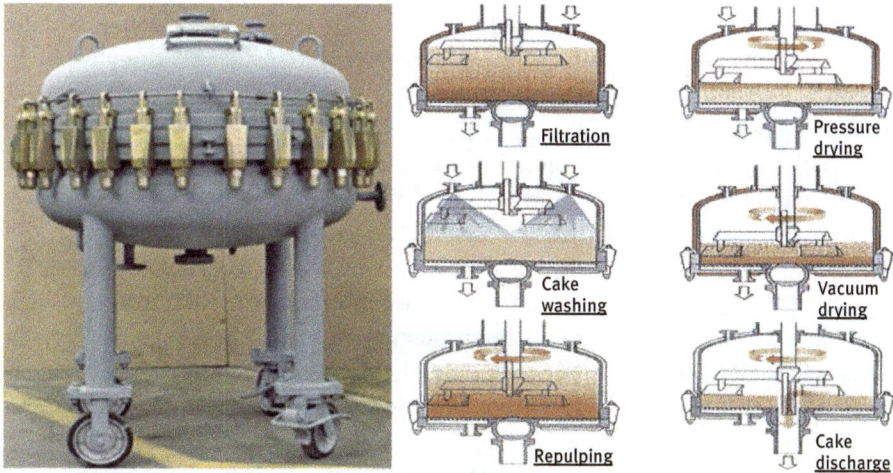

Fig. 2.82: Nutsche filter: (a) Photo [43] and (b) schematics of operation [44].

sequence starts with filtration per se when the filter is charged with slurry and the pressure is applied. In the washing stage, the wash liquid is introduced over the cake displacing the mother liquor. In the drying stage, air or gas purges the cake until the desired drying level. The final step is the cake discharge and in some instances washing the cloth or woven mesh screen with water to remove any cake residue.

Washing is essentially a replacement of the mother liqueur with pure water, with desorption or exchange of certain undesirable ions for those that could be decomposable during calcinations. Thus, for example, a known catalyst poison like Cl^- is washed away by water or replaced with NO_3^-.

Drying, which is the elimination of the solvent (usually water) from the pores of a solid, is done at temperatures higher than critical temperature of water or by evaporation under vacuum at low temperatures (from –50 °C to –5 °C). Inappropriate drying can result in non-uniform distribution along the catalyst grain of an active component not strongly bound to the surface. Slow drying or freeze drying can be used to avoid such problems. Drying of powders with the channels above 10 μm is faster than for smaller ones. The drying rate depends on the catalyst shape being faster for spherical particles than for cylindrical ones.

The objective of calcinations is to develop a well-defined structure for the active agents/supports, modify the support structure by, for example, transforming γ-alumina to α-alumina, adjust the texture and pore volume and to provide mechanical resistance.

Calcination means heating without the formation of a liquid-phase and is a further heat treatment beyond drying. It can determine the properties of the final catalyst, as many processes are occurring, such as decomposition of the precursor, spreading, sintering, and formation of new phases from the active material and the support. Calcination is specific for each system and depends on the gas-phase. Although air is often used, inert gases or controlled water vapor atmosphere could be also applied.

Typically, however, calcination is carried out in air at temperatures higher than those used in the catalytic reaction or catalyst regeneration.

In a chain-like belt, calciner gases can circulate readily through the belt and the catalyst of a fairly larger size in the form of extrudates or pellets. Because for catalyst shaping, lubricants are added to aid extrusion, they are burned out during calcination, which might lead to excessive temperature increase. A remedy in such cases is to limit the oxygen content or alternatively to perform calcination in a temperature-programmed mode to control oxidation of combustible materials present in the catalyst.

The word "calcination" is used because of historical reasons, as burning of calcium carbonate (limestone) to calcium oxide (quick lime) proceeds with removal of CO_2 (i.e., a process obviously beyond drying). A limestone calciner (rotary kiln) is shown in Fig. 2.83.

Apart from loss of chemically bonded H_2O or CO_2 several other processes occur during calcination or heating, such as modification of the nature and/or structure of the phases present, generation of the active phase and stabilization of mechanical properties and texture modification through sintering. The heating temperature and atmosphere must be properly chosen to obtain phases that are stable in the reaction and regeneration conditions, at the same time avoiding excessive sintering phenomena (agglomeration of crystals) as much as possible, as they have a negative effect on the catalytic performance. In order to understand at which temperature sintering can

Fig. 2.83: Panoramic view of rotary kiln (photos of the author).

occur for a particular metal it is important to know the Tamman temperature T_T and Hüttig temperature T_H when lattice and surface atoms, respectively, become mobile. Semi-empirical relations between these temperatures and the melting temperature T_{melt}

$$T_T = 0.5\,T_{melt}, \quad T_H = 0.3 T_{melt} \tag{2.23}$$

where T (K), indicate that when the melting temperature is not as high as for some metals (Cu: 1,357 K, Ag: 1,234 K) sintering can be profound at temperatures typically used for the thermal treatment of catalysts.

Technical arrangements of drying, calcination and activation should be performed in a way that all the catalyst particles are exposed to the same conditions, which is not possible to achieve in fixed beds. One should also realize that conditions of these treatments and applied equipment in the lab and industrial scales are very different. For example, different types of moving beds are used in industry (Fig. 2.84).

Activation refers to other thermal treatments, such as reduction of metal supported catalysts or sulfidation of hydrodesulfurization catalysts, usually performed in special atmospheres in the reactor at the start-up of the unit. Therefore, activation is not considered strictly to be a preparation procedure. Reduction is mostly done with hydrogen and a higher dispersion is obtained with a higher reduction rate unless nucleation is a limiting factor. Some catalyst manufactures do not have hydrogen at their production facilities and use chemical reductants like formic acid. It should be noted that reducibility of supported metals might be different to bulk materials and the preparation of bimetallic catalysts could be complicated because of a possible preferential reduction of one metal before the other. In Fig. 2.85 hexachloroplatinate is used as an example of the complexity of calcinations and the reduction processes.

Fig. 2.84: Calcination. (From [45].)

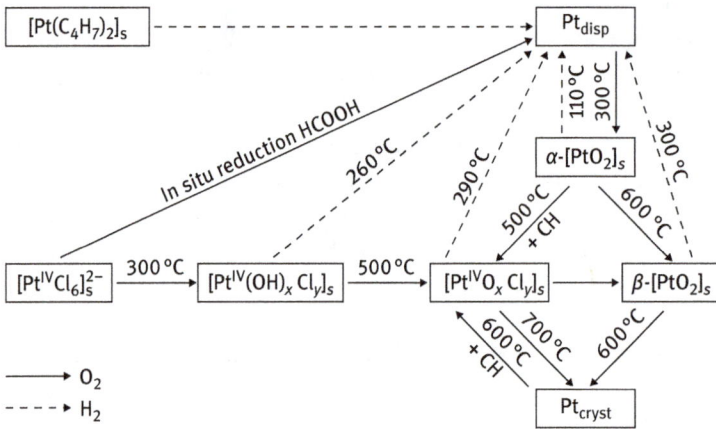

Fig. 2.85: Effect of pre-treatment on Pt/alumina.

Because of the presence of catalyst particles in technical catalysts with different location depth, crystalline size, interaction strength with the support, etc., activation rate is usually diminished as the process proceeds and some catalytic species might not be properly activated. The degree of activation (for example, degree of metal reduction) should be examined, which might not be easy when such reduction is done at a plant site in a catalytic reactor.

2.3.1.3 Forming

When a powder in a water suspension is fed through a nozzle that is spraying small droplets into hot air, such spray-drying process results in 10–100 µm particles of an almost identical shape. They could, in principle, be used in fluidized bed reactors. As discussed in Chapter 1 such small particles result in large pressure drops when applied in fixed beds, therefore forming operations are required.

Granulation of humidified powder could be performed in a horizontal rotating cylinder or a pan rotating around a 45° axis. Prior to this the powder is crushed, ground (in the presence or absence of a liquid, typically water) and screened. In granulation, different-sized particles are put into motion with a spray of liquid, resulting in particles of 1–20 mm with a broad size distribution and high-pressure drop. The slurry might contain binders for better adhesion and layer-by-layer growth of particles that are predominantly spherical after granulation.

Extrusion is the most economic and commonly applied shaping technique for catalysts and supports. During extrusion a wet paste from a hopper at the top is forced through a die and the emerging ribbon that passes through holes in the die plate is cut to the desired length using a suitable device (Fig. 2.86).

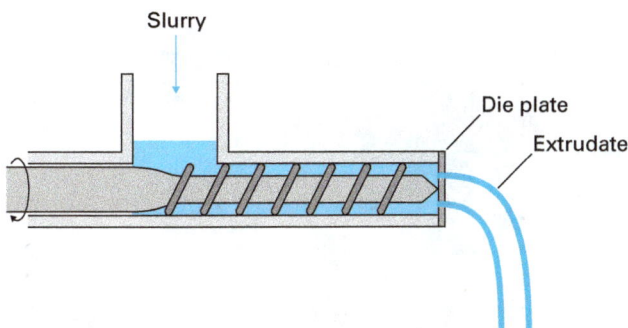

Fig. 2.86: Principle of extrusion.

Usually the catalyst powders obtained after the thermal treatments behave like sand, i.e., do not have by themselves the required moldability and plasticity, even when water is added. Various additives are used in the formulation of pastes, such as: (a) compounds for improving the rheological behavior (clays or starch); (b) binders (alumina or clays); (c) peptizing agents to de-agglomerate the particles (dilute acetic or nitric acid); and (d) combustible materials to increase the porosity (carbon black, starch, etc.).

The operating variables include mixing time, additive content, water content, aging and extrusion temperature. The quality of the extrudates also depends on the drying and calcination procedure. Special shapes (trilobates, rings, hollow cylinders, monoliths or honeycombs) can be obtained using proper dies. Press extrudates for viscous pastes and screw extrudates for less viscous ones are mainly used.

Pellets or tablets of few a mm are obtained by dry tableting, which is more expensive than extrusion. A dry powder is pressed between two punches in a press with a pressure up to 30 MPa. Small cylinders and rings are formed because of this compression. Plasticizing agents, such as graphite, stearic acid or talc, binders (alumina, clays) or porosity additives (i.e., polymer fibers) could be added during tableting to achieve the desired properties.

More details on the forming methods will be given in Section 2.3.7.

2.3.2 Unsupported metals

2.3.2.1 Colloidal catalysts

Synthesis of metal blacks can be done by reduction of aqueous solutions of their salts with hydrogen, formaldehyde formic acid and hydrazine. This procedure results in large metal particles having a surface composed mainly of face atoms. Polymer stabilized metal nanoparticles do not undergo coalescence and agglomeration (Fig. 2.87).

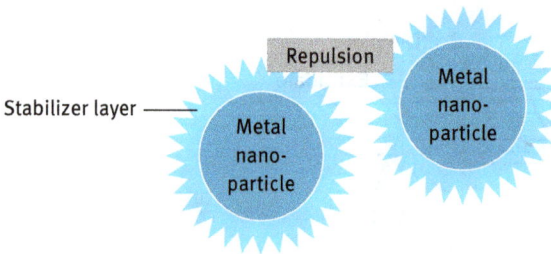

Fig. 2.87: Principles of nanodispersed metal nanoparticles.

Metal precursors, surfactants, solvents and reducing agents are needed for the synthesis of nanoparticles by colloidal methods. Typically, chlorides, nitrates, sulfates, and acetates are selected as metal salts. Moreover, two or more precursors can be applied to generate bimetallic nanoparticles.

Nanoparticles can be prepared, for example, by microemulsion methods. Microemulsion is a system containing water, oil and a surfactant, with the dispersed phase having monodispersed droplets of the size 5–100 nm. The size of water droplets in a continuous oil phase for water-in-oil microemulsions, surfactant concentration and the nature of the precipitation agent are the parameters regulating the properties of nanoparticles. Nanocolloids can be synthesized by mixing microemulsions containing the metal precursor and the precipitating agent or the reducing agent. Alternatively, the precipitating agent can be added directly into the microemulsion containing the metal precursor. Surfactants, such as various organic molecules and polymers, ensure colloidal stability of nanoparticles. The variations of the precursor, reductant, surfactants, additives and conditioning are used to control the shape and size of the nanoparticles. Cationic or anionic surfactants can be applied. Cationic cetyltrimethylammonium bromide or chloride is used together with n-hexanol as a co-surfactant. Water-to-surfactant ratio is a parameter that controls the size of particles, for example, larger particles are formed when this ratio is higher.

Poly (N-vinyl-2-pyrrolidone) (PVP) is one of the polymers that is used efficiently to control the size of transition metal nanoclusters by varying its concentration: larger concentration leads to smaller nanoclusters. Polyvinylalcohol (PVA) is used for synthesis of gold nanoparticles, which could be thereafter deposited on a required support. This method is particularly useful in synthesis of gold catalysts supported on carbon, when other preparation methods result in poor metal dispersion.

Formation of nanoparticles can be explained by simple nucleation and growth. Prior to nucleation, in a super saturation step the concentration of monomers increases up to a point of spontaneous generation of clusters. This results in a decline of monomer concentrations, preventing nucleation of new particles. The particles that are available in the solution start to grow. As growth can also occur during the nucleation phase, short nucleation phases and slow growth kinetics are required, to allow for narrow cluster size distribution.

The most favored shape with a minimum surface area and an interfacial free energy for an fcc structure of metal nanoparticles is truncated octahedron. By introducing special additives it is possible to achieve not only the size but also shape control. This is because of the different affinity of surfactants to different crystal facets hindering growth of a facet with preferential surfactant binding. The shape of Pt nanoparticles could be changed by varying the concentration of the surfactants and through a control of reduction kinetics (Fig. 2.88).

One of the challenges in the colloidal preparation method is the removal of surface capping molecules that could deteriorate catalytic activity. Determination of metal cluster size (metal dispersion) by TEM and probe gas chemisorption can thus give very different values. This is then reflected in potentially erroneous calculations of TOF. Stability of nanoparticles at high temperatures against sintering should be also ensured. The size of particles could also change during a catalytic reaction because of Ostwald ripening.

The shape purity of resulting nanoparticles still needs to be improved. In addition the size of particles is often rather large, exceeding 10–20 nm with consequently low efficiency. Moreover, the stability of shape-controlled nanocatalysts during the reduction and prevention of leaching are unresolved issues. New

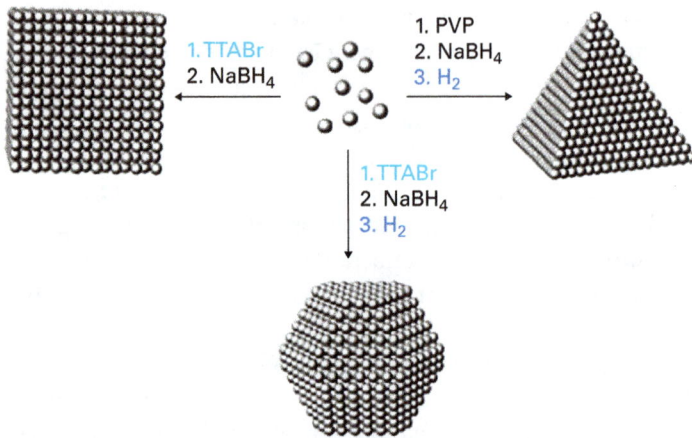

Fig. 2.88: Schematics illustrate a generic synthetic procedure for preparing Pt nanoparticles with cube, octahedron and cuboctahedron shapes (TTABr, tetradecylammonium bromide; PVP, poly-vinylpyrrolidone. (Reproduced with permission from [46]).

developments in the application of colloidal methods in catalyst preparation are related to the formation of core-shell, hollow and multi-branched (dendrimatic) mono, bi- and multi-metal nanoparticles. More details on the synthesis of precious metal nanoparticles are provided in the review by Jin [47].

2.3.2.2 Bulk metal catalysts

Only a few catalysts belong to this group of unsupported bulk metal catalysts. Synthesis of sponge catalysts has been briefly addressed before. Another important example is an iron catalyst for ammonia synthesis (Fig. 2.89). Melting of iron ore in an electric furnace occurs at 1,600 °C and during this process promoters such as alumina and silica are introduced into the furnace, while other promoters (CaO and KNO_3 or K_2CO_3) are added in the subsequent chill cast step in a shallow tray when oxygen is fed at 1,600–2,000 °C.

The melt is first cooled down and then the catalyst is crushed, separated from the undersized particles and screened. Oversized particles are further crushed, while smaller than required particles are recycled to the casting stage or melted in another furnace at 1,600–1,800 °C. This melt is cooled, then crushed and sieved into several size fractions (for example, 6–10 mm; 1.5–3.0 mm, etc.) depending on the operation requirements. The irregular granules of iron catalysts are formed using this method when the melt is passed through a hole and is put in contact with a cooling liquid. The catalyst mix crystallizes rapidly when cooled below a temperature of 570 °C to form a solid solution of wustite and alumina in the magnetite crystal lattice with a rather small surface area (1–2 m^2 g^{-1}).

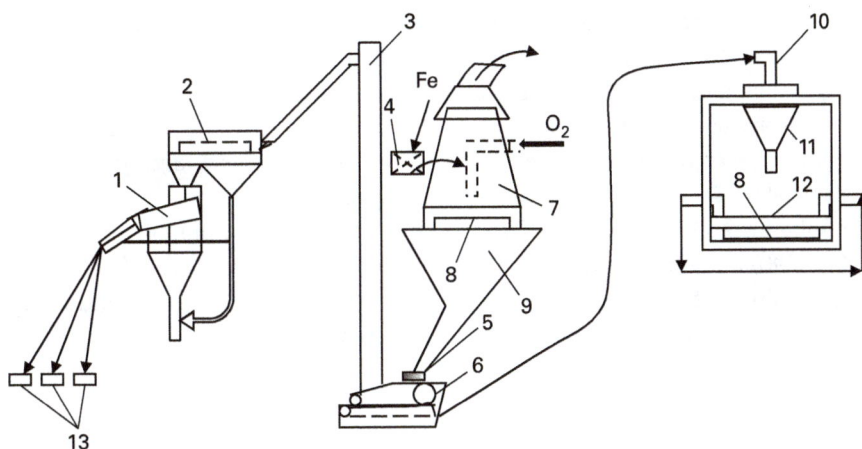

Fig. 2.89: Scheme for iron oxide synthesis, 1, screen; 2, drum; 3, lift; 4, furnace; 5, feeder; 6, crushing; 7, oxidation; 8, cooling; 9, bunker; 10, cyclone; 11, bunker of undersized fines; 12, furnace; 13, storage of catalyst. (Redrawn from [48]).

In general, parameters such as holding time, gas phase nature, efficiency of mechanical mixing and homogeneity of temperature in the melt are important to achieve homogeneity of the melt composition. For alloy catalysts, the final composition is also controlled during cooling the melt by regulating the rate of cooling, crystallization kinetics and uniformity of temperature across the bulk of the melt. Slow cooling results in the thermodynamic equilibrium composition, while rapid cooling (> 100 K/s) gives metastable amorphous, glassy phases. The latter can be transformed into nanocrystalline materials, which may be also metastable in composition and catalytically active. Such metastability is desired for fused iron oxide catalysts where unique crystallographic states cannot be prepared by the precipitation method followed by calcination. Rapid cooling also influences the formation of smaller crystallites.

2.3.3 Preparation of bulk oxides by precipitation

Precipitation of metal salts or hydroxides is a widespread method to produce bulk oxides allowing for more uniform mixing and uniform distribution of active species. Methanol and Fischer-Tropsch syntheses catalysts are examples of catalysts prepared by precipitation. This method gives more control over pore size and pore size distribution as it is not limited to the formation of a support. Thus, catalysts of different size and shape could be prepared. Nitrates are usually the preferred precursors because of their high solubility in water, availability and comparatively low price. However, careful control of NOx emissions is required. Other salts, such as sulfates and chlorides could be also applied, although their disposal could be challenging. As mentioned above, in special cases in industry more expensive organic compounds are used for the manufacturing of supports. In addition to mixing components well on an atomic scale, precipitation can afford high loadings at relatively high dispersion. At the same time, low loadings give low dispersion.

Despite the apparent superficial simplicity of the method, a number of parameters (Fig. 2.90) influence the quality of the final product, therefore even if the chemical composition of a particular catalyst could be the same for several industrial manufacturers, properties could be still very different.

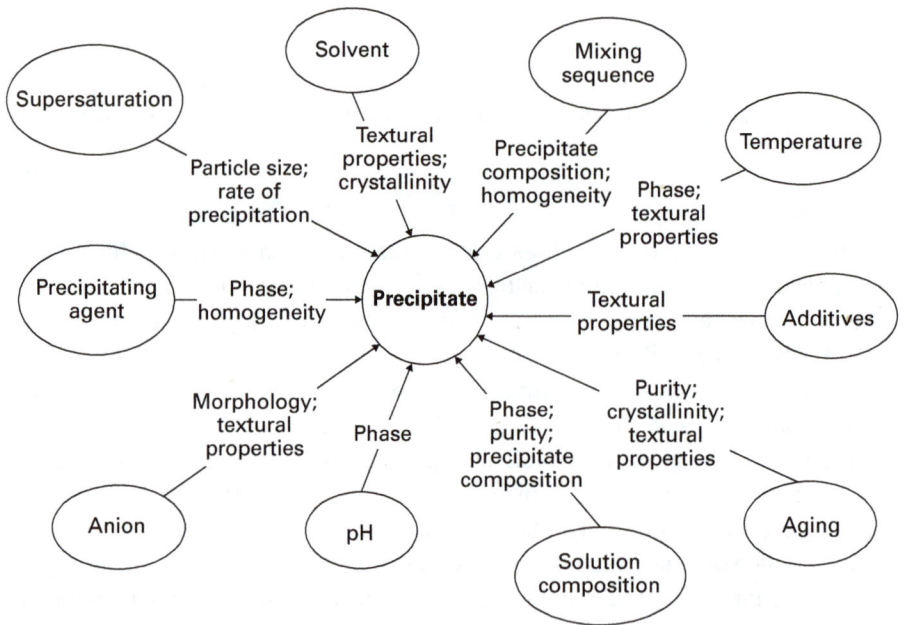

Fig. 2.90: Parameters influencing precipitation. (Reproduced with permission from [38]).

Synthesis complexity also explains why the scaling up of precipitation is difficult. For the methanol synthesis Cu/ZnO catalyst, changing precipitation conditions many result in different phases with varying catalytic activity. The precipitation method requires large consumption of chemicals and gives high amounts of process water.

For easily soluble compounds a classical theory of crystallization from oversaturated solutions can be applied in order to understand the mechanism of precipitation. After forming a supersaturated solution precipitation is initiated by physical or chemical means, resulting in nucleation followed by the crystal growth. The precipitate formed after aging and modification undergoes spray-drying or is first filtered and then dried. The dried material is further calcined and then shaped. These operations could be also performed in the reverse order.

The processes occurring during precipitation are shown in Fig. 2.91.

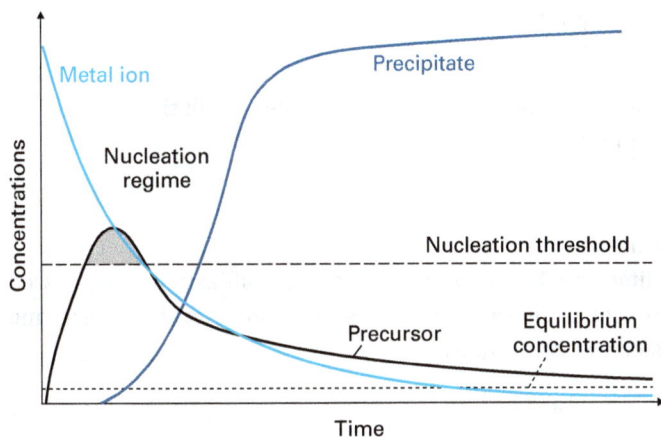

Fig. 2.91: Dependence of concentration on time during precipitation.

During dissolution of a salt the following processes occur:

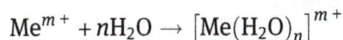

$$Me^{m+} + nH_2O \rightarrow \left[Me(H_2O)_n\right]^{m+}$$

$$\left[Me(H_2O)_n\right]^{m+} + pH_2O \rightarrow \left[Me(H_2O)_{n-p}(OH)\right]^{m-p} + pH_3O^+$$

Hydrolysis products are polymerized into complexes of $[Me(OH)m]_n^{p+}$ type, where n depends on hydrolysis conditions and the nature of the metal. Intensive mixing improves the rate of dissolution. Another important parameter is temperature. For most substrates, such as chlorites, nitrates, ammonium salts, solubility in water improves with temperature increases. However, several salts, such as sodium sulfates and carbonates display maximum solubility as a function of temperature.

Precipitation, happening after dissolution of the precursors, consists of nucleation and crystal growth. The catalytically active phase is typically metastable, thus crystallization, which is enhanced at low temperatures, should be performed at conditions far from equilibrium.

In general, kinetics of solid–solid phase transformations tends to be much more slower than liquid–solid transformations because of higher activation energy related, among other reasons, to larger solid–solid interfacial energies, smaller changes of free energy per unit volume and slower solid-state diffusion. As a consequence, solid–solid phase transformations rarely reach equilibrium, resulting in the formation of metastable phases as mentioned earlier. Nucleation and growth depend not only on temperature and supersaturation, but also on impurities.

Supersaturation must first be achieved independent of from which phase crystals can be grown. The degree of supersaturation is expressed by the concentration difference Δc:

$$\Delta c = c - c^* \tag{2.24}$$

where c is the actual solution concentration and c^* the equilibrium saturation value. The supersaturation ratio S defined as

$$S = c/c^* \tag{2.25}$$

is commonly used in explaining crystallization.

The fundamental dimensionless driving force for crystallization is expressed through the chemical potential difference of a given substance in the transferring (solution) and the transferred (solid) states

$$\Delta\mu/RT = \ln(a/a^*) = \ln S \tag{2.26}$$

where a^* is the activity of a saturated solution, implying that

$$S = \exp(\Delta\mu/RT) \tag{2.27}$$

Solubility–supersolubility diagram (Fig. 2.92) is often used to illustrate how concentration depends on temperature.

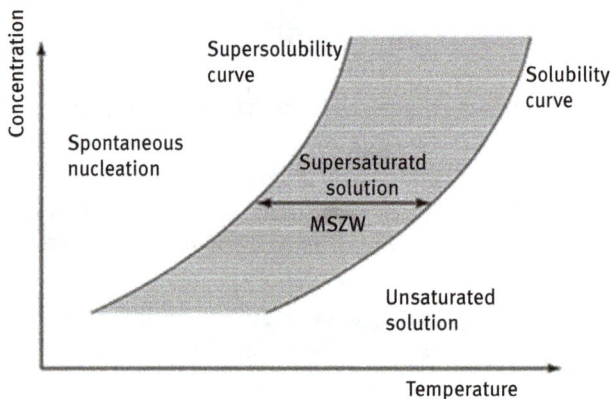

Fig. 2.92: Solubility–supersolubility diagram with MSZW denoting the metastable zone width.

While position of the lower equilibrium solubility curve can be accurately determined, location of the upper supersolubility curve is less clear as it is influenced by many factors such as the rate of supersaturation, agitation intensity, presence of crystals or impurities. The metastable zone width visible in Fig. 2.92 is thus characteristic of a particular crystallization system.

Nucleation may occur by means of homogeneous or heterogeneous mechanisms, either through primary (absence) or secondary route (presence of crystals). In primary homogeneous nucleation, nuclei are generated from the solution, while for heterogeneous nucleation, the solid surfaces are involved.

For homogeneous nucleation, the most likely shapes are spheres as this shape minimizes the amount of interfacial area per unit volume. The total Gibbs energy change, ΔG_{tot}, involved in forming a spherical nucleus of radius r is defined through

$$\Delta G_{tot} = \Delta G_{volume} + \Delta G_{int} = \frac{4}{3}\pi r^3 \Delta G_V + 4\pi r^2 \gamma \tag{2.28}$$

where ΔG_V is the free energy per unit volume released upon creating the new second-phase particle. This term is negative to allow a phase transition. The interfacial energy per unit area γ is positive as energy is expanded when an interface is created. This term is associated with the creation of new interfacial area.

The total free energy change caused by agglomeration can be expressed as a function of the nuclei number (Fig. 2.93).

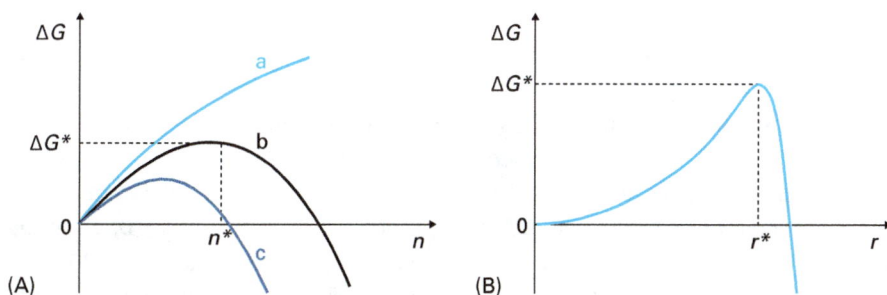

Fig. 2.93: Dependence of Gibbs energy as a function of nuclei precursors: (A) amount and (B) radius.

Fig. 2.93(A) demonstrates the influence of nuclei number on Gibbs energy, which is composed of the difference in free energy between the solution and bulk species, free energy change related to formation of the interface as in eq. (2.28) and other changes, for example, because of strain or the presence of impurities. Obviously, for agglomeration to occur the Gibbs energy should be negative, which happens when the critical number of nuclei is reached. This number n^* depends on the supersaturation ratio S. Without saturation (case a in Fig. 2.93A) nuclei are not formed. For higher supersaturation (compare cases b and c in Fig. 2.93A) there are more nuclei with a smaller size, resulting at the end in a precipitate with a smaller size. Therefore, in order to increase the number of nuclei, concentrated solutions of precursors should be used.

An important concept in crystallization is the critical crystal size r^* (0.5–5 nm) illustrated in Fig. 2.93B, starting from which there is a fast growth of nuclei resulting in a large number of crystals of different sizes. For lower r^* the degree of supersaturation should be higher. Beyond the critical nucleus size r^*, the total free energy of the system is decreasing and it is energetically unfavorable for nuclei to continue growing, while above the critical size ($r > r^*$), there is a continuation in growth.

The critical nucleus size r^* can be determined by taking the derivative of eq. (2.28) and setting it equal to zero

$$\frac{\partial \Delta G_{tot}}{\partial r} = 0 = \frac{\partial \frac{4}{3}\pi r^3}{\partial r}\Delta G_V + \frac{\partial 4\pi r^2}{\partial r}\gamma = 4\pi(r^*)^2\Delta G_V + 8\pi r^*\gamma \tag{2.29}$$

Giving

$$r^* = -2\gamma/\Delta G_V \tag{2.30}$$

When nucleation occurs heterogeneously at specific sites (surfaces, interfaces or grain boundaries), the total free energy associated with the nucleating phase is

$$\Delta G_{tot} = \Delta G_{volume} + \Delta G_{int} = \frac{4}{3}\pi r^3 \Delta G_V f(\theta) + 4\pi r^2 \gamma f(\theta) \tag{2.31}$$

which is different from homogeneous nucleation by a factor $f(\theta)$, accounting for geometry of the cap compared to a sphere.

The critical nucleus size r^*_{het} for heterogeneous nucleation calculated in a similar fashion as for homogeneous nucleation, that is, setting $\partial \Delta G_{tot}/\partial r = 0$, gives the same expression $r^*_{het} = -2\gamma/\Delta G_V$ as for homogeneous nucleation (i.e., eq. (2.30)).

For heterogeneous nucleation, activation barrier, ΔG^*_{het}, is different from homogeneous nucleation and is defined as

$$\Delta G_{tot} = \Delta G_{volume} + \Delta G_{int} = \frac{4}{3}\pi(r^*)^3 \Delta G_V f(\theta) + 4\pi(r^*)^2 \gamma f(\theta) =$$

$$= \frac{4}{3}\pi\left(-\frac{2\gamma}{\Delta G_V}\right)^3 \Delta G_V f(\theta) + 4\pi\left(-\frac{2\gamma}{\Delta G_V}\right)^2 \gamma f(\theta) = \frac{16\pi\gamma^3}{3(\Delta G_V)^2}f(\theta) \tag{2.32}$$

Concentration of nuclei (n^*) that can be formed at any given temperature T is calculated through the change in ΔG^*:

$$n^*_{het} = n_0 \exp\left(-N_A \Delta G^*_{het}/RT\right) \tag{2.33}$$

Because ΔG^*_{het} strongly decreases with decreasing θ, n^*_{het} increases exponentially with decreasing θ explaining why concentration of viable heterogeneous nuclei is orders of magnitude higher than the corresponding value for homogeneous nuclei.

The size of the critical nuclei is often expressed through the Gibbs-Kelvin equation [49]:

$$r^* = \left(\frac{3n^*v}{4\pi}\right)^{\frac{1}{3}} = \frac{2\gamma v}{kT \ln S} \tag{2.34}$$

where n is the number of precursor species, v is the molecular volume of precursor, γ is the interfacial energy, and S is the supersaturation ratio.

From eq. (2.35) it can be concluded that the size of the critical nuclei depends on $\ln S$. It is, however, difficult to control nucleation by alteration of the supersaturation ratio. A more efficient way is through the interfacial energy (surface tension).

The size of the nuclei is smaller for higher supersaturation and smaller surface tension (interfacial energy), and the latter could be affected by solution pH, ionic strength through changing nature and the concentration of electrolytes or adsorption of ions.

The size distribution of the particles depends on the duration of the nucleation period; shorter nucleation results in more monodispersed particles, while if the nucleation period is long a wide particle size distribution is obtained.

Although nucleation is decisive, growth, limited by surface reactions, is also important because the crystal size depends on the ratio of rates $r_{nucleation}$ and r_{growth}. In most crystal growth processes, not only diffusion but also surface reactions are important. As a result, the crystal growth kinetics often depends not only on the crystal size but also on the surface structure. The growth rate can be represented with the first-order kinetics in the solute concentration, while the nucleation rate is defined as proportional to the supersaturation concentration and the number of critical nuclei, corresponding to supersaturation.

Rapid precipitation can be initiated by adding seeds to the solution increasing the concentration of the solution at the interface. High stirring favors formation of small crystals, because the concentration at the interface is increasing, while higher temperature increases the crystal size.

Crystallization is complicated by the so-called Ostwald ripening. Solubility of small crystals is higher than of the larger particles. Concentration in the solution in equilibrium with a solid particle depends on the particle size. The larger the particle the lower the equilibrium concentration, thus particles larger than equilibrium size will grow, while for particles smaller than the equilibrium size at a certain concentration the solution is not saturated and the particles dissolve. Thus, during precipitation in addition to nucleation and crystal growth larger crystals are also formed at the expense of the smaller ones.

Nucleation and growth are not separated in time as once few viable nuclei are formed they start to grow. The Johnson–Mehl equation describes the overall transformation rate:

$$F(t) = 1 - \exp\left(-(\pi/3)\, \dot{N}\, G^3\, t^4 \right) \qquad (2.35)$$

Relating the overall fraction of the transformed material as a function of time [$F(t)$] with the nucleation (N) and the growth (G) rate considered to be time independent. The third power in the linear growth rate follows because of three-dimensional (3D, spherical) growth resulting in the transformed volume increased in proportion to G^3.

Equation (2.35) corresponds to sigmoidal transformations, when $F(t)$ first increases exponentially in time slowing down thereafter and finally asymptotically approaching complete transformation. A simplified version of the Johnson–Mehl equation, known as the Avrami or Avrami–Erofeev equation, is very often used in practice:

$$F(t) = 1 - e^{-kt^n} \qquad (2.36)$$

where n is known as the Avrami exponent and k is a parameter comprising both the nucleation and growth rates associated with the phase transformation.

From the discussion presented above it is possible to identify some key engineering issues related to precipitation. As concentration is changing during the process quality of the product can vary even for precipitation of a single component. This can be overcome by precipitation in a buffer solution of an electrolyte with pH corresponding to the pH of precipitation. An alternative method is continuous precipitation when pH and concentration is constant when the precursors are fed into the reactor and the precipitated suspension is removed continuously. Application of the proper solvent and pH also influences the crystal shape. Habitat (or crystal shape) can be important for either technological reasons (improving rheological properties, downstream operations, handling, etc.) or from the viewpoint of catalytic behavior. The rate of cooling or evaporation, the degree of supersaturation, crystallization temperature and presence of impurities can influence the crystal shape. In particular, adsorption of impurities blocking the crystal surface may change the crystal shape or decrease the growth rate. Therefore, impurities can be removed prior to crystallization or alternatively can be added influencing the crystal habit. Moreover, the precipitate can contain some impurities or ions that might influence catalytic performance. A remedy is to apply decomposable ions, such as nitrates, oxalates or ammonium.

Supersaturation degree can be different in different parts of the vessel. In order to avoid local supersaturation at the points when the solutions are mixed and achieve homogeneous precipitation, it is possible to add another component (for example, urea), which keeps the solution homogeneous at least for some time.

For multicomponent precipitates it is even more difficult to achieve homogeneity of the precipitate. Because of different solubilities, solid-phase composition could be different at the beginning and at the end of precipitation. For example, precipitation of hydroxides from a mixture does not occur simultaneously, but depends on the pH. Tab. 2.4 contains values of pH at which hydroxides start to precipitate for 0.01 M solutions and corresponding values of equilibrium constants $K_L = [Me^{n+}][OH^-]^n$ are also given for the equilibrium $Me(OH)_n \leftrightarrow [Me^{n+}] + [OH^-]^n$. For precipitation to occur the term $[Me^{n+}][OH^-]^n$ should exceed K_L. Thus, during neutralization of acidic solutions, when mixing is sufficient, formation will be observed in the solid phase of a hydroxide with a lower value of precipitation pH. Co-precipitation of hydroxides with similar precipitation pH results in the generation of complex crystal structures, inclusion of one hydroxide into another one or adsorption of one hydroxide onto another one, which is formed first, etc.

Tab. 2.4: Dependence of precipitation on pH for hydroxides for 0.01 M solutions.

Hydroxide	pH	K_L
$Mg(OH)_2$	10.4	$(2 \div 0.6) \times 10^{-11}$
$Mn(OH)_2$	8.6	1.9×10^{-13}
$Ni(OH)_2$	7.7	$10^{-15} \div 10^{-18}$
$Zn(OH)_2$	5.2	10^{-17}
$Cu(OH)_2$	5.1	2.2×10^{-20}
$Cr(OH)_3$	4.6	6.3×10^{-31}
$Al(OH)_3$	3.9	5.1×10^{-33}
$In(OH)_3$	3.7	1.2×10^{-33}
$Fe(OH)_3$	2.3	6.3×10^{-38}

(From [50].)

For low soluble compounds the mechanism of crystal formation is different from the classical method and follows orientation attachment when larger crystals are formed by crystallographically oriented assembly of smaller nanocrystals.

Aging follows nucleation and growth stages and allows the system to approach thermodynamic stability. During aging there could be changes in the particle size caused by dissolution-crystallization and aggregation, changes in the crystal phase (for example, titania from anatase to rutile), crystallization of amorphous phases and changes in morphology.

Washing, drying and calcination are typical preparation steps that are performed after aging. A flow scheme of industrial-scale catalyst preparation by precipitation is given in Fig. 2.94.

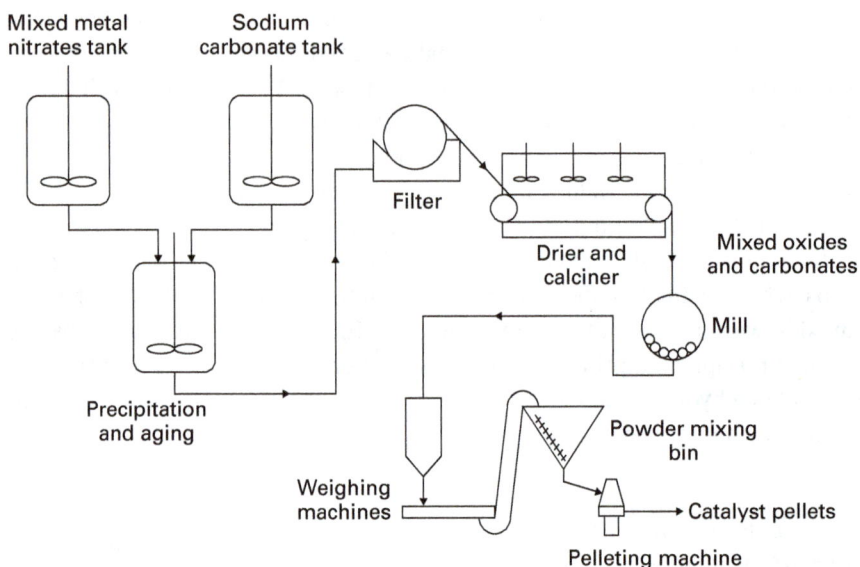

Preparation of catalysts by precipitation. (From [51].)

This scheme is rather general including the most important unit operations needed to prepare bulk oxides type of catalysts by precipitation. It is worth considering a specific case of copper- zinc-aluminum oxides catalyst used for methanol synthesis.

Synthesis of this catalyst includes several steps, namely the preparation of solutions, mixing, filtration, washing, drying, grinding, pelletizing, calcinations and final packaging. Precursors of high purity are needed to make a catalyst with minimal amount of impurities, giving highly dispersed or amorphous precipitates and subsequently catalysts with developed surface area. The scheme of the catalyst preparation is presented in Fig. 2.95.

Water solutions of the corresponding salts of copper and aluminum together with sodium carbonate are fed into the reactor in (pos. 5), where 20–40 nm crystals of copper carbonate are formed mixed with aluminum hydroxide particles of a size < 15 nm. After the addition of zinc nitrate, small crystals of zinc carbonate are generated that are included in the copper carbonate crystals, preventing the growth of the latter. After filtration (pos. 6) from the mother liqueur, drying and calcination (pos. 7), the crystals are ground (pos. 8) resulting in small particles. Graphite (pos. 9) is then added to these particles, which after additional grinding (pos. 11) are tabletized (pos. 12). The amount of graphite in the final catalyst is below 2 wt%. Reduction (pos. 15) results in the formation of copper crystallites incorporated in the matrix of Al_2O_3-$ZnAl_2O_4$ spinel. The presence of zinc diminishes dehydrogenation ability of the catalyst.

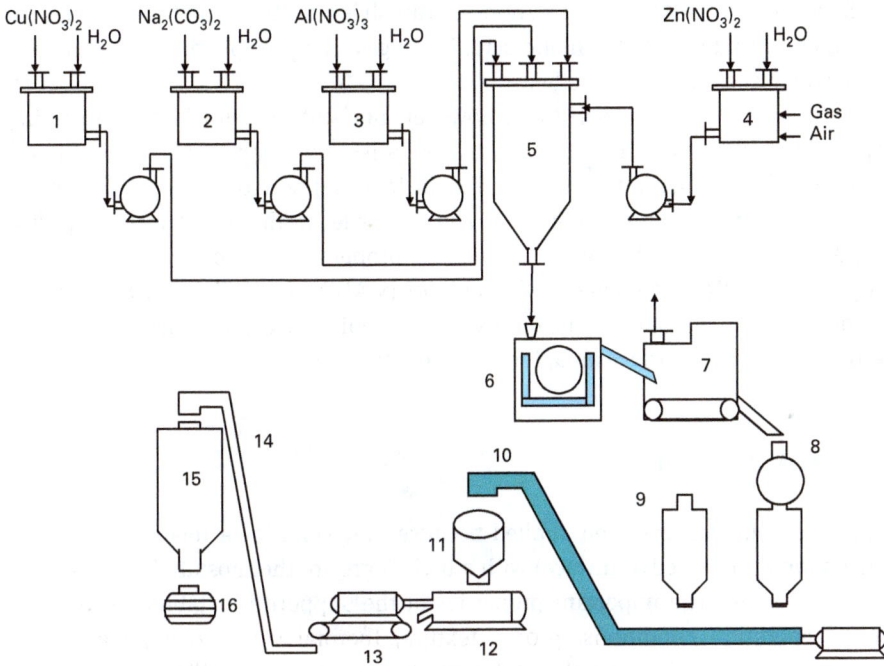

Fig. 2.95: Flow scheme of Cu-Zn-Al oxide catalyst preparation: 1, 4 vessels for preparation of water solutions; 5, reactor; 6, filter; 7, drying and calcination unit; 8, grinder; 9, vessel with graphite; 10, mixing and conveyer; 11, additional grinding unit; 12, tabletizing machine; 13 and 14, conveyers; 15, reactor for reduction (or oxidation); 16, vessels for the final catalyst. (From [52]).

2.3.4 Heteropoly acids

Heteropoly acids (HPA) or polyoxometalates are an interesting class of bulk acid catalysts with the anion $[X_xM_mO_y]^{(-q)}$ balanced by protons. In heteropoly acids X stands for a hetero atom, such as B, Al, Si, Ge and P, As or Fe, Mn, Co, Cu and Zn. M indicates addenda atoms (Mo, W, V, Ta, Nb, Os), which are attached to hetero atoms through oxygen atoms and q is the charge, varying from −3 to −28. Addenda atoms typically have high charge (+5 or +6), small size with ionic radius in the range 0.53Å–0.70Å, an expandable coordination number from 4 to 6 and the ability to form double bonds with unshared oxygen atoms.

Well-known Keggin structures $H_nXM_{12}O_{40}$ ($H_4X^{n+}M_{12}O_{40}$, with X = Si, Ge; M = Mo, W and $H_3X^{n+}M_{12}O_{40}$, with X = P, As; M = Mo, W, for example, $H_3PMo_{12}O_{40}$ and $H_3PW_{12}O_{40}$) can possess such qualities as good thermal stability, high acidity and high oxidizing ability and could be used as heterogeneous acid catalysis.

Synthesis of $H_3PW_{12}O_{40}$ can be done by dissolving Na_2WO_4 and Na_2HPO_4 in boiling water and adding hydrochloric acid, leading to the precipitation of $H_3PW_{12}O_{40} \cdot nH_2O$, which is extracted with ether.

The thermal stability of HPA is, however, not sufficient for high temperature applications, moreover the surface area of bulk heteropoly acids is rather low ($2\text{–}10$ m^2 g^{-1}) and therefore activity is insufficient. Supported heteropoly acids can be used to overcome these limitations. The key issues in the use of such supported HPA are the retention of polyanion structure, proper dispersion and fixation to the support. Typically, the choice of the support ($\gamma\text{-}Al_2O_3$, MgO, ZrO_2, TiO_2, activated carbon, zeolites, mesoporous materials, clays or mixed oxides) is determined by a need to have a high surface area, inertness and thermal stability.

2.3.5 Catalyst supports

Supported catalysts are often applied because they combine a relatively high dispersion (amount of active surface) with a high degree of thermostability of the catalytic component. The important properties of the support are stability at reaction and regeneration conditions, proper texture, thermal conductivity, mechanical strength and low costs. Typically, different phases of alumina, silica, titania, zirconia, magnesia, zinc oxide and active carbon are used.

The support should also enable production of shaped particles (mm range) where small easily sintered crystals of the active phase (nm range) do not coalesce. Pre-shaping of supports is an attractive option. However, care must be taken that dispersion of the catalytic components is not modified in the following steps. With powdered supports, the intimate mixing during deposition of the catalytic components is easily realized during the first step, however, the following operations, in which the grains are transformed into their required shape with desired porosity, are more difficult and the dispersion may not be uniform.

2.3.5.1 Carbon

Active or activated carbon, together with alumina and silica, is mainly used as a catalyst support. Activated carbon is preferred in applications where properties like inertness, lower coke propensity compared to silica and alumina, stability at a wide range of pH and temperatures, high adsorption capacity, wide variety of textural properties, are of importance. Carbon supported noble metal catalysts are typically applied in various liquid-phase hydrogenation reactions at rather low temperatures.

In oil refining the application of activated carbons is usually limited by chemical reactivity in the presence of oxygen. This means that conventional methods of catalyst regeneration by burning off the coke away are not applicable. Moreover, active carbons are mechanically weak, which prevents their applications in fixed-bed reactors and leads to generation of fines in slurry systems. Conversely, in order to recover the active phase (expensive precious metals for example) from spent catalysts, carbon can be easily burned away. It also should be noted that although conventional carbons are used mainly as powders, different forms, such as extrudates, cloths and fibers could be prepared. Finally, the costs of carbon supports are usually lower than other conventional support materials.

Activated carbon is produced from carbonaceous (for example, carbon rich) source materials such as nutshells, wood and coal by physical or chemical activation. In the former case carbonization is used.

Once during a scientific presentation a PhD candidate, who happened to be the author of this book, got a question related to selection of the support, which was active carbon. The person who posed the questions was wondering why this particular support was used as it was active in the reaction in addition to the metal on it. That person was assuming that the term "active" was related to catalytic activity of the support and not the activation procedure during preparation.

Materials with carbon content are pyrolyzed at temperatures in the range 600–900 °C, in the absence of air (usually in inert atmospheres with gases like argon or nitrogen) (Fig. 2.96). This process is followed by activation/oxidation as carbonized material is exposed to oxidizing atmospheres (carbon dioxide and/or steam at 400–600 °C). Some carbon is burned away to give the porous structure. Chemical activation means impregnation with, for instance, phosphoric acid or zinc chloride, and carbonization at temperatures in the range of 400–600 °C. Carbonization/activation steps proceed simultaneously. Activating chemicals remain in the structure after carbonization. After washing, the final activated charcoal is produced. Activated carbons have a much higher specific surface area than other types of supports ($> 1,000$ m^2 g^{-1}).

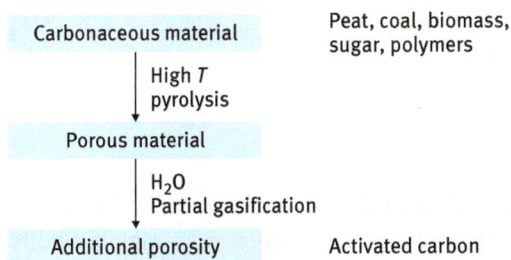

Fig. 2.96: Preparation of activated carbon.

All activated carbons have a porous structure, usually with a relatively small amount of chemically bonded heteroatoms (mainly oxygen and hydrogen). In addition, activated carbon may contain up to 15% of mineral matter (the nature and amount is a function of the precursor), which is usually given as ash content.

The adsorptive properties of activated carbon are determined not only by its porous structure but also by its chemical composition. In graphite, with a highly oriented structure, the adsorption takes place mainly by the dispersion component of the van der Waals forces, but the random ordering of the imperfect aromatic sheets in activated carbon result in incompletely saturated valences and unpaired electrons, and this will influence the adsorption behavior, especially for polar or polarizable molecules. In addition activated carbon is associated with such heteroatoms as oxygen and nitrogen (derived either from the starting material, activation process or post-treatment) and with the inorganic ash components. The presence of oxygen and hydrogen in surface groups can be up to 30 mol% H and 15 mol% O, which has a large effect on the adsorptive properties of the activated carbon. Some functional groups in carbon supports are given in Fig. 2.97.

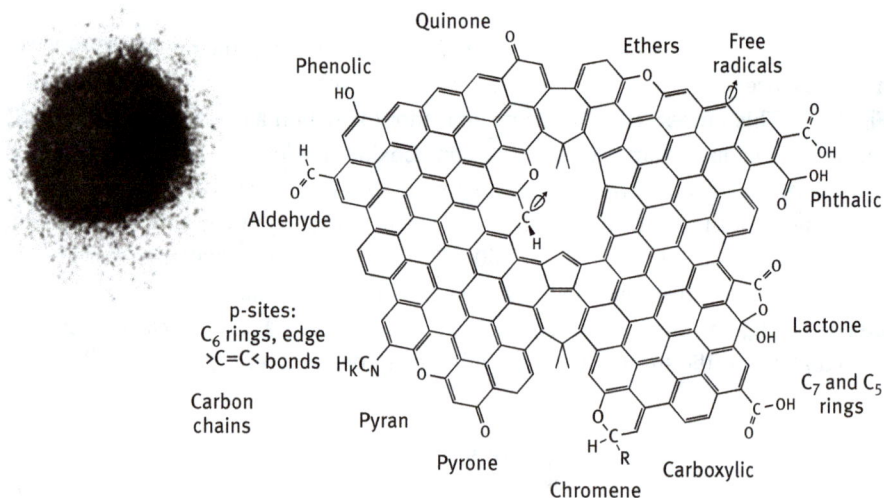

Fig. 2.97: Functional groups in activated carbon.

Oxidation with hydrogen peroxide and nitric acid results in the formation of C–O bonds, lactones, quinine, and carboxyl-carbonate structures with higher acidity generated by treatment with HNO_3. Therefore, even if carbon is hydrophobic the chemical nature could be modified to increase hydrophilicity.

In some cases, the high surface area of the carbon support may be detrimental if the active catalytic phase is confined in narrow micropores that are not accessible to the reactant molecules. This is important in processes where large molecules are involved, for example, in liquid-phase reactions of fatty acids when diffusion of reactants and products may be hindered by the narrow porosity.

Because activated carbons are lacking purity and could be hardly tunable materials with difficult to reproduce properties (even having the same surface area) several attempts were made in academia and industry to synthesize a carbon support that will be reproducible and mechanically stable. Thus, there is a growing interest in making catalyst supports from carbon nanotubes and carbon nanofibers, which could be produced in reproducible ways and make it possible to tailor the properties of these materials.

Carbon nanotubes (CNTs) that are categorized as single-walled (SWNTs) and multi-walled nanotubes (MWNTs) possess a cylindrical nanostructure with a length-to-diameter ratio of millions to one. They can be viewed as rolled, long, hollow structures with the walls formed by one-atom-thick sheets of carbon. High thermal conductivity and special mechanical and electrical properties made them a subject of intense research. The chemical bonding of nanotubes is composed of sp^2 bonds, as they are stronger than sp^3 bonds and thus result in materials with substantial mechanical strength. Single-walled nanotubes (SWNT) have a diameter of close to 1 nm. Multi-walled nanotubes (MWNT) consist of multiple rolled layers of graphene (Fig. 2.98).

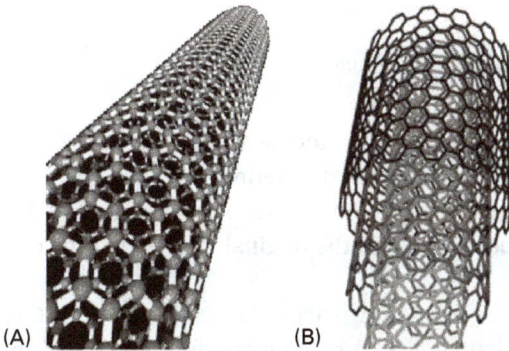

(A)　　　　(B)

Fig. 2.98: Structure of (A) single wall- and (B) multiwall nanotubes.

Carbon nanotubes are prepared by catalytic vapor phase deposition of carbon that uses an initial layer of metal catalyst particles, most commonly nickel, iron or cobalt. The diameter of the nanotubes, which is controlled by annealing or plasma etching of a metal layer, is related to the size of the metal particles. Nanotubes grow at the metal catalyst sites where the carbon-containing gas (acetylene, ethylene or methane) is dissociated on the surface and formed carbon is transported to the edges of the catalyst particle, where the nanotubes are formed with the metal particles staying either as the nanotube base (Fig. 2.99, top) or rest at the tip of the growing nanotube (Fig. 2.99, bottom) depending on the adhesion.

Extrusion or root growth

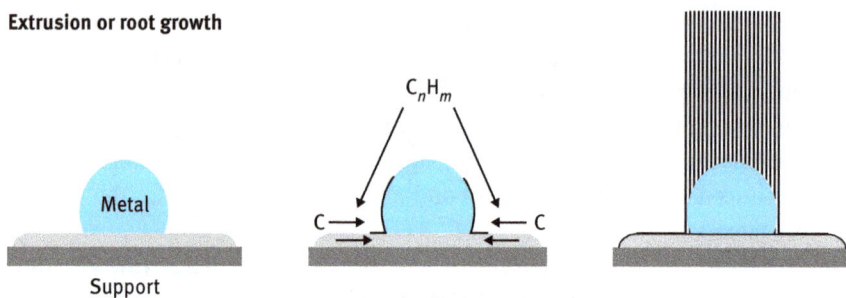

Tip growth

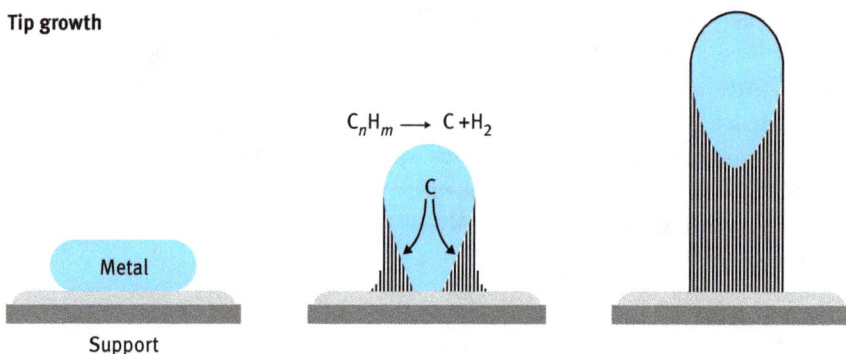

Fig. 2.99: Mechanisms of nanotubes growth. (Reproduced from [53]).

When CNT are used as catalyst supports they provide a spatial restriction on the metal particles by adjusting the CNT diameter and sintering.

Removal of the initial catalytic material could be an issue in this method. If this is done by an acid treatment (to remove alumina) the original structure of the carbon nanotubes could be destroyed.

Process-wise CNT can be synthesized in a flow reactor (Fig. 2.100) at ~700 °C to which a carbon-containing gas is fed along with a process gas (for example, nitrogen or hydrogen).

When plasma is generated by applying a strong electric field using plasma enhanced chemical vapor deposition, the nanotubes grow in the direction of the electric field.

Fluidized bed reactors are used industrially for CNT preparation. The fluidized bed pilot plant of 10 t year^{-1} developed by Arkema (Fig. 2.101A) for the production of Graphistrength™ MWNT allowed for the synthesis of materials as highly entangled bundles (Fig. 2.101B) with a size of several hundreds μm.

Similarly a fluidized bed reactor was used for the production of Baytubes® (Fig. 2.102) resulting in agglomerates of multiwall nanotubes, where each tube (diameter ~15 nm) comprises several graphite layers. The process, with an announced annual capacity of 3,000 t, is claimed to provide high carbon yields and high space-time.

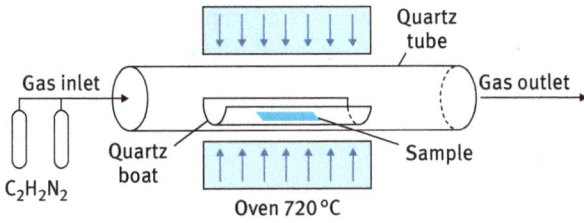

Fig. 2.100: Scheme of CNT synthesis in a flow reactor.

(A)   (B)

Fig. 2.101: Production of CNT at Arkema. (A) Pilot plant photo, (B) SEM image of MWNT bundles. (From [54]).

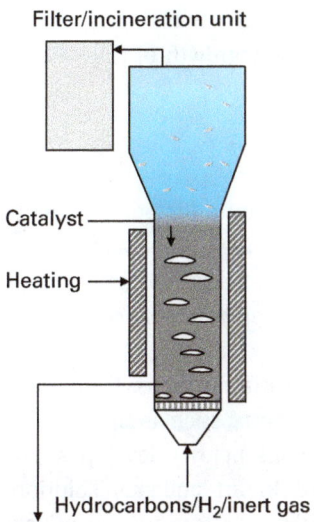

Fig. 2.102: CNT production in a fluidized bed (redrawn from [55]).

Carbon nanofibers (CNFs) are another material made of graphene layers, which are high surface area materials ($\sim$200 m^2 g^{-1}). They are synthesized in a similar way to CNT using the catalytic growth methodology with iron, cobalt and nickel as the catalytically active metals. Graphical layers of different orientation can be formed, for example, the fishbone type (Fig. 2.103).

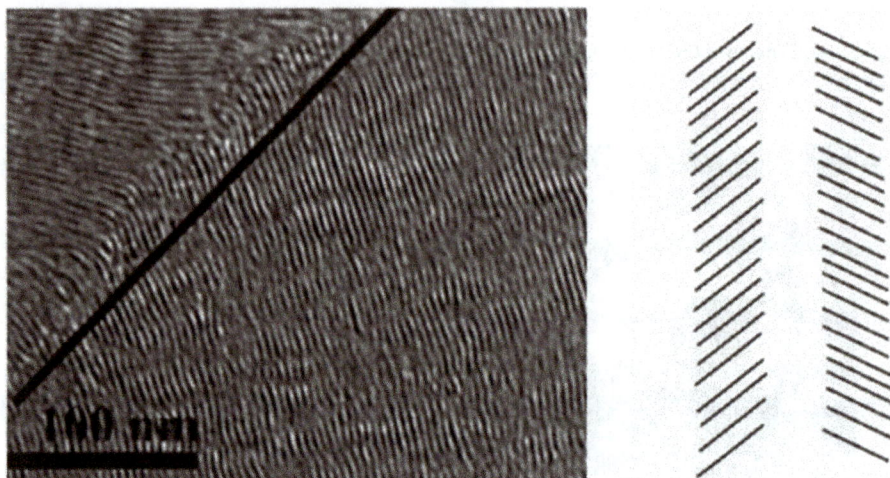

Fig. 2.103: Electron microscopy of fishbone nanofibers. (Reproduced with permission from [56]). Copyright (2012) American Chemical Society.

The thickness of the fibers can be controlled by the metal particle size, while the graphite plane orientation depends on the growth temperature and the metal type (parallel or fishbone types are formed with iron or nickel, respectively). The growth rate influences the strength of the fibers, for example, slow growth of thick fibers leads to strong particles.

2.3.5.2 Deposition of metals on carbon and surface charge of supports

Carbon is essentially hydrophobic in nature. It has low affinity for polar solvents such as water and high affinity for non-polar solvents (acetone). When a metal, for example, platinum, is deposited from water solutions of hexachloroplatinic acid, it will be, therefore, located primarily at the outer surface. Application of acetone as a solvent allows more uniform distribution along the catalyst grain.

Another important factor is the presence of oxygen surface groups. Because of these surface groups carbon supports (as well as other supports, such as silica or alumina, which will be discussed in the subsequent sections below) develop a pH-dependent surface charge (Fig. 2.104A) when placed in an aqueous solution. Dissociation of the acidic groups on the surface of a particle will lead to a negatively

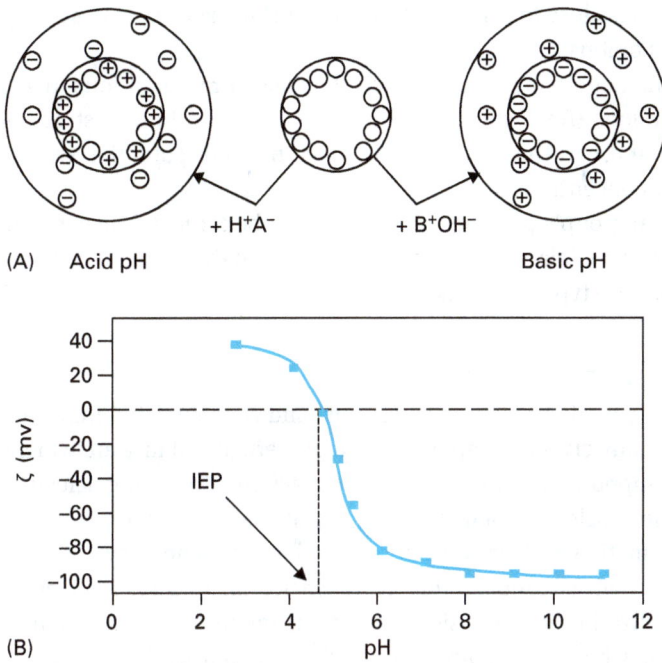

Fig. 2.104: (A) Surface polarization as a function of the solution pH. (B) Dependence of zeta potential on solution pH for Norit carbon.

charged surface. A decrease of pH in the case of a negatively charged surface will reduce the surface charge to zero (isoelectric point or point of zero charge). The basic groups will attract a positive charge from the solution and an increase of pH will counterbalance it. As the surface of the support is charged depending on the pH it can accept either cations or anions. Thus, a special care should be taken during catalyst preparation regarding the charge of the surface and the charge of species to be deposited.

Figure 2.104(B) illustrates how the zeta potential of activated carbon Norit depends on the solution pH.

Because of the surface charge there are more ions of the opposite charge in the interfacial region around the particle. The electric double layer consists of two zones: an inner region or Stern layer where the ions are strongly bound and the outer layer (diffuse) with less firmly bound ions. A boundary in the diffuse layer separates ions (which move when a particle moves, for example, because of gravity), from those ions that stay in the bulk. Zeta potential reflects the potential at this boundary. High values of zeta potential typically mean that the particles will repel each other. As clearly seen from Fig. 2.104B zeta potential displays strong dependence on pH in aqueous solutions. Figure 2.104B also illustrates that the position of the isoelectric point (IEP) is close to pH = 4.5. Below this pH the surface is positively charged and will accept anions.

The thickness of the double layer depends on the concentration of ions in solution. Some inorganic ions (for instance contaminants in the solution) strongly adsorb on the surface and can thus strongly influence the zeta potential even if they are present in low concentrations.

Although the oxygen group's presence is beneficial for better metal dispersion, it worsens reducibility of metals in a carbon supported catalyst, where carbon-active phase interactions are typically weak.

2.3.5.3 Transition metal oxides

Transition metal oxides (e.g., titania, chromia, zirconia and vanadia) exhibiting variable oxidation states, many crystallographic forms and thermal stability are widely used as catalysts and supports in different catalytic reactions, including selective oxidation, hydrogenation, selective dehydrogenation, alkylation, aldol condensation, carbonylation, amination and ammoxidation reactions to name a few. Surface acidity and basicity as well as cationic/anionic vacancies make transition metal oxides attractive for utilization in catalysis. A particular interesting application is related to photocatalysts when titania, which is a wide band gap semiconductor, is often implied for various reactions including solar water splitting.

Titanium dioxide TiO_2, or titania, as a naturally occurring oxide of titanium, is used in the form of nanoparticles, as a pigment in paints and in sun protection creams. It is typically prepared by aqueous hydrolysis of titanium salts or flame hydrolysis of $TiCl_4$. Two crystalline forms of titania – anatase or rutile – comprising chains of TiO_6 octahedra are obtained depending on the preparation conditions. Both rutile and anatase have tetragonal crystalline structure containing, respectively, 6 and 12 atoms per unit cell. Anatase is formed as a metastable phase with high surface area and can be then slowly transformed into a thermodynamically more stable rutile.

Flame hydrolysis of titanium chloride leads to anatase with a surface area 40–80 $m^2\,g^{-1}$ and mean pore diameter close to 50 nm. This phase is more stable at high T and thus more suitable for catalytic applications. Aqueous hydrolysis of titanium salts gives a support with the surface area of 190–200 $m^2\,g^{-1}$ and a pore diameter of 15–20 nm. Contrary to silica and alumina, which are insulators, titania is a semiconductor and thus can be used as a photocatalyst. Moreover because of the possibility of ready reduction of Ti^{+4} to Ti^{3+} the partially reduced titanium oxide species can decorate supported metal nanoparticles, significantly changing the catalytic properties because of strong metal support interactions.

There is a long history of using an aerosol flame method for the production of titania (spherical particles of predominantly anatase) for various photocatalytic reactions (such as titania marketed by Degussa as P25). It is not only titania that can be produced by this method but also other supports (alumina or silica) with non-porous structure and high external surface areas, which includes exposure to high temperatures and cooling rates. The resulting materials have good thermal stability.

Aerosol flame synthesis can be classified into vapor-fed and liquid-fed variants. In the vapor-fed methods applied for the synthesis of fumed silica and titania, volatile precursors (such as chlorides) are evaporated and fed into the flame. For synthesis of titania, the flame is needed to ignite the process, while a hydrogen/oxygen flame can also assist the process as in fumed silica manufacturing. The metal precursor is first converted into the metal oxide, which undergoes nucleation from the gas-phase. The schematic diagram is presented in Fig. 2.105A. A drawback of the method is associated with limited availability of a volatile precursor at a competitive price. In flame or flame-assisted spray pyrolysis liquid precursors are introduced as either non-combustible (FASP, Fig. 2.105B) or combustible (FSP, Fig. 2.105C). The main advantage of this method is the formation of nanosized particles with possible size control through the precursor-solvent composition.

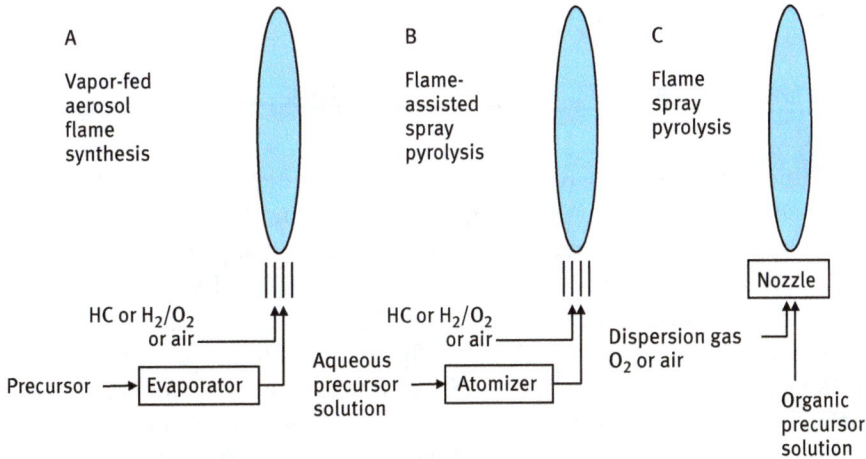

Fig. 2.105: Various flame configurations. (Redrawn from [57]).

The schematic picture of the particles formation is illustrated in Fig. 2.106.

For other common catalyst supports, such as silica or alumina, the flame methods result in highly aggregated structures. In general, the final particle size of the solid material depends on the droplet size, gas velocity of the spray or the residence time in the reactor and quenching efficiency. The flame process is influenced by a number of parameters, such as gas and liquid-feed rates, type and concentration of the precursor and solvent, solvent-to-fuel ratio. For instance, nucleation and coagulation rates are increased at high liquid-feed rates, which results in enhanced

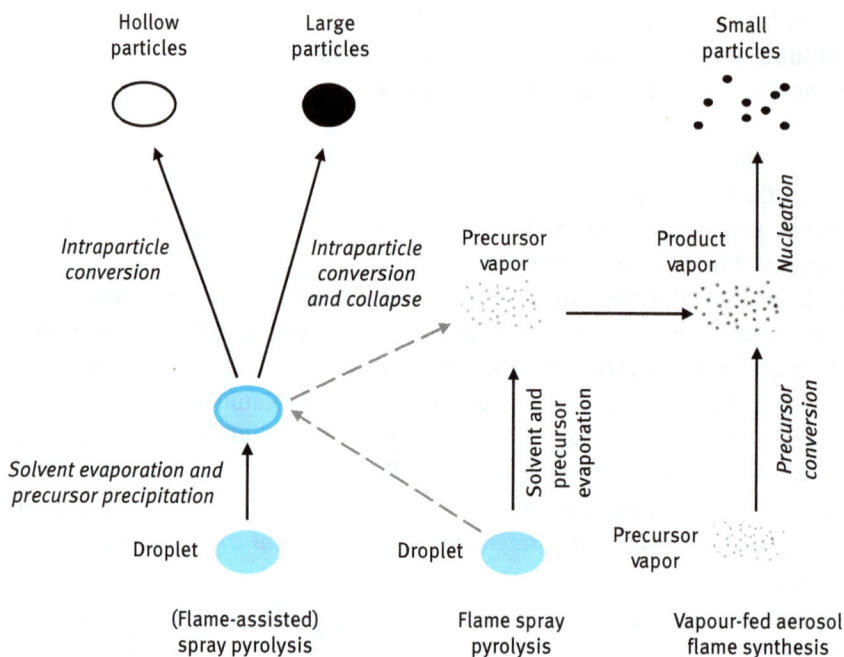

Fig. 2.106: Schematics of the particle formation. (Redrawn from [57]).

particle growth. Conversely, a reduction of the overall particle growth is achieved at lower flame temperatures and thus allows for faster cooling. This in turn is possible at high dispersion gas-flow rates. However, because of the complexity of flame methods, precise control over the size is not easy and the particles exhibit generally a wide particle size distribution.

Moreover, complete crystallization is difficult to achieve in conditions of very short residence times at high temperatures and fast cooling rates. This can result in partially amorphous materials restricting the use of flame methods for synthesis of catalyst supports when high crystallinity is needed.

As flame pyrolysis methods give materials with high thermal stability, in addition to titania such supports such as ceria are also produced by these methods. Such thermo-stability is important for automotive exhaust applications.

Ceria along with zirconia are bifunctional catalysts and contain both acid and base sites, which obviously make them attractive in bifunctional heterogeneous catalytic reactions as well as in catalytic conversion of CO_2 to value-added chemicals. In particular, appealing are the high oxygen mobility and redox property of CeO_2.

Zirconia is formed mainly in tetragonal and monoclinic crystallographic phases. The former phase is metastable, possessing a high surface area, while the monoclinic form is thermodynamically stable.

2.3.5.4 Silica

The amorphous silica supports are typically used in applications that require relatively low temperatures, as it has lower thermal stability when compared to alumina. Moreover, under exposure to steam volatile hydroxides can be formed at elevated temperatures. By its chemical nature, silica is a weak acid and is more resistant than alumina to acidic media.

Two different methods are used for preparation of silica supports: flame hydrolysis giving fumed silica and sol-gel precipitation leading to silica gels.

Flamed or fumed silica is a very pure material with nanometer sized dense particles (40–50 nm) not possessing micropores. The surface area is up to 300 m^2 g^{-1} and the diameter of pores is above 7 nm. This type of silica is manufactured by flame hydrolysis of SiCl$_4$, which is first formed by carbo chlorination of a macroscopic oxide:

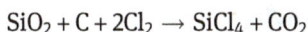

$$SiO_2 + C + 2Cl_2 \rightarrow SiCl_4 + CO_2$$

Silicon tetrachloride then reacts in a flame reactor where (in addition to SiCl$_4$) hydrogen and oxygen (air) are fed, leading to the following reactions: $2H_2O + O_2 \rightarrow 2H_2O, SiCl_4 + 2H_2O \rightarrow SiO_2 + 4HCl$. Nanometer sized silica particles are separated from HCl in a cyclone followed by deacidification of the silica surface with steam and air in a fluidized bed downstream of the cyclone (Fig. 2.107). Formed silica has a low density and needs to be compacted prior to forming operations in a compactor, located downstream of a fluidized bed reactor and a hopper. Process parameters such as flame temperature, content of the substrate, hydrogen to oxygen ratio and the residence time allows control of the size and aggregation of silica.

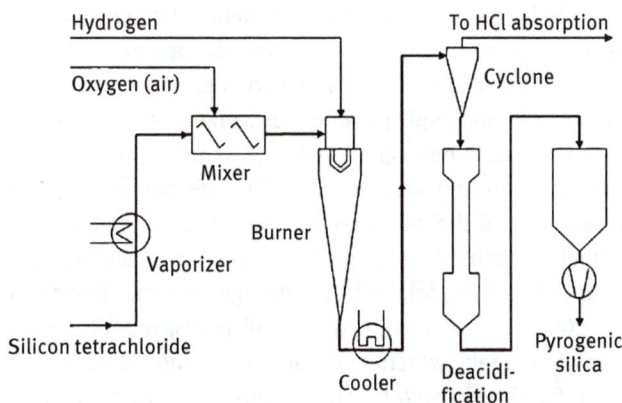

Fig. 2.107: A flow scheme of fumed silica production. (From [51].)

In the sol-gel route the alkaline solution of sodium or potassium silicate (for example, the pH of water glass for sodium silicate is 12) is mixed with sulfuric acid, which triggers the formation of first $Si(OH)_4$ monomer units, followed by polymerization to a colloidal solution (sol) where silanol (Si–OH) groups and Si–O–Si bonds are formed. Minimization of –OH and maximization of Si–O–Si bonds is the driving thermodynamic force for the formation of oligomers and cyclic species, small particles and eventually larger particles. Colloidal solutions are formed by micelles, which are separated because of repelling electric charges on their surface and in the solution, prohibiting coagulation. Sodium ions play a critical role in controlling agglomeration of micelles. At some point the sol undergoes gelation – the formation of a 3D hydrogel (Fig. 2.108). Gelation depends on the micelle concentration, temperature, ionic strength and especially pH.

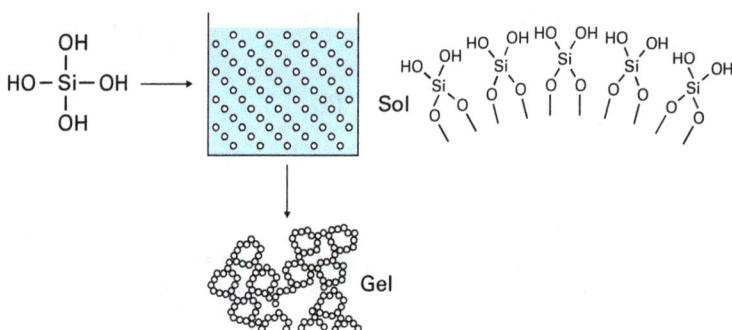

Fig. 2.108: Formation of a gel.

The key factors are the rates of hydrolysis –M–O–Na(R) + H_2O → –MOH + Na(R) OH condensation –MOH + XOM → M–OM + XOH, where X = H, R. R indicates that precursors in sol-gel synthesis could be not only salts such as metal silicates, but also alkoxides. If hydrolysis is faster than condensation then the material is less branched, while slower hydrolysis leads to a highly branched one.

The gelation rate is slow at low and high pH and can range from minutes to days. The density of hydrogel, which is a metastable polymer, increases with an increase of the initial salt concentration and gelation rate. The subsequent steps in the preparation of silica are washing of the gel to remove sodium and then drying. The drying is critical in producing a material that could be used as a catalyst support. Dried material can occupy 5% of the original hydrated gel volume. Drying at moderate temperatures (150–200 °C) leads to the collapse of pores, drastic reduction of porosity and formation of xerogels, which lack macropores and the sufficient strength for them to be used in fixed-bed reactors. These xerogels are first milled to the desired size and mixed with binders prior to shaping. An alternative to this procedure is drying under supercritical conditions leading to aerogels. First a less polar then water solvent (i.e., acetone) is used to wash away water and then acetone itself is washed away with high-pressure liquid carbon dioxide. After heating

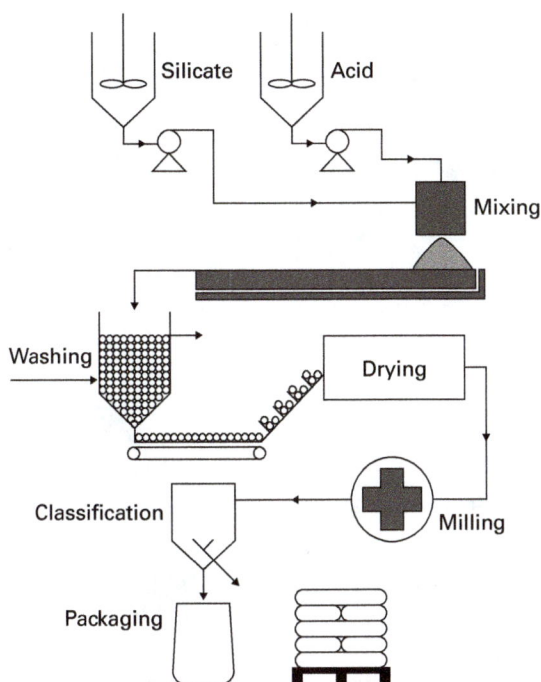

Fig. 2.109: Preparation of silica by sol-gel method. (From [51].)

beyond the critical point and gradually releasing pressure a dried product is formed. The flow scheme of silica preparation is illustrated in Fig. 2.109.

After drying at a low temperature the concentration of silanol groups is high (4–5 OH per nm^2) and the material is hydrophilic. Heating at a high temperature results in dehydroxylation, and thus a decrease in concentration leading to hydrophobic surfaces. The surface area of commercially available silica supports varies between 400–700 m^2 g^{-1} with an average pore diameter of 2.5–5 nm depending on the preparation conditions.

The surface charge of silica depends on pH (Fig. 2.110). At pH equal to the point zero charge (pH ~2) the surface is neutral with –SiOH groups, while at acidic and basic pH the main surface groups are –Si–OH$_2^+$ and –SiO$^-$, respectively.

In addition to colloidal and fumed silica, another type of silica – kieselguhr – found an application as a support for vanadium pentoxide catalysts, which are applied in oxidation of SO$_2$ to SO$_3$. Diatomaceous earth (diatomite) kieselguhr is a form of silica composed of the siliceous shells of unicellular aquatic plants of microscopic size. This material is heat resistant and has been used as an insulator, as a component in toothpaste and as an abrasive in metal polishes. Kieselguhr, with a surface area of 20–40 m^2 g^{-1} and a broad range of pore sizes, contains small amounts of alumina and iron as part of the skeletal structure.

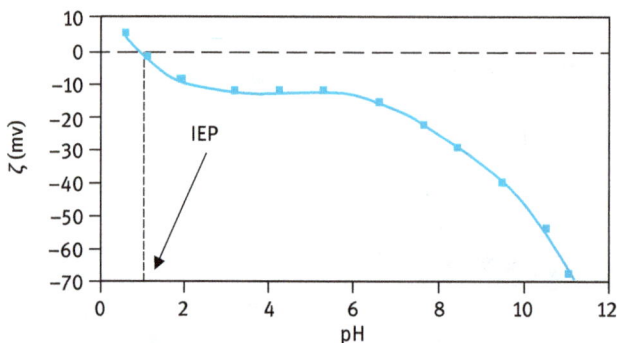

Fig. 2.110: Zeta potential as function of pH for silica.

2.3.5.5 Alumina

Although a variety of alumina exists, distinguishable by XRD, only a few phases are applied as catalyst supports, such as non-porous low surface area crystalline α-alumina, and porous amorphous η and γ-alumina. Annual world production of alumina is approximately 45 million tons, over 90% of which is used in the manufacture of aluminum metal.

Aluminum is extracted from bauxite, an ore containing aluminum hydroxide, silica and other oxides by treating it with sodium hydroxide. Formed sodium aluminate $Na_2O \cdot Al_2O_3$ (or $Na_2Al_2O_4$) undergoes crystallization that leads to different gels depending on the pH. This is done by adding nitric acid to sodium aluminate. When precipitation occurs at $8 < pH < 11$ bayerite ($Al(OH)_3$) is formed, which upon aging at high pH gives gibbsite gel with the same chemical formula of aluminum hydroxide. Precipitation at a lower pH ($6 < pH < 8$) results in crystalline boehmite gel AlOOH. Different calcination procedures result in various phases of alumina (Fig. 2.111).

In cases when the presence of alkali might influence catalytic performance, an alternative to the above mentioned route should be used. Alkali-free alumina could be prepared from aluminum sulfates (Fig. 2.111) or aluminum alcoholates. Both alkali and alkali-free methods result in high amounts of acids and bases needed (2–4 t per ton of alumina), which are cumbersome to regenerate. γ-Alumina has a surface area in the range 50–300 $m^2\ g^{-1}$ and pores between 5–15 nm. Rather high thermal and mechanical stability also makes alumina a popular support for various applications. The surface of alumina contains Brønsted acid sites (H^+ donors), Brønsted basic sites (acceptors of protons) and Lewis acid sites (electron acceptors). After dehydroxylation the surface of alumina contains mainly Lewis acidity,

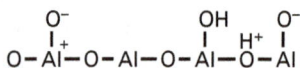

$$
\begin{array}{ccccc}
O^- & & OH & & O^- \\
| \, + & & | & H^+ & | \\
O-Al-O-Al-O-Al-O-Al
\end{array}
$$

Lewis site ···· Basic site ············ Brønsted acid sites

```
Al-sulfate solution                                    Na-aluminate solution
   Al₂O₃.3SO₃                                              Na₂O.Al₂O₃
     pH < 2                                                  pH > 12
        +                                                       +
      Base                                                    Acid
        ↓                                                       ↓
```

At 3 < pH < 6 · precipitation of micro crystalline boehmite gel AlO(OH)

At 6 < pH < 8 · precipitation of crystalline boehmite gel AlO(OH)

At 8 < pH < 11 · precipitation of bayerite gel Al(OH)₃

Gibbsite gel Al(OH)₃

Heating above 770 K   Aging at pH 8 and 353 K   Aging at high pH

Amorphous Al₂O₃

γ-Al₂O₃   η-Al₂O₃   χ-Al₂O₃
Spinel structure: cubic close packing

Heating at about 1,170–1,270 K

θ-Al₂O₃   δ-Al₂O₃   α-Al₂O₃

Heating above 1,370 K

α-Al₂O₃
Corund structure: hexagonal close packing

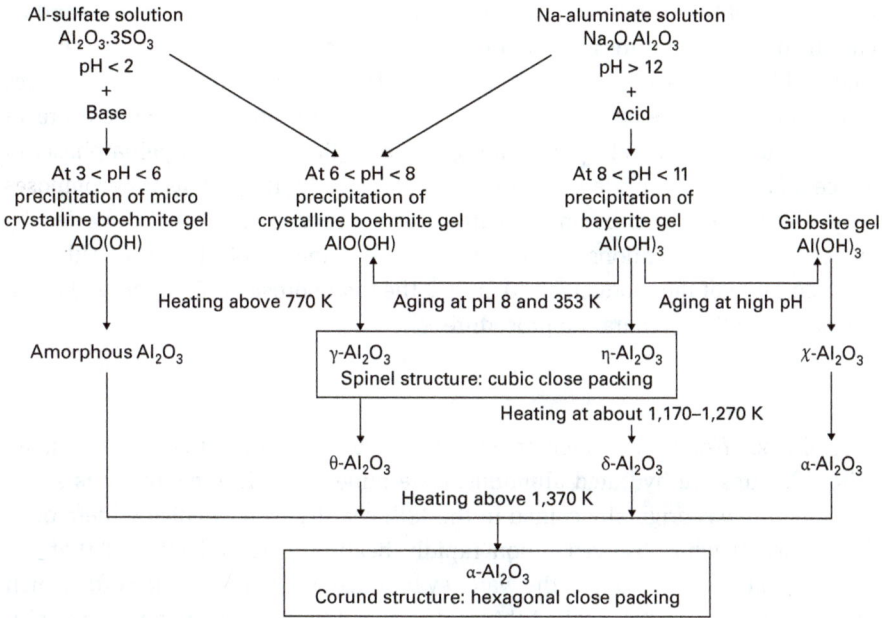

Fig. 2.111: Preparation of different alumina phases. (From [51].)

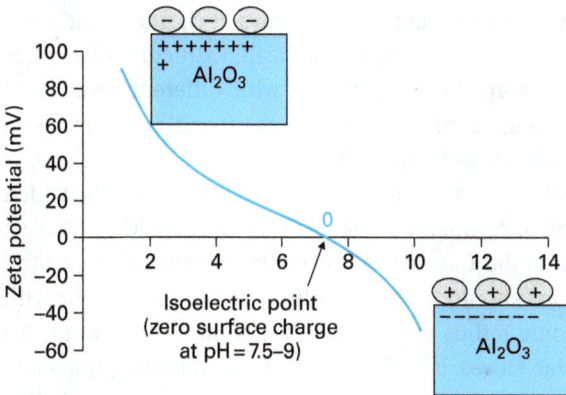

Fig. 2.112: Zeta potential dependence on pH.

In water solutions, the surface charge depends on pH, with the isoelectric point close to 7 (Fig. 2.112).

Heat treatment of γ-alumina at temperatures above 1,370 K results in the formation of first θ alumina and then α-alumina (corundum). Simultaneously, the primary particles of γ-alumina grow in size to 70 nm for α-alumina concomitant with the reduction of the surface area (few $m^2 g^{-1}$). The latter phase of alumina found its niche in high temperature applications, such as steam reforming of natural gas, which will be discussed in detail in Chapter 4. The use of additives such as oxides of chromium, iron, molybdenum or HF allows for lowering the temperature of phase transformations and even avoiding the formation of the θ alumina phase. In practice ammonium fluoride is used instead of hydrofluoric acid, as it decomposes to NH_3 and HF. HF adsorbs on the surface of γ -alumina at ~700 °C followed by polymorphic transformations of γ -alumina to α-alumina at 700–1,000 °C. Although the surface area of the product is ~10 $m^2 g^{-1}$, the total porosity (~0.55 $cm^3 g^{-1}$) could be preserved in this preparation procedure.

2.3.5.6 Zeolites

A special type of supports, which are widely used as catalysts, are solid acids called zeolites. Zeolites are hydrated aluminosilicate minerals with a microporous structure. The term was originally coined in the 18th century by a Swedish mineralogist A. F. Cronstedt who observed, upon rapidly heating a natural mineral that the stones began to dance about as the water evaporated. Using the Greek words which mean "stone that boils", he called this material zeolite. Zeolites are crystalline aluminosilicates containing pores and cavities of molecular dimensions. There are close to 50 natural zeolites and almost triple this amount of synthetic zeolites that do not occur in nature. The synthetic zeolites are among the most widely used sorbents, catalysts and ion exchange materials in the world. Zeolite crystals are porous on a molecular scale, their structures revealing regular arrays of channels and cavities (~3–15 Å), creating a nanoscale labyrinth. The complicated structure of zeolites is constructed from several secondary building blocks with different number of units, possessing 6, 8, 10 or 12-membered rings of T-atoms (Si, Al). Properties of some zeolites are presented in Tab. 2.5 and Fig. 2.113.

Channels contain water molecules and alkali metal cations (M^I, M^{II}), which can be ion-exchanged. The general formula of zeolites is $M^I M^{II}_{0.5}[(AlO_2)_x*(SiO_2)_y*(H_2O)_z]$. In addition to Al, silicon in the T atom position could be replaced, for example, by P, Ti, Ga, Ge, Hf or Zr. Ions should have a coordination number of four with the respect to oxygen and an ionic radius that fits into the zeolitic framework. An interesting example of molecular sieves is VPI-5, which is an aluminophosphate with a pore aperture of 1.2 nm. Such materials are electrically neutral and thus largely catalytically inactive.

The key properties of zeolites are size and shape selectivity, together with the potential for strong acidity. In zeolites SiO_4 and AlO_4 tetrahedra linked through common oxygen atoms and cations (such as Na^+) are needed to neutralize the charge difference between Al^{3+} and Si^{4+}. Brønsted acidity appears when Si^{4+} is replaced by Al^{3+} and the charge is balanced by proton, localized in vicinity of Al. Zeolites could thus have acidities compared to mineral acids. Acidity depends on the Si/Al ratio. With an

Tab. 2.5: Properties of zeolites (From [29]).

Common name	Three-letter code	Channels	Window or channel diameter (nm)	Pore volume $(cm^3\,g^{-1})$	Si/Al
Zeolite A	LTA	3D, Cages connected by windows, cubic	0.45	0.30	1
Zeolite X	FAU	3D, Cages connected by windows, tetragonal	0.75	0.35	1–1.5
Zeolite Y	FAU	3D, Cages connected by windows, tetragonal	0.75	0.35	>2.5
Mordenite	MOR	1D, two straight, parallel channels	0.70; 0.65	0.20	>5
Zeolite L	LTL	1D, one straight channel			
ZSM-5	MFI	2D, one straight, one sinusoidal, mutually perpendicular channels	0.55	0.15	>10
Zeolite Beta	BEA	3D, two straight, one sinusoidal, perpendicular channels	0.76; 0.64	0.25	>5

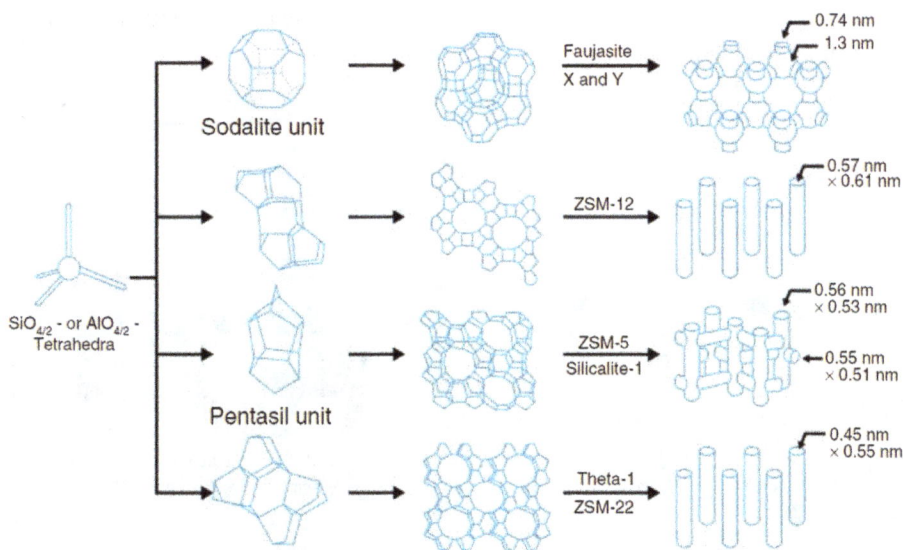

Fig. 2.113: Various types of zeolites with different numbers of T-atoms (copyright [58]).

increase of Si content, Lewis acidity decreases and is zero for silica. Brønsted acidity (zero for alumina) starts to be seen at Si/Al ~0.3, has a maximum at Si/Al ~3, and is zero for silica. An empirical Lowensteins' rule determines the structure of linkages in zeolites, stating that Al-O-Al linkages are forbidden and thus Si/Al ratio must be above or equal to unity. In fact, Si/Al ratio varies from 1 (LTA) to infinity (pure SiO_2).

Zeolites with a low aluminum content are hydrophobic (and vice versa). Zeolites are thermally stable and if deactivated could be regenerated by simply burning the carbon deposits. Stability of zeolites increases with Si/Al ratio.

Usually the as-synthesized forms of zeolites (typically with Na^+ balancing the charge) are not the most suitable for the catalytic purposes. Acidity can be introduced in the as-synthesized zeolite by ion-exchanging sodium cations with NH_4Cl or NH_4NO_3 to achieve an ammonium form. The ammonium form of zeolite is dried and calcined to obtain the proton form of zeolite. The exchange of extra-framework species (for example, Na^+ to NH^4 and H^+) is one of the ways that zeolites structure modifications. Other methods include dealumination by steaming or acid treatment in order to change Si/Al ratio. Removal of aluminum from the framework by, for example, hydrothermal treatment with steam at 600–900 °C leads to an ultrastable zeolite Y, industrially used as a catalyst for catalytic cracking.

Incorporation of transition metals into zeolites gives rise to bifunctional catalysis, where both metal and acid centers are involved in a catalytic reaction. Incorporation of Pt into zeolites tested in a ring opening of decalin (Fig. 2.114A), resulted in a decrease of the amount of strong acid sites while the overall number of acid sites was almost unchanged (Fig. 2.114B). As a consequence, the amount of cracking products decreased with a simultaneous increase in selectivity towards desired ring-opening products.

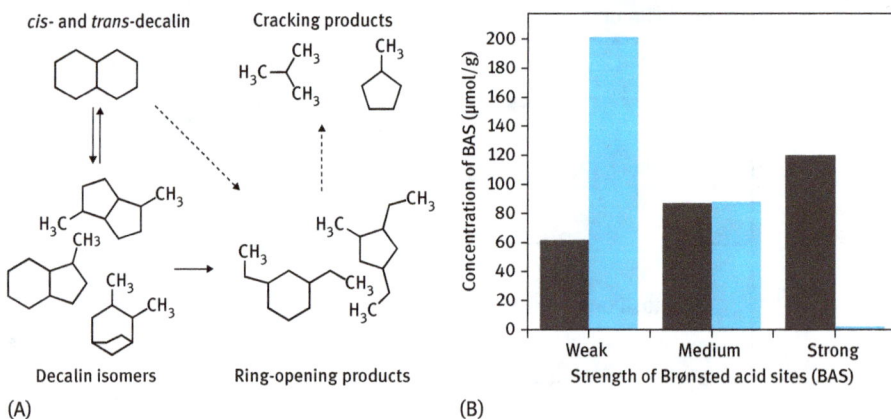

Fig. 2.114: (A) Ring opening of decalin and (B) concentration of Brønsted acid sites after incorporation of Pt into zeolites. (Data from [59]).

In general, non-uniform distribution of aluminum leads to materials with different acid strength. Acidity of zeolites is probed by mainly non-aqueous spectroscopic methods (IR spectroscopy of probe molecules) and temperature-programmed desorption of bases, giving information on the total amount of acid sites and the acid sites distribution.

For water non-sensitive materials, for example, acidic amorphous oxides or active carbons, acidity could be determined by titration of aqueous slurries of the acidic solids with a standard base.

Although catalytic activity of zeolites is often related to Brønsted acidity (for example, the presence of protons) they also contain Lewis sites (coordinatively unsaturated Al^{+3} ions).

Zeolites are microporous materials with pore diameters below 2 nm, which results in some interesting selectivity properties. Thus an important property of zeolites is shape selectivity, which refers to reactants (Fig. 2.115A), product (Fig. 2.115A) or transition state selectivity (Fig. 2.116). Fig. 2.115A illustrates that in the case of a zeolite with narrow pores, reactants with bulky substituents cannot enter the pores. For product selectivity (Fig. 2.115B) only *p*-xylene can diffuse out of the ZSM-5 channel pores, while o-xylene and *m*-xylene cannot be formed. Some transition state intermediates are too large to be accommodated within the pores/cavities of the zeolites, even though diffusion of neither the reactants nor the products is restricted. This is the case for transition state selectivity (Fig. 2.116). The diameter of the cage is important for the reactant and transition state selectivity, while diameter and the pore length are the critical parameters for the product selectivity. In some instances coking to reduce the pore size can be a method to improve shape selectivity.

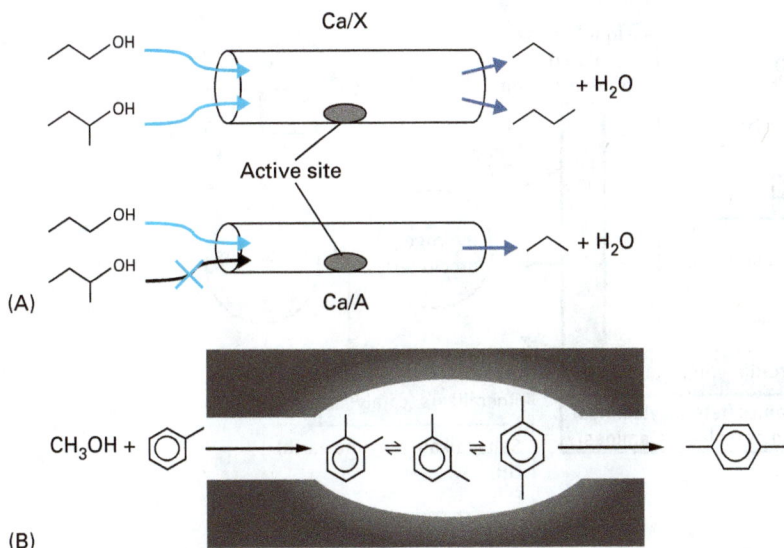

Fig. 2.115: (A) Reactant and (B) product selectivity in zeolites.

Aluminosilicate zeolites are formed by hydrothermal synthesis, typically under mild conditions (350–525 K) and autogenous pressure (above 373 K). Zeolites are synthesized by mixing a silica source (the primary building units of the framework,

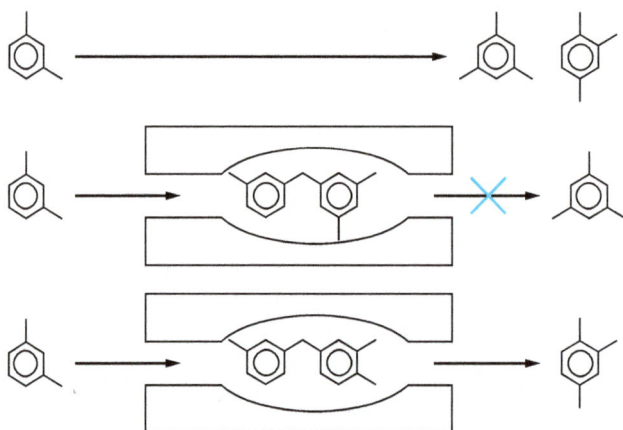

Fig. 2.116: Transition state selectivity.

i.e., SiO_2), an alumina source (origin of the framework charge), organic templates, mineralizing agents (OH^- anion, NaOH) and water (as solvent) to form an aluminosilicate gel. The process in illustrated in Fig. 2.117.

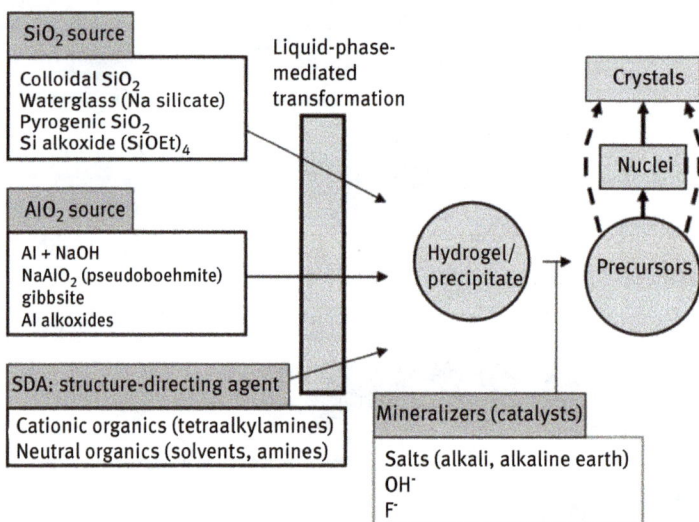

Fig. 2.117: Schematic representation of zeolitization. (Reproduced with permission from [60]).

After initial transformation of Si, Al sources and the structure directing agent (SDA) through the liquid phase mediation into a hydrogel or precipitate, the latter is converted into precursor species. Such species assemble into nuclei, which then grow in crystal according to classical crystallization. SDAs act as topology-determining

templates while transformations of the gel to precursor/nuclei are thought to be catalyzed by mineralizing agents such as alkali (alkaline earth) salts and OH^-. Figure 2.118
illustrates the crystal growth of zeolites exhibiting typical S-shaped crystallization
curves [61] with an induction period.

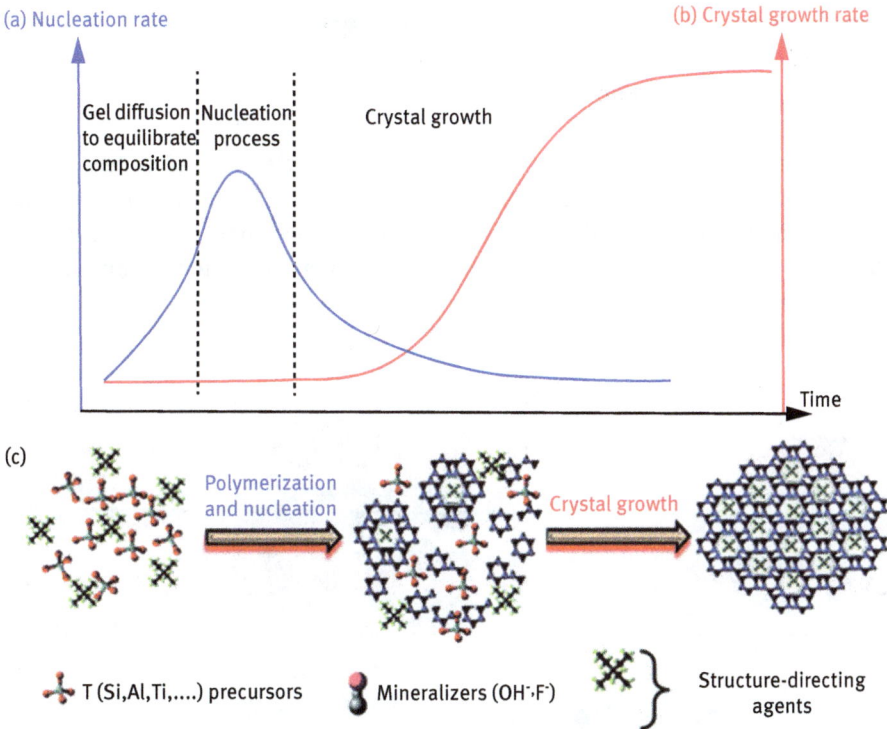

Fig. 2.118: Schematic representation of the (a) nucleation rate and (b) crystal growth rate of
zeolites described with a typical S-shaped curve and (c) related rearrangements from amorphous
particles into crystalline zeolite during the synthesis (Reproduced with permission from [61]).

Templates of the NR_4^+ type are used, such as tetra-alkylammonium hydroxide.
Alkylphosphonium cations (R_4P+) and organic complexes can be also applied.
Reactants are mixed according to a synthesis recipe. Usually, different solutions are
prepared under continuous stirring and mixed to obtain a unique thick milky white
gel. pH of the solutions is an important control parameter. Other critical parameters
are: the molar ratios (OH^-/SiO_2 and Si/Al); temperature; the sources of silica and
alumina; presence of contaminants; the order of the addition of the reactants;
aging and ripening prior to crystallization; synthesis time; the nature of the organic
templates; pre-treatment of the reactants; inclusion of special additives; presence
of seeds and stirring rate and so on.

In laboratory-scale preparations, once the gel is ready it is transferred to Teflon-lined autoclaves, where the nucleation and crystallization of the gel takes place. Condensation reactions occur faster than hydrolysis which leads to formation of nuclei and growth by the addition/condensation of monomers. The particles are negatively charged, preventing the formation of aggregates.

The Teflon-lined autoclaves are put in an oven with a constant predetermined temperature for a period of time according to the synthesis procedure. In a laboratory scale, crystallization can be performed in a static mode, under internal rotation or under rotation of the autoclaves.

When the crystallinity is plotted as a function of time, S-shapes curved are usually obtained (Fig. 2.118, 2.119), where the fraction of the transformed material F can be described by the Avrami-Erofeev equation (eq. 2.36). with n determined by data fitting. S-shaped features are caused by autocatalytic behavior at low conversions because of slow nucleation and decline of the crystal growth rate with an increase in crystallinity.

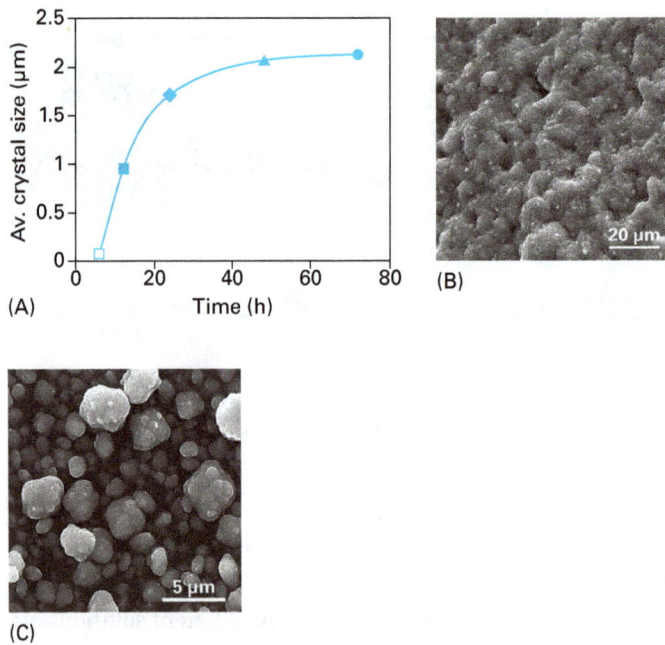

Fig. 2.119: Effect of synthesis time on ZSM-5 zeolite. (A) Average crystal size as a function of synthesis time, (B) morphology after 6 h, (C) morphology after 72 h. (Reproduced with permission from [62]).

An increase of the synthesis temperature at constant alkalinity (pH) increases nucleation and linear growth rate. The induction time in zeolite crystallization decreases rapidly with an increasing temperature, up to a certain point. Similarly, the induction time decreases and crystal growth is enhanced by an increase in pH at constant temperature. Alkalinity also has an effect on the Si:Al ratio of the zeolite.

After crystallization, the autoclaves are quenched with cold water. The formed zeolite crystals are filtered, washed and dried. Washing is required to reduce high alkalinity created by the addition of NaOH. In order to remove the organic template, zeolites are typically calcined after drying (100 °C). A calcination step is needed to remove the SDA. It is performed at 500–600 °C for several hours, depending on the zeolite. Different calcination techniques have an effect on the properties of the synthesized zeolite. After calcination, the zeolite is in the sodium form (the as-synthesized form, denoted by Na at the beginning of the name, for example, Na-ZSM-5) because it contains sodium ions to balance the framework charge induced by Al as T-atoms. Acidity can be introduced in the as-synthesized zeolite by ion-exchanging sodium cations with NH_4Cl or NH_4NO_3 to achieve an ammonium form. The ammonium form of zeolite is dried and calcined to obtain the proton form of zeolite.

A flow sheet for the preparation of a sodium form of zeolites is given in Fig. 2.120.

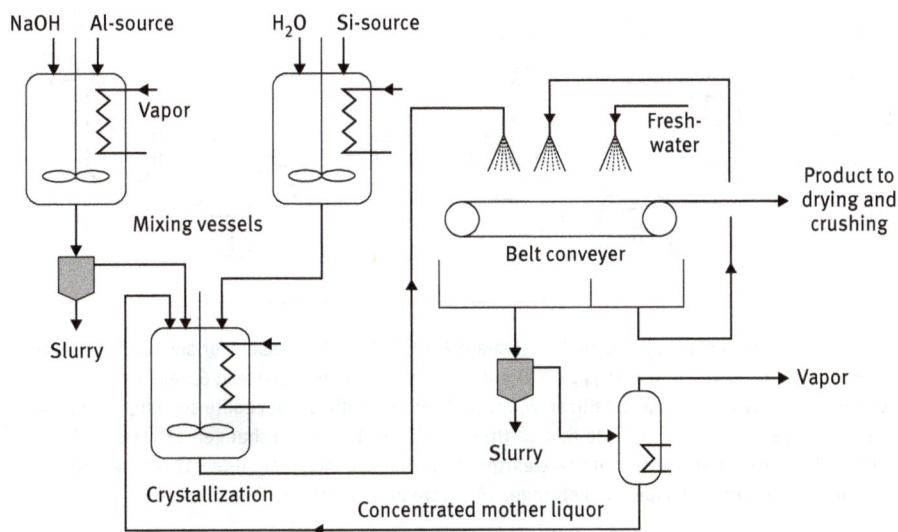

Fig. 2.120: Preparation of a sodium form of zeolites. (From [51]).

The catalyst in the final process always requires further fine-tuning by a number of secondary treatments. It is worth considering as an example the production of fluid catalytic cracking catalysts. Detailed description of these catalysts is given in Chapter 4. Typical commercial cracking catalysts are a mixture of rare earth-Y zeolite and SiO_2–Al_2O_3 amorphous aluminosilicate as a binder providing the required mechanical stability. Small pore ZSM-5 is applied as a co-catalyst with rare earths Y zeolite to increase the octane number of gasoline. The sodium form of zeolites Y is first produced followed by an exchange of sodium with rare earth ions, which is an example of a secondary treatment. Although there is a clear advantage to using binders the drawbacks are associated with blocking of the zeolites pores by the binder and longer diffusion paths. The flow scheme for the production of rare earth zeolites Y is given in Fig. 2.121.

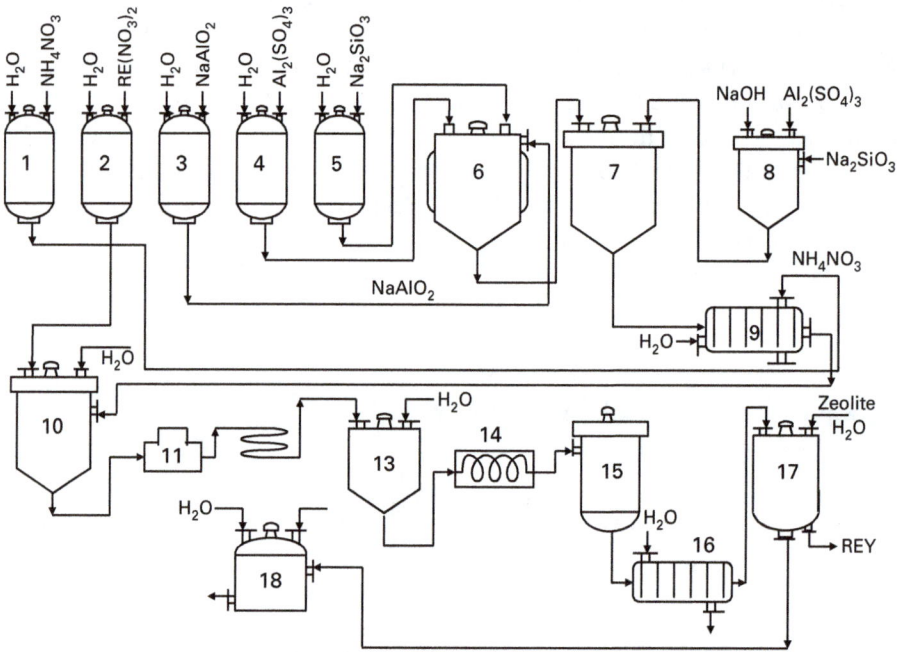

Fig. 2.121: Synthesis of rare earth zeolite. (From [52].) 1, 2, 3, 4, 5 = vessels for ammonium nitrate, rare earth nitrate, $NaAlO_2$, $Al_2(SO_4)_3$ 18 H_2O and sodium silicate respectively; 6, reactor for sodium aluminosilicate synthesis; 7,crystallizer; 8, crystallizer for synthesis of zeolite seeds; 9, filter; 10, vessel for ion exchange of NH_4^+ into rare earths; 11, milling; 12, heat exchanger; 13, vessel of additional ion exchange of NH_4^+ into rare earths; 14 cooler; 15, mixer; 16, filter; 17, vessel for repulping of zeolite; 18, suspension storage. RE, a rare earth element.

Water solutions of ammonium nitrate, rare earth nitrate, $NaAlO_2$, sodium silicate and aluminum sulfate are prepared in vessels 1–5. Sodium aluminosilicate gel is obtained in vessel 6 and a solution of $NaAlO_2$ is added. After cooling, the mixture is routed to a crystallizer in 7 to which a seed of zeolites Y is added from vessel 8 in the quantities corresponding to 1–3 wt%. Mixing of the Na-Y zeolites seed with the reacting mixture in vessel 7 lasts for ~1 h, thereafter hydrothermal synthesis is performed for ~24 h at 371 K followed by cooling to room temperature. After filtration (pos. 9) the mother liquor is separated and the zeolites are washed with water and ammonium nitrate to ion exchange sodium for ammonium. This is followed by primary ion exchange with rare earth (Ln^{3+}) nitrates in vessel 10. After milling (pos. 11) and additional ion exchange (pos. 13) at an elevated temperature (up to 473 K) the zeolite is cooled (pos. 14), filtered (pos. 16) and transported to a vessel 17. From this vessel it is taken as rare earth zeolite Y or routed to another vessel for storing in the form of suspension (pos. 18) to be used in the synthesis of, for example, microspherical FCC catalysts. The flow scheme of the microspherical catalyst synthesis is given in Fig. 2.122.

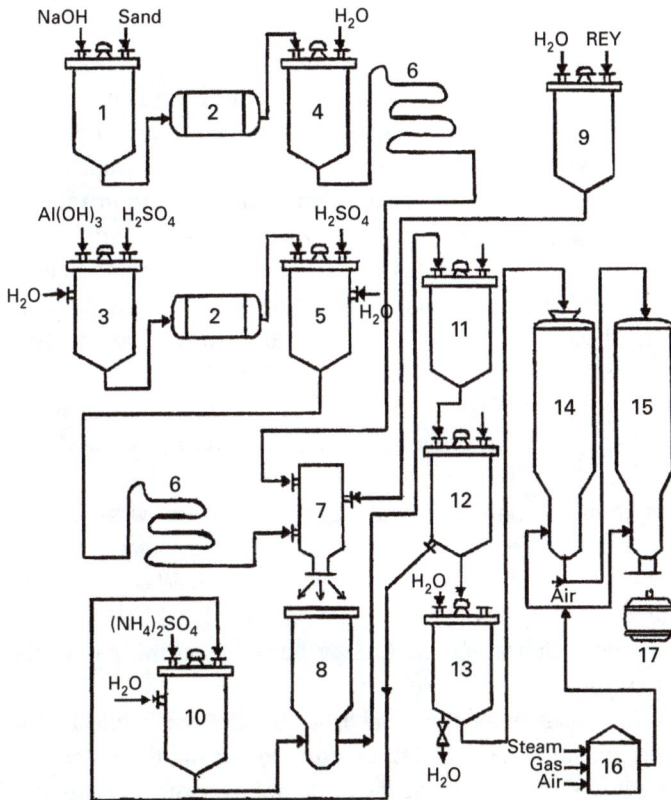

Fig. 2.122: Microspherical catalyst synthesis. (From [52].) 1, vessel for water glass synthesis; 2, filter; 3, vessel for the preparation of aluminum sulfate solution; 4, vessel for the preparation of water glass solution; 5, vessel for preparation of aluminum sulfate solution with sulfuric acid; 6, coolers; 7, mixer for preparation of the sol of zeolitic aluminosilica gel; 8, column for preparation of microspherical gel particles; 9, vessel for rare earth zeolite suspension; 10, vessel for water solution of ammonium sulfate; 11, vessel for gel syneresis; 12, vessel for gel activation; 13, vessel for washing out microspherical gel particles from sulfate and other ions; 14, drying column; 15, column for calcination; 16, burner for flue gases generation; 17, storage.

In vessel 1 (Fig. 2.122) sand, water and sodium hydroxide are introduced followed by the preparation of water glass at 98–110 °C, which is filtered to remove unreacted sand particles (pos.2) and transported to vessel 4 where the solution of a needed concentration is prepared. Aluminum sulfate is first prepared in vessel 3, then filtered and transported to vessel 5, where a working solution of aluminum sulfate is prepared by adding sulfuric acid. This is necessary as otherwise the solution containing the required amount of alumina is too basic, which will result in the generation of a zeolite with overly large pores and lower bulk density. Vessel 9 contains water suspension of rare earth zeolite Y. Solutions from vessels 4, 5 and 9 are routed to a mixer-sprayer 7, where water glass, aluminum sulfate and the rare earth zeolite Y suspension are mixed at the ratio to assure pH 7.5–8.6. Microdroplets of 2–5 nm leave vessel 7 and enter the column 8, which is filled (~3 m) with transformer oil. The microdroplets pass through the column at ~ 10–12 °C to form the gel. The large column diameter (1.5 m) is needed to prevent sticking of the gel to the vessel walls. A droplet passes through the transformer oil in 8–11 s, while 5–8 s are needed for coagulation. The coagulation rate increases with temperature increase and concentration of the parent solutions and depends on pH and presence of electrolytes. As mentioned above, an acid is added to the mixture to ensure the required coagulation rate. Passing through oil, a gel of microspherical particles is formed that contains 90% water and 10% dry substance. Syneresis (expulsion of a liquid from a gel) is done in vessel 11 at 40–50 °C for 6–24 h. Replacement of sodium ions by ammonium occurs in vessel 12 at 50–80 °C for another 6–12 h. In order to prevent removal of Al^{3+} and rare earth ions during washing (pos. 13) they are added in small amounts to the washing step while sodium, ammonium and sulfate ions are washed out. Washed gel particles can be treated with water solutions of surfactants to prevent the collapse of the gel structure during drying (pos. 14). Washing is followed by drying of the zeolite (pos. 14) and calcination (pos. 15). The flue gases needed for drying and calcination are generated by burning natural gas in the presence of steam and air (pos. 16).

During drying the flue gases enter the dryer from the bottom, while zeolite suspension falls in the opposite direction. Temperature along the dryer changes from ~315–325 °C at the top to 475–485 °C at the bottom. This process leads to a material with ~8–10% water. Calcination can be performed in a fluidized bed (Fig. 2.123). The vertical vessel for calcination has an extended upper part, while the bottom part has an inlet for flue gases. The amount of flue gases is controlled to allow fluidization, while at the same time preventing carryover of smaller particles. Calcination starts at 600–650 °C for 10 h, followed by a decrease in temperature of the flue gases leaving the burner. At 250–300 °C the fuel is shut off, while air is still added. The gradual decrease of the catalyst temperature for 4–6 h to 80–90 °C allows the transfer of the catalyst to metallic drums for storage.

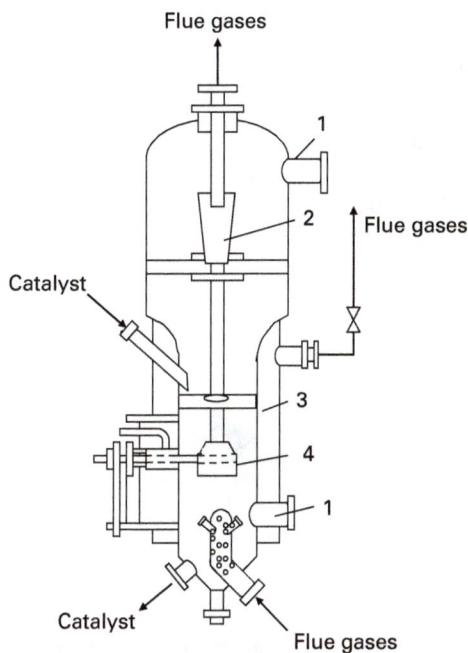

Fig. 2.123: Fluidized bed calcination unit. (From [48].)

2.3.5.7 Metal organic frameworks

These materials with a very high surface area ($1,800–3,000$ m^2 g^{-1}) have inorganic building blocks interconnected with organic linkers. The organic units are typically mono-, di-, tri-, or tetravalent ligands. The structure and properties of the metal organic framework (MOF) are influenced by the choice of metal and the linker. Similar to zeolites, MOFs exhibit shape selectivity with more flexibility of chemical composition.

MOFs are synthesized by hydrothermal or solvothermal methods, although, contrary to zeolites, organic ligands remain intact. A variety of options for post-synthetic modification of MOFs is thus available. One example of the use of MOFs in catalysis are 2D, square-grid MOFs containing single Pd(II) ions as nodes and 2-hydroxypyrimidinolates as struts (Fig. 2.124). If all the Pd centers are catalytically active in reactions such as Suzuki C–C coupling, when redox oscillations of the metal nodes between Pd(+2) and Pd(0) intermediates are needed to accomplish the reactions, these changes in the coordination number would lead to destabilization and potential destruction of the original framework.

For hydrogenation reactions on MOFs it is also possible that reactions occur on the surface of MOF-encapsulated palladium nanoparticles.

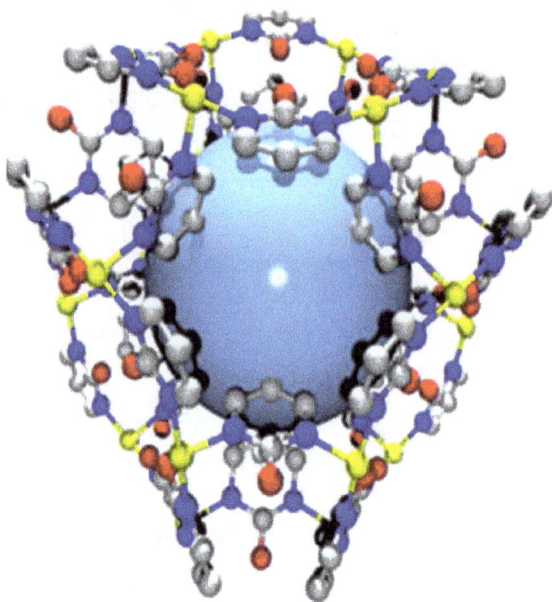

Fig. 2.124: A palladium 2-hydroxypyrimidinolate metal organic framework (MOF) used to catalyze efficient oxidation of allylic alcohols to aldehydes. (From [63].) blue, N; red, O; yellow, Pd; gray, C; light blue, Cl.

Besides high costs, the drawbacks of MOFs are air/moisture-sensitivity and scaling up issues associated with their low-temperature stability, which limits the application of MOFs in various catalytic reactions.

2.3.5.8 Mesoporous materials

Because of the limitations caused by zeolites having pores below 1.1–1.5 nm there is an interest in materials with characteristics similar to zeolites in terms of acidity but with larger pores. Such materials could separate and can catalyze reactions with larger molecules, which are otherwise not able to penetrate into the pores of zeolites. From the point of view of synthesis the size of the templating molecule determines whether ordered mesoporous materials or zeolites are formed. Isolated short alkyl chain quaternary ions lead to formation of microporous materials, while self-assembled surfactant structures with long alkyl chain quaternary ions (such as $C_{16}H_{33}(CH_3)_3NH_3^+ Br^-$) the direct formation of mesoporous molecular sieves (Fig. 2.125). In solution, surfactants can self-assemble to form micelles, rods, sheets and 3D structures. Synthesis of mesoporous materials is similar to zeolites, as it is also conducted in basic media.

One of the most famous examples of mesoporous materials is MCM-41 silicate, which is highly ordered and even gives a diffraction pattern although the glass-like

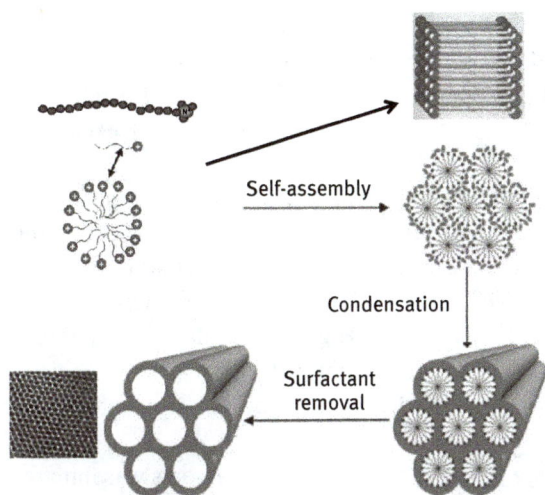

Fig. 2.125: Formation of mesoporous materials. (Modified from [64].)

silicate walls are not crystalline. The pore size distribution in amorphous MCM-41 is usually quite narrow although not as tightly defined as for zeolites.

Acidity of MCM-41 is much less than of zeolites, which along with rather expensive templates somehow hinders potential use of this material as a catalyst despite hundreds of published scientific papers.

Hexagonal mesoporous silica SBA-15 and cubic cage SBA-16 are two other interesting materials that belong to the group of mesoporous materials. Hexagonal mesoporous silica SBA-15 is synthesized in acidic media (pH 1–2) using a nonionic tri-block copolymer, such as pluronic P123 $(PEO)_{20}(PPO)_{70}(PEO)_{20}$ or HO $(CH_2CH_2O)_{20}$–$(CH_2CH(CH_3)O)_{70}$–$(CH_2CH_2O)_{20}H$ containing poly(oxyethylene) PEO and poly(oxypropylene) PPO units. Cylindrical micelles are formed with silica particles assembling around the template with hydrophobic heads and hydrophilic tails.

SBA-15 has thicker pore walls than MCM-41 and possesses monodisperse pores of 5–20 nm, while microporosity can be also introduced. Thermal and hydrothermal stability is also higher than for MCM-41. Synthesis of mesoporous SBA-15 proceeds through dissolving tri-block copolymer, preparation of gel in acidic media, ripening of the gel at 373 K for 24 h, filtration, washing with distilled water, drying at 373 K and calcination at 813 K.

2.3.5.9 Layered compounds
Layered compounds can be categorized into three classes: (a) neutral ones (e.g., brucite, phosphates and chalcogenides), (b) compounds containing negatively charged layers with compensating cations in the interlayer space (e.g., montmorillonite) and

(c) materials containing positively charged layers with compensating anions in the interlayer space, such as layered double hydroxides (e.g., hydrotalcites).

A specific feature of cations or anions in the interlayer space of this relatively high surface area layered materials is related to their ability to ion exchange. Another feature is related to swelling properties.

Cationic clays, which may exhibit both Brønsted and Lewis acid sites, contain two types of sheets comprising $Si(O,OH)_4$ tetrahedron and $M(O,OH)_6$ octahedron ($M = Al^{3+}$, Mg^{2+}, Fe^{3+} or Fe^{2+}) as building units. They are typically synthesized from minerals possessing negatively charged aluminosilicate layers compensated with interlayer cations. The Brønsted acidity is due to OH groups, while Lewis acidity is associated with Al^{3+} substituting Si^{4+} in tetrahedral sheets. Another parameter influencing hydrophilic–hydrophobic properties is the type of an exchangeable cation.

Pillaring of clays to develop large pore and strongly acidic materials was initiated by replacing the interlayer exchangeable cations first in smectites with tetraalkylammonium ions. Organic pillared clay minerals are, however, thermally unstable at high temperatures (above 250 °C), leading to the interlayer collapse. Below the decomposition temperature, organic pillared clays can be, however, applied in catalysis and adsorbents. Intercalation of clay minerals with inorganic species allows to generate pillared clays, which retain their micro- and mesoporosity after heating above 300 °C.

Natural and modified clays have found applications in various acid-catalyzed reactions, including cracking and alkylation of aromatics, as well as in bulk and fine chemical synthesis.

Hydrotalcites or anionic clays (Fig. 2.126) are layered double hydroxides of general formula $[M^{2+}_{(1-x)}M^{3+}_x(OH)_2]^{x+}(A^{n-}_{x/n})\, yH_2O$, where M^{2+} is typically Mg (or Ni, Zn, Cu), and M^{3+} is typically Al, but can be Mn, Cr, Co or Fe, while A is any anion (e.g., CO_3^{2-}, NO_3^-, F^- and Cl^-). They are formed in nature by weathering of basalts or from precipitation in saline water. A typical example of a hydrotalcite is $Mg_6\,Al_2CO_3\,(OH)_{16}\cdot4(H_2O)$. The main application of hydrotalcites is in base-catalyzed organic reactions.

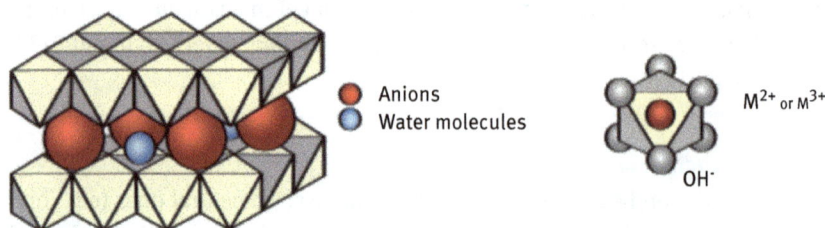

Anions
Water molecules
M^{2+} or M^{3+}
OH⁻

Fig. 2.126: Structure of hydrotalcite.

2.3.5.10 Ceramic and metallic monoliths and foams as catalyst supports

Ceramic and metallic monoliths are typically used as catalyst supports [65]. Ceramic monoliths are made from cordierite $2MgO \times 2Al_2O_3 \times 5SiO_2$ or mullite $3Al_2O_3 \times SiO_2$ by extrusion with subsequent firing or by corrugation. This is followed by washcoating with γ-, θ- and δ-alumina, enhancing the original surface area of just few m^2/g to approximately 100 m^2/g. After washcoating again firing at an appropriate temperature is performed. The components, having a similar thermal coefficient of expansion, are able to withstand high operation temperature (1,200–1,300 °C) as well as heating and cooling cycles without disintegrating. Typical cordierite honeycombs contain 400–600 channels per square inch with the wall thickness of approximately 0.15 mm.

Metallic monoliths, made of Fe-Cr-Al alloy by corrugation, have thinner walls (ca. 0.05 mm), better warm-up properties and are more mechanically strong than cordierite. Corrugation is followed by rolling up or folding into monoliths of required size and shape. A suitable oxide layer should be formed to allow adhesion of the alumina washcoat.

As already mentioned previously the surface area of cordierite is low and the active metal cannot be deposited directly on the monolith. Therefore, there is a need for applying wash-coat to increase the surface area and to deposit the active component, which is typically silica or alumina, on the wash-coat. A wash-coat thickness varies in the range of 15–100 μm providing 15–30 $m^2\ g^{-1}$ surface area. A colloidal solution, for example, of bohemite alumina, can be used for wash-coating.

In addition to alumina, the slurry contains stabilizers and binders. Stabilizers such as ceria, zirconia and barium oxide are added to the washcoat to prevent phase changes at the higher temperatures or to improve catalytic performance. In three-way automotive exhaust catalysts, cerium oxide acts as an oxygen sink absorbing it during lean operation and promotes the water-gas shift reaction generating hydrogen and thereby improving catalyst performance under rich conditions. The same function of oxygen absorption under lean conditions is provided by barium oxide. Zirconia in the washcoat composition stabilizes and serves in the three-way catalysts as a support for rhodium.

To ensure chemical and physical binding of the washcoat of the honeycomb surface and filling the large macropores of the ceramics, alumina in slurry should be ground, providing a size distribution mainly in the range of 1–10 mm that is compatible with the pores of the ceramic wall.

The pH is usually controlled when preparing the colloidal solution and the active components can be incorporated both directly in the slurry or after the washcoating. Typically, the slurry contains 10–20 wt% of the solid material depending on the monolith cell density. The laboratory procedure of wash-coating comprises first dipping the monolith into the wash-coat slurry with forced flow. The wash-coat slurry contains typically more than the calculated desired amount of the wash-coat. Dipping is followed by blowing the extra slurry out of the channels, drying and calcination of the monolith. If the desired amount of the wash-coat (usually 20 wt%) is not reached the procedure is repeated.

Often wash-coating is non-uniform along the channel unless appropriate drying methods are applied (Fig. 2.127).

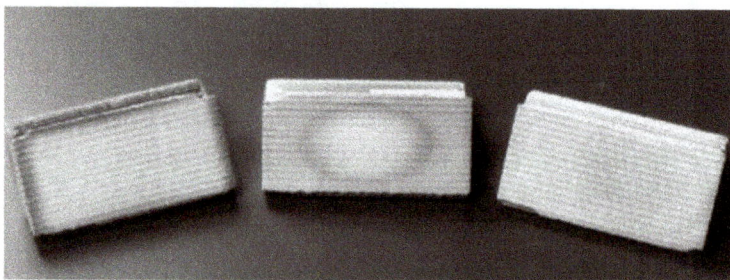

Fig. 2.127: Photograph of the cross section of monolithic catalysts showing the impregnation profiles obtained after static drying (left), freeze drying for 1 h, followed by static drying (center), and freeze drying for 24 h (right). (Reproduced with permission from [66]).

In general, it can be stated that detailed information on the preparation of monoliths at the commercial scale is typically not released by the catalyst manufacturers keeping it proprietary.

Metallic or ceramic foam are special types of structured catalysts with the advantages, that a shape required to fit the reactor size can easily be produced using a template (Fig. 2.128).

Fig. 2.128: Metallic foams. (From [67].)

Ceramic foams have several other important advantages besides the ability to match the shape and size of the reactor, such as much lower than packed beds' pressure drop and much larger external surfaces allowing higher effectiveness factors and better heat transfer. The latter is important in avoiding local overheating and hot spots. Ceramic foam catalysts can be thus advantageous in case of strongly exothermic or endothermic reactions. The open pore structure (Fig. 2.128) can also help to control selectivity.

Lower solid loading when compared with packed beds (i.e., 10–20% vs 45–55%, respectively) inevitably gives a lower productivity per reactor volume. This should be compensated by better mass transfer within the catalyst layer, larger loading or even more active catalytic phase.

Because of a more complex structure and a laborious procedure to prepare foams discussed below, the costs of manufacturing foams are apparently higher than that of more conventional catalyst shapes.

Complications in utilization of ceramic foam catalysts are related to their brittle character and thus a potential breakage during manufacture and loading into a reactor. Moreover, long lengths of preformed ceramic foam might be troublesome to charge in not perfectly straight reactor tubes.

Ceramic foams have lower pressure drop compared to packed beds, and compared to conventional monoliths the flow pattern in gas–liquid system is closer to that of packed beds.

Metallic foams could be either closed-cell or open-cell materials. The most common way to prepare closed-cell foams is to create gas bubbles in molten metal by gas injection. Stabilization agents are introduced into the melt to prevent pore coalescence. The size of the pores, or cells, is usually 1–8 mm. Open-celled metal foams are usually replicas that use open-celled polyurethane foams as a skeleton when first physical vapor deposition of metal is made onto reticulated polyurethane foam followed by burn-off from the polymer foam.

Metal foams could be based on aluminum, nickel, chromium, iron or alloys. Because of the large surface area, high thermal conductivity, mechanical durability and corrosion resistance metal foams can be used as catalyst supports offering increased heat and mass transfer and reduced diffusion resistance. Even if these materials have a potential to replace some conventional pellets or packing materials the introduction of metallic foams into industry requires more quantitative data on heat transfer coefficient, temperature profiles, stability and especially catalytic data to support more widespread use of these foams.

Ceramic and carbon forms are also produced commercially. Mechanical strength of these foams is lower compared to metal and cordierite monoliths. Polyurethane foam templates with different pore size (Fig. 2.129) can be used for the preparation of ceramic foams enabling unlimited shape and size combinations.

In order to prepare a ceramic foam block (Fig. 2.9) a defined piece of template is dipped into the initial water solution, containing, for example, alumina as the main compound and additives such as magnesia, titania and polyvinylalcohol. Uniform infiltration of the polymer by the aqueous slurry, composed of the ceramic particles with the 0.1–10 µm diameter particles along with wetting agents, dispersion

Fig. 2.129: Ceramic foams: polyurethane templates.

stabilizers and viscosity modifiers, is challenging. For better adherence, the polymer foam might be pretreated (etched). Improved wettability needed for better quality of the ceramic foam can be achieved by applying mechanical pressure or by vacuum processing. Shear thinning of the fluid (i.e., the non-Newtonian behavior with viscosity decreases under shear strain) is necessary for polymer template coating.

Low-viscosity suspensions are mainly used in dip coating and excess slurry is removed by blowing air through the foam or by squeezing and kneading. The wet foam is then dried and calcined in air at temperatures of 1,000–1,700 °C. Repeated slurry coating is carried out with the green form for adjustment of the strut thickness or filling cracks for crack filling or for functionalization of ceramic foams. Recoating might be carried out with the so-called green foam prior to calcination or after an intermediate sintering of the ceramic foam.

The laboratory-scale preparation procedure thus includes dipping, drying, re-dipping, squeezing, centrifuging and air blowing followed by calcination of the impregnated template. During the latter stage parameters such as ramping, temperature and time are important in preparing the desired material. During the calcination the template is burned and a replica of the template structure remains consisting of α-Al_2O_3 and additives. Firing is a delicate process and can be accompanied by a foam block bending. The simplified scheme of the replica method for synthesis of foams is presented in (Fig. 2.130).

The final steps in the preparation of the support are first wash-coating the foam with a high surface area support material and then deposition of the active component, which will be discussed in Section 2.3.6. The wash-coating of foams is principally made in the same way as the wash-coating for cordierite monoliths.

2.3.6 Supported catalysts

In supported catalysts, catalytically active compounds are attached to a support with a large specific surface area. Industrial preparation of catalysts is considered to be strategic and thus details of preparation are typically not disclosed or are

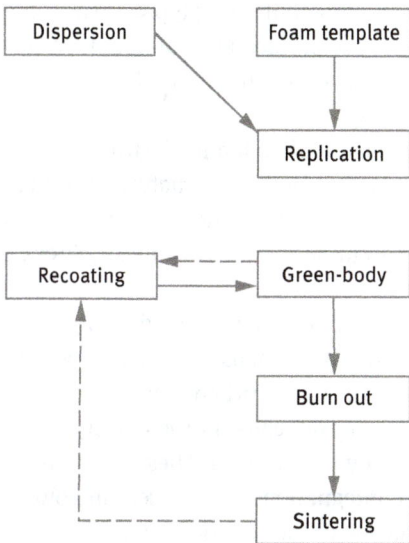

Fig. 2.130: Processing scheme for foam manufacturing (Reproduced with permission from [68]).

protected by patents. Patents typically feature broad claims and thus the exact preparation procedures applied in industry are not revealed. Deposition of the precursor on the support surface can be done by using aqueous solutions and "wet" methods of ion exchange, adsorption, precipitation and impregnation. Alternatively, dry methods (i.e., atomic layer deposition) involving the gas-solid interface could be applied. Initial deposition is followed by transformation of the precursor into the required active phase (metal, oxide, sulfide). An overview of the methods applied for preparation of supported metals is given in Tab. 2.6.

Tab. 2.6: An overview of the methods for preparation of supported metals (Reproduced with permission from [58]).

Synthesis method	Metal loading	Metal dispersion	Advantage	Challenge
Colloidal synthesis	Low–medium	Medium	Controlled size/ shape	Anchoring on support
Chemical vapor deposition	Low	Monolayer	Monolayer control	Special equipment needed
Ion adsorption	Low	High	Atomic dispersion	Gradients over support
Deposition precipitation	High	High–medium	Reproducible	Limited applicability
Coprecipitation	Very high	Medium	Very high loadings	Catalyst activation
Impregnation and drying	High	Medium	Widely applicable	Limited reproducibility

Placement of a support in contact with a solution containing a precursor leads to electrostatic interactions, chemical binding to the surface and formation of surface compounds. In addition, the support could undergo dissolution, at least partially.

An ion exchange method involves the replacement of an ion in electrostatic contact with the support by another ion. Because of the equilibrium nature of ion exchange a single operation might not be sufficient and several successive operations should be adopted. Alternatively, the exchange solution is fed continuously through the solid bed.

Adsorption of an active precursor on charged surfaces is typically done from liquid and the precursor should be selected to have a charge opposite to that of the support. The latter depends on the isoelectric point of the oxide, pH and ionic strength of the solution. The distribution of the active precursor depends on the density of the adsorption sites on the surface, which is affected by the pH level. These methods involve a subsequent washing step, contrary to impregnation, when a certain volume of a precursor containing solution is put in contact with the support.

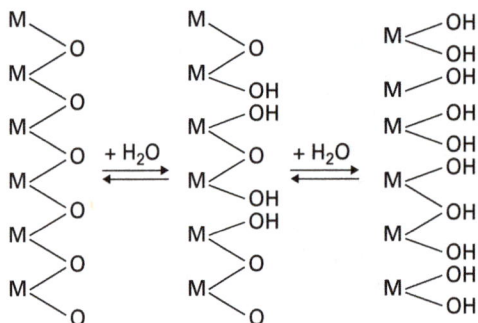

Fig. 2.131: Reaction of water with support hydroxyl groups.

2.3.6.1 Adsorption

Reaction of water with catalyst supports leads to hydroxylated surfaces (Fig. 2.131) and formation of $M-O^-$ or $M-OH_2^+$ species.

The surface is thus charged depending on the solution pH. Surface charge is equal to zero (isoelectric point or point of zero charge) at some particular value of pH depending on the support pH. An increase in the pH levels results in the hydrolysis reaction of the surface OH groups shifting equilibrium from left to the right hand side on Fig. 2.131. At a pH larger than the pzc the surface is negatively charged, welcoming the adsorption of cations. For example, amine complexes $[Cu(NH_3)_4]^{2+}$ are very suitable for adsorption on silica at an elevated pH, when the surface is negatively charged.

The pzc of γ-alumina is between 7 and 9, so it can adsorb cations and anions. Special care should be taken during catalyst preparation to the charge of the surface and the charge of species to be deposited. For example, at low pH when the alumina surface is positively charged platinum can be adsorbed from hexachloroplatinic acid $H_2[PtCl_6]$ as anions $[PtCl_6]^{2-}$. Fixation of anions on the surface is typically fast and does not limit the overall kinetics. In the case of platinum deposition on alumina from $H_2[PtCl_6]$, because of strong interactions between the complex and the surface, platinum tends to adsorb strongly and rapidly, leading to heterogeneous distribution of the metal on the support. In order to avoid this, a competitor (HCl) is added in industrial impregnations to ensure homogeneous distribution of platinum. The conventional interpretation is that effective concentration of the platinum precursor in the solution is influenced by the following equilibrium with Cl^- playing the key role:

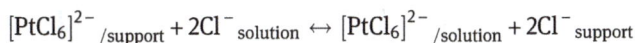

$$[PtCl_6]^{2-}{}_{/support} + 2Cl^-{}_{solution} \leftrightarrow [PtCl_6]^{2-}{}_{/solution} + 2Cl^-{}_{support}$$

An alternative explanation [69] assumes that the protons in HCl are needed to charge the surface sufficiently for adsorption to occur.

In addition, HCl acts as a stabilizing agent for the coordination sphere of platinum, as otherwise $[PtCl_6]^{2-}$ can lose several chloride ligands.

A successful recipe for one support will not necessarily work for another one. Thus at pH 8 platinum could be adsorbed on silica from the cation using $[Pt(NH_3)_4](NO_3)_2$, while the same precursor will be inefficient for alumina (Fig. 2.132). Isoelectric points of some typical supports are given in Tab. 2.7.

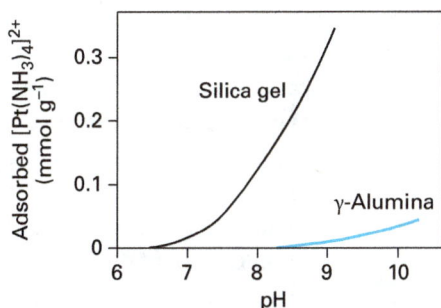

Fig. 2.132: Adsorption of $[Pt(NH_3)_4]^{2+}$ from the solution on the surface of silica gel and γ-Al$_2$O$_3$. (Redrawn from [70].).

The discussion above was mainly concerned with hydroxyl groups on the oxide supports, however, the same careful attitude towards adsorption is valid for carbons supports, where the surfaces are also charged because of the presence of various functions groups, such as carboxylic ones (Fig. 2.133). Thus $H_2[PtCl_6]$ is used at

Tab. 2.7: PZC of supports.

Support	SiO$_2$	CeO$_2$	TiO$_2$	Fe$_2$O$_3$	ZrO$_2$	Al$_2$O$_3$	ZnO	Co$_3$O$_4$	MgO
pzc	2	6	6	6.5–6.9	6.7	7.5–9	8.5–10	7.5	12.1–12.7

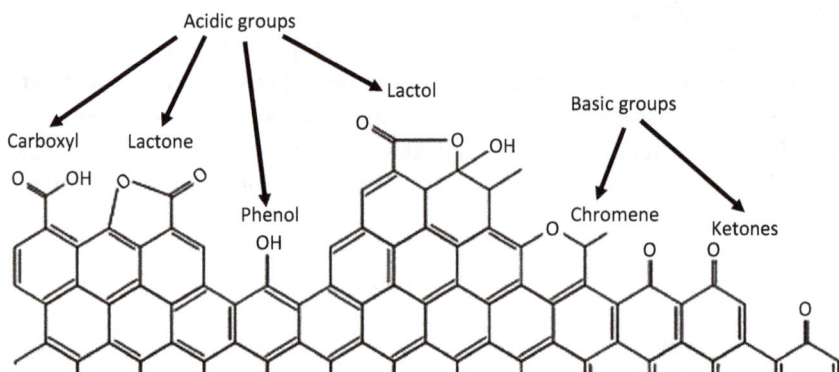

Fig. 2.133: Functional groups on carbon support.

low pH on positively charged surfaces, while [Pt(NH$_3$)$_4$](NO$_3$)$_2$ should be applied at high pH.

Point of zero charge of the support and pH are thus important parameters influencing the electrostatic interactions between the metal complex and the support. Regulation of pzc could be done by changing temperature (pzc is increasing for alumina and decreasing for silica with temperature rise) and/or selective doping of the support surface. In selective doping the dopant should not deactivate the catalyst. As an example of selective doping, in alumina modifications with Mg^{+2} or F$^-$ ions change pzc in the opposite directions. Even if pH is an important parameter, it should be noted that measurements and control of pH are difficult tasks. Changes in pH as mentioned above can cause dissolution of support and in addition can change metal speciation, which leads to adsorption of other types of species. Such changes in chemistry of adsorption should be thus taken into account during catalyst preparation both at the experimental level influencing post-treatment and at the theoretical level of modeling.

The type of ligand, in particular its charge and bulkiness, influences the nature and the strength of the metal complex-support interactions. For example, shielding a bulky ligand can decrease the interactions between the metal complex and the charged surface. Moreover, a hydrated complex could be too large to enter the support pores.

Several metal precursors are suitable for catalyst synthesis. It should be noted that metal salts (Tab. 2.8) are typically more expensive than the bulk metals. As chloride containing precursors could suppress catalyst activity, thus nitrate precursors are applied or alternatively efficient washing methods should be used to wash away chloride.

Tab. 2.8: Metal cations and anions used in catalyst preparation.

Metal	Anion	Cation
Cobalt		$Co(NH_3)_4^{++}$
Nickel		$Ni(NH_3)_4^{++}$
Copper		$Cu(NH_3)_4^{++}$
Ruthenium		$Ru(NH_3)_5Cl^{++}$
Rhodium	$RhCl_6^{3-}$	$Rh(NH_3)_5Cl^{++}$
Palladium	$PdCl_4^{2-}$	$Pd(NH_3)_4^{++}$
Silver		$Ag(NH_3)_2^{++}$
Iridium	$IrCl_6^{2-}$	$Ir(NH_3)_5Cl^{++}$
Platinum	$PtCl_6^{2-}$	$Pt(NH_3)_4++$
Gold	$AuCl_4^-$	

In general, formation of different species with different charges should be modeled as a function of pH and other concentrations (Fig. 2.134). The surface ionization reactions are described by two constants (Fig. 2.135), which along with the surface density of hydroxyls and the dielectric constant, characterize the charging properties of an oxide defined through $pzc = [0.5\,(pK_1 + pK_2)]$.

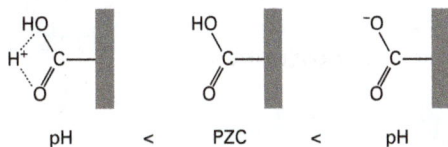

Fig. 2.134: Surface charge dependence on pH for carbon supports with carboxylic groups.

In order to model adsorption behavior (K_{ads}) a simple Langmuir type adsorption model for the metal adsorption capacity is often applied

$$a = a_\infty \frac{c_p K}{c_p K + 1} \qquad (2.37)$$

where c_p is equilibrium concentration, K is adsorption equilibrium constant and a_∞ is maximum sorption capacity. As a typical example, adsorption isotherms of H_2PtCl_6 on carbons are presented in Fig. 2.136.

Adsorption models available in the literature very seldom consider lateral interactions between adsorbed ions, site heterogeneity and oxide dissolution. Such dissolution could occur when deposition of the active agent happens at a pH far from

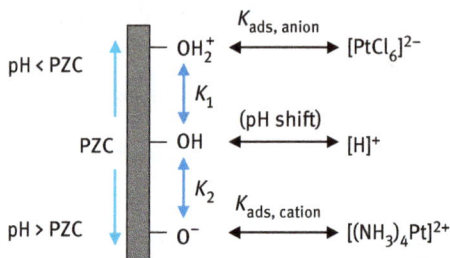

Fig. 2.135: A schematic presentation of the mechanism of electrostatic adsorption of platinum. (From [71].)

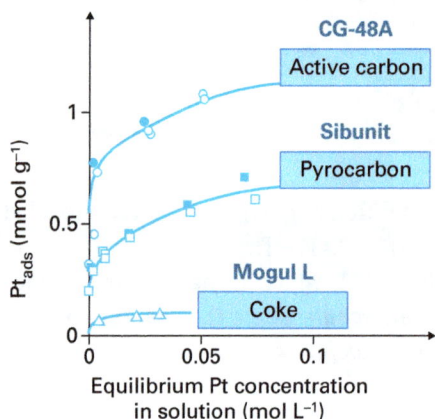

Fig. 2.136: Adsorption isotherms of hexachloroplatinic acids on carbon (From [50]).

the pzc of the support, i.e. the point of maximal stability of the solid because of minimum solubility.

Dissolution of the support could also be beneficial to a certain extent. New hydroxyl groups created because of silica dissolution in a basic medium act as adsorption sites, resulting in an increase of cations adsorption on silica at pH higher than 10.5.

It should be noted that not only the desired ion but also the associated counterion could interact with the support. Moreover, the counter ions can determine the precursor solubility.

2.3.6.2 Impregnation and drying

Impregnation, which is in a way similar to adsorption, could be done by contacting a certain volume of solution with a precursor and the support (Fig. 2.76). The method is used to obtain higher loading than in simple adsorption and to apply active precursors that do not adsorb easily on support. In this preparation method the solution of the precursor is broken down into small discontinuous elements presented in the pore of the support by gradually evaporating the solvent. The wetting time could be modified by changing the temperature of impregnation, which has an influence not only on pzc but also on the precursor solubility and viscosity of the solution. Mass transfer of the impregnated compound within the pores during impregnation and drying determines the concentration profiles. Solubility of the precursor in the solution limits the maximum loading, thus successive impregnations with drying and even calcinations between these impregnations are applied to increase the loading, as will be discussed separately in Section 2.3.6.3.

In wet impregnation an excess of solution is used and after a certain time the solid is separated and the excess solvent is removed by drying. In this non-exothermal method the pores are filled with the neat solvent prior to impregnation. When this solvent saturated support is introduced into the solution containing the precursor there is a concentration gradient between the external solution and the pores driving the precursor salt in the pores.

In dry methods (pore volume impregnation or incipient wetness) the volume of the solution containing the precursor corresponds exactly or approximately to the volume of the pores. One of the disadvantages of the method is broad particle size distribution. If pH is not adjusted and, for example, impregnation is done at conditions when pH is close to pzc, only few hydroxyl groups are available and non-adsorbing metal particles might agglomerate giving large crystallites.

In dry methods a replacement of the solid-gas interface by a solid liquid one causes a significant decrease in the free enthalpy. This results in heat release, which in general is of minor importance for impregnation. In some special cases this heat release might influence impregnation by provoking unwanted precipitation, limiting solubility or modifying the surface.

When solution is sucked into pores by capillary forces, air bubbles are trapped and could burst the support and lower the mechanical strength. This is caused by considerable forces exerted on the pore walls being in contact with these bubbles during dry impregnation. The reason is in the high pressure (several MPa) inside the compressed bubbles defined by the Young-Laplace law:

$$\Delta P = \frac{2\gamma}{r} \tag{2.38}$$

where ΔP is the pressure difference inside and outside the bubbles, γ is the liquid-gas interfacial tension and r is the liquid-gas meniscus radius.

In order to prevent or limit the bursting of the support during impregnation, a surfactant could be added to the solution or impregnation could be done in vacuum, complicating the whole procedure.

Time needed for the liquid to fill the pores is very short (a few seconds) in dry impregnation, although in practice longer operation times are used. Contrary to capillary impregnation in wet methods (diffusional impregnation) migration of the precursor into the pores is much longer and distribution of the solute is governed by diffusion into the pores (Fick law can be applied) and adsorption on the support (typically Langmuir model is used). As already mentioned above, adsorption depends on the adsorption capacity and the equilibrium constant of adsorption.

Phenomena during wet and dry impregnation are illustrated in Fig. 2.137. The characteristic time t needed to attain equilibrium in wet impregnation is proportional to the square of a pellet radius, thus is increasing with the pellet size and is far from instantaneous but requires several hours.

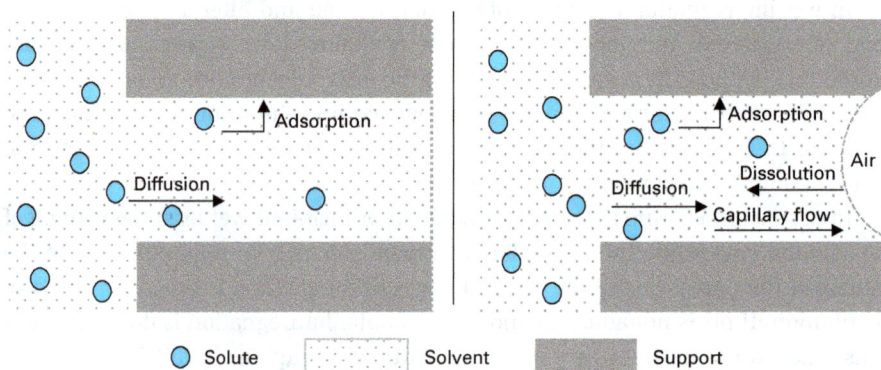

Fig. 2.137: Phenomena of transport involved in (left) wet impregnation and (right) dry impregnation. The solute migrates into the pore from the left to the right of the figures. (Redrawn from [72]).

In dry impregnation pressure-driven capillary flow of the solution inside the empty pores is described by the Darcy's law:

$$N_{l,s} = -C_{l,s} \frac{KK_{l,eff}}{\eta_l} \Delta P_l \qquad (2.39)$$

where $K_{l,eff}$ is the relative permeability of the liquid-phase, η is its viscosity, K the intrinsic permeability, and P_l is the liquid-phase pressure that is equal to the local gas pressure minus the capillary pressure P_C. Influence of viscosity and concentration on the diffusion is complex. High viscosity associated with high concentrations prevents the precursor diffusion according to eq. (2.39), while at the same time a high concentration favors the diffusion of the solute toward the center of the pellet.

Quantification of the conditions leading to profiling during adsorption and impregnation can be done by considering the characteristic time t_C at which the solution completely fills the pore volume:

$$t_C = \frac{8r^2\eta}{r_C\gamma\cos\theta} \tag{2.40}$$

where r is the grain radius, θ is the wetting angle, γ is the liquid-gas interfacial tension, η is the liquid viscosity, r_c is the pore radius. Uniform distribution can be achieved when the impregnation time is a priori longer than the time t_c of capillary impregnation. Otherwise distribution will be non-uniform.

For wet impregnation the characteristic impregnation time t_D is given by:

$$t_D = \frac{r^2(1+P)\tau}{D_M\varepsilon} \tag{2.41}$$

where τ is the pore tortuosity, D_M is the molecular diffusion coefficient of a component in the solution, τ is the porosity and P is the ratio between adsorbed and non-adsorbed amounts of the deposited compound in the support volume, i.e.,:

$$P = \frac{a}{g-a} \tag{2.42}$$

In eq. (2.42) a is the amount of a compound adsorbed and g is the overall amount of this compound located in the pores (trapped or adsorbed). When $P > 1$ the precursor is mainly adsorbed and the metal deposition can be defined as adsorption, while at $P < 1$ the role of adsorption is much less prominent. In practice contribution of adsorption and impregnation is comparable.

Uniform distribution is achieved when the impregnation time is substantially larger than the characteristic impregnation time t_D, which as follows from eq. (2.41) depends at $P \ll 1$ on the grain radius, pore structure and diffusion coefficient. When $P \gg 1$ the overall process might be limited by adsorption. For strongly adsorbing compounds the characteristic impregnation time could be so high that uniform distribution cannot be reached.

Otherwise, for weakly adsorbing compounds long impregnation times will lead to uniform distribution (Fig. 2.138). Such distribution is needed for slow reactions occurring in the kinetic regime without diffusional limitations.

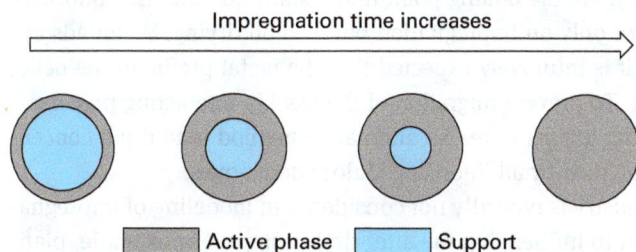

Fig. 2.138: Influence of impregnation time on active phase profile.

Increase of the precursor concentration in principle increases the penetration depth and leads to more uniform distribution. This approach is not, however, very efficient as it gives low P values and requires unnecessary high quantities of the active compound.

Another option to improve uniformity of the active phase profiling is to diminish the grain size, which obviously has technical limitations because of other requirements such as pressure drop in the reactor. Uniform precursor distribution is also achieved when precursors and competitors (HCl or acetic acid) are equally interacting with the surface.

Finally, temperature of impregnation is usually of minor importance regarding precursor distribution.

Egg-shell distribution is applied for diffusional limited reactions when the internal part of a grain is not utilized in catalysis. In general, in order to get an egg-shell distribution, one of the options is to use a precursor, which should be strongly adsorbed during impregnation. Impregnation with a very viscous solution can be recommended as an alternative technique.

An interesting option is egg-yolk distribution, when the active phase is concentrated in the center of the grain. Such distribution is obtained when there is a competitor interacting more strongly with the surface than the precursor. It could be useful for slow reactions when the particles undergo severe attrition. For example, synthesis of platinum on alumina catalysts could be considered with hexachloroplatinic acid as a precursor. In the presence of oxalic or citric acids, which adsorb stronger than the platinum precursor, adsorption of $H_2[PtCl_6]$ at the outer layer is prevented, forcing it to move further towards the grain center.

Finally an egg-white distribution with the location of the active phase away from the outer surface and the center can be used when the center of the grain is not used because of diffusional limitations, while the outer surface without the active phase is needed to protect the catalyst because of attrition or poisoning. Such distribution could be achieved by, for example, using oxalic acid prior to impregnation and then regulating the concentration of competitor and impregnation time.

Elimination of the solvent after impregnation is done by drying, or in other words heating the solvent up to the boiling point under static conditions. Uniformity of distribution depends not only on impregnation but also on drying. When adsorption is weak or moderate it is intuitively expected that the metal profile in the pellet can change during drying. To prevent migration of the weakly interacting precursor, drying can be done at room temperature. An alternative method is to dry a concentrated, viscous solution obtained by addition of cellulose derivatives.

Although drying chemistry is typically not considered in modeling of impregnation, it might be important in influencing the after-drying steps. For example, platinum complexes adsorbed on the oxide surface undergo further transformations at the drying stage. When high loaded $[PtCl_6]^{2-}$ on alumina is dried at room temperature two chloro-ligands are replaced by the two hydroxyl groups of the support, changing the complex-support bonding. When dried at higher temperatures the hydrolyzed platinum complexes could lead to polynuclear hydroxo complexes, which require reduction at higher temperatures.

Changes in the heating rates also govern drying and can lead to non-uniform distribution. The slow and fast drying regimes are presented in Fig. 2.139A. Compared with slow drying, fast drying, usually associated with high heating rate and drying temperature, tends to block all movements of the solute toward the outside of the pellet.

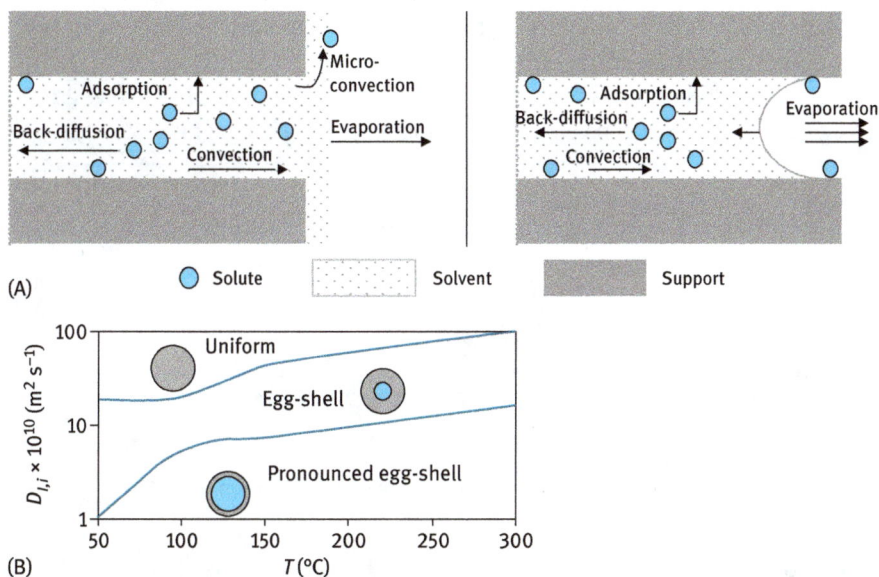

Fig. 2.139: (A) Phenomena of transport involved in (left) slow and (right) fast drying. The *solvent* migrates from the left to the right of the figures. Microconvection is not mentioned on the right side for the sake of clarity. (Redrawn from [73].) (B) Contour map of the final metal profile in the diffusion coefficient – drying temperature plane. (Reproduced with permission from [73]).

During drying the reactions between the precursor and the support could be non-uniform because of different solvent (water) amounts and non-isothermicity along the pore leading to a position-dependent particle size distribution.

It is apparently clear from the discussion above that impregnation is a complex process depending on a number of interrelated parameters (concentrations, pH, viscosity, presence of counterions and competitors, solvent). As operational parameters, pH adjustment, introduction of ligands or organic additives and impregnation duration could be used. In the drying step temperature, heating rate, gas flow and pressure, as well as drying time could be also altered to influence the resulting supported catalyst.

Because of the complexity of impregnation, which should be considered together with drying, only a few studies have been conducted with a detailed engineering analysis. As an example of an engineering approach to study impregnation and drying, the work on parametric investigation of impregnation and drying of supported catalysts by Liu et al. [73] should be mentioned. For drying, the mass balances of the drying medium (air) and the solvent (water) were considered in addition to metal dissolved in the liquid and adsorbed on the support. Simulations for the case of drying with uniform initial conditions demonstrated that the effective diffusion coefficient and the drying temperature have the most significant impact on the metal redistribution during drying. The most pronounced egg-shell profile can be observed at the point of the highest drying temperature and the smallest effective diffusion coefficient (Fig. 2.139B).

The rate of drying also influences the profile of the active phase (Fig. 2.140). Thus egg-shell profiles can be enhanced by increasing the drying rate.

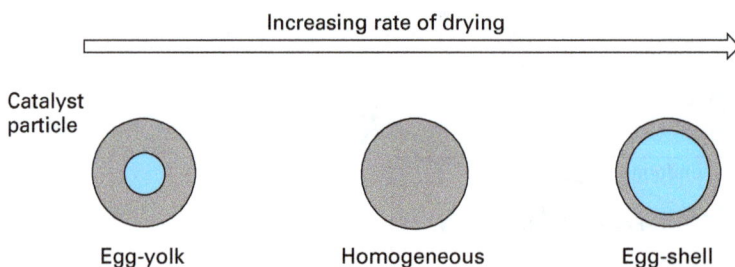

Fig. 2.140: Influence of the drying rate on the active phase profile. (Redrawn from [74]).

2.3.6.3 Multiple adsorption and impregnation

As the required amount of the active precursor cannot necessarily be dissolved in a liquid volume used in pore volume impregnation, high metal loadings could not be always achieved and multiple impregnations are needed.

Let us first analyze the mass balance for impregnation following the treatment presented by Pakhomov [50]. The initial amount of the active component precursor G_0 can be calculated as:

$$G_0 = V_0 c_0 \tag{2.43}$$

where V_0 is the initial volume of the solution with concentration c_0.

After the impregnation time τ a certain amount of the precursor g is trapped in the pores of the volume V_Σ and the concentration of the precursor in the solution changes to c_τ. Thus:

$$G_\tau = g + (V_0 - V_\Sigma) c_\tau \tag{2.44}$$

For wet impregnation in the diffusional regime and combining eqns. (2.43) and (2.44) one gets an expression for g:

$$g = V_\Sigma c_0 \frac{n}{n+1} \qquad (2.45)$$

where $n = V_0/V_\Sigma$ is the excess of the impregnation solution compared to the pore volume. For dry impregnation:

$$g = V_\Sigma c_0 \qquad (2.46)$$

The maximal amount of the active compound that could be loaded on the support during one impregnation is determined by pore volume V_Σ and the solubility c_s of the precursor:

$$g = V_\Sigma c_s \qquad (2.47)$$

Multiple impregnations are usually needed when the desired amount of the active compound is rather high (above 20 wt%).

It should be noted that since the amount of the available pore volume is related to the volume of the precursor (bulky organic salts or hydrates), this volume could be already filled after the first impregnation, while the amount of the active compound is below the requirement. In such cases thermal treatment is needed between subsequent impregnation steps to form an intermediate (for example, oxides from metal salts). Obviously, as multiple impregnations complicate catalyst preparation, the number of impregnations should be as small as possible.

Volume of either the intermediate compound or the metal after the first impregnation (followed by thermal treatment) is:

$$\Delta V = \frac{g_{1M}}{\rho} = \frac{g_{1A}}{\rho} \frac{M}{A} = g_1 \alpha \qquad (2.48)$$

ρ is density, A/M is metal content in intermediate, A is metal mass, M is mass of the intermediate, g_{1M} or g_{1A} is mass of intermediate or metal after first impregnation and $\alpha = M/A\rho$.

Free volume of pores after the first impregnation is given by

$$V_1 = V_0 - \Delta V = V_0 - g_1\alpha = V_0(1 - c_0\alpha) \qquad (2.49)$$

Free volume after i impregnation for uniform distribution and no pore blockage can be expressed in the following way

$$V_i = V_0(1 - c_0\alpha)^i = V_0 - g\alpha \qquad (2.50)$$

The value of α could be easily calculated for the intermediate compounds when density is known or experimentally determined through estimation of the pore volumes after impregnations. When α is known all the required parameters to perform multiple impregnations could be calculated, such as the number of impregnations i needed to introduce g grams of metal per gram of support at c_0:

$$i = \frac{lg\left(1 - \frac{g\alpha}{V_0}\right)}{lg(1 - c_0\alpha)} \tag{2.51}$$

the amount of metal g loaded in i impregnations:

$$g = V_0 - \frac{(1 - \alpha c_0)^{i-1}}{\alpha} \tag{2.52}$$

the volume of solution needed for i impregnations:

$$V_p = V_0 \sum_{1}^{i-1} (1 - c_0 a)^{i-1} \tag{2.53}$$

and the minimum volume of pores needed to introduce the needed amount of metal g:

$$V_{0, min} = \frac{g\alpha}{1 - (1 - c_0\alpha)^i} \tag{2.54}$$

2.3.6.4 Industrial implementation of impregnation

Industrial-scale impregnation is done in both common type reactors and in dedicated equipment. The conveyor type of wet impregnation equipment (Fig. 2.141) has been used for large-scale impregnation. Conceptually, in this method the catalyst is loaded in buckets that are moving down into the impregnation basin. The impregnation time depends on the conveyor speed and the length of the basin. The concentration of the impregnation solution is kept constant.

For dry impregnation the impregnating solution is sprayed over the support (Fig. 2.142), which is rotating.

 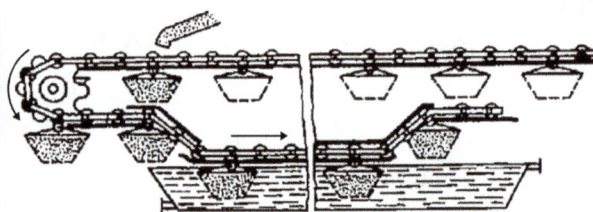

(A) (B)

Fig. 2.141: Conveyor wet impregnation. (A), from [75], (B) from [48]. Explanations are given in the text.

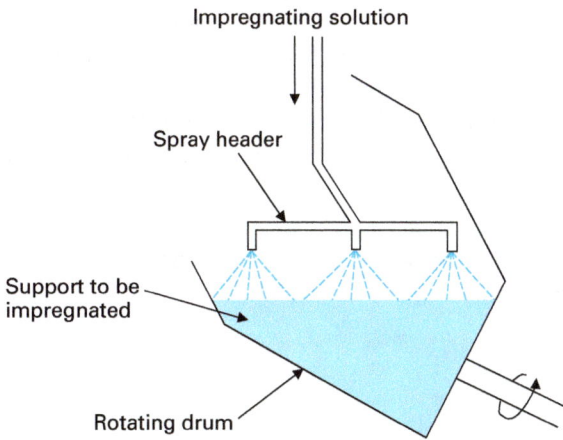

Fig. 2.142: Incipient wetness impregnation. (Redrawn from [38].)

There are also reactors operating batch-wise, which allow impregnation and drying and calcination in the same reactor. The reactor with the volume of 0.25–10 m³ can have a capacity of 100–2,000 kg/day, operating with air as a drying and calcination agent at, respectively, 80–180 °C and up to 600 °C. An example of synthesis of zinc acetate on a carbon support is illustrated in Fig. 2.143. Activated carbon is charged through the pipe pos. 1. For a reactor diameter of 2 m, the height of the support bed is approximately 0.6–0.7 m. Prior to impregnation, the support is dried by air through the distribution grid 4. The bed is fluidized by adding

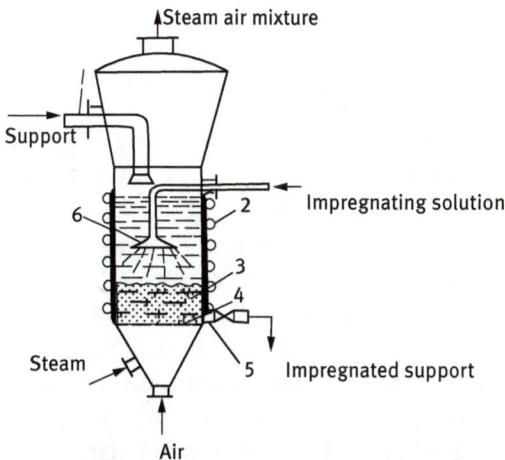

Fig. 2.143: Reactor combining impregnation, drying and calcination [48]. 1, pneumatic tube for loading the support; 2, heating elements; 3, fluidized bed of the support; 4, distribution grid; 5, discharging outlet; 6, spraying.

approximately 20–30% more air than the fluidization threshold. After drying of the support, the solution of zinc acetate is sprayed through distributer (pos. 6). Intensive mixing guarantees efficient impregnation. Air steam mixture containing dust is taken from the reactor top. Reactor is heated by steam through spiral elements pos. 2. After impregnation in the same reactor, the catalyst is matured, dried in air at 70–110 °C and unloaded through pos. 5.

This reactor operates batch-wise. To enhance throughput, two reactors for impregnation can be employed (Fig. 2.144), one undergoing impregnation while the support is unloaded from the second one.

Fig. 2.144: An installation with two reactors for impregnation [76].

As an example, for catalyst preparation by impregnation, a supported silver catalyst for synthesis of formaldehyde by oxidation of methanol will be discussed. The synthesis of formaldehyde can be performed in a fixed-bed reactor with silver catalysts. The global production of formaldehyde is divided by this process and alternative oxidative dehydrogenation at a lower temperature on Fe–Mo oxide. More details are provided in chapter 4. Because the methanol and air mixture is flammable, a safe process should operate either below or above the flammability limits (6.7–36.5% methanol in dry air at 100 °C). Oxidative dehydrogenation of methanol on silver is organized above the higher limit at 600–700 °C, allowing to achieve the yields of formaldehyde exceeding 85%.

The classical method of catalyst preparation by impregnation developed in the 1950s is still used, although at the beginning of the twenty-first century an alternative electrolytic preparation method was commercialized based on the discovery that the crystal growth could be controlled by the creation of a silver-diamine complex during the electrochemical synthesis step.

In the classical method silver (30–35 wt%) on pumice is prepared by impregnation of pumice by silver nitrate (Fig. 2.145).

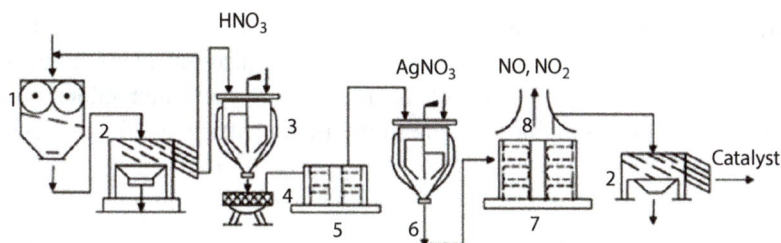

Fig. 2.145: Preparation of silver catalyst by impregnation, 1 – grinding, 2 – sieving, 3 and 6 reactors with impeller and heating jackets. 4 – filter, 5 – dryer, 7 – calcination. (Redrawn from [48].)

The large grains of the support are ground (pos.1) and sieved (pos.2). The smaller and larger fractions are disposed and returned to grinding, respectively, while the medium fraction (2–5 mm) is introduced in the reactor with a heating jacket (pos. 3). Nitric acid is added into the reactor 3 at 60–70 °C for 7–8 h order to remove iron containing impurities

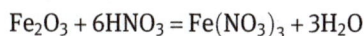

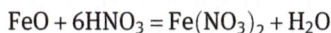

$$Fe_2O_3 + 6HNO_3 = Fe(NO_3)_3 + 3H_2O$$

$$FeO + 6HNO_3 = Fe(NO_3)_2 + H_2O$$

that otherwise lead to cracking and coking.

After filtering out and washing with water the support is dried at 100–110 °C and then impregnated in another reactor (pos. 6) with a solution of silver nitrate, with simultaneous evaporation of water at 100 °C. After impregnation that results in deposition of silver mainly in the pore mouth and on the external surface, the catalyst is first thermally treated at 400–500 °C for 1–2 h to remove NO_x followed by calcination at approximately 650 for 2–3 h. The level of NO_x emissions is approximately 0.4 ton per 1 ton of catalyst, while an approximate feedstock consumption for production of 1 ton of silver on pumice catalyst is as follows: pumice – 630 kg, nitric acid (47%) – 800 kg, hydrochloric acid – 475 kg, silver nitrate – 740 kg. The example above has mainly educational purposes as in industrial practice mainly silver gauze's are applied.

2.3.6.5 Precipitation
This method has two main variants. The first one is used when cheap catalysts are produced and the aim is to optimize the activity per volume unit. It relies on

co-precipitation of the support and the active component which gives a material that is then dried, calcined and if needed reduced. Details of this method are covered in a section devoted to the precipitation of bulk oxides (Section 2.3.3), as the $CuZnOA_2O_3$ (Fig. 2.95) catalyst described in that section can be also considered as copper and zinc oxide supported on alumina. Precipitation is caused by adding a precipitation agent such as sodium hydroxide or bicarbonate.

A short description of the nucleation phenomena was provided in Section 2.3.3. Thermodynamic analysis of deposition-precipitation should also consider the interfacial energy with the support. Because of the presence of such interactions the total free energy also includes enthalpy of the support-nanoparticles (NP) interface:

$$\Delta G^{total} = \Delta H_{NP} + \Delta H_{\frac{NP}{sup}} - T\Delta S \tag{2.55}$$

and can be significantly lowered. This in turn results in a lower concentration of the active precursor at which the crystallization rate can be measured. Solubility plots including the solubility and supersolubility curves are presented in Fig. 2.92. The solubility curve (saturation concentration) separates liquid from the solid, indicating that solubility increases with temperature. There is a concentration (supersolubility, or nucleation threshold) where the precipitation becomes abruptly measurable. Between these two lines of solubility and supersolubility precipitation does not occur within the bulk and happens exclusively on the support. Keeping the concentration of the catalytically active precursor in this range is the basis of the deposition-precipitation method.

This second method of precipitation, originally developed for high loading of the active phase, employs pre-existing supports and is preferred when the active phase precursors are expensive and deposition of a nm size active phase is needed. Obviously, precipitation in the bulk should be avoided. As mentioned above, metal hydroxides or carbonates precipitate on the support after the addition of a base to increase pH.

In order to achieve homogeneous precipitation, pH should be homogeneous (difficult to achieve during scaling up) and the concentration of the active precursor should be below the supersolubility curves.

Such homogeneous solution can be obtained in the case of deposition-precipitation with urea that slowly hydrolyzes at 70–90 °C according to the following reactions:

$$(NH_2)_2CO = NH_4^+ + CNO^-;$$

$$CNO^- + 3H_2O = NH_4^+ + HCO^- + OH^-;$$

$$CNO^- + 2H_2O = NH_4^+ + CO_2 + 2OH^-.$$

The rate of urea hydrolysis does not depend on pH as it is of first-order towards urea. The amount of urea is typically 1.5–2.5 times above stoichiometric because of ammonia volatilization.

It should be also mentioned that although urea hydrolysis takes a relatively long time there is a possibility to thoroughly mix the support suspension at a temperature where there is no reaction. This allows for preparing highly concentrated solutions. Large amounts of highly viscous suspensions of support can therefore be used with urea, separating the mixing of the ingredients for catalyst production and the generation of precipitating agents. This is especially important for the scaling up of catalyst preparation as efficient stirring is difficult to achieve in large vessels.

Deposition-precipitation with urea gives a constant level of supersaturation by maintaining a uniform concentration of OH^- in both the bulk and pore solutions. The precipitated phase is a hydroxide of the active metal. Supported Mb, V, Fe, Ni, Cu and gold catalysts could be prepared by deposition-precipitation with urea.

The drawbacks of the method are the possible formations of cyanate and nitrite in the waste water and release of NO.

Obviously, properties of the support are important in the deposition-precipitation process. Thus, large particles could be formed in the case of neutral support surfaces close to the support pzc, neutral metal complexes, lack of hydroxyl groups on the support surface at low pH or dissolution of the support at low pH. In addition, the following lead to large particles: the repulsions occurring between the anion and the negatively charged, deprotonated support surface at pH above the pzc of the support, the presence of chloride causing agglomeration of metal particles, low stability and slow adsorption rates of some metal complexes, such as $Al(OH)_4^-$.

Besides the strength of interactions between the metal precursor and the support, maturation time and redispersion also have an impact on the final nanoparticle size.

Changing the pH level of a solution of an active precursor, changing the valency of a dissolved active precursor, or decreasing the concentration of a compound forming soluble complexes with the active precursor are procedures that can be used in deposition-precipitation. Many deposition-precipitation procedures involve an increase in the pH level of a suspension of the support in a solution of the active precursor. As a significant number of catalytically important metal compounds dissolve as oxy-anions, raising the pH level cannot be used to bring about the precipitation of such metal ions, and consequently a procedure has been developed to reduce the metal in the oxy-anions to a lower valency. The metal ions of a lower valency exhibit a basic behavior and are thus soluble at low pH levels and precipitate at high pH levels.

For the supports with high pzc the use of the deposition-precipitation (DP) method is challenging. The DP method is also not very easy to perform for hydrophobic carbon. As the nature of adsorption sites for oxides and for carbon is different, the same method of pH adjustment cannot always be applied. Zero potential of silica does not change when increasing the pH from 3 to 7, whereas these changes are significant for carbon in the range of pH 2–5. Carbon surface is thus oxidized to efficiently increase the number of carboxylic groups and subsequently the loading. A particular case is preparation of carbon supported gold catalysts, which could be easily reduced during deposition, leading to large clusters of gold. In order to avoid it, gold colloidal sols are prepared with polyvinyl alcohol (PVA) followed by a reduction with NaBH$_4$, drying, and finally deposition onto the carbon support.

A simple thermodynamic model [77] that takes into account the number of hydroxyl groups on oxide surfaces during the preparation of supported catalyst particles by deposition-precipitation with urea assumes that the interfacial energy can be represented by two terms:

$$\gamma_{ss} = \gamma_{SS}^{0} + \gamma_{SS}^{electrostatic} = \gamma_{SS}^{0} + \gamma_{SS}^{0}\lambda\frac{(pH - pzc)}{pzc} \tag{2.56}$$

one γ_{SS}^{0} describing pH independent non-electrostatic interfacial energy, while the other $\gamma_{SS}^{electrostatic}$ denotes electrostatic energy between a nanocluster and the charged surface.

The latter is proportional (with λ being the cluster size independent proportionality constant) to the relative approach of pH to the point zero charge $\Delta z = (pH - pzc)/pzc$. The latter parameter could be either positive or negative (for example, for supports with high pzc, such as magnesia). This simple approach allows for predicting dependence of the cluster size on the pH at which precipitation was performed. The enthalpy for nanoparticles is a function of the number of nanoparticles multiplied by the surface free energy and the atomic volume of bulk metal. The number of nanoparticles for a hemispherical cluster can be written as a cubical equation with respect to the particle radius r. The minimal total energy with respect to this radius can be derived by taking into account the above mentioned correlations via taking a derivative $\frac{\Delta G^{total}}{dr} = 0$ giving finally:

$$r^3 - \frac{p_2}{\Delta z}r - \frac{p_2}{\Delta z} = 0 \tag{2.57}$$

With:

$$p_1 = 3\frac{(\beta RT - \gamma_{SS}^{0})}{\gamma_{SS}^{0}\pi\lambda}, p_2 = -2\frac{4\alpha\gamma\Omega}{\gamma_{SS}^{0}\pi\lambda} \tag{2.58}$$

where α is a dimensionless coefficient of an order of unity (α can be equal to 1 only for spherical particles), Ω is the atomic volume of the bulk metal, R is the gas constant, β is a dimensionless constant ($\beta \approx$ 5–8). Despite the apparent simplicity of eq. (2.57), modeling of the cluster size dependence as a function of Δz (Fig. 2.146) shows clearly that the proposed model can be used as a tool for the estimation of the metal particle size on a given support surface.

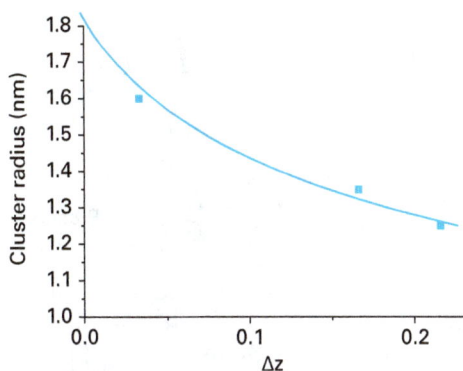

Fig. 2.146: Dependence of cluster radius on Δz for gold catalysts on titania prepared by deposition-precipitation with urea at different pH. (Redrawn from [77].)

2.3.6.6 Atomic layer deposition

Dry methods of catalyst preparation, such as chemical vapour deposition (CVD) and atomic layer deposition (ALD), being inherently more expensive than wet methods, have their advantages resulting in catalysts with controlled metal dispersion. In fact the dispersion could be very high at significant loadings. In CVD, volatile precursors are deposited in a "cold zone" directly from the gas-phase and thin films are formed on the support surface by thermal decomposition or reactions of gas compounds.

ALD uses alternating cycles of self-limiting reactions and deposition occurs in a layer-by-layer fashion. Sequential exposure of the support with surface hydroxyl groups to, for example, trimethyl aluminum $Al(CH_3)_3$ and water, results first in the formation of O-Al linkages and the generation of methane in the gas-phase (Fig. 2.147). When no surface hydroxyl groups are available the reaction terminates. The surface methyl groups can react, however, with water producing methane and another set of O-Al linkages.

Typically a constant stream of an inert gas (N_2) is flowing through a chamber at a pressure of ~0.1–10 Torr and the precursor vapors are injected into this carrier gas. A purge period is introduced between each precursor dose to avoid mixing of the chemicals, which can give non-self-limited growth. The film thickness is very precisely controlled by the number of precursor cycles rather than the deposition time, as is the case for conventional CVD processes.

For growing thin films, reaction cycles with the metal and the non-metal reactants require just seconds being performed at the same temperature. For porous materials, longer reactant exposures (hours) are needed. Moreover, longer times allow different reaction temperatures to be used for the two reactants. Organometallic compounds and metal halides are most commonly applied as metal reactants. The deposition of noble metals by ALD requires a long time because of the difficulties in removing the

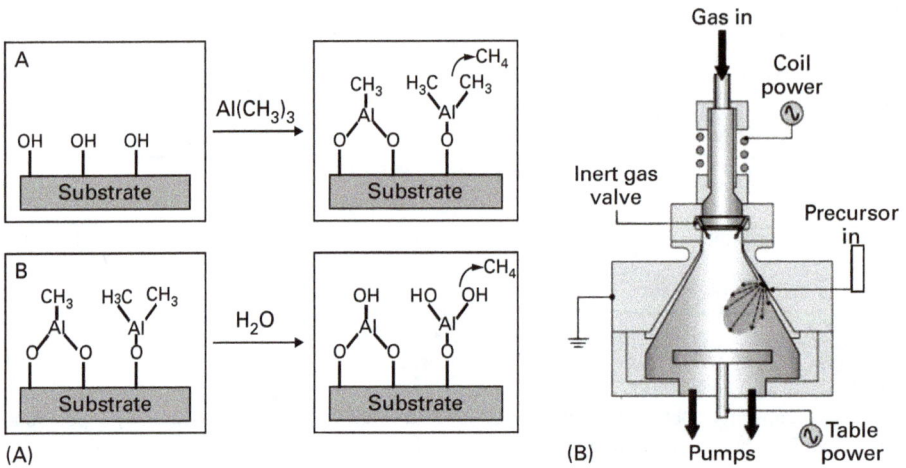

Fig. 2.147: Illustration of surface chemistry for alumina ALD. (A) (From [78]), (B) Figure courtesy of [79].

ligands from the noble metal precursor bound to an oxide surface. Poor nucleation and the ability of noble metal atoms to agglomerate results in the formation of highly dispersed metal clusters of nanometer size.

Desirable characteristics for ALD precursors include high volatility, good thermal stability at the ALD growth temperatures, and high reactivity with the other compound used for the film growth.

2.3.7 Catalyst forming operations

Several post-treatment and forming operations, which are performed during catalyst preparation, were briefly discussed in Section 2.3.1.2. Catalyst forming, which is seldom considered in academic research, is, however, extremely important for the preparation of catalysts influencing the final performance. Information about catalyst forming is typically proprietary and primarily empirical. Only a few reviews on catalyst forming are available, such as Schüth and Hesse [80]. At the same time, it is apparently clear that solid engineering approaches are needed in this field, otherwise unit operations in catalyst preparation will resemble those used centuries ago (Fig. 2.148).

Spray-drying, extrusion, pelleting and granulation are the most important catalyst forming methods to give different catalyst shapes, which define the pressure drop in a reactor. The properties of a primary solid in terms of activity and selectivity should be maintained. In addition to the primary porosity, compaction creates additional porosity that is important for mass transfer. Porogenes are added during forming, and are burned when the catalyst is calcined. These porogens that help to

Fig. 2.148: Unit operations. (A) Georg Bauer (Agricola), "De Re Metallica" (1566). (B, C) Catalyst preparation operations. Photo courtesy of Dr. I. Simakova.

improve porosity could have adverse effects. The distribution of particle size in the powder that is used for compaction can be beneficial, as the fines fill the voids and, after agglomeration, improve the strength of the large particles. Mixing the particles with different sizes could be challenging as they tend to be accumulated in different parts of a reactor (smaller at the top). Moreover, in some cases excessive fine powder leads to pilling problems.

Forces of varying strength ranging from weak and medium (van der Waals, electrostatic, capillary forces) to strong bonds (covalent, siloxane bridges) are involved in the agglomeration of solid particles. Even the weak van der Waals forces of very small particles could be of importance when dispersion forces are prominent for many points of contacts between particles.

2.3.7.1 Spray-drying

Spray drying is a method for producing attrition-resistant microspherical catalyst particles. In the spray dryer, the slurry comprising the catalyst particles, suspended in an aqueous sol or hydrogel of a binder (e.g., silica–alumina, alumina, clays), is sprayed into a hot gas flow. The binders are used if shaping is otherwise not possible in the spray-drying process.

Spray-drying can be used to generate the catalysts in the final shape or make powders that will be further processed with another method. Spray-drying process results in 10–100 μm of almost identical shape particles, which could be used in processes operating in fluidized bed reactors, such as fluidized catalytic cracking.

During spray-drying a powder slurry in water suspension (10–30 wt%), prepared by milling (for example, the filter cake to the size of several hundred nanometers), is fed through a nozzle that sprays small droplets into hot air. The formation of microspheres occurs when the spray falls downward in the dryer, the binder particles undergo gelation and the solvent evaporates. This results in embedding of the catalyst particles into a matrix of cross-linked particles.

In high gas temperatures calcination can even occur. This process can be repeated, that is, spray-dried particles can be processed the second time to furnish them with a protective shell.

The nozzle (or atomizer) generating the spray is very important in spray-drying, defining the diameter and size distribution of the droplets. The choice of the spray nozzle construction material is important, especially in the presence of abrasive components in the slurry.

Rotary atomizers with a hollow wheel rotate at high speed (up to 20,000 revolutions min^{-1}) to spray the slurry mechanically which leads to fine to medium coarse powders. Another parameter is the flow direction, which could be cocurrent, counter-current, or a mixed flow. The air distribution is needed to ensure homogeneous flow across the chamber. In a cocurrent operation mode the hottest air contacts the droplets at a stage when the majority of the water is still present, giving the most desired profile.

The particle size distribution is adjusted through control of the atomizer wheel speed. Coarse grained powders can be produced using a low speed atomizer wheel operation.

For conditions when the inlet drying temperature is restricted to below 550 °C, roof air dispersers are used (Fig. 2.149). When high outlet temperatures are needed, part of the exhaust air can be recycled to preheat the drying air passing to the drying chamber, improving the overall heat efficiency.

Catalyst formulations containing organic solvents are dried in an inert gas closed cycle plants.

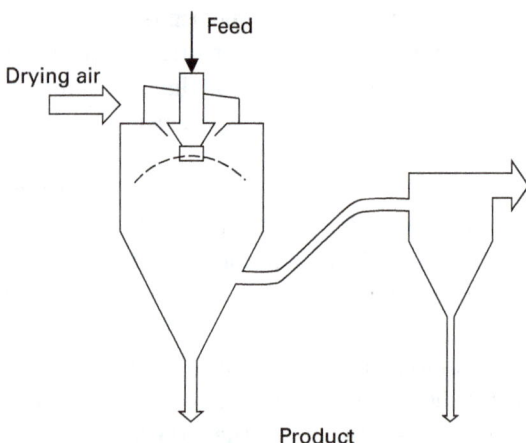

Fig. 2.149: Spray dryer with a rotary atomizer and a roof air disperser. (From [81].)

There are several advantages of rotary atomizers, such as rather narrow particle size distributions, requirements of high-pressure pumps only when very thick pastes are sprayed, resistance to plugging, ability to handle relatively abrasive slurries, and high capacities. The design is, however, more mechanically demanding and the size of the catalyst particles is tightly bound to the diameter of the drying chamber. Dependence of the mean particle size as a function of the chamber diameter is discussed by Roberie et al. [82] and demonstrates that, for example, FCC catalysts (65–80 μm) require a particular chamber diameter (2.8–4 m).

2.3.7.2 Size reduction and granulation

Both types of operation (size reduction and size enlargement) are used in catalyst preparation.

Ball mills and agitated media mills can be used to produce powders of very high fineness from dry or wetted feedstock even down to the submicron range. For wet grinding, the powder is suspended in a liquid upstream milling and then fed in such form to the mill. Ball mills (Fig. 2.150) are equipped with grinding media. The bed of balls or grinding beads is put into motion by rotation of the grinding chamber or by an agitator. The feed material is comminuted between the grinding media because of impact and shear forces. Ball milling can be organized in a continuous

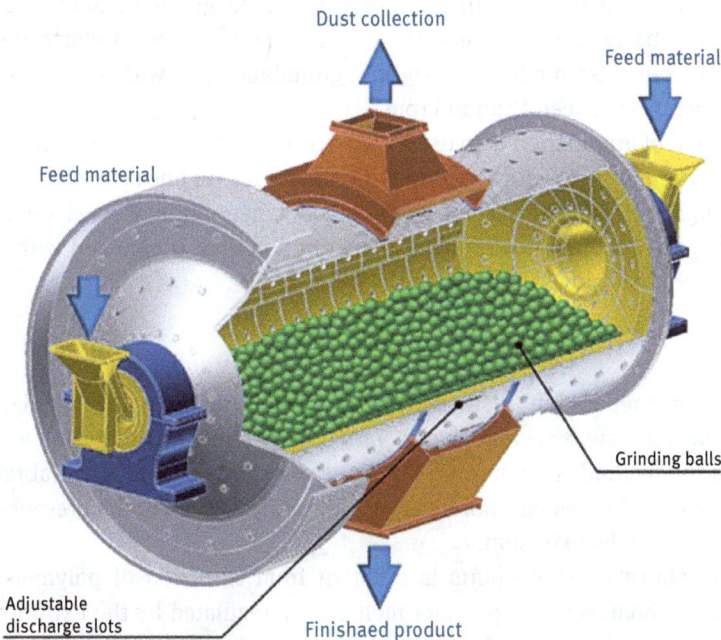

Fig. 2.150: A scheme of ball milling [83].

mode, when the product is pumped to the mill, flows through the agitated bed of grinding beads and finally exits the mill through a suitable discharge. The volume of the grinding beads can be approximately 70–80% of the mill volume. The size of the mills used in catalyst manufacturing is approximately 1–3 m³.

Size enlargement by granulation, which is conceptually similar to a snowball formation, could be used for large-scale production of support spheres, giving particles of irregular shape and size. Moreover, particles are less dense and strong compared to tableting and extrusion. During granulation, size enlargement occurs by wet tumbling-growth agglomeration. Adhesion between wetted particles originates from capillary forces or can be additionally stabilized by cross-linking if the sprayed liquid contains binders, which can either undergo hydrolysis and condensation reactions or gelation. Better mechanical strength provided by the inorganic binders is achieved at the expense of the pore size. Alternatively, application of organic binders necessitates their removal by calcination, leading thereby to lower mechanical strength.

Granulation can be done by placing the powder, which could be pre-wetted, on a tilted (45–55° against the floor) rotating pan (Fig. 2.151). The particle growth and agglomeration occur while the pan is rotating. After reaching a certain size the particles fall over the rim, which could be higher than 1 m for pans with large diameters of several meters. In addition to growth, several other processes proceed during granulation, such as wetting and nucleation prior to agglomeration; and also attrition and breakage, which decrease the size of the formed granules. Rotational speed is adjusted to be 25–40% less than the critical rotation speed at which the centrifugal force is compensating for the weight. A granulation pan with 3 m diameter tilted to 50° operates between 12 and 15 rpm.

Another method of granulation is to use a cylindrical tilted rotated drum (diameter 0.5–1 m, length up to 3 m) with the granulation mass fed on one side and then transported through the drum. This mode of operation, where tilting is used solely for transportation, gives a broader granule size distribution than is obtained with a granulation pan, requiring downstream granulation.

2.3.7.3 Extrusion

Compared to other preparation methods, the extrusion process affords high throughput at relatively low costs and gives a variety of possible extrudate shapes. The downside of the method is a non-uniform shape of extrudates and lower abrasion resistance compared to pellets. Moreover, not all catalysts even in the presence of binders can be shaped by extrusion.

Extrusion of catalysts and supports is different from extrusion of polymers, when properties of a homogeneous polymer melt can be regulated by the extruder temperature. Because pastes for catalyst extrusion are highly concentrated dispersions, their behavior is determined by rheological characteristics. The pore

Fig. 2.151: A granulation pan.

structure and mechanical stability of extrudates are determined by the properties of the paste and extrusion conditions. For catalytic applications, a certain pore structure is needed to allow transport of the reactants to the active sites. Typically, extrudates contain large transport pores of 300–600 nm in addition to mesopores (10–25 nm) being different from materials prepared by tabletting as the latter have mainly monomodal distribution of mesopores.

During extrusion a wet paste from a hopper at the top is forced through a die, and the emerging ribbon passing through holes in the die plate is cut to the desired length using a suitable device. The pressure that is developed in the screw extruder as the paste moves toward the die is affected by the screw geometry and the paste rheology.

After extrusion, the formed extrudates are either broken because of gravity giving a non-uniform sized material or cut-off, for example, by a rotating blade. The cut-off leads to very similar extrudates.

The effect of the extrusion pressure on the compaction in extrusion is, however, smaller than in pelletizing, because in the former case the particles are suspended in a liquid. Adjustment of the liquid-to-solid ratio in the extrusion paste is thus used not only to influence the apparent viscosity of the paste, which obviously decreases when there is more liquid in the paste, but also to vary the porosity of extrudates. On the other hand, too high liquid content will prevent the formation of extrudates after leaving the die. Even if plasticity can be improved by addition of plasticizers, flow of suspensions is characterized by strong non-Newtonian flow. While understanding of suspension flow is not yet sufficient to predict extrusion and control the final properties of catalyst extrudates; nevertheless, rheology is important in analyzing extrusion processes and will be briefly discussed below.

Flow of suspensions can be described by a single property, namely, the shear viscosity. Shear is an action of stress when forces are applied causing two contiguous parts of a body to slide relative to each other in a direction parallel to their plane of contact. Figure 2.152 illustrates deformations produced between parallel plates when the upper plate of surface area A is moved in response to force F. The displacement of a plane layer (dx) over the separation between layers (dy) is termed the shear (dx/dy) acting on the fluid.

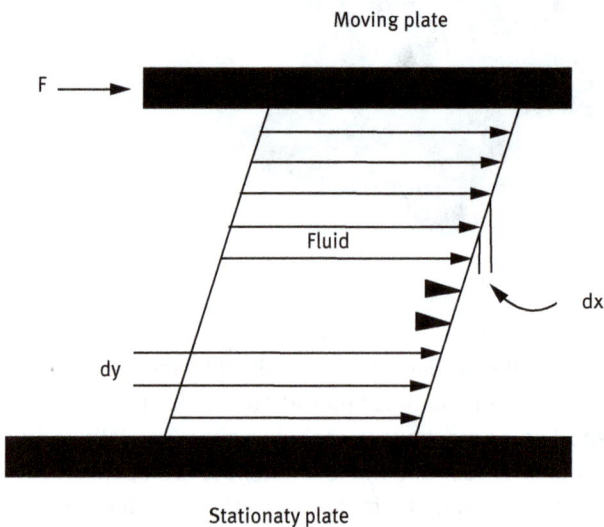

Fig. 2.152: Illustration of deformations between parallel plates.

Depending on the material, there are differences in the behavior when the shear stress is removed. While a solid will tend to return to its original shape, a plastic material may show only partial recovery. The force per area in the plane termed the shear stress (τ) is expressed as

$$\tau = F/A \tag{2.59}$$

According to the Newton's law of viscosity, the shear stress is proportional to the derivative dV/dy termed the shear rate $\dot{\gamma}$:

$$F/A \propto dV/dy \propto \dot{y} \qquad (2.60)$$

giving dependence of shear stress on shear rate

$$\tau = \eta \, \dot{y} \qquad (2.61)$$

where η is the coefficient of viscosity having units of mass·length^{-1}·time^{-1}. In SI units, τ is in Pa, $\dot{y}$ in s^{-1}; thus, η is expressed in Pa·s. The coefficient of viscosity, being the ratio of the shear stress to the shear rate, represents the internal resistance to flow of the fluid. Equation (2.61) is applied to Newtonian fluids when viscosity (temperature dependent) does not change with the relative flow velocity. For the large class of fluids relevant for chemical product technology, this does not hold and viscosity changes depending on the shear rate

$$\tau = \eta(\dot{y}) \, \dot{y} \qquad (2.62)$$

The rheograms for non-Newtonian fluid can often be described by an empirical power-law function called the Ostwald-de Waele equation

$$\tau = k \left(\frac{dy}{dt} \right)^{n} \qquad (2.63)$$

where k is the consistency index. Materials such as concentrated dispersions used in preparation of catalysts often exhibit the so-called shear thinning flow, which is distinctly different from the Newtonian behavior (Fig. 2.153). For shear thinning (pseudoplastic) $n < 1$, while for the Newtonian flow $n = 1$. Shear thickening (dilatant) fluid has n exceeding unity ($n > 1$).

Figure 2.153A illustrates a parabolic profile for the Newtonian liquid. Deviations for shear thinning liquids from the ideal behavior are attributed to a low viscosity near the wall, where the shear rate is high and a high viscosity in the middle. The velocity profile is more flat resembling a plug flow. The opposite behavior with a sharp profile is observed for shear thickening fluids. Dependence of viscosity on the shear rate is shown in Fig. 2.153B. Shear viscosity decreases or increases with increasing shear rate, leading to, respectively, shear thinning and shear thickening.

The entire flow curve for pseudoplastic behavior can often be fitted using the Reiner–Philippoff equation:

$$\tau = - \dot{y} \left(\eta_{\infty} + \frac{\eta_0 - \eta_{\infty}}{1 + \tau^2/A} \right) \qquad (2.64)$$

where A is an adjustable parameter, η_0 is viscosity at $\dot{y} \to 0$, η_{∞} is viscosity at $\dot{y} \to \infty$.

An often used model of Herschel–Bulkley is the power-law type of equation with the addition of a yield stress

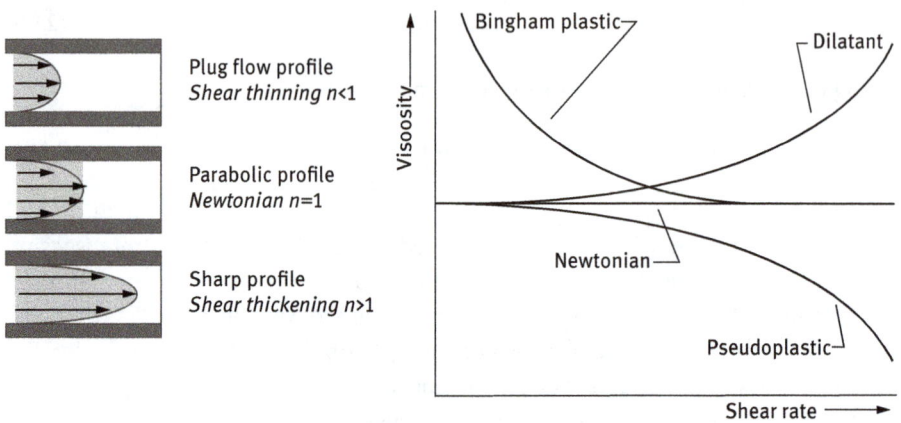

Fig. 2.153: Behavior of suspensions. (A) Flow profiles. Coefficient n appears in eq. (2.63) and (B) typical curves of viscosity versus shear rate [84].

$$\tau = \tau_0 + k(\dot{\gamma})^n \tag{2.65}$$

where τ, τ_0 and $\dot{\gamma}$ are shear stress, shear yield stress and shear rate; k is the consistency constant and n is the flow index. In this model, viscosity is expressed as

$$\eta = \frac{\tau_0}{|\dot{\gamma}|} + k|\dot{\gamma}|^{n-1} \tag{2.66}$$

Binders can also influence the flow behavior, altering the velocity profile in the paste, thereby deteriorating extrudated properties due to inadequate adhesion during extrusion. The desired product during extrusion (high shear rates) is achieved for pseudoplastic pastes with low viscosity, since after leaving the die at a low shear rate the viscosity is increased leading to stable extrudates. Most catalyst pastes are, however, prone to shear deformation. The flow behavior of molding masses is well described by Ostwald–de Waele equation (eq. (2.63)). The flow behavior index n in eq. (2.63) indicates pseudoplastic fluids being for simple shapes at most 0.7, while for preparation of honeycomb monoliths it is below 0.3.

Because extrusion of catalysts and supports is in essence defined by the flow of a suspension inside an extruder, the ability to make an extrudate is determined by the geometry of the extruder, chemical properties of the screw, wall and die materials, which should ensure minimal adhesion of the paste, most importantly by the properties of the paste.

The value of the flow behavior index is related to the velocity profile. There is a radial profile of flow velocity in the die channel as the flow velocity decreases with an increasing distance from the flow axis. If the velocity gradient is significant, the defects will be generated after a granule leaves a die channel because of the velocity equalization. Subsequently, efficient molding will not be possible. A small flow behavior index corresponds to less significant velocity gradients and an ideal plug flow is achieved when the index is approaching zero.

Molding pastes have non-Newtonian flow behavior and during their movement through dye channels, different velocity profiles are observed (Fig. 2.154), determining the tensile strength of shaped catalysts. Obviously, in case of substantial velocity profiles in the radial direction, produced grains will have macrodefects, originating from improper "layering" of the material.

Fig. 2.154: Flow deformation profile during extrusion (From [85]).

Pastes that are too viscous can block the extruder, while the opposite leads to unstable extrudates. Therefore, an important characteristic of suspensions used in catalyst extrusion is viscosity, which can be expressed by the Einstein equation as a function of the solid content at low concentrations:

$$\eta_r = 1 + 2.5\varphi \tag{2.67}$$

where φ is the solid fraction. At higher concentration, the data for relative viscosity appear to follow more of an exponential function of concentration according to the Mooney equation:

$$\eta_r = \exp\left(\frac{\xi\phi}{1 - k\phi}\right) \tag{2.68}$$

where ζ is a coefficient determined by fitting, constant k is the self-crowding factor, being in the range between 1.35 and 1.91. The self-crowding factor is inversely proportional to the maximum concentration. The latter has a limit because of the exponential character of the curve. The self-crowding factor at maximum concentration does not allow the movement of neighboring particles as the suspension acts like a solid.

Krieger–Dougherty semi-empirical equation describes the concentration dependence of the viscosity:

$$\eta_r = \left(1 - \frac{\phi}{\phi_{max}}\right)^{-K\varphi_{max}} \tag{2.69}$$

where φ_{max} is the maximum packing fraction or the volume fraction at which the zero shear viscosity diverges. Parameter K is equal to 2.5 for spherical particles, while deviation from this value is related to shape changes of the particles. Equation (2.69) reduces to the Einstein relation at low particle concentration.

An alternative approach of Brodnyan–Mooney was developed for ellipsoidal particles

$$\eta_r = \eta_o \exp\left[\frac{2.5\phi + 0.399(p-1)^{1.48}\phi}{1-k\phi}\right] \tag{2.70}$$

where k is the self-crowding factor.

The treatment above considered suspensions of hard spheres without colloidal or thermodynamic interactions. Repulsive surface forces stabilize dispersions and prevent aggregation, as the particles are kept away at a certain distance. Electrostatic or steric repulsion results thus in an excluded volume inaccessible to other particles, which can be expressed as follows:

$$\varphi_{eff} = \varphi\left(\frac{a_{eff}}{a}\right)^3 \tag{2.71}$$

where a_{eff} is the effective particle radius, which is defined as half the distance to which centers of two particles can approach each other under colloidal forces. Rheology of colloids in many cases is similar to hard sphere dispersions being quantitatively described as a hard sphere system with the concept of effective radius. Equation (2.71) implies that particle size is a parameter that influences φ_{eff}. An increase in the particle radius a at constant φ and a constant range of the repulsive colloidal interactions corresponds to a decrease in φ_{eff}. This means that dispersions with the same φ but different a may exist in different phases having a strong impact on the shear rate-dependent viscosity.

Not only the size but also the particle size distribution can influence the phase behavior of colloidal dispersions. A significant viscosity decrease is typically observed by mixing particles with different sizes when the particle volume fraction φ exceeds 0.5. This effect is more pronounced with increasing φ and the ratio between large and small particles χ. When repulsive colloidal interactions become relevant, the viscosity passes through a minimum increasing again when χ is increased at a constant total particle concentration and a fixed fraction of small particles.

Attractive particle interactions can lead to large compact aggregates and rapid phase separation. Alternatively, loose aggregates attract water and increase an effective volume fraction φ_{eff}. This results in an increase in the zero-shear viscosity. Upon an increase in the shear rate, there is a gradual floc breakdown and/or their alignment in the flow direction, and as a result a decrease in viscosity.

The extrusion process is influenced by the paste composition: water content (typically between 20 and 40 wt%); presence of peptizing agents (to adjust the colloid-chemical properties of the paste); binders; plasticizers; lubricants and porogenes. It is also important that no large solid particles are present for the extrusion of complex shapes. The operating variables also include mixing time, aging and extrusion temperature.

While elevation of temperature is positive, improving relative deformations from the viewpoint of technology, the process becomes much more energy intensive and unnecessarily complex. Moreover, temperature fluctuations can influence mechanical and rheological properties of the suspension, primarily the optimal water content. Therefore, extrusion of highly concentrated suspensions is carried out at ambient temperature.

The paste should be plastic enough to be able to pass through the die and obtain the desired shape. After extrusion, the extrudate should not have any visible or microdefects having adequate properties for further processing (transport, drying, calcination, etc.). The coagulation structure of the paste, on the one hand, should be strong enough ensuring adequate rheological properties and retention of the shape after leaving the die, and on the other hand, too strong coagulation structure implies higher energy needed for extrusion.

As already mentioned an important parameter influencing properties of the extrudates is the size of the particles in the suspension, which should be small enough enhancing plasticity. An energy-intensive operation, ball milling, is typically applied to reduce the size of the solid phase.

When the amount of water is much lower than the solid pore volume, it might be difficult to process efficiently such dry pastes. Settling of the pastes after extruder can be, on the contrary, troublesome for a significant water content (Fig. 2.155). The amount of water depends on the surface lyophilic and adsorption properties of the solid phase and its specific surface area. For example, the optimum amount of water in mass% for an alumina support using 20% nitric acid solution as a peptizing agent is 22–23%, while somewhat higher amount of water (28–29 mass%) is required when PVA is used instead of nitric acid. A catalyst for

Fig. 2.155: Extrusion of too wet paste (courtesy of Dr. Z. Vajglova).

oxidation of SO_2 to SO_3 based on vanadia on silica when processed with sulfuric acid as a peptizing agent needs as high as 37–38 mass% of water.

Porogenes (carbon black, starch and sawdust) are added to the paste to create porosity, which is generated during calcination. Besides porogenes, plasticizers and lubricants are also burned away during calcination, contributing to porosity generation. Such porosity while beneficial for mass transfer can deteriorate mechanical stability.

Modification of shaping masses, when the particle sizes are in the (sub)micrometer range, from the colloidal viewpoint is controlled by zeta-potential. These systems that are used in extrusion could be modified, for instance, by the addition of peptizing agents (e.g., nitric, formic or acetic acids) influencing colloid–chemical interactions between particles and assuring deviation of pH from the point zero charge. Otherwise , the liquid phase might start to agglomerate. Corrosivity of peptizing agents (e.g., nitric acid) should be considered when selecting them. In addition, utilization of nitric acid during extrusion might influence acidity and can give NO_x during calcination. To mitigate these emissions, a selective catalytic reduction of NO_x should be installed, obviously increasing the catalyst production costs. As a consequence, application of electrolytes as peptizing agents is justified when other methods to regulate rheological properties of suspension pastes are not successful.

Plasticizers or their mixtures applied to improve the paste rheological behavior are selected from a wide variety of plasticizers, including clays, starch, sugars, cellulose derivatives, polyethylene glycol and PVA. The amount of plasticizers should not be too high (up to 3–5 mass%), corresponding to covering the available surface of the solid and allowing the desired changes in rheological properties. Higher concentration might not be beneficial from the rheological viewpoint. Selection of rheology improvers should also consider potential adhesion of the suspension paste to the extruder material, which should be minimized.

Formation of shaped catalysts implies cross-linking between particles in the suspension, which can be achieved either by condensation of surface hydroxyls, resulting in oxygen bridges, or by sintering. In the absence of suitable surface hydroxyl groups in many catalysts or catalyst precursors, addition of inorganic binders allows interparticle cross-linking. Binders, such as hydroxide forms of alumina, silica–alumina, clays or silica sols applied together with the catalytic phase result in cross-linking at sufficiently low temperatures avoiding thereby damage of the catalytically active material. Binders required during extrusion make the extrudates strong enough, preventing a potential collapse.

Typically, inorganic binders are utilized because organic ones will be burned away at the calcination step at 500–600 °C, which also results in the burnout of organic additives like plasticizers.

In many cases, the catalytic pastes cannot be shaped without adding binders. When alumina is added in the form of boehmite or pseudoboehmite, it is transformed into transition alumina during calcination. Gelation of silica sols and agglomeration of delaminated fragments of clays are responsible for the stability of extrudates when these binders are used.

Introduction of binders can influence catalytic activity sometimes in a negative way. In such a case, a compromise should be made between the rheological properties and mechanical stability on one hand, and catalytic behavior, on the other. For example, introduction of aluminosilicates for extrusion of monolithic blocks based on titania improves the rheological properties of the suspension paste at the expense of the catalytic properties.

A special care should be taken on the surface properties of binders, which can be themselves catalytically active. In production of traditional ceramics, this is a rare case. In production of catalysts by extrusion, there are, however, several examples when the main active component and the binder influence each other, which has a consequence in terms of physicochemical properties of the final catalyst. Figure 2.156 shows temperature-programmed desorption of ammonia from fresh forms of H-ZSM-5 zeolite and the material with aluminum phosphate as a AlPO binder.

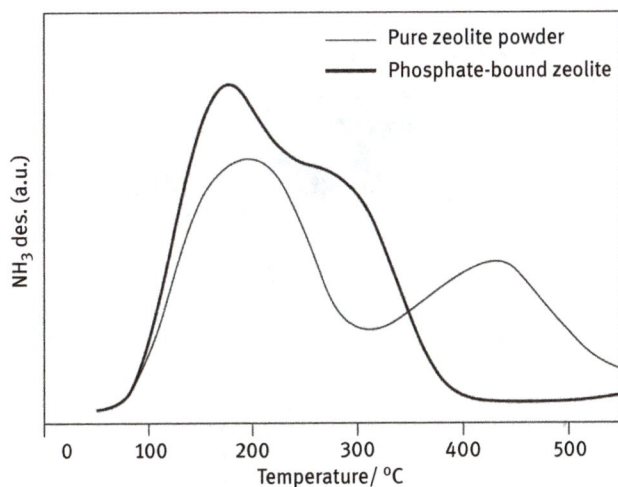

Figure 2.156: Ammonia TPD of H-ZSM-5 powder after calcination and $AlPO_4$-bound extrudate containing the same zeolite. Reproduced with permission from [86].

There are evident changes in acidity after addition of a binder resulting in another distribution of sites in terms of acidic strength. Such an example of the binder influences on the acidic strength, and thus catalytic properties can be interpreted by the phosphor migration from the binder to the zeolite crystal structure.

Application of binders can result in nonuniformity of the active component distribution. It implies that there could be zones with a high concentration of the active phase, and consequently, higher rate, maybe local overheating and appearance of zones which are controlled by mass transfer rather than kinetics.

Special shapes (trilobates, rings, hollow cylinders, monoliths or honeycombs) can be obtained using proper dies (Fig. 2.157).

Fig. 2.157: Shapes of extruded catalysts.

Internal cavities can be formed when, for instance, a two-piece aluminum extrusion die set is used. In a set with separated parts, the piece at the right of Fig. 2.158 is used for forming the internal cavity.

Fig. 2.158: A two-piece aluminum extrusion die set to form extrudates with an internal cavity [87].

The quality of the extrudates depends not only on extrusion but also on downstream drying and calcination. These steps require special attention in the case of larger structures such as extruded monoliths, which are applied in SCR of NO_x at power and waste incineration plants or the destruction of dioxins generated during incinerating municipal wastes. The active phase V_2O_5 is deposited on titania (anatase) support. To improve the mechanical strength, silicoaluminates or glass fibers are added as mechanical promoters (3.5 wt%). The suspension for extrusion contains clays (ca 6.5 wt%), water and small amounts of methyl hydroxyethyl cellulose and polyethylene glycol as lubricants. Drying of the extruded monolith must be slow enough to prevent ruptures and cracks. Calcination is performed at approximately 500 ± 600 °C.

Screw extruders with short screw length to barrel diameter are used for catalysts requiring less viscous pastes. For more viscous pastes, press extruders with a rotating pressing cylinder are utilized due to shorter distance at which the pressing force is applied. Figure 2.159 shows that this distance is just between two cylinders, the pressing and the screening ones. Because of lower shear deformation, the flow behavior of the paste is less critical than in screw extruders. The latter allow, however, production of extrudates with more complex shapes.

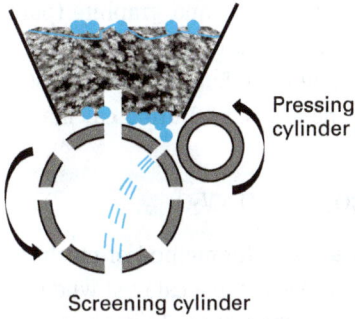

Fig. 2.159: Extruder press.

As an example of mechanical mixing and extrusion during catalyst preparation, the synthesis of sulfuric acid catalyst of the SVD type (on diatomite) is illustrated in Fig. 2.160.

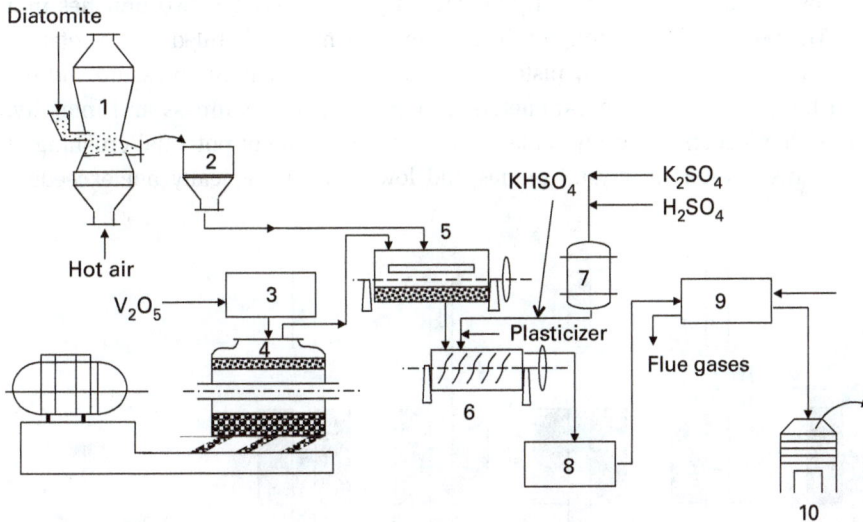

Fig. 2.160: Production of sulfuric acid catalyst. (Redrawn from [48].) 1, dryer; 2, bunker; 3, grinder; 4, vibration milling; 5, ball milling; 6, extruder; 7, reactor for potassium bisulfate; 8, dryer; 9, calcination; 10, sieve.

The active component in this catalyst is vanadium pentoxide promoted with sulfates of alkali metals, potassium and sodium. The support is diatomite, which is silica with some minor presence of other oxides such as Fe_2O_3 and Al_2O_3. The catalyst performance depends very much on the type of diatomite.

The support is dried in a spray dryer at 120–150 °C up to a humidity of 5–7%. The technical quality vanadium pentoxide (80% of V_2O_5) is ground up, first to 2–3 mm (pos. 3) and then to ~200 μm (pos. 4). After mixing and ball milling (pos. 5) the solid phase, containing diatomite and vanadium pentoxide, is extruded with the addition of potassium hydrosulfate (promoter), stearic acid (plasticizer) and graphite (porogenes) directly to the extruder. Potassium bisulfate is prepared in the reactor (pos. 7) by contacting potassium sulfate with sulfuric acid. The extrudates contain 26–30% water and are thus dried at 120 °C (pos. 8) and calcined (pos. 9). During calcination the following reactions occur

$$2KHSO_4 = K_2S_2O_7 + H_2O; \; K_2S_2O_7 + V_2O_5 = K_2S_2O_7 \cdot V_2O_5$$

In this example the supported catalyst was produced by a dry method of mechanical mixing with subsequent calcination. This procedure leads to a catalyst when vanadium is present as a complex sulfated salt mixture, which is in the molten state during actual catalysis.

2.3.7.4 Tableting

Pellets or tablets of high shape accuracy, much better than in extrusion, are obtained by dry tableting when a dry powder is pressed between two punches in a press. The powder flows through a feeder mechanism, and is filled into a rotating die, where it is compressed by pistons (Fig. 2.161). The maximum pressure and the rate of the pressure are the parameters influencing pellet hardness and integrity. High mechanical stress during tableting has the downside of potential crushing of crystals, dislocations of crystal planes and low porosity, typically not exceeding

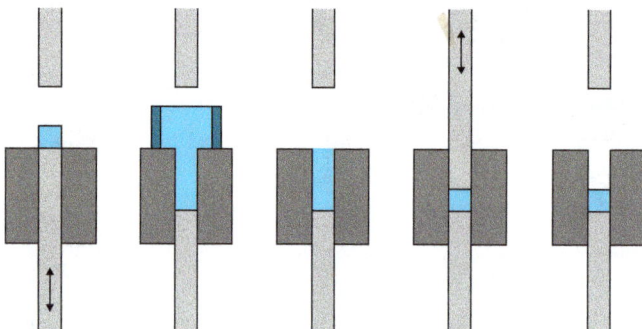

Fig. 2.161: A tableting machine cycle.

30%, thereby the formulation can contain porosity improvers if porosity is important for the final application.

At the start of the cycle, the lower piston is in the upper position. The pellet made in the previous cycle is pushed out as the die passes under the feeder mechanism. The upper piston moves down by a fixed length, which defines the dimension of the pellet. Although usually simple cylinders with almost the same height and diameter are produced, it is also possible to form ring-shape tablets with special dies and punches.

The important parameter in tableting is plastic deformation, which depends on the applied stress, overall time and rate of compression, and the time for the material to be under maximum stress. The elastic component of the response to the applied pressure stored in the material is relaxed after the pressure release. A too high elastic component gives unstable tablets.

Small amounts of liquid or solid additives may be used to prevent dust formation and improve compaction by lubrication. Because in tableting dry feeds are used, thus lubricants or pilling aids such graphite, PVA, polyethylene, talc, silicates, aluminates, to name a few, are added to enable the powder to flow. One of the most frequently used additives is graphite at concentrations between 0.5 and 1%. If for some reasons application of graphite can deteriorate catalyst performance, it should be removed by careful oxidation.

In general, the feed for tableting should have flow properties of sand. Mechanical stability after compression can be improved by adding binders. Moreover, high mechanical stability is obtained when the feed consists of mixed particle sizes. If the feed is not sufficiently dense prior to tableting, pre-compacting is applied.

Different geometries can be achieved during tabletting including tablets per se when the height is low than the diameter or cylinders when the opposite holds (height > diameter). Moreover, cylinders with holes or ribs on the external surface as well as their combination can be manufactured.

An example of nickel-molybdenum alumina catalyst preparation is shown in Fig. 2.162. Suspension of aluminum hydroxide and water solutions of NH_4MoO_4 and nitric acid are prepared respectively in vessels (pos. 1–3) and then mixed in a reactor (pos. 4). Thereafter the tablets are produced from this mixture (pos. 6). After drying at room temperature (pos. 7) for 24 h the tablets are first dried at 383–423 K for 12 h and then calcined at 803 K for 8 h. After grinding the tablets to the powder size, the powder is impregnated with $Ni(NO_3)_2$ for 1 h in the reactor (pos. 4). Subsequent tableting, drying and calcination give the final catalyst that contains 13–15 wt% MoO_3, 3–4 wt% NiO, 0.05–0.1 wt% Na_2O, 0.2–0.3 wt% Fe_2O_3 and alumina (the rest). Approximately 30 kg of NO_x is released per ton of catalyst.

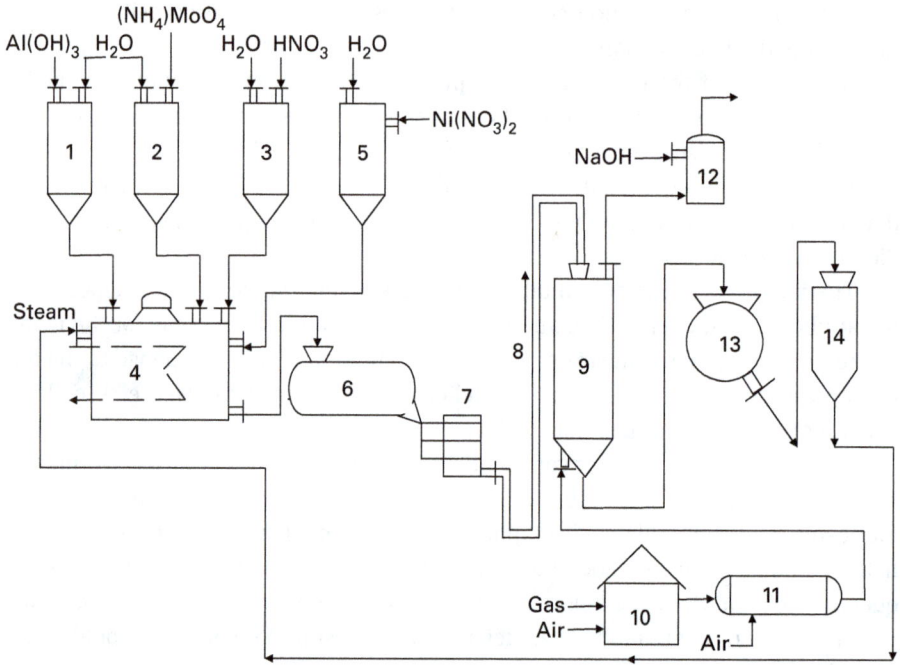

Fig. 2.162: Production of nickel-molybdenum alumina catalyst. (From [48].) 1, 2, 3, 5, vessels for Al(OH)$_3$ suspension and water solutions of NH$_4$MoO$_4$, HNO$_3$, Ni(NO$_3$)$_2$; 4, reactor; 6, tableting machine; 7, oven; 8, lift; 9, calcination unit; 10, burner for flue gases generation; 11, heat exchanger; 12, absorber with caustic solution; 13, milling; 14, intermediate vessel.

2.3.7.5 3D printing in catalytic technology

The 3D printing process is a promising manufacturing method, which can in the future be applied to many different catalytic materials, allowing careful design of targeted structures. The method is suitable for large-scale manufacturing as well as for catalyst development being applicable to lab-scale operations. 3D printing allows accurate control of various sophisticated structures, including monolithic catalysts with a complex preparation procedure. Not surprisingly, additive manufacturing tools (e.g., 3D printing) have been mainly explored for shaping 3D catalyst bodies such as structured monoliths with improved mass and heat transfer characteristics. Figure 2.163 illustrates a comparison between an extruded monolith and a robocast structure with FCC (Fig. 2.163C) lattice. In robocasting, highly concentrated suspensions of the material and a solvent are deposited in a layer-wise fashion through a syringe-type nozzle. The minimum printable feature size in robocasting is typically 30–500 μm.

Besides fabricating structured monoliths, geometrically complex reaction vessels and microreactors with integrated catalysts can be manufactured using 3D printing,

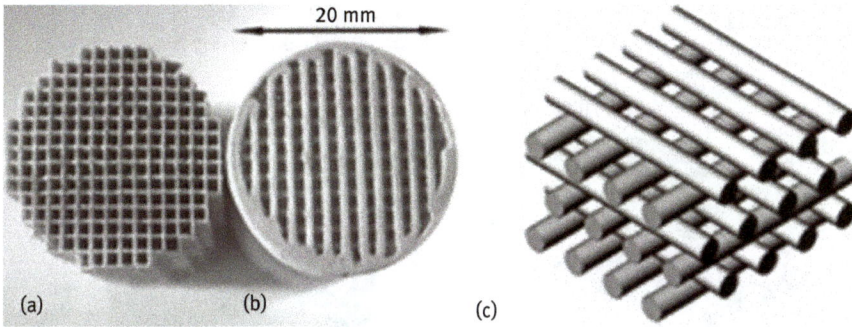

Fig. 2.163: Monoliths of similar outer dimensions and surface areas: (a) extruded honeycomb, (b) robocast material with (c) FCC lattice (Reproduced from [88]).

when the structure of the target material can be first designed through computer-aided design.

The most suitable methods for catalyst preparation include (Fig. 2.164) fused deposition modeling (FDM), direct ink writing (DIW), stereolithography (SLA) and selective laser sintering (SLS).

In SLS a laser is used as the power source to sinter a powdered material, while in SLA layer-by-layer deposition is done using photochemical processes. DIW (robocasting) is an extrusion-based printing process similar to FDM with the difference that contrary to the latter process in DIW, the material is extruded directly without melting or solidification.

Similar to extrusion processes described earlier, a low viscosity of the printing ink is required for maintaining fluidity while passing through the nozzle. Thereafter, the material should maintain the shape on the printing structure, which implies high viscosity. Pseudoplastic fluids used in robocasting typically require utilization of a solvent, necessitating post treatment drying and sintering. In addition to solvents, mixtures for robocasting of more complex materials, such as zeolites, should contain different additives aiding extrusion (e.g., clays and plasticizing organic binders).

Future advances in 3D printing of catalysts require [89]:

- better control of the micro/mesostructures of printed catalytic materials by improving the accuracy;
- improvement of the feedstock materials to diminish processing costs;
- structural optimization of 3D-printed materials, including pore size, pore structure and active phase distribution;
- better preservation of the 3D-printed structures after post-treatment processes.

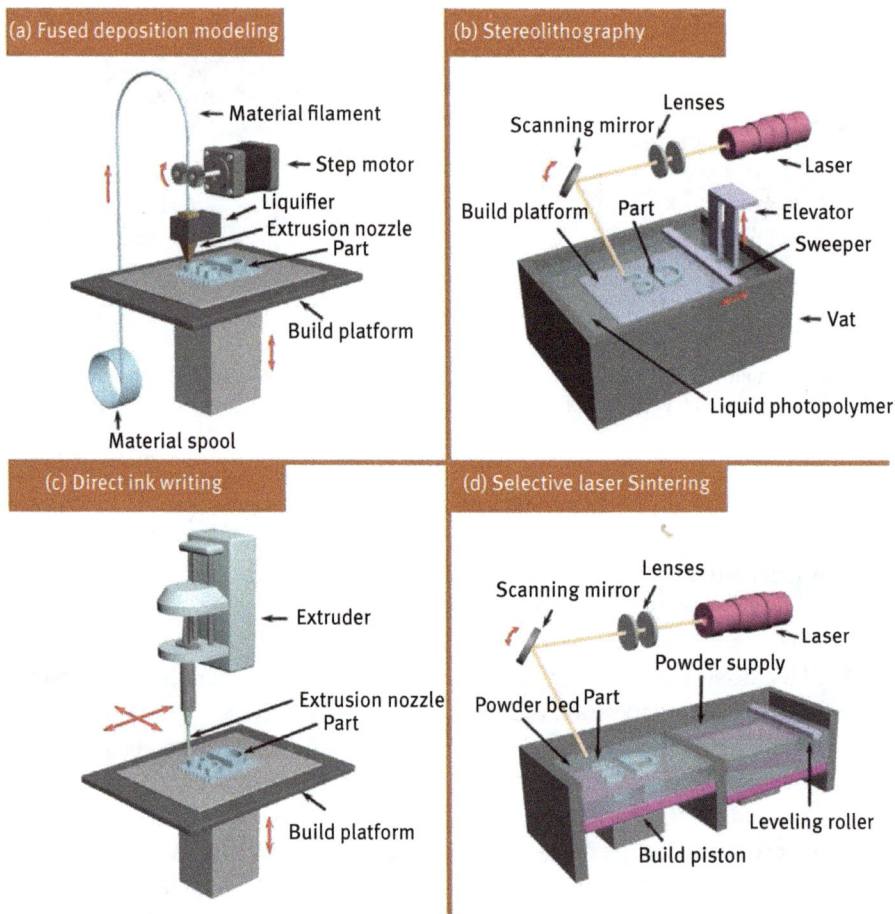

Figure 2.164: Some 3D printing processes that are useful for catalyst preparation: (a) fused deposition modeling; (b) stereolithography; (c) direct ink writing; and (d) selective laser sintering. (From [89]).

2.3.7.6 Scaling-up catalyst preparation

The typical strategy in academic work on catalyst design is related to the use of catalyst particles of such a size that diffusion limitations are suppressed. Of less concern are the preparation methods and selection of precursors, which are feasible for use on an industrial scale. For example, metal chlorides could be used for impregnation on a laboratory scale, while nitrates are preferred industrially. Catalyst reduction in academia is usually carried out by molecular hydrogen, while some catalyst manufactures prefer reduction with chemical reagents. There is therefore a difference in catalytic materials prepared on a small scale and those used for piloting and eventual production. For example, the type of the support or the preparation method should not necessarily be the same. Commercial raw materials are used for catalyst preparation in a large scale instead of pure chemicals on a laboratory scale. Water is the most used solvent, while the use of organic solvents is uncommon.

In lab scale, costs of chemicals are rarely considered, while cost-effective raw materials should be selected for industrial production. Moreover, the corrosive nature of raw materials, possibility of emissions and their mitigation, efficient disposal of effluents, wastewater purification along with process water recovery should be carefully considered, limiting catalyst preparation technology. Low slurry concentration during precipitation can be, for example, applicable in lab-scale recipes but will be challenging from the production technology viewpoint.

Preparation of catalysts for larger than lab scale is not straightforward as parameters such as stirring rate, heating and cooling rates, heat transfer, flow pattern, drying and calcination conditions can be different. In particular, heating and cooling rates are much slower when the scale is increasing. This influences chemical and physical process in catalyst preparation, for instance, crystallization by hydrothermal synthesis.

In addition, large reactors have more prominent spatial inhomogeneity in temperature, pH and concentration. As the size increases, significant variations in the local temperature profiles are possible during drying and calcination. Proper equipment design is thus essential to minimize different spatial variations. Larger production volume also requires equipment capable of handling such volumes.

Preparation methods used in industry are mainly kept as trade secrets or cover a broad range of parameters, making it difficult to reveal the exact preparations details. Very few flow schemes for catalyst preparations can be found in the open literature.

In general, scaling up the catalyst preparation procedure is not an easy task. In pilot scale and industrial fixed-bed reactors, usually larger catalyst particles are used than in the laboratory scale where the aim is to avoid problems with the pressure drop. This can result in changes not only in the reactant conversion but also in the product selectivity, which is known to be influenced by internal and external diffusion.

Lab-scale academic research, aimed primarily at publications in high-impact factor journals, does not typically consider a necessity of a high crush strength required for applications in fixed bed reactors, when also a minimally allowed pressure drop should be achieved. This requirement implies higher compaction during forming at the expense of porosity as well as elevated heat and mass transfer within the catalyst particle. The latter can obviously influence activity and selectivity. Not only the value of the crushing strength is of importance, but also the distribution in strength, in particular percentage of catalyst particles with low strength should be minimal, as otherwise the particles will break giving high pressure drop.

Another important issue is the presence of a binder, for example, for extrusion. Introduction of a binder can influence the catalyst performance, thus such differences should be recognized when progressing from laboratory to pilot and industrial scale with kinetic and diffusion modeling.

Catalyst stability is an important issue that is often not in the focus of experimental and theoretical studies at the laboratory scale. The necessity to develop materials and structures resistant to deactivation also requires some alterations on the material level, resulting in a difference between catalysts tested in a lab and on an industrial scale. In conclusion the scale-up of the catalyst preparation methods from a laboratory scale to pilot and industrial scales is challenging, which calls for more widespread involvement of the theories of physical chemistry and chemical engineering in catalyst preparation.

References

[1] Rouquerol, F., Rouquerol, J., Sing, K.S.W. (1999) Adsorption by Powders and Porous Solids. San Diego, CA: Academic Press, pp. 1–25.

[2] Leofanti, G., Padovan, M., Tozzola, G., Venturelli, B. (1998) Surface area and pore texture of catalysts. Catal. Today 41: 207.

[3] http://www.vermeer.net/pub/communication/downloads/future-perspectives-in-cata.pdf.

[4] https://pubs.acs.org/doi/10.1021/acscatal.8b01708

[5] Catalysts: Combinatorial Catalysis, 2005, DOI: 10.1016/B0-12-369401-9/01128-1, Encyclopedia of Condensed Matter Physics, A. Hagemeyer, A. Volpe

[6] Rodemerck, U., Linke, D. (2009) High throughput experimentation, in. Synthesis of solid catalysts, ed. K.P. de Jong., Wiley, Weinheim.

[7] Li, H., Zhang, Z., Liu, Z. (2017); Application of Artificial Neural Networks for catalysis: A Review, Catalysts 7: 306; doi:10.3390/catal7100306. https://onlinelibrary.wiley.com/doi/full/10.1002/aic.16198

[8] Medford, A.J. Ross Kunz, M., Sarah M. Ewing, S.M., Borders, T., Fushimi, R. (2018), Extracting knowledge from data through catalysis informatics, ACS Catal. 8: 7403-7429. https://pubs.acs.org/doi/10.1021/acscatal.8b01708

[9] Rothenberg, G. (2008) Data mining in catalysis: Separating knowledge from garbage, Catal. Today, 138:2-10. https://www.sciencedirect.com/science/article/pii/S092058610800062X

[10] Goldsmith, B.R., Esterhuizen, J., Liu, J.-X., Bartel, C. J., Sutton, C. (2018) Machine learning for heterogeneous catalyst design and discovery, AIChe J, 64: 2311-2323. https://pubs.acs.org/doi/10.1021/acscatal.8b01708

[11] Condon, J.B. (2006) Surface Area and Porosity Determination from Physisorption, Measurements and Theory. Elsevier.

[12] Dubinin, M.M. (1960) The potential theory of adsorption of gases and vapors for adsorbents with energetically nonuniform surfaces. Chem. Rev. 60: 235–241.

[13] Kalantar Neyestanaki, A., Mäki-Arvela, P., Backman, H., Karhu, H., Salmi, T., Väyrynen, J., Murzin, D.Yu. (2003) Kinetics and stereoselectivity of o-xylene hydrogenation over Pd/Al_2O_3. Mol. Catal. A. Chem. 193: 237–250.

[14] Bayraktar, O., Kugler, E.L. (2002) Characterization of coke on equilibrium fluid catalytic cracking catalysts by temperature-programmed oxidation. Appl. Catal. A. Gen. 233: 197–213.

[15] Sahin, S., Mäki-Arvela, P., Tessonier, J.P., Villa, A., Reiche, S., Wrabetz, S., Su, D.S., Schlögl, R., Salmi, T., Murzin, D.Yu. (2011) Palladium catalysts supported on N-functionalized hollow vapour-grown carbon nanofibers: the effect of the basic support and catalyst reduction temperature. Appl. Catal. A. Gen. 408: 137–147.

[16] Li, L., Lin, J., Li, X., Wang, A., Wang, X., Zhang, T (2016) Adsorption/reaction energetics measured by microcalorimetry and correlated with reactivity on supported catalysts: A review. Chin. J. Catal. 37: 2039–2052.

[17] Simakova, I. (2010) Catalytic transformations of fatty acid derivatives for food, oleochemicals and fuels over carbon supported platinum group metals. PhD thesis. Åbo Akademi University, Turku, Finland.

[18] Sharma, R., Bisen, D.P., Shukla, U., Sharma, B.G. (2012) X-ray diffraction: a powerful method of characterizing nanomaterials. Recent Res. Sci. Techn. 4: 77–79.

[19] Davis, S.M. (1990) Effect of surface roughness on particle size predictions from photoemission results. J. Catal. 122: 240–246.

[20] Eränen, K., Klingstedt, F., Arve, K., Lindfors, L.-E., Murzin, D.Yu. (2004) On the mechanism of the selective catalytic reduction of NO with higher hydrocarbons over a silver/alumina catalyst. J. Catal. 227: 328–343.

[21] Villegas Perdomo, J.I. (2006) Engineering bifunctional catalysts for hydrocarbon valorisation, PhD thesis. Åbo Akademi University, Turku, Finland.

[22] Guerrero-Perez, M.O., Banares, M.A. (2006) From conventional in situ to operando studies in Raman spectroscopy. Catal. Today 113: 48–57.

[23] Aho, A., Kumar, N., Lashkul, A., Eränen, K., Ziolek, M., Decyk, P., Salmi, T., Holmbom, B., Hupa, M., Murzin, D.Yu. (2010) Catalytic upgrading of woody biomass derived pyrolysis vapors over iron modified zeolites in a dual-fluidized bed reactor. Fuel 89: 1992–2000.

[24] Arve, K., Popov, E.A., Rönnholm, M., Klingstedt, F., Eloranta, J., Eränen, K., Murzin, D.Yu. (2004) From a fixed-bed Ag-alumina catalyst to a modified reactor design: how to enhance the crucial heterogeneous-homogeneous reactions in HC-SCR. Chem. Eng. Sci. 59: 5277–5282.

[25] US Patent US 8,017,086 B2 (2011) to Volvo Technology Corporation, Catalyst unit for reduction of NOx compounds.

[26] Fierro, G., Moretti, G., Ferraris, G., Andreozzi, G.B. (2011) A Mossbauer and structural investigation of Fe-ZSM-5 catalysts: Influence of Fe oxide nanoparticles size on the catalytic behaviour for the NO-SCR by C_3H_8. Appl. Catal. B-Environ. 102: 215.

[27] Nieminen, V., Karhu, H., Kumar, N., Heinmaa, I., Ek, P., Samoson, I., Salmi, T., Murzin, D.Yu. (2004) Physico-chemical and catalytic properties of Zr- and Cu-Zr ion-exchanged H-MCM-41. Phys. Chem. Chem. Phys. 6: 4062–4069.

[28] Niemantsverdriet, J.W. (2007) Spectroscopy in Catalysis. Weinheim, Germany: Wiley-VCH.

[29] Chorkendorff, I., Niemanstverdriet, J.W. (2003) Concepts of modern catalysis and kinetics. Weinheim, Germany: Wiley-VCH.

[30] Lauritsen, J.V., Besenbacher, F. (2006) Model catalyst surfaces investigated by scanning tunneling microscopy. Adv. Catal. 50: 97–147.

[31] Vang, R.T., Lauritsen, J.V., Lægsgaard, E., Besenbacher, F. (2008) Scanning tunneling microscopy as a tool to study catalytically relevant model systems. Chem. Soc. Rev. 37: 2191–2203.

[32] Gladden, F., Mantle, M.D., Sederman, A.J. (2006) Magnetic resonance Imaging of catalysts and catalytic processes. Adv. Catal. 50: 1–75.

[33] Karreman, M.A., Buurmans, I.L.C., Geus, J.W., Agronskaia, A.V., Ruiz-Martínez, J., et al. (2012) Integrated laser and electron microscopy correlates structure of fluid catalytic cracking particles to Brønsted acidity. Angew. Chem. Intern. Ed. 51: 1428–1431.

[34] Piccolo, L., Nassreddine, S., Toussaint, G., Geantet, C. (2010) Discussion on "A comprehensive two-dimensional gas chromatography coupled with quadrupole mass spectrometry approach for identification of C10 derivatives from decalin" by C. Flego, N. Gigantiello, W.O. Parker, Jr., V. Calemma [J. Chromatogr. A 1216 (2009) 2891].

[35] Bengaard, H.S., Norskov, J.K., Sehested, J.S., Clausen, B.S., Nielsen, L.P., et al. (2002) Steam reforming and graphite formation on Ni catalysts. J. Catal. 209: 365.

[36] Rostrup-Nielsen, J.R. (2012) Perspective of Industry on Modeling Catalysis. In: Deutschmann, O., editor. Modeling and Simulation of Heterogeneous Catalytic Reactions: From the Molecular Process to the Technical System. Weinheim, Germany: Wiley-VCH, pp. 283–301.

[37] Norskov, J.K., Abild-Pedersen, F., Studt, F., Bligaard, T. (2011) Density functional theory in surface chemistry and catalysis. Proc. Nat. Acad. Sci. 108: 937–943.

[38] Campanati, M., Fornasari, G., Vaccari, A. (2003) Fundamentals in the preparation of heterogeneous catalysts. Catal. Today 77: 299–314.

[39] Fokin, S. (1913) Catalytic oxidation reaction at high temperatures, Preliminary communication, J. Russ. Phys. Chem. Soc. 45: 286-288.

[40] Xu, Q., Pearce, G.K., Field, R.W. (2017), Pressure driven inside feed (PDI) hollow fibre filtration: Optimizing the geometry and operating parameters, J. Membrane Sci., 537: 323–336. https://www.sciencedirect.com/science/article/pii/S0376738817306403

[41] https://en.wikipedia.org/wiki/Rotary_vacuum-drum_filter#/media/File:Rotary_vacuum-drum _filter.svg

[42] https://www.tsk-g.co.jp/en/tech/industry/horizontal-belt-filter.html

[43] http://trends.directindustry.com/comber-process-technology-srl/project-34050-131310.html

[44] http://www.solidliquid-separation.com/pressurefilters/nutsche/nutsche.htm

[45] http://www.gusu.com.cn/index_e.asp

[46] An, K., Alayoglu, S., Ewers, T., Somorja, G.A. (2012) Colloid chemistry of nanocatalysts: A molecular view. J. Colloid. Interf. Sci. 373: 1.

[47] Jin, R. (2012) The impacts of nanotechnology on catalysis by precious metal nanoparticles. Nanotechnol. Rev. 1: 31–56.

[48] Mukhlenov, I.P., Dobkina, E.I., Derjuzhkina, V.I. (1989) Catalysts Technology. Leningrad: Khimia.

[49] Rusanov, A.I. (2012) The development of the fundamental concepts of surface thermodynamics, Colloid J. 74: 136–153.

[50] Pakhomov, N.A. (2011) Scientific Basis for Catalyst Preparation. Novosibirsk: Institute of Catalysis.

[51] Moulijn, J.A. http://www.tnw.tudelft.nl/fileadmin/Faculteit/TNW/Over_de_faculteit/ Afdelingen/Chemical_Engineering/Research/Catalysis_Engineering/Education/ Miscellaneous/doc/catcat03.ppt.

[52] Kolesnikov, I.M. (2004) Catalysis and Production of Catalysts. Moscow, Russia: Teknika.

[53] Bathgate, G., Iyuke, S., Kavishe, F. (2012) Comparison of straight and helical nanotube production in a swirled fluid CVD reactor. ISRN Nanotechnology. Article ID 985834, doi:10.5402/2012/985834. http://www.hindawi.com/isrn/nanotechnology/2012/985834/

[54] Bordere, S., Corpart, J.M., Bounia, El., Gaillard, N.E., Passade-Boupat, P.N., Piccione, P.M., Plee, D., Industrial production and applications of carbon nanotubes. http://www.graphis trength.com/export/sites/graphistrength/.content/medias/downloads/literature/General-in formation-on-carbon-nanotubes.pdf

[55] www.bayerbms.com

[56] Jiménez, V., Ramírez-Lucas, A., Díaz, J.A., Sánchez, P., Romero, A. (2012) CO_2 capture in different carbon materials, Environ. Sci. Technol. 46: 7407–7414.

[57] Strobel, R., Baiker, A., Pratsinis, S.E. (2006) Aerosol flame synthesis of catalysts. Adv. Powder Technol. 17: 457–480.

[58] Hanefeld, U, Lefferts, L. (Ed). (2018) Catalysis. An integrated textbook for students, Wiley, Weinheim.

[59] Kubicka, D., Kumar, N., Venäläinen, T., Karhu, H. Kubickova, I., et al. (2006) The metal-support interactions in zeolite-supported noble metals: the influence of metal crystallites on the support acidity. J. Phys. Chem. B. 110: 4937–4946.

[60] Bulut, M., Jacobs, P.A. Concepts for preparation of zeolite-based catalysts, in Synthesis of Solid Catalysts (ed. K.P. De Jong), Wiley, Weinheim, 2009, 243–376.

[61] Grand, J., Awala, H., Mintova, S. (2016) Mechanism of zeolites crystal growth: new findings and open questions, CrystEngCom, 18: 650–664.

[62] Kumar, N., Nieminen, V., Demirkan, K., Salmi, T., Murzin, D.Yu. et al. (2002) Effect of synthesis time and mode of stirring on physico-chemical and catalytic properties of ZSM-5 zeolite catalysts. Appl. Catal. A-Gen. 235: 113.

[63] Fraile, J.M., Herrerías, C.I. (2011) CAFC9: 9th congress on catalysis applied to fine chemicals. Platinum Met. Rev. 55: 13-19.

[64] http://www.chem.strath.ac.uk/people/academic/lorraine_gibson/research/sorbents.

[65] Twigg, M.V., Richardson, J.T. (2002) Theory and applications of ceramic foam catalysts, Chem. Eng. Res. Des. 80: 183–189.

[66] Vergunst, T., Kapteijn, F., Moulijn, J.A. (2001) Monolithic catalysts - non-uniform active phase distribution by impregnation. Appl. Cat. A: Gen. 213: 179–187.

[67] http://www.selee.com/Selee_Corporation_Metal_Foam.php.

[68] Fey, T., Betke, U., Rannabauer, S., Scheffle, M. (2017) Reticulated replica ceramic foams: processing, functionalization, and characterization, Adv. Eng. Mat. 19: 170036.

[69] Spieker, W.A., Regalbuto, J.R. (2001) A fundamental model of platinum impregnation onto alumina. Chem. Eng. Sci. 56: 3491–3504.

[70] Geus, J.W., van Veen, J.R. (1999) Preparation of supported catalysts, Chapter 10, In: Moulijn, J.A., van Leeuwen, P.W.N.M., van Santen, R.A., editors. Catalysis, An Integrated Approach to Homogeneous, Heterogeneous and Industrial Catalysis. Amsterdam, The Netherlands: Elsevier, p. 467.

[71] Jiao, L., Regalbuto, J.R. (2008) The synthesis of highly dispersed noble and base metals on silica via strong electrostatic adsorption: I. Amorphous silica. J. Catal. 260: 329–341.

[72] Marceau, E., Carrier, X., Che M. (2009) Impregnation and drying. In: de Jong, K.P., editor. Synthesis of Solid Catalysts. Weinheim, Germany: Wiley-VCH.

[73] Liu, X., Khinast, J.G., Glasser, B.J. (2008) Parametric investigation of impregnation and drying of supported catalysts. Chem. Eng. Sci. 63: 4517–4530.

[74] Hagen, J. (2006) Industrial Catalysis. Weinheim, Germany: Wiley-VCH.

[75] Stillwell, W.D. (1957) Preformed catalysts and techniques of tableting. Ind. Eng. Chem. 49: 245–249.

[76] https://prodselmash.ru/specializirovannoe-oborudovanie/ustanovki-dla-vypuska-vysokotehnologicnoj-produkcii/61

[77] Murzin, D.Yu., Simakova, O.A., Simakova, I.L., Parmon, V.N. (2011) Thermodynamic analysis of the cluster size evolution in catalyst preparation by deposition-precipitation. React. Kinet. Catal. Let. 104: 259–266.

[78] Lu, J., Lei, Y., Elam J.W. Atomic layer deposition of noble metals – new developments in nanostructured catalysts. Noble Metals. doi: 10.5772/33082.

[79] www.oxford-instruments.com.

[80] Schüth, F., Hesse, M. (2008) Catalyst forming. In: Ertl, G., Knözinger, H., Schüth, F., Weitkamp, J., editors. Handbook of Heterogeneous Catalysis. Weinheim, Germany: Wiley-VCH. 676–699.

[81] http://www.niroinc.com/food_chemical/spray_dryer_selection.asp.

[82] Roberie, T., Hildebrandt, D., Creighton, J., Gilson, J.P. (2002) Preparation of zeolite catalysts, Chapter 3, in Zeolites for Cleaner Technologies. Guisnet, M.J., Gilson, P. Editors. London, UK: Imperial College Press, 57-73. Available from http://www.granmix.ru/equip/.

[83] https://media.licdn.com/dms/image/C5612AQFeVY3UvI9UJA/article-inline_image-shrink _1500_2232/0?e=1568851200&v=beta&t=YNcarijTqHhqPeqqFZf3Ci4JGqS4bUPVvYs8DSIkk_c

[84] http://www.thermopedia.com/content/5408/762NNFFig1.gif

[85] Devyatkov, S. Yu., Zinnurova, A.A., Aho, A., Kronlund, D., Peltonen, J., Kuzichkin, N. V., Lisitsyn, N. V., Murzin, D.Yu. (2016) Shaping of sulfated zirconia catalysts by extrusion: understanding the role of binders, Ind. Eng. Chem. Res., 55: 6595.

[86] Freiding, J., Patcas, F.-C., Kraushaar-Czarnetzki, B. (2007) Extrusion of zeolites: Properties of catalysts with a novel aluminium phosphate sinter matrix, Appl. Catal. A Gen. 328: 210.

[87] http://upload.wikimedia.org/wikipedia/commons/thumb/2/21/Al_extrusion_die_set.jpg/ 440px-Al_extrusion_die_set.jpg.

[88] Stuecker, J. N., Miller, J. E., Ferrizz, R. E., Mudd, J.E., Cesarano, J. (2004) Advanced support structures for enhanced catalytic activity, Ind. Eng. Chem. Res., 431: 51–55.

[89] Zhou, X., Liu, C.-J., (2017) Three-dimensional printing for catalytic applications: Current status and perspectives, Adv. Funct. Mater. 27, 1701134. https://onlinelibrary.wiley.com/doi/ 10.1002/adfm.201701134.

3 Engineering reactions

3.1 Introduction

In order to understand a process it is important to have a model describing the factors that control the process outcome. Knowledge of reaction thermodynamics and kinetics is thus required and physico-chemical understanding of catalytic processes is essential for proper reactor and process design. According to the very definition of catalysis, it is a kinetic phenomenon, and thus reliable kinetic models, which describe the rates and equilibria of catalytic reactions, are of vital importance for solving the problems in mathematical modeling, design and intensification of chemical processes.

Moreover, the performance of catalytic processes depends on how the reaction is conducted and the physical state of reactants and equipment (for example, the reactor). Understanding of hydrodynamics, mass transfer and heat transfer along with reaction kinetics is needed in order to properly select conditions, equipment and operation mode (batch, semi-batch or continuous). It should also be remembered that while the intrinsic rate constants do not change with scale and equipment, rates of physical processes are scale- and equipment-dependent thus severely influencing the outcome of a particular heterogeneous catalyst reaction in terms of activity, conversion or selectivity.

As heterogeneous catalytic reactions inherently require several phases, interphase mass transfer is always present, increasing in complexity when the reaction system is multiphase, which comprises not only the solid catalysts but reactants in both gas and liquid (or several liquid) phases. In addition to influencing the reaction rate, the mass transfer can influence selectivity in complex reactions, as mentioned above and addressed in detail below. The heat transfer depends on the reaction rates, which could be concentration dependent.

Proper engineering on catalytic reactions needs understanding of the process chemistry and reaction mechanism but also data on physico-chemical properties of reactants, phase behavior, reaction thermodynamics and kinetics, catalyst deactivation, mass transfer and heat transfer.

3.2 Thermodynamics

Evaluation of thermodynamics is the starting point in reaction engineering, as it addresses the overall feasibility of the reaction and defines the composition of the reaction mixture at equilibrium and thus the maximum conversion and selectivity. In addition, the reaction rate depends on the approach to equilibrium, or the distance

https://doi.org/10.1515/9783110614435-003

of the reaction mixture composition to the equilibrium one, as at equilibrium the rate of the forward and reverse reactions is equal to each other.

Thermodynamic conditions for equilibrium system:

$$aA + bB = cC + dD \tag{3.1}$$

can be expressed through the Gibbs energy, which must be stationary. This means that the derivative of G with respect to the extent of reaction, ξ, is zero, or that the sum of chemical potentials μ (partial molar Gibbs energy) of the products and reactants are equal to each other:

$$a\mu_A + b\mu_B = c\mu_C + d\mu_D \tag{3.2}$$

where chemical potentials of reagents depend on activity:

$$\mu_i + \mu_i^0 = RT \ln a. \tag{3.3}$$

In eq. (3.3) μ_i^0 is the standard chemical potential. If a mixture is not at equilibrium, the driving force for the reaction to approach equilibrium, leading to subsequent changes in composition, is the minimization of the excess Gibbs energy (or Helmholtz energy at constant volume reactions). The equilibrium constant for eq. (3.1):

$$K = \frac{[A]^a [B]^b}{[C]^c [D]^d} \tag{3.4}$$

is related to the standard Gibbs energy change for the reaction by the equation:

$$\Delta G_T^0 = - RT \ln K_{eq} \tag{3.5}$$

where R is the universal gas constant and T the temperature.

The Gibbs energy is defined through reaction enthalpy ΔH_T^0 and entropy ΔS_T^0

$$\Delta G_T^0 = \Delta H_T^0 - T \Delta S_T^0 \tag{3.6}$$

The temperature dependence of the standard Gibbs energy change of the reaction is given by:

$$\frac{\partial \Delta G^0}{\partial T} = \frac{\Delta G^0 - \Delta H^0}{T} \tag{3.7}$$

After rearrangement and division by RT, we arrive at an expression relating reaction enthalpy and equilibrium constant:

$$\frac{\Delta H^0}{RT^2} = -\left(\frac{1}{RT} \frac{\partial \Delta G^0}{\partial T} - \frac{\Delta G^0}{RT^2} \right) = - \frac{\partial \left(\frac{\Delta G^0}{RT} \right)}{\partial T} = \frac{\partial (\ln K)}{\partial T} \tag{3.8}$$

The standard enthalpy changes are given by:

$$\Delta H_T^0 = \sum v_i \Delta H_{298K}^0 (\text{prod}) - \sum v_i \Delta H_{298K}^0 (\text{react}) + \int_{298K}^{T} \left(\sum_{i=1}^{N_i} v_i c_{p,i} \right) dT \tag{3.9}$$

where $c_{p,i}$ is the temperature-dependent heat capacity, which for gases is given by:

$$c_{p,i} = a_i' + b_i' T + c_i' T^2 + d_i' T^{-2} \tag{3.10}$$

In eq. (3.10) a, b, c and d are coefficients. The rates of forward and reverse reactions are related through thermodynamics in the following way. Expressing the rates of the forward and the reverse reactions, respectively:

$$r_+ = k_+ \prod_I C_I^{\alpha_+, i}, r_- = k_- \prod_I C_I^{\alpha_-, i} \tag{3.11}$$

the ratio of rates is:

$$\frac{r_+}{r_-} = \frac{k_+}{k_-} \prod_I C_I^{\alpha_+, i - \alpha_-, i} = K \prod_I C_I^{\alpha_+, i - \alpha_-, i} \tag{3.12}$$

with the equilibrium constant defined as $K = k_+/k_-$. In eq. (3.11) $\alpha_{+,i}$ stands for reaction order in compound I in the forward direction. Equation (3.12) can be rearranged to the De Donder equation:

$$\frac{r_+}{r_-} = e^{A/RT} \tag{3.13}$$

where A is the reaction affinity:

$$A = RT \ln K \prod_I C_I^{\alpha_+, i - \alpha_-, i} \tag{3.14}$$

which is defined as the derivative of the Gibbs free energy with respect to the extent of the reaction:

$$A = \left(-\frac{\partial G}{\partial \xi} \right)_{T,p} \tag{3.15}$$

The relationship between the overall process rate and the rate in the forward reactions is also expressed through reaction affinity:

$$r = r_+ - r_- = r_+ \left(1 - e^{\frac{-A}{RT}} \right) = r_+ \left(1 - \varphi \frac{(C)}{K} \right) \tag{3.16}$$

where r is the overall rate, r_+ is the rate of the forward reaction, $\varphi(C)$ is a function of only reactants concentrations and K is the equilibrium constant. For a reversible reaction in eq. (3.1), assuming that stoichiometry corresponds to molecularity, the rate is given by:

$$r = k_+ C_A^a C_B^b - k_- C_C^c C_D^d = k_+ C_A^a C_B^b \left(1 - \frac{k_- C_C^c C_D^d}{k_+ C_A^a C_B^b} \right) = r_+ \left(1 - \frac{1}{K} \frac{C_C^c C_D^d}{C_A^a C_B^b} \right) \tag{3.17}$$

where C_A, etc. are partial pressures (in the case of gas-phase reactions) or concentrations (for liquid-phase reactions).

3.3 Kinetics

3.3.1 Definitions

Chemical kinetics studies the rates of chemical reactions and deals with experimental measurements in batch, semi-batch or continuous reactors. It addresses how the reaction rates depend on concentrations, temperature, nature of catalyst, pH and solvents, to mention a few reaction parameters.

Experimental kinetic investigations are one of the powerful ways to reveal the reaction mechanisms. The following problems can be solved by using a kinetic model:

- choosing the catalyst and comparing the selectivity and activity of various catalysts and their performance under optimum conditions for each catalyst;
- determination of the main and by-products formed during the process;
- determination of the optimum sizes and structures of catalyst grains and the necessary amount of the catalyst to achieve the specified values of conversion and selectivity;
- elucidation of the dynamics of the process and the determination of the behavior of the reactor under unsteady state conditions;
- revealing the influence of mass and heat transfer processes on the chemical reaction rates and product selectivity as well as the determination of the kinetic region of the process;
- selecting the type of a reactor and structure of the contact unit that provides the best approach to the optimum conditions.

Kinetic data thus forms the basis of catalysis, allowing the design and modeling of catalytic reactors. In addition to these practical aspects, kinetic analysis is also important in the elucidation of catalytic reaction mechanisms.

Formulation of the reaction mechanism needed for kinetic modeling is difficult and should be based on all available information about the reaction chemistry, kinetic experiments in steady and non-steady-state conditions, physical measurements, surface science data and theory.

Accurate kinetic measurements with high precision are needed as large devia-tions in the values of the experimentally measured rates pose serious obstacles for quantitative considerations. Such deviations could be caused by variations in the feedstock, inhomogeneity of catalyst samples, alterations of reaction parameters dur-ing a reaction, influence of heat and mass transfer. Another necessary feature is the ability to generate maximum kinetic information in the minimum time. The analysis of products as well as reactor lay-out should preferably be as easy as possible.

It should be noted that on many occasions kinetic calculations are based on erro-neous data and interpretations. In Fig. 3.1, conversion after a certain period of time reaches 100% independent of the catalyst applied. For example, reported values of 100% conversion after 24 h and corresponding turnover frequency (TOF) values de-fined by eq . (1.3) are thus often misleading, because catalysts have very different activ-ity. There is also an apparent danger in using the values of initial rates, as the reaction rates are not obtained directly from measurements. As an illustration, in Fig. 3.1 a case is presented when the calculation of the initial rate based on the first measured experi-mental point for the most active catalyst obviously gives an overestimation.

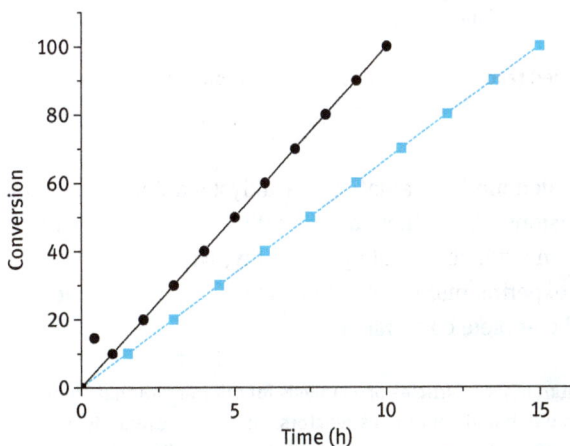

Fig. 3.1: Concentration vs time dependence for difference catalysts (see explanations in the text).

For simple kinetics it is possible to apply algebraic equations for batch reactors. Differential equations representing mass balances can be also analytically integrated rather easily for simple cases.

In heterogeneous catalysis, turnover frequency is often reported as an indicator of catalyst activity. In organometallic catalysis, turnover number (TON) is used, which has a different meaning, that is, the number of moles of substrate that a mole of catalyst can convert until it is deactivated completely. Reporting TON as the number of transformations during a certain period of time without proving that the catalytic activity is completely lost would be thus misleading.

Another common mistake regarding interpretation of deactivation is illustrated in Fig. 3.2. It is sometimes considered that when conversion is 100% in fixed-bed reactors and the activity decline is visible only after a certain time-on-stream, there is no deactivation prior to that point. In fact, deactivation is also present even when conversion is 100%, but the whole catalyst bed is not required to reach the full conversion and the catalyst could be much more active initially. Stability of catalysts should be thus analyzed at intermediate conversion levels, which can be done by decreasing the residence time (i.e., higher flow or less catalyst).

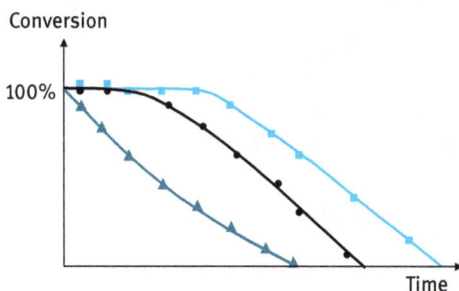

Fig. 3.2: Experimental data in a fixed-bed reactor showing an initial 100% conversion that declines after some time-on-stream.

For reactions performed in the batch mode, evaluation of catalyst stability is done by catalyst recycling. In many occasions, the authors are reporting conversion of 100% after a certain period of time as an evidence of catalyst stability, which is misleading as elaborated earlier. In batch experiments, the reaction times or catalyst amounts should be thus selected to avoid complete conversion.

The author is reviewing a significant number of articles on catalysis for different journals reporting kinetic measurements mainly in batch and continuous reactors. There are several details of experimental procedures, which are very often not reported. Few of them are listed below and can be even used by the readers as a checklist, while preparing their next papers. I hope that it will help in smoothening the reviewing process as the reviewers, including myself, will be less frustrated asking the same questions over and over again.

One of the key questions is related to a potential impact of mass and heat transfer limitations. Differences between different catalysts could be masked if the reaction is performed under a significant influence of mass transfer. Simple diagnostic tests or more elaborative calculations can be done to evaluate mass transfer limitations. In particular, to assess the internal mass transfer (i.e., inside the catalyst grain), knowing the size of the catalyst particles is critical. For batch slurry reactors, contrary to fixed-bed reactors, the size of catalyst grains (in the micron range) is, however, frequently missing in scientific publications. Seldom calculations of the catalyst effectiveness factor (will be discussed in Section 3.5) are done. The liquid load in a batchwise-operated slurry reactor should be approximately 0.3–0.5 of the total reactor volume to allow efficient energy dissipation. In many instances, the load is much lower, making the reviewers wondering if the impeller is really in the liquid, but not above it. It is not enough to report the stirring speed to assess the external mass transfer limitations, as it is the relative ratio between the mass transfer to the external catalyst surface to reaction kinetics which is important, rather than the absolute values of the mass transfer coefficient.

In many instances, the catalyst and the reactant are introduced simultaneously and then the mixture is heated to the reaction temperature. Apparently, things are happening during heating and should be properly accounted for. A better option is to heat the reactant and catalyst separately.

When sampling of the reaction mixture is done, the volume of samples and details on the presence of the catalyst in the samples are rarely reported. Depending on how the samples are taken, the catalyst bulk density can, however, change, influencing interpretation of the kinetic data.

Often three-phase reactions are performed under hydrogen or oxygen. Only the overall pressure is often reported without specifying if this was an overpressure (manometer reading) or the total pressure and what was the vapor pressure of the solvent at the reaction temperature. It thus remains a mystery what was the pressure of the reacting gas, which is important for evaluation of the reaction order and reaction kinetics.

In papers describing experiments in a continuous mode quite often, it is not clear at which temperature the flow rates or residence times are reported (i.e., room or reaction temperature) and how gas expansion was taken into account in calculation of selectivity.

Another typical mistake is not to report the mass balance (or carbon balance) closure.

An essential feature for catalytic reactions is the readiness in reduction/activation of heterogeneous catalysts and the possibility to use them in the required geometrical form. As activity deterioration takes place in many catalytic reactions (contrary to the definition of catalysis) there should be always activity control in measurements of catalytic kinetics.

A measure of catalytic activity is the reaction rate, which is defined through the extent of the reaction. The change in the extent of reaction (number of chemical transformations divided by the Avogadro number) is given by $d\xi = dn_B/v_B$, where v_B is the stoichiometric number of any reaction entity B (reactant or product) and n_B is the corresponding amount.

This extensive property $d\xi/dt$ is measured in moles and cannot be considered as reaction rate, since it is proportional to the reactor size.

For a homogeneous reaction, when the rate changes with time and is not uniform over a reactor volume v, the reaction rate is:

$$r = \frac{\partial^2 \xi}{\partial t \partial v} \tag{3.18}$$

while for the constant reactor volume it is defined as:

$$r_i = \frac{1}{v_i} \frac{dC_i}{dt} \tag{3.19}$$

where i is the reactant or product with a corresponding stoichiometric coefficient v_i.

For a heterogeneous reaction occurring over a reaction space S (catalyst surface, volume weight, or number of active sites) the rate expression is given by:

$$r = \frac{\partial^2 \xi}{\partial t \partial S} \tag{3.20}$$

leading to further simplifications when the rate is uniform across the surface:

$$r = \frac{1}{S} \frac{\partial \xi}{\partial t} \tag{3.21}$$

Rate laws express how the rate depends on concentration and rarely follows the overall stoichiometry as given in eq. (3.1). In fact, reaction molecularity (the number of species that must collide to produce the reaction) determines the form of a rate equation. Elementary reactions occur when the rate law can be written from its molecularity and when the kinetics depend only on the number of reactant molecules in that step. For elementary reactions the reaction orders have integral values typically equal to 1 and 2 (Fig. 3.3), or occasionally 3 for tri-molecular reactions.

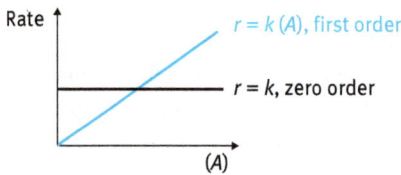

Fig. 3.3: Reaction kinetics of first and zero order.

Reaction orders, which for a particular reaction can be fractional ($r_A = -kc_A^n$) indicating a complex reaction mechanism, could be determined using logarithmic plots:

$$\ln\left(\frac{-dc_A}{dt}\right) = \ln k + n \ln c_A \tag{3.22}$$

with the reaction order corresponding to the slope.

The temperature dependence of the reaction constants is expressed by the Arrhenius equation:

$$k = A e^{\frac{-E_a}{RT}} \tag{3.23}$$

where the pre-exponential factor A weakly depends on temperature, which is often neglected. The Arrhenius equation should be applied for elementary reactions, and thus typically apparent values of activation energies are reported for multistep heterogeneous catalytic reactions. Such values could be easily determined by taking the natural logarithms of both sides of the eq. (3.23):

$$\ln k = \ln A - \frac{E_a}{R}\left(\frac{1}{T}\right)$$ (3.24)

A plot of $\ln k$ vs $\frac{1}{T}$ gives a straight line with a slope of $-\frac{Ea}{R}$ and an intercept of $\ln A$.

3.3.2 Reaction mechanism

A mechanism of a catalytic reaction comprises several elementary steps. A linear mechanism contains steps that contain only one adsorbed species on both sides of these elementary steps, while non-linear steps might have more than one species.

A proposal for a reaction mechanism should correspond to the principle of microscopic reversibility, which implies that reactions proceed in both directions through the same set of elementary steps and intermediates on the same type of catalytically active centers. A rate-limiting step reaction, if any, should be the same in both directions.

Derivation of kinetic equations can be substantially simplified using the steady-state approximation originally developed by Bodenstein. In general, a reaction is considered to be at steady-state if the concentrations of all species in each element of the reaction space (surface coverage of reactive intermediates) do not change in time (Fig. 3.4).

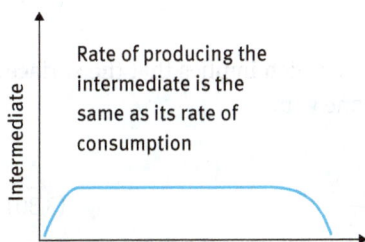

Fig. 3.4: Concentration of a reaction intermediate as a function of reaction time.

According to Bodenstein there are intermediates in chemical reactions that are present in "inferior" amounts, for example, in much less concentrations than the major species in the mechanism. Under these conditions, the rate of concentration changes of an intermediate can be considered negligible. The application of the steady-state approximation to heterogeneous catalysis does not require that the surface concentrations are low, but implies that they do not change with time:

$$\frac{d\theta}{dt} = 0$$ (3.25)

Another formulation of the steady-state approximation resulting from eq. (3.25) is that all the steps in a catalytic reaction, occurring at steady-state, are equal to each other.

To illustrate the steady-state approximation, an isomerization of A to B will be considered:

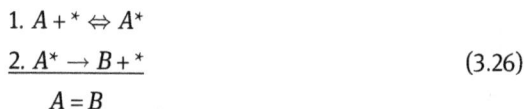

$$
\begin{aligned}
&1.\ A + * \Leftrightarrow A^* \\
&2.\ A^* \rightarrow B + * \\
&\quad A = B
\end{aligned}
\tag{3.26}
$$

where $*$ is the surface site and the second step is irreversible. The rates for elementary reactions 1 and 2 in (3.26) are given by:

$$
r_1 = r_{+1} - r_{-1} = k_{+1} C_A \theta_v - k_{-1} \theta_A, \qquad r_2 = r_{+2} = k_{+2} \theta_A
\tag{3.27}
$$

where θ_v and θ_A is the fraction (coverage) of vacant sites and adsorbed A, respectively. The overall rate of the product formation is equal to the rate of the second step in the reaction mechanism. In the steady-state hypothesis the rate is:

$$
r_{A^*} = \frac{d\theta_A}{dt} = r_{+1} - r_{-1} - r_{+2} \approx 0
\tag{3.28}
$$

leading to the dependence of the coverage of adsorbed A on the fraction of vacant sites:

$$
\theta_A = \frac{k_{+1} C_A}{k_{-1} + k_{+2}} \theta_v
\tag{3.29}
$$

Taking into account the balance equation $\theta_A + \theta_v = 1$ which implies that the surface sites are either vacant or occupied by adsorbed A, one gets:

$$
\theta_A = \frac{\dfrac{k_{+1} C_A}{k_{-1} + k_{+2}}}{\left(1 + \dfrac{k_{+1} C_A}{k_{-1} + k_{+2}}\right)}
\tag{3.30}
$$

With this balance equation it is implicitly considered that the rate is proportional to the amount of catalytic sites.

The rate of product B formation in the steady-state is equal to the rate of each step, which could be taken as the rate of the second step for simplicity of derivation:

$$
-r_A = r_B = k_{+2} \theta_A = \frac{\dfrac{k_{+2} k_{+1} C_A}{k_{-1} + k_{+2}}}{\left(1 + \dfrac{k_{+1} C_A}{k_{-1} + k_{+2}}\right)}
\tag{3.31}
$$

Quite often, adsorption and desorption [for example, step 1 in eq. (3.26)] are considered to be much faster than the reaction step. It implies that $k_{-1} > k_{+2}$ and subsequently:

$$r = \frac{\dfrac{k_{+2}k_{+1}C_A}{k_{-1}}}{1 + \dfrac{k_{+1}C_A}{k_{-1}}} = \frac{k_{+2}K_A C_A}{1 + K_A C_A} \tag{3.32}$$

with $k_{+1}/k_{-1} = K_1 = K_A = K_{A,0}\exp\left(\left[\dfrac{(-\Delta H_{ads})}{RT}\right]\right)$, where $(-\Delta H_{ads})$ is the heat of adsorption of a molecule, A, on the active catalyst surface. The exothermic nature of adsorption ($\Delta H < 0$) implies that K_A decreases with temperature.

When the rates of a step in both directions are significantly higher than the rates of some other steps, such step is considered to be at quasi-equilibrium (Fig. 3.5). This is because $r_+/r_+ \approx 1$. In heterogeneous catalysis, adsorption and desorption are usually considered to be much faster than the steps of surface reactions, making it possible to apply adsorption isotherms (for example Langmuir isotherm) directly to these steps.

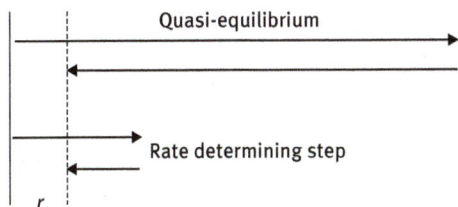

Fig. 3.5: Illustration of quasi-equilibrium and rate-determining step concept.

Two extreme limits are possible for eq. (3.32), the first when concentrations are low and $K_A C_A \ll 1$, then the denominator is close to unity, that is, $(1 + K_A C_A) \sim 1$ The rate is then proportional to the reactant concentration: $r \sim k_{+2} K_A C_A$, as a first-order reaction with respect to concentration of A with an apparent first-order rate constant, k' = $k_{+2}K_A$ (Fig. 3.6). This is the low concentration or pressure (or weak binding i.e., small K) limit: under these conditions the steady-state surface coverage, θ, of the reactant molecule is very small.

The other limit is at high concentrations, $K_A C_A \gg 1$; then $(1 + K_A C_A) \sim K_A C_A$ and the rate is $r \sim k_{+2}$, giving a zero order reaction with respect to the concentration of A. This is the high concentration (or strong binding, i.e., large K) limit. Under these conditions the steady-state surface coverage, θ, of the reactant molecule is almost unity. The rate shows the same concentration variation as surface coverage in the Langmuir adsorption isotherm.

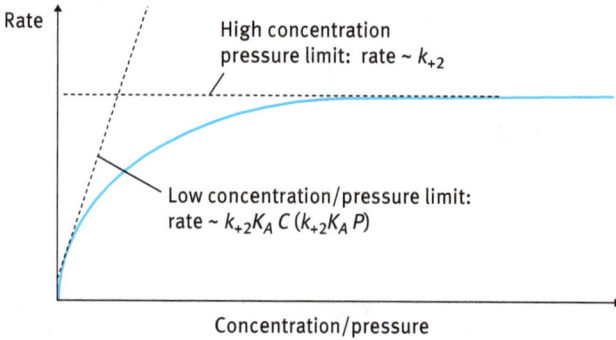

Fig. 3.6: Concentration/Pressure dependence of reaction rate (eq. 3.32).

The law of mass action applied to processes on surfaces implies that rate expressions and adsorption isotherms contain concentrations. For gas-phase reactions, it is tempting to utilize partial pressures rather than the concentrations as shown in Fig. 3.6. Strictly speaking from the ideal gas law $PV = nRT$ and thus $C = P/RT$ giving instead of eq. (3.32)

$$r = = \frac{k_{+2}K_A P_A / RT}{1 + K_A P_A / RT} = \frac{k_{+2}K'_A P_A}{1 + K'_A P_A} \tag{3.33}$$

The adsorption equilibrium constant defined through pressures K_A' is thus different from the adsorption constant expressed through concentrations.

A gas-phase reaction can proceed according to the Langmuir-Hinshelwood mechanism:

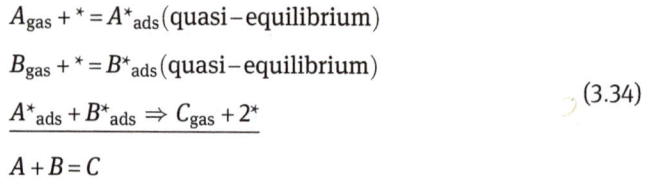

$$A_{\text{gas}} + ^* = A^*_{\text{ads}} \,(\text{quasi}-\text{equilibrium})$$
$$B_{\text{gas}} + ^* = B^*_{\text{ads}} \,(\text{quasi}-\text{equilibrium})$$
$$\underline{A^*_{\text{ads}} + B^*_{\text{ads}} \Rightarrow C_{\text{gas}} + 2^*} \tag{3.34}$$
$$A + B = C$$

In this mechanism the surface reaction between the two adsorbed species competing on the surface is the rate-determining step. The reaction rate can be easily written as:

$$r = k\theta_A \theta_B = k \frac{K_A C_A}{1 + K_A C_A + K_B C_B} \frac{K_B C_B}{1 + K_A C_A + K_B C_B} = \frac{k K_A C_A K_B C_B}{(1 + K_A C_A + K_B C_B)^2} \tag{3.35}$$

When $K_A C_A \ll 1$ and $K_B C_B \ll 1$ both coverage θ_A and θ_B are very low, and the rate is:

$$r = k K_A C_A K_B C_B \tag{3.36}$$

meaning that the reaction is first-order in both reactants. If $K_A C_A \ll 1 \ll K_B C_B$, the coverage of A is low, while the surface is predominantly covered by B ($\theta_A \to 0$, $\theta_B \to 1$), and the rate is becoming first-order in A, but negative first-order in B:

$$r = \frac{k K_A C_A}{K_B C_B} \tag{3.37}$$

Depending upon the concentration (partial pressure) and binding strength of the reactants, a given model for the reaction scheme can give rise to a variety of apparent kinetics. This highlights the dangers inherent in the reverse process – namely trying to use kinetic data to obtain information about the reaction mechanism.

If the product C adsorbs on the catalyst surface the rate equation is becoming slightly more complicated, as the coverage of C should be included in the denominator:

$$r = \frac{k K_A C_A K_B C_B}{(1 + K_A C_A + K_B C_B + K_C C_C)^2} \tag{3.38}$$

If the reaction is reversible, with the rate determining step $A^{\star}_{\text{ads}} + B^{\star}_{\text{ads}} \Leftrightarrow C^{\star}_{\text{ads}} + {}^{\star}$ then:

$$r = k_+ \theta_A \theta_B - k_- \theta_C \theta_* = \frac{k_+ K_A C_A K_B C_B}{(1 + K_A C_A + K_B C_B + K_C C_C)^2} - \frac{k_- K_C C_C}{(1 + K_A C_A + K_B C_B + K_C C_C)^2}$$

$$= \frac{k_+ K_A C_A K_B C_B \left(1 - \frac{k_- K_C C_C}{k_+ K_A C_A K_B C_B} \right)}{(1 + K_A C_A + K_B C_B + K_C C_C)^2} = r_+ \left(1 - \frac{1}{K} \frac{C_C}{C_A C_B} \right) \tag{3.39}$$

Another possibility is that a molecule from the gas-phase reacts with an adsorbed molecule without adsorbing itself on the surface (for example, the Eley-Rideal process). For the mechanism:

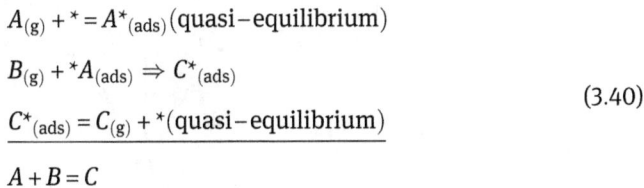

$$A_{(g)} + {}^* = A^*_{(\text{ads})} \, (\text{quasi-equilibrium})$$

$$B_{(g)} + {}^* A_{(\text{ads})} \Rightarrow C^*_{(\text{ads})}$$

$$\underline{C^*_{(\text{ads})} = C_{(g)} + {}^* (\text{quasi-equilibrium})} \tag{3.40}$$

$$A + B = C$$

the reaction rate for an irreversible reaction is given by:

$$r = k_+ \theta_A C_B = \frac{k_+ K_A C_A C_B}{1 + K_A C_A + K_C C_C} = \frac{k_+ K'_A P_A P_B}{1 + K'_A P_A + K'_C P_C} \tag{3.41}$$

For low pressures $K'_A P_A \ll 1$ and the rate is first-order in both reactants, similar to the Langmuir-Hinshelwood mechanism. However, the rate cannot have negative order dependence as followed from Langmuir-Hinshelwood treatment.

Monomolecular isomerization in the adsorbed layer is a simplified case of the mechanism (3. 40)

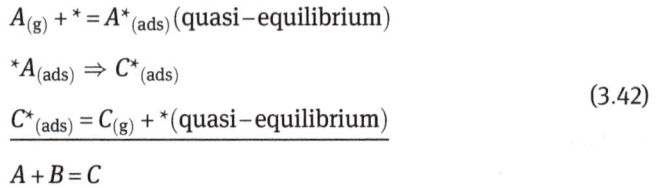

$$A_{(g)} + * = A^*_{(ads)} \, (\text{quasi}-\text{equilibrium})$$

$$*A_{(ads)} \Rightarrow C^*_{(ads)}$$

$$\underline{C^*_{(ads)} = C_{(g)} + * (\text{quasi}-\text{equilibrium})} \tag{3.42}$$

$$A + B = C$$

With the corresponding rate equation

$$r = k_+ \, \theta_A = \frac{k_+ K_A C_A}{1 + K_A C_A + K_C C_C} \tag{3.43}$$

The energy diagram for mechanism (3.42) is shown in Fig. 3.7, illustrating that the observed (apparent activation energy) is different from the activation energy of the isomerization in the adsorbed state per se. The concept of activation energy will be elaborated in section 3.4.4.

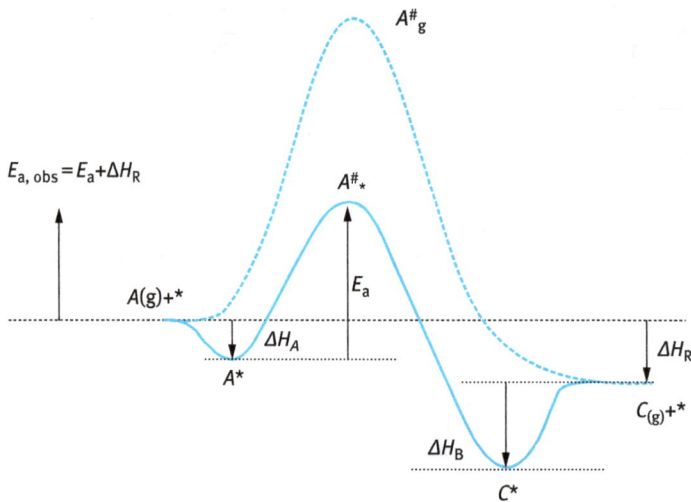

Fig. 3.7: Energy diagram for the reaction of A to C. For illustration purposes, a noncatalytic reaction is also shown.

As mentioned earlier, a standard choice in kinetic modeling is to use concentrations, although for gas-phase kinetics it is common to apply partial pressures, often neglecting a more complex behavior of the adsorption constant on temperature. This approach is straightforward for ideal gases. In the case of high pressures and deviations from ideal gas law it is, however, recommended that gas-phase fugacities in rate equations are used instead of partial pressures. For liquid-phase systems concentrations are usually used, although for strongly non-ideal solutions they should be replaced by activities, which could be rather complicated.

At this point, some generalization of kinetic models is possible. For a reaction $A + B = C + D$ the reaction rate takes a form:

$$r = \frac{\text{kinetic factor* driving force}}{(\text{adsorption term})^n} \tag{3.44}$$

where the adsorption term includes $K_A C_A$, etc. for molecular adsorption or $(K_A C_A)^{1/2}$ for adsorption with dissociation and the power in the denominator corresponds to the number of species in the rate-determining steps, while the driving force is $(1 - P_C P_D / K_{eq} P_A P_B)$ or $\left(1 - C_C C_D / K_{eq} C_A C_B\right)$.

The classical application of Langmuir kinetics to uniform surfaces can be extended to the reactions that occur on two types of sites in close proximity. The reaction proceeds via interactions between species adsorbed on such distinct sites. Examples of such reactions are various oxidation reactions on metal oxides, which contain two types of sites, metal atoms sites and oxygen atom sites, each having a specific function. Another example is hydrogenolysis of haloarenes where spillover hydrogen reacts with an organic molecule adsorbed on the metal surface. The mechanism is

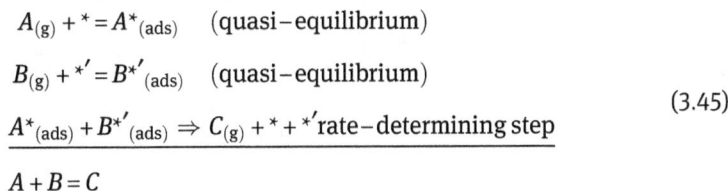

$$A_{(g)} + * = A^*_{(ads)} \quad \text{(quasi–equilibrium)}$$

$$B_{(g)} + *' = B^{*'}_{(ads)} \quad \text{(quasi–equilibrium)}$$

$$\underline{A^*_{(ads)} + B^{*'}_{(ads)} \Rightarrow C_{(g)} + * + *' \text{rate–determining step}} \tag{3.45}$$

$$A + B = C$$

Similarly, it can be written for the reaction rate:

$$r = k \theta_A \theta'_B = k \frac{K_A C_A}{1 + K_A C_A} \frac{K_B C_B}{1 + K_B C_B} \tag{3.46}$$

Dual site formalism can be easily extended to other cases, such as dissociative adsorption of for example B_2 with stepwise addition where the first addition step is rate limiting:

$$r = k \frac{K_A C_A}{1 + K_A C_A} \frac{\sqrt{K_B C_{B_2}}}{1 + \sqrt{K_B C_{B_2}}} \tag{3.47}$$

Models with semi-competitive adsorption could be also applied. The principle of semi-competitive adsorption is that the large organic molecules always leave some empty space in between them even at complete coverage, i.e., vacant interstitial sites, where the smaller molecule (typically hydrogen and oxygen) can adsorb. Compared to classical models, the semi-competitive adsorption model includes one additional parameter, namely the maximum coverage of the organic molecule. Some examples of rate equations for various catalytic mechanisms are given in Tab. 3.1.

Tab 3.1: Rate expressions for catalytic gas-phase reactions with various mechanisms (rds – rate-determining step). In the rate expressions, pressures are used instead of concentrations.

Surface reactions	Rate expression
1.$A + * = A*$ 2. $A* \Rightarrow B*$ r.d.s. 3. $B* = B + *$	$r_A = \dfrac{k_2 K_1 \left(P_A - \frac{P_B}{K_{eq}}\right)}{1 + K_1 P_A + K_3^{-1} P_B}$
1. $A + * \Rightarrow A*$ r.d.s. 2. $A* = B*$ 3. $B* = B + *$	$r_A = \dfrac{k_1 \left(P_A - \frac{P_B}{K_{eq}}\right)}{1 + (1 + 1/K_2) K_3^{-1} P_B}$
1.$A + * = A*$ 2.$A* = B*$ 3.$B* \Rightarrow B + *$ r.d.s.	$r_A = \dfrac{k_3 K_1 K_2 \left(P_A - \frac{P_B}{K_{eq}}\right)}{1 + (1 + K_2) K_1 P_A}$
1.$A + * = A$ 2.$B + * = B$ 3.$A* + B* \Rightarrow C* + *$ r.d.s. 4.$C* = C + *$	$r_A = \dfrac{k_3 K_1 P_A K_2 P_B}{\left(1 + K_1 P_A + K_2 P_B + K_4^{-1} P_C\right)^2}$
1.$A + * = A$ 2.$A* + B_{(g)} \Rightarrow C* + D_{(g)}$ r.d.s. 3.$C* = C + *$	$r_A = \dfrac{k_2 K_1 P_A P_B}{1 + K_1 P_A + K_3^{-1} P_C}$
1.$A_2 + *' = 2A*'$ 2.$B + * = B*$ 3.$A*' + B* \Rightarrow AB* + *'$ r.d.s. 4.$A*' + AB* \Rightarrow C* + *'$ (fast) 5.$C* = C + *$	$r_A = \dfrac{k_3 \sqrt{K_1 P_{A_2}} K_2 P_B}{\left(1 + \sqrt{K_1 P_{A_2}} + K_2 P_B + K_5^{-1} P_C\right)^2}$
1.$A + * = A*$ 2.$A* + * \Rightarrow B* + *$ r.d.s. 3.$B* = B + *$	$r_A = \dfrac{k_2 K_1 P_A}{\left(1 + K_1 P_A + K_3^{-1} P_B\right)^2}$

The two-step sequence is a frequent case in heterogeneous catalysis and can be presented in the form:

$$1. \ * + A_1 \leftrightarrow *I + B_1$$

$$2. \ *I + A_2 \leftrightarrow * + B_2 \tag{3.48}$$

$$A_1 + A_2 \leftrightarrow B_1 + B_2$$

where $*I$ is the surface intermediate. The mechanism assumes binding of the reactant A_1 in the first step, giving an adsorbed intermediate I and releasing the product

B_1. In the second step the surface intermediate reacts with the reactant A_2 leading to a vacant site and the product B_2. The rate of the steps is given by:

$$r_1 = k_{+1}{}^{[*]}C_{A_1}, r_{-1} = k_{-1}[*I]C_{B_1}, r_2 = k_{+2}[*I]C_{A_2}, r_{-2} = k_{-2}{}^{[*]}C_{B_2} \tag{3.49}$$

The steady-state approximation gives the relationship between * and *I:

$$*I = \frac{k_{+2}C_{A_2} + k_{-1}C_{B_1}}{k_{+2}C_{A_2} + k_{-1}C_{B_1}} \tag{3.50}$$

The conservation equation for the catalytic sites is:

$$[*] + [*I] = [*] = 1 \tag{3.51}$$

Applying the steady-state principle for mechanism (3.48) and eliminating the coverage of the intermediate I, it can be demonstrated that the reaction rate is equal to:

$$r = \frac{k_{+1}C_{A_1}k_{+2}C_{A_2} - k_{-1}C_{B_1}k_{-2}C_{B_2}}{k_{+1}C_{A_1} + k_{+2}C_{A_2} + k_{-1}C_{B_1} + k_{-2}C_{B_2}} = \frac{\omega_{+1}\omega_{+2} - \omega_{-1}\omega_{-2}}{\omega_{+1} + \omega_{+2} + \omega_{-1} + \omega_{-2}} \tag{3.52}$$

where ω_{+1}, etc. are frequencies of steps ($\omega_{+1} = k_{+1}C_{A_1}$). In eq. (3.52) the rate is calculated per catalyst mass. Otherwise the rate expression contains the catalyst concentration:

$$r = C_{cat}\frac{k_{+1}C_{A_1}k_{+2}C_{A_2} - k_{-1}C_{B_1}k_{-2}C_{B_2}}{k_{+1}C_{A_1} + k_{+2}C_{A_2} + k_{-1}C_{B_1} + k_{-2}C_{B_2}} \tag{3.53}$$

Equation (3.53) contains liquid-phase concentrations, while for gas-phase reactions, partial pressures are often used. The water-gas shift (WGS) reaction can be, for example, represented by the two-step sequence:

$$* + H_2O \leftrightarrow *O + H_2$$
$$*O + CO \leftrightarrow * + CO_2 \tag{3.54}$$
$$H_2O + CO = H_2 + CO_2$$

The rate expression for WGS reaction can be written:

$$r = \frac{k_{+1}P_{H_2O}k_{+2}P_{CO} - k_{-1}P_{H_2}k_{-2}P_{CO_2}}{k_{+1}P_{H_2O} + k_{+2}P_{CO} + k_{-1}P_{H_2} + k_{-2}P_{CO_2}} \tag{3.55}$$

In the treatment above it was assumed that the number of adsorption sites on which a molecule is adsorbed is equal to unity, which is not necessarily the case, especially in organic catalysis (Fig. 3.8).

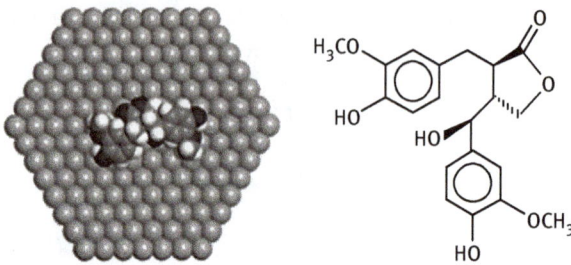

Fig. 3.8: The lignin hydroxymatairesinol (HMR) depending on the adsorption structure covers ~ 10–20 Pd surface atoms.

As an example, the hydrogenolysis reaction of a hydrocarbon (HC) will be considered here. For a uniform surface (where all the sites are equal and there is no interaction between adsorbed species), the following mechanism could be proposed:

$$1.\ HC + z^* \rightleftharpoons HC_{z^*}$$
$$2.\ H_2 + 2^* \rightleftharpoons 2H_*$$
$$3.\ HC_{z^*} + H_* \rightarrow \text{Products}$$

(3.56)

In (3.56) z is the number of free neighbor potential sites required for hydrocarbon adsorption, steps 1 and 2 represent adsorption steps, while step 3 reflects carbon-carbon bonds rupture, which could be followed by other fast steps leading to final reaction products. Quasi-equilibria for the adsorption steps of hydrocarbon and hydrogen give:

$$K_1 = \frac{\theta_{HC}}{P_{HC}\theta_*^z}, K_2 = \frac{\theta_H^2}{P_{H_2}\theta_*^2}$$

(3.57)

where θ_H denotes the coverage of hydrogen, θ_{HC} the coverage of hydrocarbon, θ_* the fraction of vacant sites and P_{HC} and P_{H_2} are the partial pressures of hydrocarbon and hydrogen, respectively. The rate of step 3 is expressed by:

$$r_1 = k_3 \theta_{HC}\theta_H$$

(3.58)

which when combined with expressions for adsorption gives:

$$r_1 = k_3 K_1 K_2^{0.5} P_{HC} P_{H_2}^{0.5} \theta_*^{z+1}$$

(3.59)

The site balance equation:

$$1 = \theta_{HC} + \theta_H + \theta_* = K_1 P_{HC}\theta_*^z + K_2^{0.5} P_{H_2}^{0.5}\theta_* + \theta_*$$

(3.60)

should be solved numerically (except for $z = 1$) together with eq. (3.59) during parameter estimation and comparison with experimental data.

Although the steady-state regime can only be realized in open systems, the same steady-state approximation is usually applied for closed systems (batch reactors), as relaxation times (time to reach steady-state) are shorter than times of one turnover.

3.4 Kinetics of complex reactions

3.4.1 Theory of complex reactions kinetics

Catalytic reactions are complex, which means that there is a set of different elementary reactions occurring jointly. These reactions are related to each other, with some participating species in common. For example, hydrogenation of pentene involving the formation of complexes with a catalyst (where z is the surface site, very often used along with other signs to indicate surface sites, such as *) is described by the reaction:

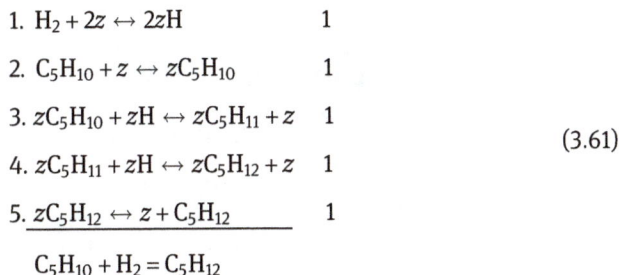

$$
\begin{aligned}
&1.\ H_2 + 2z \leftrightarrow 2zH && 1 \\
&2.\ C_5H_{10} + z \leftrightarrow zC_5H_{10} && 1 \\
&3.\ zC_5H_{10} + zH \leftrightarrow zC_5H_{11} + z && 1 \\
&4.\ zC_5H_{11} + zH \leftrightarrow zC_5H_{12} + z && 1 \\
&\underline{5.\ zC_5H_{12} \leftrightarrow z + C_5H_{12} \qquad\qquad\quad 1} \\
&\ \ \ C_5H_{10} + H_2 = C_5H_{12}
\end{aligned}
\tag{3.61}
$$

The latter equation is the overall chemical equation, which includes only reactants and reaction products, while the chemical equations of elementary reactions grouped into steps in (3.61) include surface intermediates that do not appear in the overall equations. For reversible reactions steps consist of a pair (forward and reverse) of elementary reactions, while only one elementary step is needed for an irreversible reaction. If the rates of some elementary reactions are sufficiently large compared to the rate of the complex reaction as a whole, several elementary reactions are grouped into one more complex step.

Overall reaction equations are linear combinations of chemical equations of steps. They are obtained by the addition of chemical equations of steps multiplied by stoichiometric numbers (positive, negative or zero). The numbers, introduced by Horiuti, must be chosen in such a way that the overall equations contain no intermediates.

Stoichiometric (Horiuti) numbers could be also fractional, for example in SO_2 oxidation:

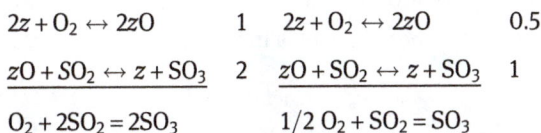

$$
\begin{array}{llll}
2z + O_2 \leftrightarrow 2zO & 1 & \quad 2z + O_2 \leftrightarrow 2zO & 0.5 \\
\underline{zO + SO_2 \leftrightarrow z + SO_3} & 2 & \quad \underline{zO + SO_2 \leftrightarrow z + SO_3} & 1 \\
O_2 + 2SO_2 = 2SO_3 & & \quad 1/2\,O_2 + SO_2 = SO_3 &
\end{array}
$$

The overall equation of the scheme on the right side is obtained from the overall equation of the left one when multiplied by 0.5. Such an operation cannot be performed for the equations of steps, because reaction $z + 1/2O_2 \leftrightarrow zO$ cannot occur. The overall chemical reaction $1/2O_2 + SO_2 = SO_3$ only describes stoichiometry, but not the reaction mechanism. The reaction route after Horiuti is defined as a set of stoichiometric numbers of the steps that produces an overall reaction equation. Routes must be essentially different and it is impossible to obtain one route from another through multiplication by a number, although their respective overall equations can be identical.

For an example of this, the synthesis of water from elements will be considered:

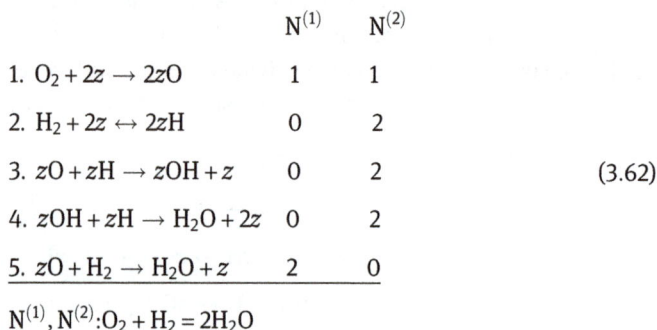

$$
\begin{array}{lcc}
 & N^{(1)} & N^{(2)} \\
\text{1. } O_2 + 2z \rightarrow 2zO & 1 & 1 \\
\text{2. } H_2 + 2z \leftrightarrow 2zH & 0 & 2 \\
\text{3. } zO + zH \rightarrow zOH + z & 0 & 2 \\
\text{4. } zOH + zH \rightarrow H_2O + 2z & 0 & 2 \\
\text{5. } zO + H_2 \rightarrow H_2O + z & 2 & 0 \\
\end{array}
\tag{3.62}
$$

$$N^{(1)}, N^{(2)}: O_2 + H_2 = 2H_2O$$

The mechanism consists of two routes, the first once corresponding to an Eley-Rideal mechanism, when oxygen is adsorbed with dissociation and reacts with dihydrogen from the fluid phase (steps 1 and 5, respectively). The second route comprises dissociative adsorption of dioxygen and dihydrogen (steps 1 and 2) and the subsequent two consecutive surface reaction steps (steps 3 and 4). Such reactions are usually considered to be of a Langmuir-Hinshelwood type.

> The research proposal I wrote in the late 1980s contained the term "Langmuir-Hinshelwood kinetics". Prof. M. Temkin asked if I had ever read the original publications of Irving Langmuir and advised me to do so when the reply was negative. It then became apparent to me that Langmuir in fact considered both cases (for example for CO oxidation): the co-adsorption of both reactants (called nowadays Langmuir-Hinshelwood mechanism) and the reaction of the second compound from the fluid phase, while the first is adsorbed (the Eley-Rideal mechanism).

The route $N^{(2)}$ in eq. (3.62) cannot be obtained by multiplication of $N^{(1)}$ with a certain number, which means that the routes $N^{(1)}$ are $N^{(2)}$ different (as apparent from their different mechanism), although their overall equations are identical.

The stoichiometric number of a step s in a route $N^{(3)}$ $v_s^{(3)}$ could be defined as:

$$v_s^{(3)} = C_1 v_s^{(1)} + C_2 v_s^{(2)} \tag{3.63}$$

where C_1 and C_2 are some arbitrary numbers. Any linear combination of routes of a given reaction (infinite combinations) will also be a route of the reaction:

$$N_S^{(3)} = C_1 N_S^{(1)} + C_2 N_S^{(2)} \tag{3.64}$$

Reaction routes thus form the vector space, in terms of linear algebra. A set of such routes, when no route can be represented as a linear combination of others, is called the basis of routes, which are linearly independent. Although the basis of routes can be chosen in different ways the number of basis routes for a given reaction mechanism is determined in a unique way, being the dimension of the space of the routes.

There are also relationships between intermediates that give balance or conservation equations. For heterogeneous catalytic reactions, a balance equation relates the surface coverage of adsorbed species and vacant sites. If there are two types of sites on the surface there would be two balance equations. Another example of the balance (conservation or link) equation could be overall neutrality of the surface if adsorbed intermediates are charged species.

Following Horiuti the number of basic routes P is determined by:

$$P = S - I \tag{3.65}$$

where S is the number of stages and I is the number of independent intermediates in the sense used in the Gibbs phase rule. A similar expression was proposed by Temkin:

$$P = S + W - J \tag{3.66}$$

where P is the number of routes, W is the number of balance equations, J is the number of intermediates (including vacant sites) and S is the number of steps.

The rate along the basic route $N^{(p)}$ is denoted as $r^{(p)}$. According to its definition there are $v_s^{(p)} r^{(p)}$ runs of the sth stage per unit time per unit reaction space. Here $v_s^{(p)}$ is the stoichiometric number of the sth step along the route $N^{(p)}$. For a one-route mechanism for a steady-state reaction the overall reaction rate is expressed as:

$$r = \frac{r_{+s} - r_{-s}}{v_s} = \frac{r_{+1} - r_{-1}}{v_1} = \frac{r_{+2} - r_{-2}}{v_2} = \frac{r_{+3} - r_{-3}}{v_3} = \frac{r_{+4} - r_{-4}}{v_4} = \ldots \tag{3.67}$$

For the mechanism (3.62) the derivation could be simplified assuming equilibrium adsorption of dioxygen and dihydrogen with dissociation, giving:

$$\theta_0 = \sqrt{K_1 P_{O_2}} \theta_v, \theta_H = \sqrt{K_2 P_{H_2}} \theta_v \tag{3.68}$$

where θ_v is the fraction of vacant sites. From the steady-state for zOH ($r_3 = r_4$):

$$\theta_{OH} = \frac{k_3}{k_4} \theta_0 = \frac{k_3}{k_4} \sqrt{K_1 P_{O_2}} \theta_v \tag{3.69}$$

Taking into account the balance equation, an explicit expression for the fraction of vacant sites could be obtained as:

$$\theta_v = \frac{1}{1 + \sqrt{K_1 P_{O_2}}\left(1 + \frac{k_3}{k_4}\right) + \sqrt{K_2 P_{H_2}}} \tag{3.70}$$

The reaction rate is equal to the sum of the rates along two routes:

$$r = r^I + r^{II} = \frac{r_3}{2} + \frac{r_5}{2} = \frac{k_3 \sqrt{K_1 P_{O_2}} \sqrt{K_2 P_{H_2}}}{2\left(1 + \sqrt{K_1 P_{O_2}}\left(1 + \frac{k_3}{k_4}\right) + \sqrt{K_2 P_{H_2}}\right)^2} \tag{3.71}$$

$$+ \frac{k_5 P_{H_2} \sqrt{K_1 P_{O_2}}}{2\left(1 + \sqrt{K_1 P_{O_2}}\left(1 + \frac{k_3}{k_4}\right) + \sqrt{K_2 P_{H_2}}\right)}$$

In (3.62) there were steps with a stoichiometric number equal to zero, which means that these steps do not take part in a particular route. In fact a step can have a stoichiometric number equal to zero along all the routes. For example when solvent Y adsorbs on the surface a mechanism can take the form:

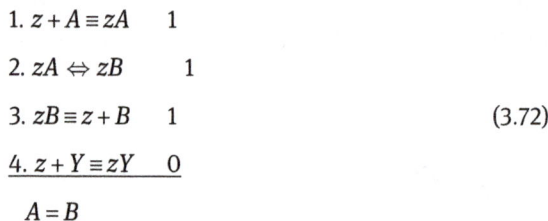

$$
\begin{aligned}
&1.\ z + A \equiv zA &&1 \\
&2.\ zA \Leftrightarrow zB &&1 \\
&3.\ zB \equiv z + B &&1 \\
&4.\ z + Y \equiv zY &&0 \\
\hline
&A = B
\end{aligned}
\tag{3.72}
$$

The stoichiometric number of step 4 in eq. (3.72) is zero, which is logical as the solvent does not take part in the chemical transformations. At the same time, the solvent concentration can influence the reaction rate by blocking some catalytic sites, which will be otherwise accessible to reactants, resulting in the following rate expression:

$$r = \frac{k_{+2} K_A C_A - k_{-2} K_B C_B}{1 + K_A C_A + K_B C_B + K_y C_y} \tag{3.73}$$

One of the papers I submitted to the Journal of Catalysis in the early 1990s contained an analysis of a mechanism similar to eq. (3.72). The referee responded that the "stoichiometric number in chemistry can never be zero." Apparently the reviewer, who was supposed to be a specialist in catalytic kinetics, was not at all familiar with the theory of complex reactions, developed by Horiuti and Temkin in the 1960s!

A reaction mechanism can be rather complex, comprising several routes and types of sites. For example, a mechanistic model describing the experimental data of

L-arabinose oxidation over Au/Al_2O_3 catalyst with oxygen in the aqueous phase [1] can be considered as:

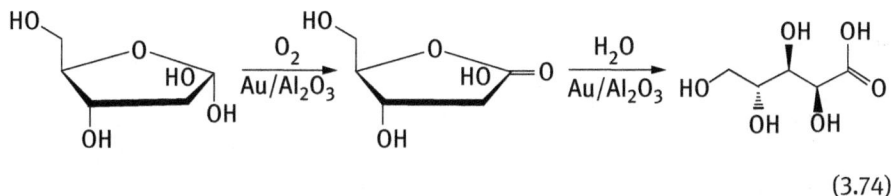

$$(3.74)$$

This was based on the concept of oxidative dehydrogenation:

		$N^{(1)}$	$N^{(2)}$	$N^{(3)}$	$N^{(4)}$
1.	$A + {}^* \equiv A^*$	0	0	0	0
2.	$A + {}^{*'} \equiv A^{*'}$	2	0	0	0
3.	$O_2 + 2^* \equiv 2O^*$	1	0	0	0
4.	$O^* + {}^{*'} \equiv O^{*'} + {}^*$	2	1	1	1
5.	$A^{*'} + O^{*'} \rightarrow B^{*'} + H_2O + {}^{*'}$	2	0	0	0
6.	$B^{*'} + OH^{-*'} \rightarrow D^{*'} + {}^{*'}$	0	1	0	0
6a.	$D^{*'} + H^+ \rightarrow (fast)C + {}^{*'}$	0	1	0	0
7.	$O^* + H_2O + 2e^- \rightarrow 2OH^- + {}^*$	0	0	-1	1
8.	$OH^- + {}^* \rightarrow O^* + H^+ + 2e^-$	0	1	0	2
9.	$O^{*'} + H_2O + 2e^- \rightarrow 2OH^- + {}^{*'}$	0	1	1	1
10.	$OH^- + {}^* \equiv OH^{-*}$	0	0	0	0
11.	$OH^- + {}^{*'} \equiv OH^{-*'}$	0	1	0	0
12.	$B^{*'} \equiv B + {}^{*'}$	2	-1	0	0

$N^{(1)}$: $2A + O_2 = 2B + 2H_2O$, $N^{(2)}$: $B + H_2O = C$, $N^{(3)}$: $0 = 0$, $N^{(4)}$: $2H_2O = 2OH^- + 2H^+$.

$$(3.75)$$

where * is the active sites on the edges of the catalyst, $^{*'}$ is the active sites on the faces, A is L-arabinose, B is arabinolactone, C is arabinonic acid and D is the intermediate. The elementary steps above can be described by four reaction routes. It was assumed that L-arabinose can adsorb on both the faces and the edges of gold clusters (steps 1 and 2, respectively). Oxygen is adsorbed on the edges with dissociation (steps 3), thereafter migrating to the faces (steps 4). The sugar oxidation involves dehydrogenation of the adsorbed L-arabinose to arabinolactone and oxygen reduction in step 5. The product, arabinonic acid, is formed in step 6 and 6a via the intermediate species

D, while step 7 and 8 account for electron transfer, which involves oxygen adsorbed on the edges. Finally, as the oxidation rate was found to be dependent on pH, adsorption of OH^- on the faces was also included in the model (step 11). Electron transfer involving adsorption of OH^- on the edges was also included in the model (step 9). In addition, step 10 accounts for direct adsorption of OH^- on the edges without electron transfer. The oxidative dehydrogenation of *L*-arabinose takes place in two steps (5 and 6) via the intermediate arabinolactone *B*. The intermediate product, arabinolactone, has been found only in small amounts in the reaction mixture (step 12). It is assumed to react much faster than *L*-arabinose and, as a reactive intermediate, it will have very low surface coverage.

One of the routes (i.e., $N^{(3)}$) is a so-called empty route, as it does not result in any chemical transformation.

According to the theory of complex reactions developed by Horiuti and Temkin, the rate of the *s* step is obtained by summation over all *P* basic routes, hence:

$$r_{|s|} = r_s - r_{-s} = \sum_{p=1}^{P} v_s^{(p)} r^{(p)} \tag{3.76}$$

where r_s and r_{-s} are the rates of the forward and reverse elementary reactions comprising the stage, $r^{(p)}$ is the rate along the basic route $N^{(p)}$, $v_s^{(p)}$ is the stoichiometric number of the *sth* step along the route $N^{(p)}$. For example the rate of step 5 in (3.75) is:

$$r_{|5|} = r_5 - r_{-5} = 2 * r^{(1)} + 0 * r^{(2)} + 0 * r^{(3)} + 0 * r^{(4)} \tag{3.77}$$

which looks trivial at first. However, such analysis is very helpful for multiroute mechanisms.

If for some steps $r_s \approx r_{-s} \succ\succ r_{|s|}$ the steps are considered as "fast" (for example, steps 1, 2, 3, 4 in eq. 3.75) or quasi-equilibrium steps (as illustrated in Fig. 3.4). Strictly equilibrium steps are also possible, when the stoichiometric numbers of a step for all basic routes are zero, then $r_s = r_{-s}$. Reversible adsorption of a solvent [eq. (3.72)] may serve as an example.

For non-linear steps there could be several solutions for (3.76). Sometimes the steady- state solution cannot be reached and the reaction system starts to oscillate.

For a one-route mechanism for a steady-state reaction the overall reaction rate follows immediately from (3.75):

$$r = \frac{r_{+s} - r_{-s}}{v_s} = \frac{r_{+1} - r_{-1}}{v_1} = \frac{r_{+2} - r_{-2}}{v_2} = \frac{r_{+3} - r_{-3}}{v_3} = \frac{r_{+4} - r_{-4}}{v_4} = \ldots \tag{3.78}$$

General expressions for the reaction rate can be constructed knowing the expressions for the rate of steps if the following equation is used:

$$\prod_{i=1}^{s} r_{+i} - \prod_{i=1}^{s} r_{-i} = (r_{+1} - r_{-1})r_{+2} \ldots r_{+s} + r_{-1}(r_{+2} - r_{-2})r_{+3} \ldots r_{+s} + \ldots$$

$$+ r_{-1} r_{-2} \ldots (r_{+s} - r_{-s}) \tag{3.79}$$

The numbers of steps in (3.79) are chosen in an arbitrary way. For a four-step reaction one arrives at:

$$\prod_{i=1}^{4} r_{+i} - \prod_{i=1}^{4} r_{-i} = (r_{+1} - r_{-1})r_{+2}\ldots r_{+4} + r_{-1}(r_{+2} - r_{-2})r_{+3}\ldots r_{+4} + \ldots$$
$$+ r_{-1}r_{-2}\ldots(r_{+4} - r_{-4}) \tag{3.80}$$

which could be transformed into:

$$r_{+1}r_{+2}r_{+3}r_{+4} - r_{-1}r_{-2}r_{-3}r_{-4} = r_{|1|}r_{+2}r_{+3}r_{+4} + r_{-1}r_{|2|}r_{+3}r_{+4} + \ldots r_{-1}r_{-2}r_{-3}r_{|4|} \tag{3.81}$$

As steady-state the global rate is expressed by the rate of a step and its stoichiometric number is then:

$$r_{+1}r_{+2}r_{+3}r_{+4} - r_{-1}r_{-2}r_{-3}r_{-4} = (v_1 r)r_{+2}r_{+3}r_{+4} + r_{-1}(v_2 r)r_{+3}r_{+4}$$
$$+ \ldots + r_{-1}r_{-2}r_{-3}(v_3 r) \tag{3.82}$$

giving the rate expression as:

$$r = \frac{r_{+1}r_{+2}r_{+3}r_{+4} - r_{-1}r_{-2}r_{-3}r_{-4}}{v_1 r_{+2}r_{+3}r_{+4} + r_{-1}v_2 r_{+3}r_{+4} + r_{-1}r_{-2}v_3 r_{+4} + r_{-1}r_{-2}r_{-3}v_3} \tag{3.83}$$

which, in more general form, is as follows:

$$r = r_+ - r_- = \frac{\displaystyle\prod_{i=1}^{s} r_{+i}}{v_1 r_{+2}\ldots r_{+s} + r_{-1}v_2 r_{+3}\ldots r_{+s} + r_{-1}r_{-2}\ldots v_s}$$

$$- \frac{\displaystyle\prod_{i=1}^{s} r_{-i}}{v_1 r_{+2}\ldots r_{+s} + r_{-1}v_2 r_{+3}\ldots r_{+s} + r_{-1}r_{-2}\ldots v_s} \tag{3.84}$$

$$= \frac{\displaystyle\prod_{i=1}^{s} r_{+i} - \prod_{i=1}^{s} r_{-i}}{v_1 r_{+2}\ldots r_{+s} + r_{-1}v_2 r_{+3}\ldots r_{+s} + r_{-1}r_{-2}\ldots v_s}$$

The ratio of rates in the forward and reverse directions is thus expressed as:

$$\frac{r_-}{r_+} = \frac{r_{-1}r_{-2}r_{-3}\ldots r_{-s}}{r_1 r_2 r_3 \ldots r_s} \tag{3.85}$$

Note that such analysis is valid for linear mechanisms.

For a two-step catalytic sequence (eq. 3.48) and when numbering the step 1 as the first one and the step 2 as the second, it follows from (3.84) that:

$$r = \frac{k_{+1}^{[*]}C_{A_1}k_{+2}^{[*I]}C_{A_2} - k_{-1}^{[*I]}C_{B_1}k_{-2}^{[*]}C_{B_2}}{k_{+2}[*I]C_{A_2} + k_{-1}^{[*I]}C_{B_1}} \tag{3.86}$$

or after rearranging:

$$[*] = r\frac{k_{+2}C_{A_2} + k_{-1}C_{B_1}}{k_{+1}C_{A_1}k_{+2}C_{A_2} + k_{-1}C_{B_1}k_{-2}C_{B_2}} \tag{3.87}$$

If numbering step 1 as the second one and step 2 as the first, another expression for the rate and coverage of the intermediate I follows from eq. (3.84) as:

$$r = \frac{k_{+2}[*I]C_{A_2}k_{+1}[*]C_{A_1} - k_{-2}[*]C_{B_2}k_{-1}[*I]C_{B_1}}{k_{+1}[*]C_{A_1} + k_{-2}[*]C_{B_2}} ; \tag{3.88}$$

$$[*I] = r\frac{k_{+1}C_{A_1} + k_{-2}C_{B_2}}{k_{+1}C_{A_1}k_{+2}C_{A_2} - k_{-1}C_{B_1}k_{-2}C_{B_2}} \tag{3.89}$$

From the balance equation:

$$[*]^{tot} = [*I] + [*] = r\frac{k_{+1}C_{A_1} + k_{+2}C_{A_2} + k_{-1}C_{B_1} + k_{-2}C_{B_2}}{k_{+1}C_{A_1}k_{+2}C_{A_2} - k_{-1}C_{B_1}k_{-2}C_{B_2}} \tag{3.90}$$

the reaction rate can be easily obtained as:

$$r = \frac{k_{+1}C_{A_1}k_{+2}C_{A_2} - k_{-1}C_{B_1}k_{-2}C_{B_2}}{k_{+1}C_{A_1} + k_{+2}C_{A_2} + k_{-1}C_{B_1} + k_{-2}C_{B_2}} = \frac{\omega_{+1}\omega_{+2} - \omega_{-1}\omega_{-2}}{\omega_{+1} + \omega_{+2} + \omega_{-1} + \omega_{-2}} \tag{3.91}$$

For a more complex case of three steps:

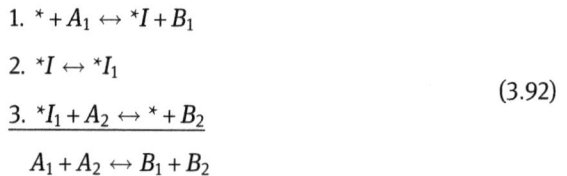

$$\begin{array}{l} 1.\ * + A_1 \leftrightarrow *I + B_1 \\[4pt] 2.\ *I \leftrightarrow *I_1 \\[4pt] 3.\ \underline{*I_1 + A_2 \leftrightarrow * + B_2} \\[4pt] A_1 + A_2 \leftrightarrow B_1 + B_2 \end{array} \tag{3.92}$$

the reaction rate can be derived either using the steady-state approximation leading to:

$$r = \frac{\omega_{+1}\omega_{+2}\omega_{+3} - \omega_{-1}\omega_{-2}\omega_{-3}}{\omega_{+2}\omega_{+3} + \omega_{-3}\omega_{+2} + \omega_{-3}\omega_{-2} + \omega_{+3}\omega_{+1} + \omega_{-1}\omega_{+3} + \omega_{-1}\omega_{-3} + \omega_{+1}\omega_{+2} + \omega_{-2}\omega_{+1} + \omega_{-2}\omega_{-1}} \tag{3.93}$$

or directly from (3.84) when changing the counting order in the reaction mechanism (3.92) for steps 1–3. The terms in the denominator for (3.84) are obtained when counting starts from the first, second or the third step, respectively:

$$s = 1, r_{+2}r_{+3}; r_{-3}r_{+2}; r_{-3}r_{-2}$$

$$s = 2, r_{+3}r_{+1}; r_{-1}r_{+3}; r_{-1}r_{-3}$$

$$s = 3, r_{+1}r_{+2}; r_{-2}r_{+1}; r_{-2}r_{-1}$$

which in combination with other terms gives eq. (3.93).

A Mars-van-Krevelen catalytic oxidation mechanism involving oxygen from the lattice can be used to illustrate the three-step mechanism:

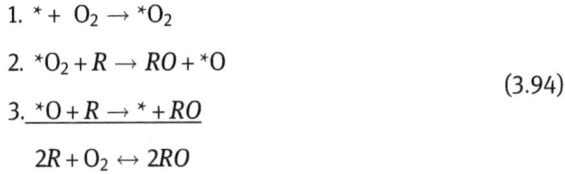

$$1. \; * + O_2 \rightarrow *O_2$$
$$2. \; *O_2 + R \rightarrow RO + *O$$
$$3. \; \underline{*O + R \rightarrow * + RO}$$
$$2R + O_2 \leftrightarrow 2RO$$

(3.94)

For the mechanism with three irreversible steps the equation is simply:

$$r_{O_2} = \frac{r_R}{2} = \frac{\omega_{+1}\omega_{+2}\omega_{+3}}{\omega_{+2}\omega_{+3} + \omega_{+3}\omega_{+1} + \omega_{+1}\omega_{+2}}$$
$$= \frac{k_{+1}k_{+2}k_{+3}P_{O_2}P_R}{k_{+2}k_{+3}P_R + k_{+1}(k_{+2}+k_{+3})P_{O_2}}$$

(3.95)

In a similar fashion using pressure instead of concentrations for gas-phase reactions the rate equation for a four-step linear mechanism (simplified mechanism for methane steam reforming):

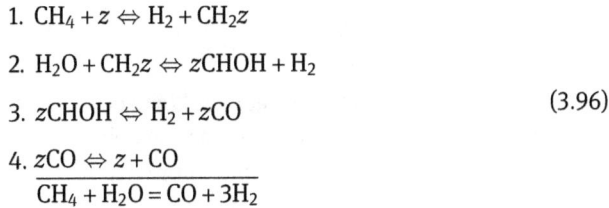

$$1. \; CH_4 + z \leftrightarrow H_2 + CH_2 z$$
$$2. \; H_2O + CH_2 z \leftrightarrow zCHOH + H_2$$
$$3. \; zCHOH \leftrightarrow H_2 + zCO$$
$$4. \; \underline{zCO \leftrightarrow z + CO}$$
$$CH_4 + H_2O = CO + 3H_2$$

(3.96)

takes the form:

$$r = \frac{k_{+1}P_{CH_4}k_{+2}P_{H_2O}k_{+3}k_{+4} - k_{-1}P_{H_2}k_{-2}P_{H_2}k_{-3}P_{H_2}k_{-4}P_{CO}}{D}$$

(3.97)

with:

$$D = k_{+2}P_{H_2O}k_{+3}k_{+4} + k_{-1}P_{H_2}k_{+3}k_{+4} + k_{-1}P_{H_2}k_{-2}P_{H_2}k_{+4} + k_{-1}P_{H_2}k_{-2}P_{H_2}k_{-3}P_{H_2} +$$
$$k_{+1}P_{CH_4}k_{+2}P_{H_2O}k_{+3} + k_{-4}P_{CO}k_{+2}P_{H_2O}k_{+3} + k_{-4}P_{CO}k_{-1}P_{H_2}k_{+3} +$$
$$k_{-4}P_{CO}k_{-1}P_{H_2}k_{-2}P_{H_2} + k_{+4}k_{+1}P_{CH_4}k_{+2}P_{H_2O} + k_{-3}P_{H_2}k_{+1}P_{CH_4}k_{+2}P_{H_2O} +$$
$$k_{-3}P_{H_2}k_{-4}P_{CO}k_{+2}P_{H_2O} + k_{-3}P_{H_2}k_{-4}P_{CO}k_{-1}P_{H_2} + k_{+3}k_{+4}k_{+1}P_{CH_4} +$$
$$k_{-2}P_{H_2}k_{+4}k_{+1}P_{CH_4} + k_{-2}P_{H_2}k_{-3}P_{H_2}k_{+1}P_{CH_4} + k_{-2}P_{H_2}k_{-3}P_{H_2}k_{-4}P_{CO}$$

(3.98)

The hydrogenation of benzene is an example of a single-route reaction mechanism with five linear steps:

1. $C_6H_6 + * \Leftrightarrow C_6H_6*$

2. $C_6H_6* + H_2 \Leftrightarrow C_6H_8*$

3. $C_6H_8* + H_2 \Leftrightarrow C_6H_{10}*$ (3.99)

4. $C_6H_{10}* + H_2 \Leftrightarrow C_6H_{12}*$

5. $\underline{C_6H_{12}* \Leftrightarrow C_6H_{12} + *}$

$C_6H_6 + 3H_2 = C_6H_{12}$

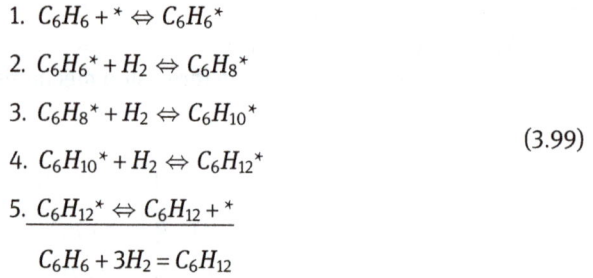

The general form of the kinetic equation is:

$$r = \frac{\omega_{+1}\omega_{+2}\omega_{+3}\omega_{+4}\omega_{+5} - \omega_{-1}\omega_{-2}\omega_{-3}\omega_{-4}\omega_{-5}}{D}$$ (3.100)

where the denominator D in its most general form consists of 25 terms. If reaction occurs far from overall equilibrium and all the steps are irreversible then after several simplifications:

$$r = \frac{\omega_{+}\omega_{+2}\omega_{+3}\omega_{+4}\omega_{+5}}{\omega_1\omega_2\omega_3\omega_4 + \omega_1\omega_2\omega_3\omega_5 + \omega_1\omega_2\omega_4\omega_5 + \omega_1\omega_3\omega_4\omega_5 + \omega_2\omega_3\omega_4\omega_5}$$

$$= \frac{P_{C_6H_6}P_{H_2}}{\left(\frac{1}{k_{+5}}\right)P_{C_6H_6}P_{H_2} + \left(\frac{1}{k_{+4}} + \frac{1}{k_{+3}} + \frac{1}{k_{+2}}\right)P_{C_6H_6} + \left(\frac{1}{k_{+1}}\right)P_{H_2}}$$ (3.101)

with the reaction order in hydrogen and benzene varying between zero and unity.

In heterogeneous catalysis the reaction mechanism is often rather complex and cannot be represented by a one-route multistep reaction sequence, as there are several routes leading to a variety of products.

If the reaction is described by P routes the rate expression is:

$$\sum_{p=1}^{P} \left(r^{(p)} \frac{V_1^{(p)}}{r_{+1}} + \frac{r_{-1}V_2^{(p)}}{r_{+1}r_{+2}} + \cdots + \frac{r_{-1}r_{-2}r_{-3} \cdots r_{-s-1}V_s^{(p)}}{r_{+1}r_{+2}r_{+3} \cdots r_{+s}} \right) = 1 - \frac{r_{-1}r_{-2}r_{-3} \cdots r_{-s}}{r_{+1}r_{+2}r_{+3} \cdots r_{+s}}$$

(3.102)

3.4.2 Relationship between thermodynamics and kinetics

Linear free energy relationships (LFER) introduced by Brønsted are widely used in homogeneous and heterogeneous catalytic reactions, relating reaction constants k with equilibrium constants K in a series of analogous elementary reactions:

$$k = gK^\alpha \quad 0 < \alpha < 1$$ (3.103)

where g and α (the Polanyi parameter) are constants. Such relationships suppose a certain relation between reaction thermodynamics and kinetics, for example, between the Gibbs free energy of activated complexes and the Gibbs energy of a reaction:

$$\Delta G^{\#} \sim \alpha \Delta G \tag{3.104}$$

Evans and Polanyi introduced this relationship to organic reactions, Semenov extended the application to chain reactions and Temkin applied LFER to heterogeneous catalysis. Equation (3.104) is often expressed through a relationship between the energy of activation and the reaction enthalpy:

$$E = E_0 + \alpha \Delta H \tag{3.105}$$

where E_0 is a constant for a reaction family.

3.4.3 Non-ideal surfaces

In the treatment of catalytic kinetics the surface is often considered to be uniform. However, adsorbed molecules change the structure of the surface layer, influencing each other, and the surface sites are not equal in the ability to bind chemisorbed molecules. This is reflected, for example, in the dependence of the heats of adsorption on coverage. The case of different adsorption sites is defined as biographical non-uniformity while the mutual influence of the adsorbed species is considered as induced non-uniformity, In the former case there is a certain distribution of properties, which can be either chaotic – when adsorption energy on a given site is independent on the neighbor site – or discrete.

The physical nature of the biographical (intrinsic or a priori) non-uniformity can be attributed to the difference in the properties of the different crystal faces and the occurrence of dislocations, defects and other disturbances.

The special Temkin model of non-uniformity assumes that each site is characterized by a definite adsorption energy of a particular substance and that there is a certain distribution of the total number of sites with this energy. The values of adsorption energy on different sites lie between a certain minimum and a certain maximal value $(\Delta G_a^{\circ})_{min} \leq (\Delta G_a^{\circ}) \leq (\Delta G_a^{\circ})_{max}$, and the number of sites dl with standard Gibbs energy of adsorption within (ΔG_a°) and $(\Delta G_a^{\circ}) + d(\Delta G_a^{\circ})$ is given by an exponential dependence:

$$dl = Ce^{\frac{\Delta G_a^{\circ}}{R\Theta}} d(\Delta G_a^{\circ}) \tag{3.106}$$

where C and Θ are constants. The values of Θ could be positive, negative or infinity, reflecting cases of Freundlich, Zeldovich-Roginski and Temkin isotherms, respectively. In case when $\Theta \to \infty$ instead of (3.106) the following distribution is valid:

$$dI = Cd(\Delta G_a^\circ) \tag{3.107}$$

The sites are numbered in the sequence of increasing ξ (decreasing adsorption strength). The relative number of sites is $s = l/L$, where L is the total number of sites ($0<s<1$). The distribution of sites function $\varphi(\xi)d\xi$, which is the number of sites with $\Delta G_a^\circ/RT$ within $\xi + d\xi$, can be defined as:

$$\xi < \xi_0 : \varphi(\xi) = 0; \xi_0 \le \zeta \le \xi_1 : \phi(\xi) = A\exp(y\xi); \xi > \xi_0 : \varphi(\xi) = 0 \tag{3.108}$$

where ξ_0 is lowest value of ξ and ξ_1 is the highest value of ξ, A is a constant, $y = T/\Theta$. For even distribution, $y = 0$, for example, $\varphi(\xi) = A$. The value of A is determined by the normalized condition:

$$\int_{-\infty}^{\infty} \varphi(\xi)d\xi = L \text{ or } \int_{\xi}^{\xi_1} \varphi(\xi)d\xi = L \tag{3.109}$$

The expressions for the constant A are

$$y \ne 0, A = \frac{yL}{e^{y\xi_1} - e^{y\xi_0}}; y = 0, A = \frac{L}{\xi_1 - \xi_0} \tag{3.110}$$

For each surface site (there are overall L sites) the reaction rate for the two-step sequence using pressures, rather than concentrations is expressed in the same way as for ideal surfaces:

$$\rho_+ = \frac{1}{L} \frac{k_{+1}P_{A_1}k_{+2}P_{A_2}}{k_{+1}P_{A_1} + k_{-1}P_{B_1} + k_{+2}P_{A_2} + k_{-2}P_{B_2}} \tag{3.111}$$

The rate in the forward direction is the sum of all the rates on each catalytic site:

$$r_+ = \int_{\xi_0}^{\xi_1} \rho_* \varphi(\xi)d\xi \tag{3.112}$$

Changing the variable from ξ to λ ($\xi = \lambda + \xi_0$) the reaction rate is:

$$r_+ = \frac{y}{e^{yf} - 1} \int_0^f \frac{k_{+1}{}^0P_{A_1}k_{+2}{}^0P_{A_2}e^{(n-m)\lambda}d\lambda}{(k_{+1}{}^0P_{A_1} + k_{-2}{}^0P_{B_2})e^{-m\lambda} + (k_{-1}{}^0P_{B_1} + k_{+2}{}^0P_{A_2})e^{n\lambda}} \tag{3.113}$$

where $n = \beta + y$; $m = \alpha - y$. After some rearrangements discussed in detail by Murzin and Salmi [2] an expression for the reaction rate is obtained, which is valid in the region of medium coverage:

$$r = \frac{y}{e^{yf} - 1} \frac{\pi}{\sin m\pi} \frac{k_{+1}{}^0P_{A_1}k_{+2}{}^0P_{A_2} - k_{-1}{}^0P_{B_1}k_{-2}{}^0P_{B_2}}{\left(k_{+1}{}^0P_{A_1} + k_{-2}{}^0P_{B_2}\right)^m \left(k_{-1}{}^0P_{B_1} + k_{+2}{}^0P_{A_2}\right)^{1-m}} \qquad (3.114)$$

For evenly non-uniform surfaces $y = 0$, $m = \alpha$ and the expression for the rate is:

$$r = \frac{1}{f} \frac{\pi}{\sin \alpha\pi} \frac{k_{+1}{}^0P_{A_1}k_{+2}{}^0P_{A_2} - k_{-1}{}^0P_{B_1}k_{-2}{}^0P_{B_2}}{\left(k_{+1}{}^0P_{A_1} + k_{-2}{}^0P_{B_2}\right)^{\alpha} \left(k^0{}_{-1}P_{B_1} + k_{+2}{}^0P_{A_2}\right)^{1-\alpha}} \qquad (3.115)$$

When both steps in the two-step sequence are irreversible a power law model is obtained:

$$r = kP_{A_1}^n P_{A_2}^{1-n} \qquad (3.116)$$

where $0 \leq n \leq 1$. A comparison between eq. (3.116) and the corresponding one for uniform surfaces:

$$r = \frac{k_{+1}P_{A_1}k_{+2}P_{A_2}}{k_{+1}P_{A_1} + k_{+2}P_{A_2}} \qquad (3.117)$$

shows the difference between the reaction kinetics on uniform and non-uniform surfaces.

These two models give qualitatively different behavior at boundary values of partial pressures. For example, according to an ideal surface model, at high partial pressures the reaction rate obeys zero order kinetics, while at low pressures the reaction order is equal to unity, in stark contrast to the mechanism for a non-ideal surface, which does not show zero order dependence in a wider range of partial pressures.

Temkin's kinetic models for intrinsic non-uniform surfaces with essentially two non-equilibrium steps were extended by Snagovskii and Avetisov to several types of reaction mechanisms. Only linear steps were considered that occur either with or without changes in the number of adsorbed species leading respectively to:

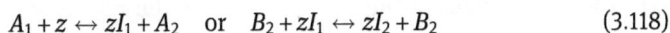

$$A_1 + z \leftrightarrow zI_1 + A_2 \quad \text{or} \quad B_2 + zI_1 \leftrightarrow zI_2 + B_2 \qquad (3.118)$$

It was assumed that the rate constants of elementary steps do not depend on the parameter ξ. The rate constants of the steps in the forward direction (for example, adsorption type) that correspond to the left side of eq. (3.118) contain the term $e^{-\alpha\xi}$, while the desorption type constants have the term $e^{\beta\xi}$. The values of the transfer coefficients α and β are considered to be the same and distribution functions are similar for all surface species. The kinetic equation for a uniform patch is expressed by:

$$\rho = \frac{1}{L} \frac{Ce^{(\beta - \alpha)\xi}}{De^{-\alpha\xi} + He^{\beta\xi}} \qquad (3.119)$$

where L is the number of surface sites, and C, D and H are some functions of the reactants partial pressures. The reaction rate on non-ideal surfaces is given by:

$$r = \frac{\pi y}{(e^{yf} - 1)\sin n\pi} \frac{C}{D^m H^n} \qquad (3.120)$$

with $m = \alpha - y$ and $n = \beta + y$. The values of C, D and H for the two-step sequences are correspondingly:

$$C = \omega_{+2}D - \omega_{-2}H, D = \omega_{+1} + \omega_{-2}, H = \omega_{-1} + \omega_{+2} \qquad (3.121)$$

For a mechanism with three steps and two surface intermediates the parameters in eq. (3.120) are:

$$C = \omega_{+3}(\omega_{+1} + \omega_{-3}) - \omega_{-3}H, D = \left(1 + \frac{\omega_{-2}}{\omega_{+2}}\right)(\omega_{-3} + \omega_{+1}),$$

$$H = \omega_{+3} + \frac{\omega_{-1}\omega_{-2}}{\omega_{+2}} \qquad (3.122)$$

The models for induced non-uniformity (presence of lateral interactions) usually consider the repulsive interactions between nearest neighbors and attractive interactions between next-nearest neighbors. Information on lateral interactions (which are not prominent at low coverage) can be extracted from high-resolution scanning tunneling microscopy data and analysis of desorption kinetics.

The simplest representation of adsorbate-adsorbate-adsorbent interactions is the Fowler-Guggenheim isotherm, which differs from the Langmuir one in the exponential term, accounting for the interactions:

$$aP = \frac{\theta}{1 - \theta} e^{v\theta} \qquad (3.123)$$

Using a similar concept the reaction rate for the region of medium coverage was derived [3] as:

$$r = \frac{T \ln U'}{\eta^2 C} \frac{k_{+1}P_{A_1}k_{+2}P_{A_2} - k_{-1}P_{B_1}k_{-2}P_{B_2}}{\left(k_{+1}P_{A_1} + k_{-2}P_{B_2}\right)^{\alpha}\left(k_{+2}P_{A_2} + k_{-1}P_{B_1}\right)^{1-\alpha}} \qquad (3.124)$$

where η is a constant and $U' = \left(k_{+1}P_{A_1} + k_{-2}P_{B_2}\right)/\left(k_{+2}P_{A_2} + k_{-1}P_{B_1}\right)$. Equation (3.124) is similar to the expression for biographically non-uniform surfaces. However, for induced non-uniform surfaces in the region of medium coverage, the concentration-dependence is slightly more complicated than the derivation for biographical non-uniform surfaces, as parameter U' is dependent on partial pressures of reactants and products.

3.4.4 Kinetic aspects of selectivity

One of the most important requirements for catalytic reactions is proper selectivity, which in a broad sense should be understood as chemo-, regio- and enantioselectivity. Selectivity is the ability of a catalyst to selectively favor one among various competitive chemical reactions. Intrinsic selectivity is associated with the chemical composition and structure of surface (support), while shape selectivity is related with pore transport limitations (see Figs 2.115, 2.116 and 3.9).

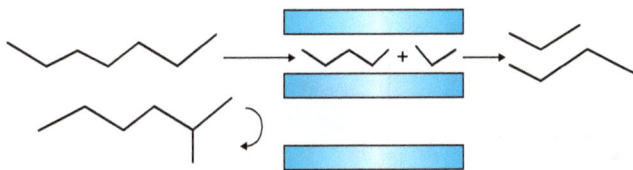

Fig. 3.9: Reactant selectivity in catalysis by zeolites.

Chemoselectivity and regioselectivity describe the ability of a catalyst to discriminate among the different and the same functional group, or several orientations, respectively. Diastereoselectivity defines control of the spatial arrangement of the functional groups in the product, while enantioselectivity is related to the catalyst ability to discriminate between mirror-image isomers or enantiomers.

Notions of differential and integral selectivity are often used. The former is the ratio between the accumulation rate of the desired product and the total rate of conversions of one of the original substances in various directions, including the formation of the desired product at a given temperature and with a given composition of the reaction mixture: $S = r_{des}/r$.

Differential selectivity relates the rates, integral selectivity is the ratio of the desired product concentration to the total concentration of all of the products, describing the course of the reaction up to a given point. In a gradientless system differential and integral selectivity coincide.

As selectivity depends on conversion, it is extremely dangerous to compare selectivity for different catalysts at just one end-point or at a certain period of time. For parallel reactions of the same kinetic order it still could be done, as selectivity for a system 1. $A \Rightarrow B$; 2. $A \Rightarrow C$ is independent of the concentration of A and therefore of conversion.

$$S_B = \frac{r_{+1}}{r_{+1} + r_{+2}} = \frac{k_{+1} K_A C_A}{k_{+1} K_A C_A + k_{+2} K_A C_A} = \frac{k_{+1}}{k_{+1} + k_{+2}} \tag{3.125}$$

At the same time, for consecutive reactions 1. $A \to C$; 2. $C \to D$ the expression for selectivity that assumes equilibrium binding of the reactants A and C is conversion dependent (Fig. 3.10):

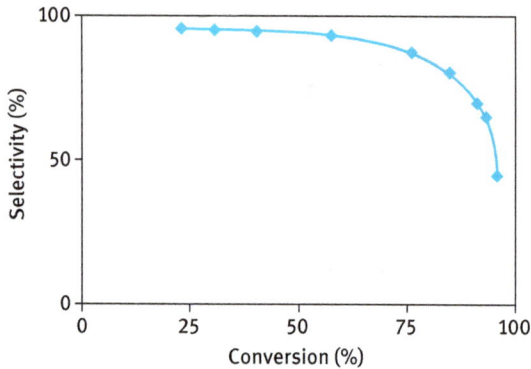

Fig. 3.10: Selectivity vs conversion dependence for consecutive reactions.

$$S_C = \frac{r_{+1} - r_{+2}}{r_{+1}} = 1 - \frac{r_{+2}}{r_{+1}} = 1 - \frac{k_{+2}K_2C_C}{k_{+1}K_1C_A} \qquad (3.126)$$

The steepness of the decline depends on $k_{+2}K_2/k_{+1}K_1$ ratio. When adsorption properties of A and C are similar, selectivity could be enhanced by applying a certain substance, which displaces C on the surface or impedes its re-adsorption. Such strategy is implied in selective hydrogenation of acetylene to ethylene when CO is added to the feed in minute (ppm) amounts.

Let us expand the kinetic analysis to a parallel-consecutive hydrogenation reaction (Fig. 3.11).

This mechanism can be visualized using the catalytic cycle approach with the vertices (nodes) and edges corresponding to the intermediates and steps, respectively. In the case of linear elementary steps the edges join the vertices, which correspond to the intermediates (adsorbed species ZA, ZB, ZC, ZD and vacant sites Z).

The rates along three routes, represented by three cycles in the graph (Fig. 3.11B) can be easily derived assuming that hydrogenation occurs in the liquid-phase by Eley-Rideal mechanism and that all reactants and products in principle could adsorb on the catalyst surface.

If deriving kinetic equations from the mechanism above, one arrives at following equations:

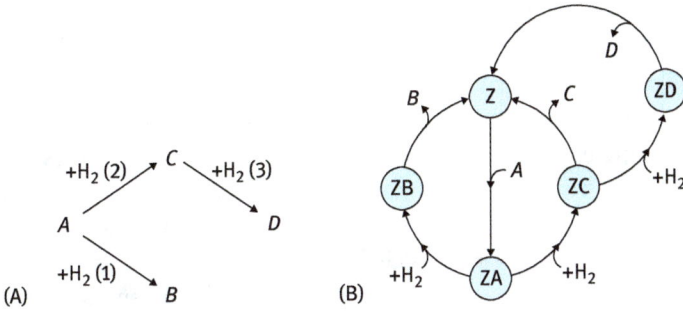

Fig. 3.11: Parallel-consecutive hydrogenation: (A) mechanism, (B) kinetic graph.

$$r_1 = \frac{k_{+1}K_A c_A c_{H_2}}{1+K_A c_A + K_B c_B + K_C c_C + K_D c_D}, r_2 = \frac{k_{+2}K_A c_A c_{H_2}}{1+K_A c_A + K_B c_B + K_C c_C + K_D c_D}$$

$$r_3 = \frac{k_{+3}K_C c_C c_{H_2}}{1+K_A c_A + K_B c_B + K_C c_C + K_D c_D} \tag{3.127}$$

where k_{+1} is rate constant of reaction 1 in the forward direction, K_A is the adsorption constant and c_A is the concentration of component (A) etc., and c_{H2} is hydrogen concentration in the liquid-phase. Equation (3.127) should be combined further to the mass balances of the components:

$$\frac{1}{\rho}\frac{dc_A}{dt} = -r_1 - r_2; \frac{1}{\rho}\frac{dc_B}{dt} = r_1; \frac{1}{\rho}\frac{dc_C}{dt} = r_2 - r_3; \frac{1}{\rho}\frac{dc_D}{dt} = r_3 \tag{3.128}$$

where ρ is the catalyst bulk density. Selectivity towards the compound C is:

$$S_C = \frac{dc_C}{dc_A} = -\frac{r_2 - r_3}{r_1 + r_2} = -\frac{k_{+2}K_A c_A - k_{+3}K_C c_C}{k_{+1}K_A c_A + k_{+2}K_A c_A}$$

$$= -\frac{k_{+2}}{k_{+1}+k_{+2}} + \frac{k_{+3}K_C c_C}{(k_{+1}+k_{+2})K_A c_A} = -L + M\frac{c_C}{c_A} \tag{3.129}$$

An analytical solution of (3.129) can be readily obtained:

$$C_C = \frac{L}{1-M}\frac{C_A{}^M}{(C_A^o)^{M-1}} - C_A \tag{3.130}$$

where C_A^o is the initial concentration of the reactant at which $C_C^o = 0$, $L = k_{+2}/(k_{+1}+k_{+2})$ and $M = k_{+3}K_C/((k_{+1}+k_{+2})K_A)$.

Equation (3.130) can be rearranged [2] to

$$S_C = \frac{L}{1-M}\frac{1}{\delta}\left[(1-\delta)^M - (1-\delta)\right] \tag{3.131}$$

where δ is conversion.

It follows from eq. (3.131) that selectivity is independent of hydrogen pressure being dependent on conversion.

Equation (3.131) implies and Fig. 3.10 illustrates that there is a substantial difference in selectivity of the intermediate on conversion. Thus, behavior of different catalysts should be compared in terms of selectivity at similar conversion levels. Selectivity values are usually more accurate than the values of reaction rates, even with comparatively large errors in these reaction rates. Selectivity analysis is thus crucial for elucidation of reaction mechanisms in complex reactions, making it possible to discriminate among rival reaction mechanisms in cases when statistical data fitting provides very similar results.

Analytic solutions are readily available for first-order reaction networks because the overall system of differential equations remains linear. For a consecutive reaction network consisting of a second-order reaction followed by the first-order reaction, an analytical solution is very complicated. In fact, the exact solutions for a mixed second-order process followed by a first-order reaction appeared only very recently [4, 5]. These expressions have a very high degree of mathematical complexity containing hypergeometric functions. They are not tractable for analysis of, for example, selectivity dependence on conversion and reaction parameters. An alternative to analytic expressions is numerical solutions, which can be rather easily obtained using commercially available software as well as freeware.

Such analysis performed in [6] demonstrated that for the reaction network comprising the second-order reaction followed by the first-order reaction just by diluting the reaction mixture 10fold selectivity to the intermediate at approximately 20% conversion can dramatically drop from 95% to 5%. Such variations in selectivity when different catalysts of different research groups are compared can incorrectly be attributed to intrinsic nature of catalysts rather than different initial substrate concentrations (solvent amounts), which often is not discussed.

The reaction mechanism with linear steps can easily be shown graphically (see Fig. 3.11B). For non-linear mechanisms the situation is somewhat different. For a simple Langmuir-Hinshelwood mechanism (eq. 3.34) visualization can be done as presented in Fig. 3.12 with the non-linear step 3 depicted in the graph as a line connecting node (*A) with (*), touching (*B) and additionally as a broken line connecting (*B) and (*).

The mechanism in eq. (3.34) contains only one route, however from Fig. 3.12 a wrong conclusion could be made about two catalytic cycles, which emphasizes that visualization of non-linear mechanisms is not straightforward. There is, however, an incentive to treat heterogeneous complex catalytic kinetics in the same spirit as homogeneous organometallic systems using the notion of catalytic cycles. Thus in the literature there are mechanisms with non-linear steps, which are incorrectly discussed in terms of catalytic cycles, giving a misleading impression of reaction mechanisms and leading to incorrect rate expressions.

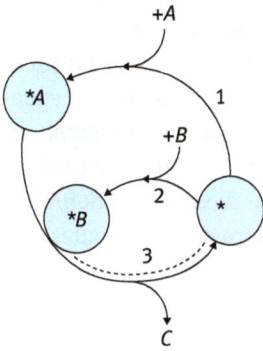

Fig. 3.12: Graph for a mechanism with a non-linear step.

A convenient way to overcome the difficulties in the visualization of reaction networks is to use the catalytic cycle approach, in a similar way to linear sequences with the number of nodes (and respective edges) corresponding to the number of steps. For example, CO oxidation with stepwise (molecular to dissociative) oxygen adsorption and two non-linear steps 2 and 4 is considered below:

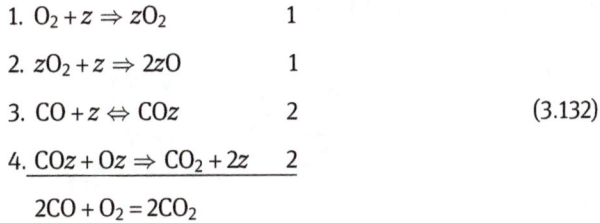

$$
\begin{array}{lll}
\text{1. } O_2 + z \Rightarrow zO_2 & 1 \\
\text{2. } zO_2 + z \Rightarrow 2zO & 1 \\
\text{3. } CO + z \Leftrightarrow COz & 2 & \text{(3.132)} \\
\text{4. } \underline{COz + Oz \Rightarrow CO_2 + 2z} & 2 \\
\phantom{\text{4. }} 2CO + O_2 = 2CO_2 &
\end{array}
$$

The graph for such mechanism can be visualized either with a conventional approach (Fig. 3.13A) or with a concept of a reaction route cycle [2] (Fig. 3.13B). In Fig. 3.13B the nodes contain all the surface species taking part in the cycle.

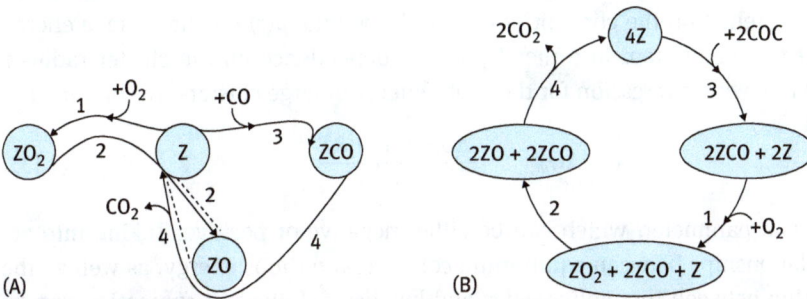

Fig. 3.13: Langmuir-Hinshelwood CO oxidation mechanism with dissociative oxygen adsorption: (A) kinetic graph and (B) catalytic cycle.

In the reaction route cycle (similar to Fig. 3.13B) the number of species in the node N_N of a cycle should at least correspond to free sites required for adsorption/reaction steps, taking into account stoichiometric numbers of these steps. This number can be calculated by considering all the free sites on the left-hand side of equations for elementary reactions multiplied by the respective stoichiometric numbers. Additionally, sites (corrected with the stoichiometric numbers of steps) that are released and appear on the right hand side of steps, preceding those adsorption/reaction steps, should be subtracted. This analysis should be performed for all routes.

3.4.5 Structure sensitivity

In the particle size regime between 1 and 20 nm, three types of reactivity change are distinguished. This is illustrated in Fig. 3.14. Some reactions are independent of particle size. The rates of other reactions, normalized per exposed metal surface atom, decrease, increase or even pass through a maximum when particle size increases. This dependence of catalytic rate on size or dispersion of catalytically active particles has been known since Boudart [7] introduced the notion of structure-sensitive and insensitive reactions in the 1960s.

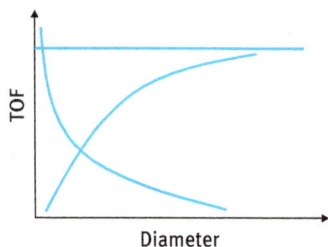

Fig. 3.14: Structure sensitivity plots.

The thermodynamic approach was used quantitatively to explain the cluster size effect [8]. It relied on the concept of chemical potential $\mu(r)$ (or the surface energy excess of the metal atom in a metal particle) dependence on the cluster radius r size. The following expression for the Gibbs energy of large clusters was obtained:

$$\Delta G_{ads,\infty} = \frac{\Delta G_{ads}(r) - \eta RT}{r} \tag{3.133}$$

where η is a parameter, which can be either negative or positive. Taking into account relationships between equilibrium constant and Gibbs energy, as well as the relationship between thermodynamics and kinetics, a following expression can be written for adsorption:

$$k(r) = gK(r)^{\alpha} = k_{\infty}e^{\frac{\alpha\eta}{T}} \qquad (3.134)$$

For the two-step sequence it was demonstrated [8] that turnover frequency $v(r)$ is:

$$v(r) = \frac{(k_{+1}P_{A_1}k_{+2}P_{A_2} - k_{-1}P_{B_1}k_{-2}P_{B_2})e^{(\alpha_2+\alpha_1-1)\frac{\eta}{T}}}{k_{1+}P_{A_1}e^{\alpha_1\frac{\eta}{T}} + k_{+2}P_{A_2}e^{(\alpha_2-1)\frac{\eta}{T}} + k_{-1}P_{B_1}e^{(\alpha_1-1)\frac{\eta}{T}} + k_{-2}P_{B_2}e^{\alpha_2\frac{\eta}{T}}} \qquad (3.135)$$

where P_{A_1}, etc., are partial pressures (for gas-phase reactions) or concentrations (for liquid-phase reactions), k_i are kinetic constants and α are the Polanyi parameters of steps. If these parameters a are equal to each other and the overall reaction is irreversible:

$$v(r) = \frac{\omega_{+2}e^{(\alpha-1)\frac{\eta}{T}}}{1 + \frac{\omega_{+2}+\omega-1}{\omega_{+1}}e^{-\frac{\eta}{T}}} = \frac{p_1 e^{(\alpha-1)\frac{\eta}{T}}}{1+p_2 e^{-\frac{\eta}{T}}} \qquad (3.136)$$

with ω_i -frequencies of steps (i.e., $\omega_{+1} = k_{+1}P_{A_1}$, etc.), $p_2 = (\omega_{+2}+\omega_{-1})/\omega_{+1}$ and $p_1 = \omega_{+2}$.

Comparison of eq. (3.136) with the experimental data demonstrated a very good correspondence of theory and experiments and even the ability to describe maxima in TOF as a function of cluster size.

The same equation can be derived [9], taking into account differences in the Gibbs adsorption energy between edges, corners and terraces:

$$\Delta G_{ads} = \Delta G_{ads,\,terraces}f_{terraces} + \Delta G_{ads,\,edfes}f_{edges} + \Delta G_{ads,\,corners}f_{corners} \qquad (3.137)$$

where $f_{terraces}$, f_{edges} denote fractions of these surface sites, whose sum is obviously equal to unity. The geometrical considerations for different particle size shapes are presented in Fig. 3.15.

Theoretical analysis of the models presented above demonstrated that, in line with experimental data, significant changes in kinetic regularities (for example, reaction orders and selectivity) could be anticipated with varying cluster size.

An alternative way of implementing the cluster size dependence in reaction kinetics is to view reaction constants as size independent. Then each type of site (terraces, corners and edges) contributes to the overall rate according to:

$$v(d) = r_{corners}f_{corners} + r_{edges}f_{edges} + r_{terraces}f_{terraces} \qquad (3.138)$$

In (3.138) it is assumed that adsorption/desorption and reaction rates on terraces, edges and corners are always the same, independent of the cluster size, which might be unrealistic especially for reactions involving large organic molecules when an ensemble of sites is required for multi-centered adsorption, and reactivity can depend on the cluster size even for terraces.

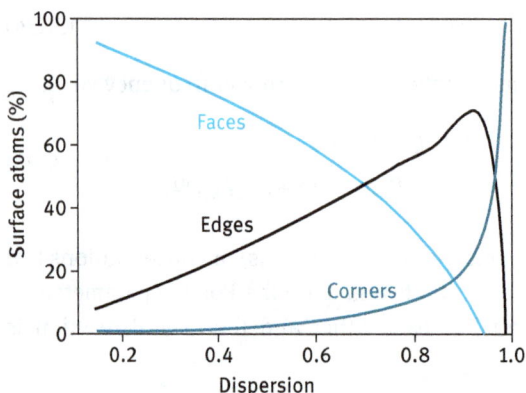

Fig. 3.15: Fraction of exposed surface atoms as a function of metal dispersion. (From [10] copyright Elsevier).

3.4.6 Mechanism-free kinetics – kinetic polynomial

For many catalytic reactions with non-linear steps, derivation of kinetic equations could be challenging. In order to avoid such difficulties Lazman and Yablonsky applied constructive algebraic geometry to non-linear kinetics that expresses the reaction rate of a complex reaction as an implicit function of concentrations and temperature. This concept of kinetic polynomial [11] has found important applications, including parameter estimation, analysis of kinetic model identifiability and finding all steady states of kinetic models. The Lazman-Yablonsky four-term rate equation for polynomial kinetics is:

$$r = \frac{k_+ \left(f_+ (c) - K_{eq}^{-1} f_- (c) \right)}{\Sigma(k, c)} N(k, c) \qquad (3.139)$$

The terms in eq. (3.139) are as follows: (1) the kinetic apparent coefficient k_+; (2) the potential term, or driving force related to the thermodynamics of the net reaction; (3) the term of resistance, i.e., the denominator, which reflects the complexity of reaction, both its multistep character and its non-linearity; finally, (4) the non-linear term $[N(k,c)]$, a polynomial in concentrations and kinetic parameters, which is caused exclusively by non-linear steps. In the case of a linear mechanism, this term vanishes. In classical kinetics of heterogeneous catalysis (LHHW equations) such term is absent.

Such phenomenological or mechanism-free expressions, although useful for some applications and for preliminary analysis of kinetic data, are in general not reliable as they do not predict reaction rates, concentration or temperature dependence outside the experimental conditions. Models based on the knowledge of elementary processes provide reliable extrapolation outside of the experimentally studied domain. In addition, the system is better understood.

3.4.7 What is behind a rate constant?

Transition state theory of surface reactions was developed independently by Laidler and Glasstone [12], Eyring [13] and Temkin [14] shortly after the appearance of the more general approach by Eyring and Polanyi. According to this theory, the reaction rate for a reaction $n_A A + n_B B + I_{ads} + J_{ads} \rightarrow C$ comprising reactants A and B from the gas phase and adsorbed species i and j is given by [14]:

$$r = \chi L \frac{k_B T}{h} \frac{q_{\neq}}{q_A q_B} \exp\left(-\frac{\Delta E}{RT}\right) [A]^{n_A} [B]^{n_B} \theta_i \theta_j \tag{3.140}$$

where k_B is the Boltzmann constant, h is the Plank constant, $q_{\neq}$ is the modified partition function of the transition state, and q_A and q_B are partition functions of the reactants, χ is the transmission coefficient, reflecting nonadiabaticity in the quantum-mechanical sense, L is the total number of sites per unit area, and n_A and n_B are reaction orders in respective compounds. The concentrations of the gas-phase molecules are given in number of molecules per volume of gas. Quite often instead of gas-phase concentrations, respective pressures are used; therefore, the equation for the rate should be modified [14]

$$r = \chi L \frac{k_B T}{h(k_B T)^{n_A + n_A}} \frac{q_{\neq}}{q_A q_B} \exp\left(-\frac{\Delta E}{RT}\right) [P_A]^{n_A} [P_B]^{n_B} \theta_i \theta_j \tag{3.141}$$

Partition functions are important because some useful properties follow from them. To calculate the values of the partition functions of the molecule, each of the motions (electronic, translational, rotational and vibrational) should be considered, keeping in mind that a molecule of n atoms has $3n$ degrees of freedom. For a molecule there are three degrees of translational freedom and $3n - 6$ and $3n - 5$ for the vibrational freedom of non-linear and linear molecules, respectively. Thus an atom ($n = 1$) has three translation degrees of freedom; a molecule of two atoms has three translational, one vibrational, and two rotational degrees of freedom. For a linear molecule of three atoms, in addition to three translational degrees of freedom, there are four vibrational and two rotational degrees of freedom, while for a non-linear molecule there will be three vibrational and three rotational degrees of freedom. There is one specific feature in the partition functions for the transition state as the one vibrational degree of freedom is replaced by the movement along the reaction coordinate. Thus the total number of degrees of freedom is $3n - 1$.

The translational partition function can be calculated knowing the particle mass, temperature and is conventionally calculated per unit volume: q_{trans}^{3D} or per dimension q_{transl}

$$q_{trans}^{3D} = V \left(\frac{(2\pi m k_B T)^{\frac{1}{2}}}{h} \right)^3 ; q_{transl} = I \frac{(2\pi m k_B T)^{\frac{1}{2}}}{h} \tag{3.142}$$

where m is the reduced mass. For example, for a molecule AB this is $m_A m_B/(m_A + m_B)$. The vibrational energy levels are usually calculated using the harmonic oscillator approximation. For a diatomic molecule, the vibrational partition function per degree of freedom is:

$$q_{vib} = \frac{1}{\left(1 - e^{-\frac{hv}{k_B T}} \right)} \tag{3.143}$$

IR spectra are applied for calculations of vibrational partition functions, clearly representing a challenge in the calculation of vibration frequency of short-lived (10^{-12} s) transition states. However, vibrational partition functions are usually equal to unity, unless the vibrations are low frequency. For calculations of rotational partition functions the moment of inertia I (for example, the molecule structure) must be known. In the case of a diatomic molecule, the rotation partition function is:

$$q_{AB} = \frac{8 k_B \pi^2 I T}{h^2} \tag{3.144}$$

The electronic partition function q_{el} is usually equal to unity.

For an indirect molecular adsorption on a surface, the adsorbed molecule is free to move across the surface, rotate and vibrate. For such precursor-mediated adsorption the pre-exponential factor in eq. (3.140) takes the form:

$$k_{ads} = \frac{(k_B T) q_{trans}^{\#2}}{h q_{trans}{}^3} \times \frac{q_{rot}^{\#} q_{vib}^{\#}}{q_{rot} q_{vib}} = \frac{(k_B T) q_{trans}^{\#2}}{h q_{trans}{}^3} \times S_{id} \tag{3.145}$$

where the former term is the collision factor and $S_{id} = (q_{rot}^{(\#)} q_{vib}^{(\#)})/(q_{rot} q_{vib})$ is the sticking coefficient for indirect adsorption. If there are no changes in the values of rotational and vibrational partition functions of the adsorbed molecule, the sticking coefficient is equal to unity. The transition state is associated, however, with the decrease of entropy, therefore the sticking coefficient is less than unity and is usually a function of surface coverage, temperature and the surface structure of the adsorbent.

For direct adsorption on solid surfaces with limited mobility of adsorbed species the pre-exponential factor is:

$$k_{ads} = \frac{k_B T}{h} \frac{q_{vib}^{\#ads}}{q_{trans}^3 q_{rot} q_{vib}} = \frac{k_B T \, q_{trans}^{\#2}}{h q_{trans}^3} \times S_d \qquad (3.146)$$

with the same collision factor as in (3.145) but a much smaller (10^{-6}–10^{-8}) sticking coefficient $S_d = \dfrac{q_{vib}^{\#ads}}{(q_{trans}^{\#2} q_{rot} q_{vib})}$ caused by considerable losses in entropy when the molecule is strongly fixed on the surface after direct adsorption.

For desorption of a molecule with similar freedom for adsorbed species and the transition state:

$$k = \frac{k_b T}{h} \frac{q_{A\#}}{q_{A^*}} \approx 10^{13} s^{-1}$$

$$\text{as} \frac{q_{A\#}}{q_{A^*}} \approx 1 \text{ and } \frac{k_b T}{h} \approx 10^{13} s^{-1}. \qquad (3.147)$$

When transition state is mobile, the prefactor for desorption is in the range 10^{14-16} s^{-1} in line with experimental data [15]. The values of prefactors for surface reactions of the type $A^* \rightarrow B^*$ are approximately 10^{13} s^{-1}, while the prefactor for dissociation of an adsorbate $AB^* \rightarrow A^* + B^*$ is related to the value of desorption prefactor of the same molecule, that is, $k_{diss} = \sim 10^{-3} k_{des}$ [15].

Transition state theory can be applied to complex reactions, theoretically predicting the values of pre-exponential factors for multistep reaction mechanisms, where because of over-parametrization the direct numerical fitting of rate constants results in values with a large error interval.

Kinetic treatment based on the theory of complex reactions – which went beyond the simplified Langmuir-Hinshelwood-Hougen-Watson approach with one rate-determining step (Tab. 3.1) – made the calculation of parameters necessary (pre-exponential factors, activation energies of elementary reactions, etc.). Such theoretical approaches started to appear in 1970s and were based on thermodynamics and transition state theory, as well as surface science experimental tools (ultra-high vacuum studies, spectroscopy) in combination with enhanced computer power. Later on they were nicely summarized, refined and popularized by J. Dumesic and co-workers [16] as microkinetic modeling. In essence, micro-kinetic modeling assembles molecular-level information obtained from quantum chemical calculations, atomistic simulations and experiments to quantify the kinetic behavior at given reaction conditions on a particular catalyst surface. Some examples will be provided in Chapter 4.

3.4.8 Apparent activation energy of complex reactions

Apparent activation energy is an often used concept for complex reactions, describing the overall dependence of a reaction on temperature. The expression for the apparent (observed) activation energy is given as follows:

$$E_{act, apparent} = -R\frac{\partial \ln r}{\partial 1/T} = RT^2 \frac{\partial \ln r}{\partial T} \tag{1.3}$$

As an example, the two-step mechanism (3.48) with two irreversible steps gives an expression for the rate

$$r = \frac{\omega_{+1}\omega_{+2}}{\omega_{+1}+\omega_{+2}} = \frac{k_{+1}C_{A_1}k_{+2}C_{A_2}}{k_{+1}C_{A_1}+k_{+2}C_{A_2}} \tag{3.148}$$

Considering eq. (1.3) the apparent activation energy can be calculated as

$$E_{act, apparent} = RT^2 \frac{\partial \ln \frac{k_{+1}C_{A_1}k_{+2}C_{A_2}}{k_{+1}C_{A_1}+k_{+2}C_{A_2}}}{\partial T} = RT^2 \frac{\partial}{\partial T}(\ln k_{+1} + \ln C_{A_1} + \ln k_{+2} + \ln C_{A_2}$$
$$- \ln(k_{+1}C_{A_1} + k_{+2}C_{A_2})) \tag{3.149}$$

Neglecting dependence of the prefactor on temperature in $\ln k_{+1}$ in comparison with the exponential term of the Arrhenius equation one gets

$$\frac{\partial}{\partial T}(\ln k^0_{+1}\exp(-E_{+1}/RT)) = \frac{\partial}{\partial T}(\ln k^0_{+1} - E_{+1}/RT) = \frac{E_{+1}}{RT^2} \tag{3.150}$$

and

$$\frac{\partial}{\partial T}(\ln k^0_{+2}\exp(-E_{+2}/RT)) = \frac{E_{+2}}{RT^2} \tag{3.151}$$

The last term in eq. (3.149) can be transformed

$$\frac{\partial}{\partial T}(\ln(k_{+1}C_{A_1} + k_{+2}C_{A_2})) = \frac{\partial \ln(k_{+1}C_{A_1}+k_{+2}C_{A_2})}{\partial(k_{+1}C_{A_1}+k_{+2}C_{A_2})}\frac{\partial(k_{+1}C_{A_1}+k_{+2}C_{A_2})}{\partial T} =$$

$$= \frac{1}{k_{+1}C_{A_1}+k_{+2}C_{A_2}}\frac{\partial(k_{+1}C_{A_1}+k_{+2}C_{A_2})}{\partial T} = \frac{C_{A_1}k^0_{+1}e^{-E_{+1}/RT}\frac{E_{+1}}{RT^2}+C_{A_2}k^0_{+2}e^{-E_{+2}/RT}\frac{E_{+2}}{RT^2}}{k_{+1}C_{A_1}+k_{+2}C_{A_2}} =$$

$$= \frac{k_{+1}C_{A_1}\frac{E_{+1}}{RT^2}+k_{+2}C_{A_2}\frac{E_{+2}}{RT^2}}{k_{+1}C_{A_1}+k_{+2}C_{A_2}} \tag{3.152}$$

The final expression for the apparent activation energy is

$$E_{a,\,app} = E_{+1} + E_{a+2} - \frac{k_{+1}C_{A_1}}{k_{+1}C_{A_1} + k_{+2}C_{A_2}} E_{+1} - \frac{k_{+2}C_{A_2}}{k_{+1}C_{A_1} + k_{+2}C_{A_2}} E_{+2} \qquad (3.153)$$

A special case of the two-step sequence is the Eley–Rideal mechanism, where the first step in the mechanism [3.48] can be considered as quasi-equilibrium, resulting in eq. (3.32). The activation energy of such reaction is illustrated in Fig. 3.16.

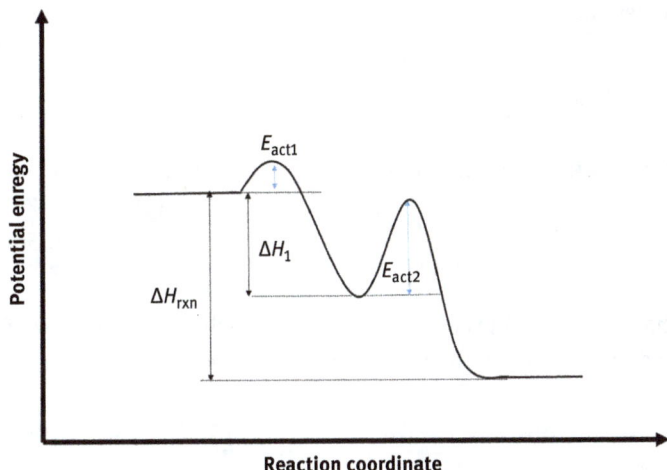

Fig. 3.16: Activation energy profile for a two-step reaction with a low value of activation energy for the binding step.

Equation (3.32) can be written as

$$r = \frac{k^o_{+2} K^o_A C_A e^{(-\Delta H_1 - E_{+2})/RT}}{1 + K^o_A C_A e^{-\Delta H_1/RT}} \qquad (3.154)$$

When binding of the substrate to the catalyst is not strong enough and the catalyst mainly exists in the form of unbound species, $K^o_A C_A e^{-\Delta H_1/RT} < < 1$, the apparent activation energy is simply $E_{act} = E_{+2} + \Delta H_1$ and can be even negative when the enthalpy of the exothermal intermediate complex formation (being negative) is larger than the activation energy of the second step. Experimentally, such negative values of activation energy were reported for a homogeneous Diels–Alder reaction of dimethylanthracene and tetracyanoethylene in the series of inert solvents [17, 18] and for the organocatalytic Mannich reaction of an imine and fluorinated ketoester catalyzed by thiourea [19]. In the latter case, theoretical calculations of the enthalpy showed that binding of the catalyst to substrate occurs without an activation barrier, and that the overall enthalpy change of the transition state is negative relative to starting substrates. In heterogeneous cata-lytic cracking of n-alkanes over ZSM-5, the observed activation energy decreases from 140 to −50 kJ/mol when the carbon chain length increases from three to 20 [20]. This dependence is a consequence of the competition between an intrinsic kinetics dependence with temperature, and a decrease in the adsorption strength and the concentration of active intermediates with temperature.

In fact, eq. (3.154) can have a maximum as a function of temperature, which can be easily demonstrated by considering the reciprocal function and taking a de-rivative of it. When the derivative is equal to zero

$$\left(\frac{1 + K_A^o C_A e^{-\Delta H_1/RT}}{k_{+2}^o K_A^o C_A e^{(-\Delta H_1 - E_{+2})/RT}} \right)' = 0 \tag{3.155}$$

an expression for the temperature at optimum is

$$T_{max} = \frac{-\Delta H_1/R}{\ln \frac{-\Delta H_1 - E_{+2}}{K_1^o C_A E_{+2}}} \tag{3.156}$$

which shows clearly that positive values of T_{max} are achieved when the enthalpy of the exothermal intermediate complex formation is larger than the activation energy of the second step.

3.4.9 Dynamic catalysis and deactivation

In the chemical industry the majority of processes are designed to be conducted in stationary conditions. Hence, it is not surprising that steady-state kinetics is mainly studied as the surface concentration of reactants is time independent. The investi-gation of reaction kinetics in gradientless reactors further simplifies the experimen-tal routine.

Although steady-state kinetic experiments give valuable information about the overall rates of catalytic reactions, they cannot unequivocally reveal the underlying reaction mechanisms because different reaction mechanisms might give similar steady-state rate equations.

From an industrial viewpoint the main drawback of kinetic models, based only on steady-state data, is that start-up and transient regimes cannot be reliably modelled. Kinetic models for non-stationary conditions should be also applied for the processes in fluidized beds, reactions in riser (reactor) – regenerator units with catalyst circulation – as well as for various environmental applications of heterogeneous catalysis, when the composition of the treated gas changes continuously.

For the construction of a kinetic model for non-stationary conditions, knowl-edge about the evolution of the concentrations of adsorbed species on the catalyst surface is needed. Under non-stationary conditions, the changes of concentration fields in time, reactor space and catalyst surface (for heterogeneous catalysis), are interrelated by complex dependencies. Therefore, knowledge about the gas and surface composition is required for experimental investigations under non-stationary conditions.

In addition, in many catalytic reactions changes in the rate constants of some elementary steps are associated with the side reactions, which are not present in catalytic cycles. When such deactivation is prominent it requires some counterbalancing at the catalyst and reactor levels, as will be illustrated in Chapter 4. The time scale of deactivation depends on the process type and can vary from a few seconds, as in fluid catalytic cracking (FCC), to several years, for example, in steam reforming of natural gas or oxidation of o-xylene, which are also discussed in Chapter 4.

The main causes of deactivation in heterogeneous catalysis are poisoning, fouling, thermal degradation (sintering, evaporation) initiated by the often high temperature, mechanical damage and corrosion/leaching by the reaction mixture (Fig. 3.17).

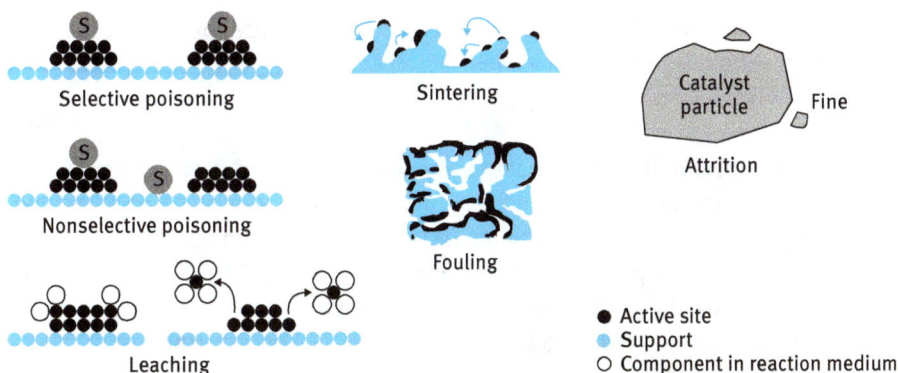

Fig. 3.17: Deactivation mechanisms. (From [21]).

Poisoning is usually caused by strong adsorption of impurities in the feed. Poisoning depends upon adsorption strength of such poisons relative to the other species competing for catalytic sites. Adsorbed poisons may not only block active sites, but change the electronic or geometric structure of the surfaces as well. Fouling is associated with covering of the catalyst surface by deposits, which are quite often hydrogen-deficient carbonaceous materials (i.e., coke), making the active sites inaccessible.

For liquid-phase reactions catalysis leaching can be prominent, with the leached metals responsible for high catalytic activity in some reactions. Such leaching is particularly problematic in oxidation catalysis because of the strong complexation and solvolytic properties of oxidants (H_2O_2, RO_2H) and/or the products (H_2O, ROH, RCO_2H, etc.).

Strong stresses of packed catalyst beds during start-ups, shut-downs and catalyst regeneration could lead to mechanical deactivation. Finally, thermal degradation (because of sintering, chemical transformations, evaporation, etc.) could cause deactivation. Of particular prominence is sintering, or, the loss of catalyst active surface caused by crystallite growth of either the support material or the active phase.

Various kinetic models to account for deactivation have been presented in the literature, of different empirical and semi-empirical natures. Such models often separate activity function a (the current reaction rate divided by the initial intrinsic rate, $a = r_t/r_0$) and deactivation:

$$-\frac{da}{dt} = f(c_i, T)\varphi(a) \tag{3.157}$$

In eq. (3.157) $f(c_i, T)$ reflects the activity on non-deactivated catalyst and $\varphi(a)$ is deactivation function. This approach of separability is applied when the reaction and deactivation rates are of different magnitudes; the reactions proceed relatively fast, while deactivation requires much longer time (hours, days or months) and in addition deactivation does not affect selectivity.

Different equations have been presented in the literature for the deactivation function, for example, a power law one:

$$\varphi(a) = a^m \tag{3.158}$$

As activity changes in fixed-bed reactors are observed with time-on-stream, deactivation is often expressed in terms of time, which is not, however, a true variable. More correctly, the deactivation function has to be expressed in terms of the deactivating agent: the coke precursor or the poison, which means that the amount of coke (or poison) on the catalyst site should be known. The determination of a rate equation for the formation of the coke precursor is thus an integral part of the kinetic study of the process.

For the formal treatment of deactivation for separable kinetics, the fraction of non-coked or non-poisoned catalyst (f) can be defined in terms of the concentration of the poison or coke on the catalyst surface (c_P) and the catalyst capacity for coke or poison ($c_{P,0}$):

$$f = \frac{c_{P,0} - c_P}{c_{P,0}} \tag{3.159}$$

Accumulation of coke (or poison) on the catalyst is given by:

$$\frac{dc_P}{dt} = R_P(c_P, p, T, WHSV) \tag{3.160}$$

where *WHSV* stands for weight hour space velocity for fixed-bed operation in continuous mode. Separating activity and deactivation, as well as assuming a power law for the dependence of R_P on coke concentration (rate of coke formation is proportional to some power (n) of unused capacity for adsorbing coke):

$$\frac{dc_P}{dt} = k(c_{P,0} - c_P)^n \phi(p, T, WHSV) \tag{3.161}$$

after integrating for constant values of $c_{P,o}$ and f the linear exponential and hyperbolic relationships between fraction of coke and time could be obtained:

$$f = \left[\frac{1}{1 + (n-1)k'c_{P,0}^{n-1}t}\right]^{\frac{1}{n-1}} \quad n \neq 1; f = \exp\left(-k'c_{P,0}^{n-1}t\right), n = 1 \tag{3.162}$$

where $k' = k\phi(p, T, WHSV)$ is only an apparent constant, as it depends on temperature, *WHSV* and the reactant pressure p.

However, when deactivation is considered as a constituent part of the reaction scheme, it gives more possibilities for clarifying deactivation mechanisms. For a scheme:

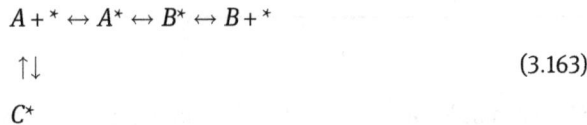

$$A + {}^* \leftrightarrow A^* \leftrightarrow B^* \leftrightarrow B + {}^*$$

$$\uparrow\downarrow \tag{3.163}$$

$$C^*$$

Where C represent adsorbed coke, deactivation r_C and self-regeneration r_{-C} proceed simultaneously with these steps being essentially slower than the reaction steps. Applying the quasi steady-state only to the main reaction $d\theta_A/dt \approx 0$ and supposing for simplification that adsorbed A is the most abundant surface intermediate, the changes in coke coverage as a function of time-on-stream are:

$$\frac{d(1-\theta_A)}{dt} = k_c\theta_A - k_{-C}(1-\theta_A) \tag{3.164}$$

where k_C and k_{-C} are deactivation and self-regeneration constants. When integrating eq. (3.164) with the boundary conditions $t = 0$, $\theta_A^0 = 1$ (i.e., the surface is initially totally covered by A), an analytical expression for the reaction rate is obtained:

$$r = r_0\theta_A = a_3 + a_1e^{-a_2t} \tag{3.165}$$

where r_0 is the reaction rate at deactivation free conditions and:

$$a_1 = \frac{k_C r_0}{k_C + k_{-C}}; \ a_2 = k_C + k_{-C}; \ a_3 = \frac{k_{-C} r_0}{k_C + k_{-C}} \tag{3.166}$$

At the infinite time the reaction rate corresponds to a_3 because of self-regeneration. The steepness of activity loss is characterized by a_2, while the sum of a_1 and a_3 corresponds to the rate at deactivation free conditions.

For sintering of the metal crystallites on a catalyst support, models for activity decline could also include certain residual activity level similarly to eq. (3.166).

Generally, it can be stated that derivation of a rate equation with deactivation can be based on the theory of complex reactions, considering deactivation as an independent route leading to coke (C) on the catalyst surface (Tab. 3.2).

Tab. 3.2: Reaction routes and kinetic equations for a mechanism including deactivation.

	$N^{(1)}$	$N^{(2)}$		
1. $A + * \rightleftharpoons A^*$	1	1	$\theta_V = \dfrac{1-\theta_{C^*}}{1+K_A C_A + K_B C_B}$	$r_A = -r_2 - |v_A| r_4$
2. $A^* \rightarrow B^*$	1	0		$r_B = r_2 - |v_B| r_4$
3. $B^* \rightleftharpoons B + *$	1	0	$r_2 = (k_{+2}K_A C - k_{-2}K_B C_B)\dfrac{1-\theta_{C^*}}{1+K_A C_A + K_B C_B}$	$r_{C^*} = r_4$
4. $A^* \rightarrow C^*$	0	1		$|v_A| = 1, |v_B| = 0$
$N^{(1)} A \rightarrow B$			$r_4 = k_{+4}K_A C_A \dfrac{1-\theta_{C^*}}{1+K_A C_A + K_B C_B}$	
$N^{(2)} A \rightarrow C^*$				

More examples are provided in the monograph [2], which covers several linear and non-linear models, including mono- and bimolecular deactivation mechanisms for reactant and product molecules as well as a reversible deactivation reaction.

Contrary to experiments in fixed-bed continuous reactors, deactivation of catalytic reactions performed in batch modes can be hidden, as the observed behavior could be similar to first-order kinetics. It is typical to then perform experiments under identical conditions but recycling the catalyst.

For the modelling of deactivation, a semi-empirical equation can be used:

$$C_R = C_R^0 - a_1 t e^{-k_d \tau} \tag{3.167}$$

where a_1 is a complex parameter (could include reactant concentration), C_R^0 is the initial concentration of reactant, k_d is the deactivation constant, t is the time, and τ is the total time starting from the beginning of the series.

In multilayer coking, as observed when organic compounds react on catalyst surfaces, several carbonaceous layers are build-up (Fig. 3.18): $R + \theta_0 \rightarrow \theta_1$, $R + \theta_1 \rightarrow \theta_2...R + \theta_{n-1} \rightarrow \theta_n$ where R denotes the reactant.

The rate of coke generation is composed from the rate of monolayer and multilayer formation:

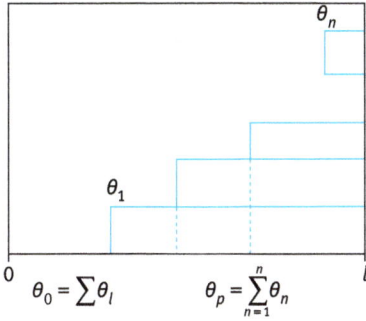

$$\theta_0 = \sum \theta_l \qquad \theta_p = \sum_{n=1}^{n} \theta_n$$

Fig. 3.18: The model for multilayer coking.

$$\frac{1}{\xi}\frac{dC_C}{dt} = k_m\theta_0 c_R + k_p c_R \sum_{n=2}^{N}\theta_{n-1} = k_m c_R(1-\theta_P) + k_p c_R(\theta_P - \theta_N) \qquad (3.168)$$

where ξ is the quantity of coke, derived from one mole of reactant and θ_p is the coverage of coke.

As activity changes in the main reaction are only caused by blocking of the active sites by coke formation in the first layer, the coverage of coke θ_p is:

$$\frac{C_m d\theta_P}{\xi\ dt} = k_m c_R(1-\theta_P) \qquad (3.169)$$

where C_m is the capacity of a monolayer, or the maximum amount of coke (gram of coke/gram of catalyst), which could be adsorbed on the catalyst for monolayer coking. Equations (3.168) and (3.169) should be solved simultaneously, which is a very demanding task. Simplifications of the model for the linear mechanism of the main reaction are $a = 1 - \theta_p$ and $\theta_N = 0$. After integration an equation relating activity to the coke concentration can be obtained:

$$\frac{C_C}{C_m} = (1-\varphi)(1-a) - \varphi\ln a \qquad (3.170)$$

where $\varphi = k_p/k_m$. Equation (3.170) includes linear and exponential dependences as special cases. If multilayer coking is not profound, then $\varphi = k_p/k_m \approx 0$ and $C_C/C_m = (1-a)$. When the deactivation constants for monolayer and multilayer coking are the same and $\varphi = k_p/k_m = 1$, then $C_C/C_m = -\ln a$ and $a = 1 - C_C/C_m$. Similar treatment could be extended to cases of non-linear deactivation and non-linear mechanisms for the main reaction.

When a poison molecule (P) deposits on a vacant surface site, $P + * \rightleftharpoons P*a$ certain part of the catalytically active sites is blocked. For the first-order kinetics in a batch reactor it would mean that instead of m (catalyst mass) an effective catalyst mass ($m^* = m - f$), should be used, where f is the amount of catalyst irreversibly blocked by the poisons. If f is independent on reactant concentration, one arrives at:

$$C = C_o e^{-k(m-f)t} \tag{3.171}$$

where $C = C_o$ at $t = 0$.

3.4.10 Mathematical treatment of experimental data

In addition to the selection and construction of experimental equipment, planning of experiments and implementing them, the procedures for obtaining kinetic parameters also involve checking the experimental data consistency of the developing kinetic models and evaluation of the kinetic data. Modern statistical methods of kinetic evaluation require parameter estimation through numerical data fitting with subsequent evaluation of the physical and statistical consistency of the parameters.

A heuristic approach to the generation of kinetic data includes experiments with varying parameters that are considered to be important, i.e., temperature, concentrations, pressures, pH, catalyst amount, volume, etc. After estimation of the rate constants by regression analysis and evaluation of the adequacy of the model, additional experiments are performed if the results are not satisfying. This is followed by another round of parameter estimation and possible simulations outside the studied parameter domain, checking if it is possible to predict the kinetic behavior outside of the studied initially domain.

In experimental design, possible experimental errors and the necessity of repeated experiments should be accounted for, and a decision on the minimum set of experiments should be made. In addition, performing experiments under non-isothermal conditions should be considered because the temperature profile can be followed on-line. For non-isothermal reactors, the experimental temperature field can be interpolated by fitting an empirical model for each temperature profile, $T = f$ (reactor length), using, for example, a polynomial model.

The structure of kinetic models used in parameter estimation includes expressions for the reaction rates $r_j = f_j(k,K,c)$, generation rates of the components $r_i = \sum v_{ij} r_j$ and the mass balances of reactors, presented in the corresponding sections of this book for some reactor types.

Differential models for batch and fixed-bed reactors can generally be described by ordinary differential equations (ODEs), such as:

$$\frac{d\hat{y}}{d\Theta} = f(y,p) \tag{3.172}$$

where Θ represents time, length, volume, etc., depending on the particular case. Partial differential equations, particularly the hyperbolic and parabolic ones, can be transformed into the form of ordinary differential equations by discretization.

Non-linear regression analysis is applied in the estimation of parameter values (typically rate and equilibrium constants, more rare mass and heat transfer coefficients), as most models in catalytic kinetics are non-linear with respect to the parameters.

The model equation is written according to $\hat{y} = f(x, p)$, where y is the dependent variable, function (f) contains independent variables (x) and parameters (p). Non-linear regression is, strictly speaking, valid for algebraic models, but it can be applied to differential models, because the solution (y) is obtained numerically from the model equation. Different criteria for minimization could be applied, including most often the sum of squares between the experimental values and model predictions:

$$Q(p) = \sum_{i=1}\sum_{s=1}\sum_{t=1}\left[\left(y_{i,s,t} - f(x_{i,s,t'}p)^2\right)\right]w_{i,s,t} \qquad (3.173)$$

where $Q(p)$ is the objective function in minimization, i denotes each component in the reaction mixture, s denotes the different data sets and t refers to the data points, $y_{i,s,t}$ is the experimental value and $f(x_{i,s,t})$ is the corresponding model prediction, $w_{i,s,j,t}$ is the weight matrix that is used if experimental scattering varies between different data. In addition to this objective function, the sum of squares of the relative errors:

$$Q(p) = \sum_{i=1}\sum_{s=1}\sum_{t=1}\left[\left(\frac{y_{i,s,t} - f(x_{i,s,t'}p)}{y_{i,s,t}}\right)^2\right]w_{i,s,t} \qquad (3.174)$$

or even the maximal deviation between the experimental and calculated parameters:

$$Q(p) = \sum_{i=1}\sum_{s=1}\sum_{t=1}\left[(y_{i,s,t} - f(x_{i,s,t'}p)\right]w_{i,s,t} \qquad (3.175)$$

could be used. Various optimization methods exist for the minimization of the objective function to be included in modern parameter estimation software. The Levenberg-Marquardt method is a rapid and efficient one provided there are good initial guess of the parameters. Otherwise, the minimization can start with a slower, but a more robust simplex algorithm, switching in the vicinity of the minimum to the Levenberg-Marquardt algorithm. The latter method belongs to the group of the gradient methods, which use the derivative vector of the objective function with respect to the parameter directions to determine the direction where this gradient changes most, i.e., the steepest descent direction. The steepest descent is not necessarily the optimum one, thus some methods therefore use the second derivative matrix of the objective function with respect to the parameters.

The main reason for deviations between experimental and predicted data originates from systematic deviations, which are caused by inadequate stoichiometry, inappropriate kinetic models and poor calibration of the analytical equipment. Analysis of residuals as a function of variables can pinpoint cases with inappropriate weighting, systematic deviations caused by an incorrect model and poor experimental design (Fig. 3.19).

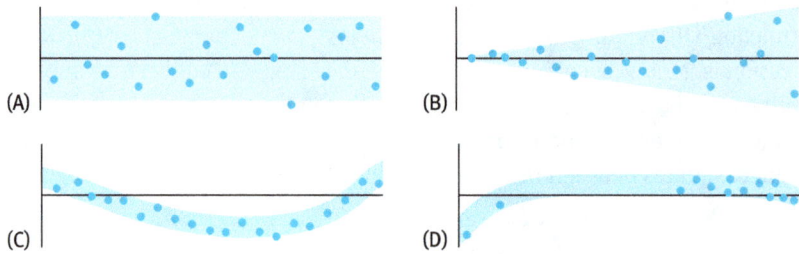

Fig. 3.19: Residuals as a function of variables: (A) adequate model with random distribution, (B) inappropriate weighting, (C) systematic deviations and incorrect model, (D) poor experimental design. From [22].

The parameter estimation results are evaluated by a standard statistical analysis. Degree of explanation can be estimated with the coefficient of determination, the R^2-value:

$$R^2 = 1 - \frac{\|y - \hat{y}\|^2}{\|y - \bar{y}\|^2} \qquad (3.176)$$

where y is the observed variable, $\hat{y}$ is the prediction of the variable and $\bar{y}$ is the average of all the data points. The estimated sum of squares is divided by the sum of squares of the simplest possible model, the average of the values. Equation (3.176) implies that R^2 values approaching 100% are desired while values exceeding 95% represent a good fit. The R^2-value does not, unfortunately, distinguish between random errors caused by scattered data, and systematic deviations caused by an inadequate model also require the sensitivity analysis.

Because the standard statistical analysis of parameters based on linearization does not always give a realistic picture of the accuracy of the parameters, it is recommended to check the objective function (Q) as a function of each parameter value (Fig. 3.20A,B)

Such sensitivity analysis provides additional information regarding accuracy and identifiability of parameters. Another option is to investigate the value of the objective function for a pair of parameters, resulting in contour plots (Fig. 3.20C).

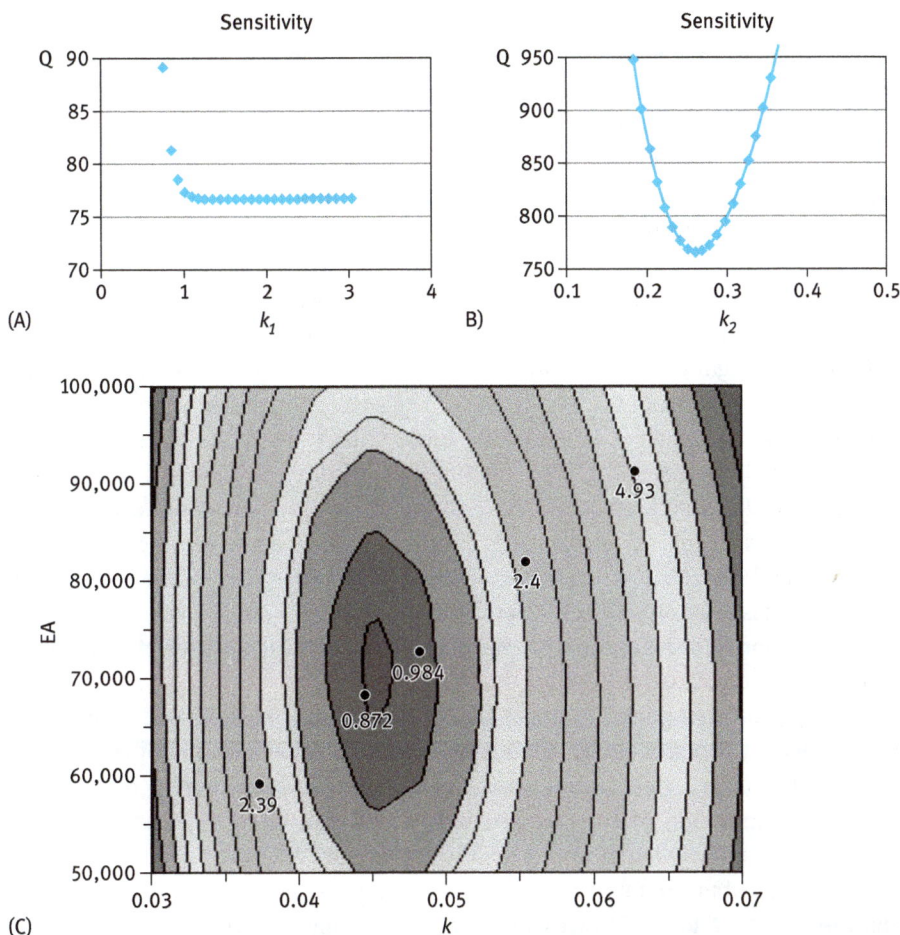

Fig. 3.20: Sensitivity plots: (*A, B*) objective function dependence on a parameter, (C) contour plots.

For a strong mutual correlation between parameters, contour plots become elongated, i.e., the numerical value of one parameter can be easily compensated by another parameter. In such cases the model cannot be used for extrapolation and the physicochemical significance of the parameters is questionable. One of the options to suppress the correlation between pre-exponential factor and activation energy is to use the modified Arrhenius equation:

$$k = k_{ave} \exp\left(\frac{-E_{act}}{R}\left[\frac{1}{T} - \frac{1}{T_{ave}}\right]\right); \text{ where } k_{ave} = k_0 \exp\left(\frac{-E_{act}}{RT_{ave}}\right) \tag{3.177}$$

Here T_{ave} is the reference temperature, chosen within the experimentally covered temperature domain, which is typically the average temperature.

In addition to strong correlations between parameters, sometimes too small or too large values of some parameters are obtained, indicating only a magnitude of the values without a reliable confidence range. Rearrangement of the model should thus be done prior to further parameter estimation. For complex reactions with many routes in addition to using microkinetic approaches, it could be recommended to perform additional experiments such as kinetic measurements with reactions intermediates as well as adsorption measurements.

Since parameters in a mechanistic model have a clear physico-chemical meaning, there are certain constraints including positive values of kinetic parameters, magnitude of rate and equilibrium constants, and activation energies. Adsorption entropies are negative because binding of the substrate to the catalyst results in the loss of entropy, therefore $0 < -\Delta S^0_{ads} < S^0_{gas}$. The adsorption enthalpy should be also negative to allow negative values of the Gibbs energy of adsorption. Moreover, as discussed in Section 3.4.7, it is possible to theoretically calculate at least some of the values of kinetic parameters based on the transition state theory even if there are difficulties in taking into account properly surface heterogeneity and lateral interactions.

Sometimes after data fitting several rival models have similar values of the objective function. Several approaches have been devised to allow a selection between these models, discussed in detail by Kapteijn et al. in [23].

Several software packages are available for modeling kinetics of chemical and biochemical systems, including catalytic programs. Athena Visual Studio can be used for modeling, optimization, estimation, optimal experimental design, and model discrimination of chemically reactive and non-reactive process systems. ModEst (from model estimation) is a software that has been designed for parameter estimation of mechanistic mathematical models as well as for experimental design, and is able to deal with explicit algebraic, implicit algebraic (systems of non-linear equations) and ordinary differential equations. This software package includes the possibility for advanced statistical analysis, including sensitivity data, contour plots and evaluation of estimated parameters accuracy by the Markov chain Monte Carlo (MCMC) method, which is based on the Bayesian approach. The MCMC method, provides a tool for the evaluation of the reliability of the model parameters by treating all the uncertainties in the data and the modelling as statistical distributions.

3.5 Mass transfer

3.5.1 Diffusion effects in heterogeneous catalysis

In any catalytic system it is not only chemical reactions that should be considered, but also mass and heat transfer effects. These effects are present inside the porous

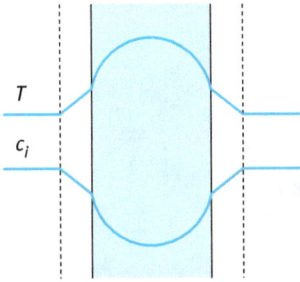

Fig. 3.21: Concentration gradients and temperature profiles for an exothermal fluid-solid reaction with interphase and intraparticle diffusion.

catalyst particles and in the surrounding fluid films, resulting in concentration gradients across the phase boundaries and within the particle (Fig. 3.21).

Because of heat and mass transfer the observed rate in a catalytic reaction (macrokinetics) is different from an intrinsic rate of a catalytic transformation (microkinetics), thus modeling of a two-phase (fluid-solid) catalytic reactor includes simultaneous reaction and diffusion in the pores of the catalyst particle. In three-phase systems (gas-liquid-solid) diffusion effects in the liquid films at the gas-liquid interphase should also be considered (Fig. 3.22).

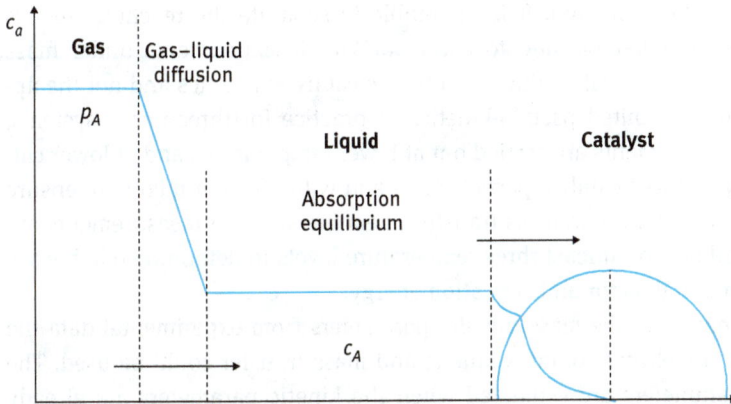

Fig. 3.22: Mass transfer processes in three-phase systems. (From [2].).

The intraparticle and interphase mass transfer coefficients display lower temperature dependence than the intrinsic rate, as shown in Fig. 3.23 and discussed quantitatively later in the book.

The design of experiments for a particular reaction system starts with defining an experimental domain, which is usually constrained by temperature, pressure, concentrations, solubilities, slurry density, mixing effects, kinetic rates, mass transfer rates, reaction enthalpies, heat transfer rates and the thermal stability of the

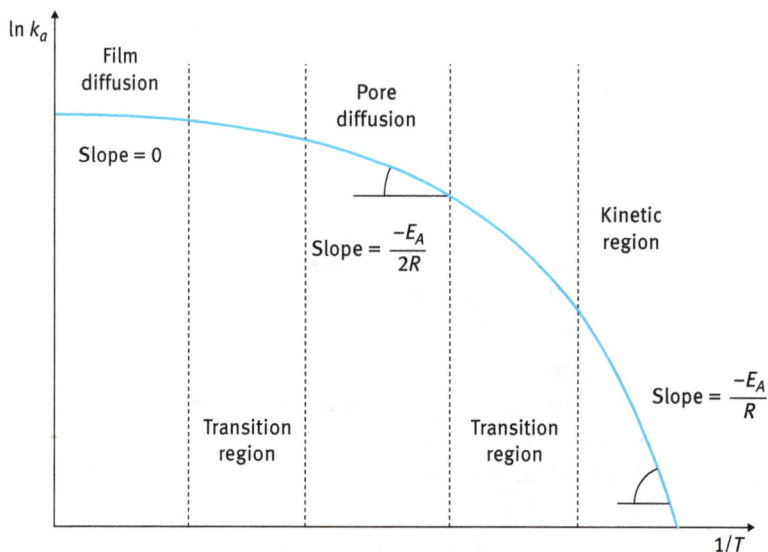

Fig. 3.23: Temperature dependence of catalytic reactions. (From [2].).

species in the reaction medium. It is preferable to separate the reaction kinetics from the transfer phenomena and to carry out kinetic experiments under mass transfer free conditions. Establishing the intrinsic catalytic kinetics and not the apparent mass transfer limited pseudo-kinetics in practice for three-phase systems, implies that the experiments are carried out at lower temperatures and at lower catalyst loadings with small catalyst particle sizes and with efficient mixing to ensure negligible influence of various mass transfer phenomena. Such measurements are preferably carried out for at least three temperature levels to determine reliably apparent pre-exponential factor and activation energy.

For determination of the mass transfer parameters from experimental data the detailed reactor model that contains kinetic and mass transfer could be used. The mass transfer parameters are estimated when the kinetic parameters are already available and are implemented as fixed parameters.

3.5.2 Reactor dependent external diffusion (interphase mass transfer, film diffusion)

First we will consider an isothermal case when there is only transfer of mass from the bulk to the external surface of the catalyst and internal diffusion does not play a role (Fig. 3.24).

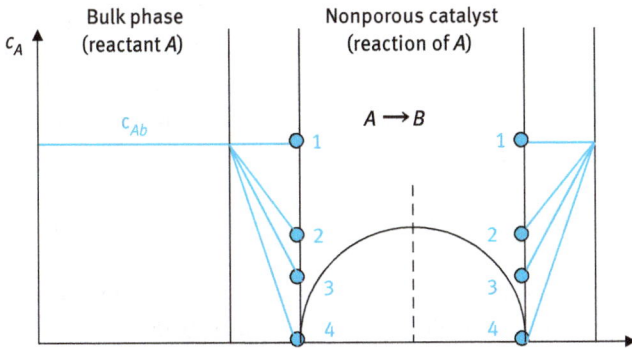

Fig. 3.24: External mass transfer. 1, no limitations by external mass transport (film diffusion); 2 and 3, external mass transfer and reaction; 4, maximum limitations by external mass transfer.

The molar flux (mol m^{-2} s^{-1}) J or number of moles i diffusing in the unit of time per unit of surface in z direction according to the Fick's law is proportional to the concentration gradient in this direction:

$$J_i = -D\frac{\partial c_i}{\partial z} \tag{3.178}$$

where D is the diffusion coefficient (m^2 s^{-1}), which for binary diffusion in gases increases with increase an in temperature and a decrease in pressure as discussed below.

The heat flow q (J m^{-2} s^{-1}) or heat transmitted per unit time per unit of surface area in z direction is determined analogously to mass flow by the Fourier's law:

$$q = -k\frac{\partial T}{\partial z} \tag{3.179}$$

and is proportional to the temperature gradient. In eq. (3.179) k is thermal conductivity of the fluid.

In the thin (boundary) layer around the solid particle the mass and heat transfer is governed by molecular diffusion and the overall mass transfer is proportional to the molecular diffusion coefficient. The molar flow rate N (mol s^{-1}) can be expressed by the following equation:

$$N = k_f A_p (c^b - c^S) \tag{3.180}$$

where A_p is the external surface area of the catalyst particle and $k_f(\beta)$ is the mass transfer coefficient in the film layer surrounding the catalyst particle, c^b concentration at the bulk, c^s concentration at the surface. At steady-state, this flux is equal to the reaction rate in the particle:

$$N = V_p r_v(c^S)$$ (3.181)

where V_p is volume of catalyst particle, and r_v is the rate per particle volume, which is related to the rate per unit mass (r_w) and particle density (p_p) in the following way: $r_v = p_p r_w$. For a first-order irreversible reaction, eqs. (3.180) and (3.181) give:

$$k_f(c^b - c^s) = \frac{V_p}{A_p} k_v c^s = \frac{1}{a'} k_v c^s$$ (3.182)

where $a' = A_p/V_p$ is volumetric external surface area. The concentration at the external surface and the expression for the reaction rates can be easily deduced:

$$r_v^{obs} = k_v c^s = k_v \frac{k_f a'}{k_v + k_f a'} c^b = \frac{1}{\frac{k_v}{k_f a'} + 1} k_v c^b = \frac{1}{1 + Da} k_v c^b = \eta_{ext} k_v c^b$$ (3.183)

This equation contains the external effectiveness factor η_{ext} which is defined as the ratio of effective (observed) rate to the intrinsic chemical rate under bulk fluid conditions $\eta_{ext} = 1/(1 + Da)$.

In eq. (3.183) Da is the Damköhler number $k_v/k_f a'$. Large values of Da correspond to strong mass transfer limitations; therefore the observed kinetics in the domain of mass transfer is of first-order. For strong external mass transfer limitations increasing catalyst activity does not influence the rate. Catalyst poisoning and deactivation might have an influence on the observed rate when catalyst activity is decreased to such extent that kinetics is becoming the limiting step.

It is apparent that the effectiveness factor depends on the mass transfer coefficient, which in turn depends on the reactor and hydrodynamic conditions, physical properties of the liquid, as well as the size of the catalyst grain.

Estimation of the catalyst effectiveness factor in eq. (3.183) requires calculation of the mass transfer coefficient, which can be done using dimensionless numbers, such as the Sherwood (Sh), the Schmidt (Sc) and the Reynolds (Re) numbers:

$$Sh = \frac{k_f d}{D}, Re = \frac{Vd}{v} = \frac{Vd\rho}{\mu}, Sc = \frac{v}{D}$$ (3.184)

where d is the characteristic length, k_f is the mass transfer coefficient, representing the ratio of total mass transfer to diffusive mass transfer, V is velocity, ρ density, μ dynamic viscosity and v kinetic viscosity.

The coefficient of mass transfer could be estimated using the mass transfer factor j_D, which is defined in the following way:

$$j_{D(M)} = \frac{Sh}{Re(Sc)^{\frac{1}{3}}} = \frac{k_f}{V}\left(\frac{\mu}{\rho D}\right)^{\frac{2}{3}} = \frac{k_f \rho}{G}\left(\frac{\mu}{\rho D}\right)^{\frac{2}{3}}$$ (3.185)

where G is the mass flow ($G = V\rho$). Experimentally established correlations between the mass transfer factor and the Reynolds number $j_D = f(Re)$ are presented in Tab. 3.3 and are graphically illustrated in Fig. 3.25. The mass transfer coefficient k_f can be calculated by taking j_D from experimental correlations and inserting it into eq. (3.185). The following procedure could be applied. If G, d and μ are known the Reynolds number can be computed, followed by calculation of the mass transfer factor, for example Fig. 3.25. Finally, by taking the known mass transfer factor and V, μ, ρ and D, the mass transfer coefficient k_f can be calculated.

Tab. 3.3: Mass and heat transfer correlations in packed beds.

Mass transfer	Heat transfer	Mass transfer factor	Range
Gases			
$Sh = \dfrac{0.357}{\varepsilon_p} Re^{0.641} Sc^{\frac{1}{3}}$	$Nu = \dfrac{0.428}{\varepsilon_p} Re^{0.641} Pr^{\frac{1}{3}}$	$0.416 < \varepsilon < 0.788$ $j_D = 0.357 Re^{-0.359}$	$3 < Re < 2000$
Liquids			
$Sh = \dfrac{0.25}{\varepsilon_p} Re^{0.69} Sc^{\frac{1}{3}}$ $Sh = \dfrac{1.09}{\varepsilon_p} Re^{\frac{1}{3}} Sc^{\frac{1}{3}}$	$Nu = \dfrac{0.30}{\varepsilon_p} Re^{0.69} Pr^{\frac{1}{3}}$ $Nu = \dfrac{1.31}{\varepsilon_p} Re^{\frac{1}{3}} Pr^{\frac{1}{3}}$	$0.35 < \varepsilon < 0.75$ $j_D = 0.25 Re^{-0.31}$ $j_D = 1.09 Re^{-0.67}$	$55 < Re < 1500$ $0.0016 < Re < 55$

ε, the interparticle void fraction of the bed of particles.

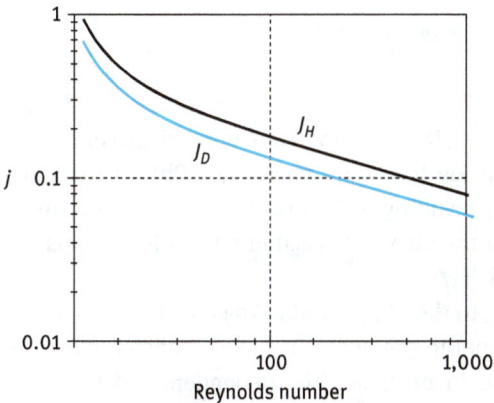

Fig. 3.25: Correlation between mass and heat transfer factors and Reynolds numbers.

It follows from eq. (3.185) that $k_f \propto j_M V$. Considering dependence of the mass transfer factor on the Reynolds number, it can be written that $j_M \propto Re^{-0.5} \propto (1/Vd_p)^{0.5}$, giving the following dependence of the mass transfer coefficient on the velocity and the diameter of catalyst particles:

$$k_f \propto V(1/Vd_p)^{0.5} \propto (V/d_p)^{0.5} \qquad (3.186)$$

Equation (3.186) is demonstrated graphically in Fig. 3.26, showing linear proportionality between the mass transfer coefficient and the square root of (V/d_p) at low values of the latter. Thus, with increasing velocity and diminishing the catalyst particle size, the impact of mass transfer on the intrinsic catalytic rate could be eliminated. Another way to calculate mass transfer coefficients is to use correlations between the Sherwood number and the Reynolds and the Schmidt numbers if the values of Re and Sc are known. The Sherwood number can be calculated using correlations in Tab. 3.3 and therefore the mass transfer coefficient can be determined.

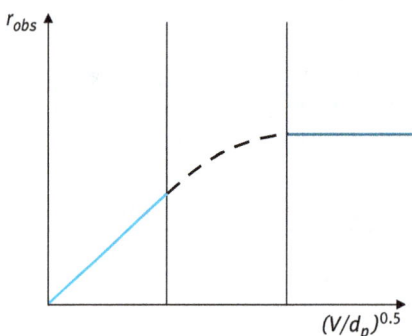

Fig. 3.26: Dependence of the observed rate on the square root of (V/d_p).

From eq. (3.185) it can be shown that $k_f \propto (D)^{2/3}$. Temperature dependence of the diffusion coefficient is expressed for diffusion in the gas-phase by a Chapman-Enskog equation, which gives $D \propto T^{3/2}$, finally resulting in $kf \propto (D)^{4/6} \propto (T^{3/2})^{2/3} \propto T$ and a very minor temperature dependence of the observed reaction rate with the apparent activation energy less than 5–10 kJ mol^{-1}.

The diffusion coefficient according to the Chapman-Enskog equation (discussed in Section 3.5.3) is inversely proportional to pressure, thus mass transfer is becoming more prominent with an increase in pressure. The dependence of the mass transfer coefficient on pressure is complicated however by the influence of pressure on other parameters.

Analogously to mass transfer, heat transfer coefficients can be computed, which are correlated in the same way as mass transfer. In fact, the Sherwood number (the ratio of convective mass transfer rate to diffusive mass transfer rate) is an analogue of the Nusselt number, which corresponds to the ratio of the total heat transfer to conductive heat transfer:

$$Nu = \frac{h_f d}{\lambda}$$

(3.187)

where h_f is the heat transfer coefficient, λ is the thermal conductivity of the fluid. Tab. 3.3 contains correlations between the Nusselt number and the Reynolds and Prandtl numbers, where the latter is defined through the specific heat capacity C_p, thermal conductivity and viscosity $Pr = C_p\mu/\lambda$. For heat transfer calculations the heat transfer factor could be also used:

$$j_H = \frac{Nu}{Re(Pr)^{\frac{1}{3}}} = \frac{h_f}{C_p G}\left(\frac{C_p \mu}{\lambda}\right)^{\frac{2}{3}}$$

(3.188)

For calculation of heat transfer coefficients, the Chilton-Colburn analogy can be applied. From Fig. 3.25 it can be seen that $j_D/j_H \approx 0.7$ and by knowing the Reynolds number first the heat transfer factor and finally the heat transfer coefficient are obtained:

$$h_f = j_H C_p^{\frac{1}{3}} G \left(\frac{\lambda}{\mu}\right)^{\frac{2}{3}}$$

(3.189)

Several correlations have been proposed in the literature for slurry reactors [24] that relate the dimensionless Sherwood number to the Reynolds and Schmidt numbers. Data collected for various slurry reactors show that the Sherwood number can be described by the following equation:

$$Sh = 1.0 Re^{\frac{1}{2}} Sc^{\frac{1}{3}}$$

(3.190)

In the case of low Reynolds numbers, dependencies in Table 3.3 should be modified to contain a correction term, accounting for mass transfer in the absence of stirring:

$$Sh = 2 + 0.4 Re_d^{0.5} Sc^{\frac{1}{3}}$$

(3.191)

The dependence of the Reynolds number on the local velocity can be established in terms of the Kolmogorov theory of turbulence. According to this theory, the only parameter needed to describe the probability distribution of the relative velocity field V_λ in a homogeneous isotropic turbulent fluid is the average energy of dissipation ε, i.e.,:

$$V_\lambda \propto (\varepsilon\lambda)^{\frac{1}{3}}$$

(3.192)

which gives the following expression for the Reynolds number:

$$\mathrm{Re}_d \propto \left(\frac{\varepsilon d}{v^3}\right)^{\frac{1}{3}} \tag{3.193}$$

finally providing a way to estimate the liquid/solid mass transfer coefficients k_{LS}:

$$k_{LS} = \left(\frac{\varepsilon D^4 \rho}{\eta d_p^2}\right)^{\frac{1}{6}} \tag{3.194}$$

where ε denotes the specific mixing power. D^0_{AB} is the mutual diffusion coefficient of solute A in solvent B, ρ is solvent density, η solvent viscosity, and d_p is the diameter of the catalyst particles. A similar equation can be used for gas-liquid mass transfer with the only difference is that instead of catalyst particle size diameter the size of bubbles should be introduced in eq. (3.194). In practice, for gas-liquid systems the gas- liquid mass transfer coefficient is difficult to decouple from the diameter of gas bubbles, because the latter is determined by the multiphase hydrodynamics. Specific correlations between $k_{Lg}a$ (where a is the interfacial area of gas per unit of volume, m^2m^{-3}) and energy of dissipation are used, whose validity is usually limited thus disallowing extrapolations beyond the ranges of these correlations.

Theoretical calculations of the maximum specific mixing power are based on the assumption that all of the energy of the impeller is dissipated in the liquid. As the power of the motor $P_{stirrer}$ (Watt) is known, the energy dissipation is defined through it and the liquid mass:

$$\varepsilon = \frac{P_{stirrer}}{\rho_L V_L} \tag{3.195}$$

where ρ_L and V_L are the liquid density and volume.

Typically, not all the power of the motor is dissipated in the liquid therefore it is advisble to determine the specific mixing power by experimentally determining the rate of mass transfer. This is done, for example, by dissolution of some solid materials (e.g. benzoic acid), whose rate is limited by mass transfer. It should be also noted that the energy is not evenly dissipated throughout the vessel, as it is much higher close to the impeller and drastically decreasing away from the impeller and leads to even dead zones without any mixing. Visualization of mixing could be done, for example, by tracer addition or using electrical resistance tomography.

For fixed-bed reactors an experimental way to determine whether external mass transfer is present is to determine the dependence of conversion α on the flow rate (Fig. 3.27). If the conversion at a fixed residence time (catalyst mass m divided by the mass flow F_A^0) in the reactor does not change with changing flow rate (and respectively, the catalyst amount), for example, $\alpha_{A,1} = \alpha_{A,2}$ at $m_1/F_{A,1}^0 = m_2/F_{A,2}^0$ (Fig. 3.27A) then external mass transfer limitations can be neglected. Experimentally is it usually

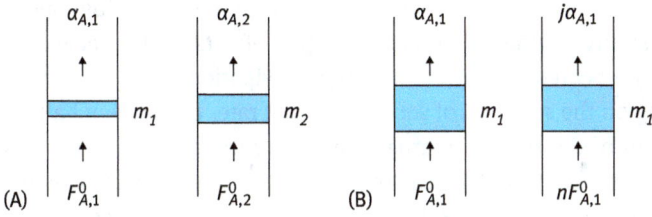

Fig. 3.27: Elucidation of mass transfer by (A) changing the flow rate and catalyst amount or (B) changing the flow rate.

easier to keep the catalyst amount the same (Fig. 3.27B). When conversion is linearly proportional to the residence time in the reactor, it is usually assumed that there are no external mass transfer limitations.

A common test to verify if mass transport controls a catalytic reaction in three-phase slurry reactors is to vary the mass of catalyst (gas-liquid mass transfer) and the rate of agitation (liquid-solid mass transfer). When the reactor productivity rate is independent on the catalyst mass the observed rate is thus governed by the gas-liquid mass transfer (Fig. 3.28A).

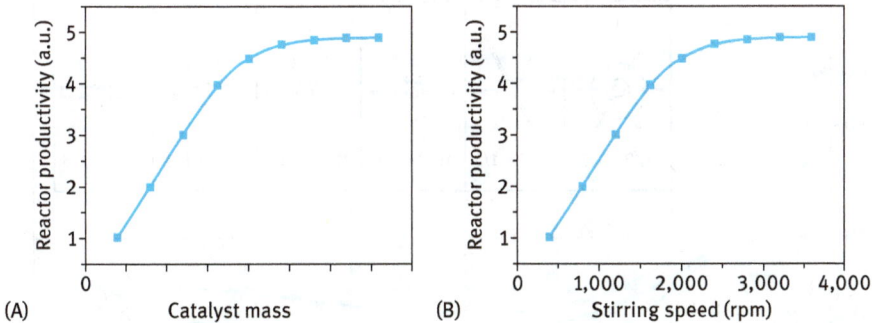

Fig. 3.28: Experimental verification of (A) gas-liquid and (B) liquid-solid mass transfer.

Agitation of the reaction mixture is usually done by stirring, although reactors in the form of shakers are also used for laboratory purposes.

The mass transfer coefficient depends on the energy dissipation, which is a function of mixing power W, often related to the stirring speed according to:

$$P = K\rho_{liquid}N^3D^5_{(stirrer)} \tag{3.196}$$

where N is the stirring speed (rev s^{-1}), K is a constant, ρ_{liquid} is the liquid density and $D_{stirrer}$ is the stirrer diameter. The dependence of the mass transfer coefficient on the stirring speed can be then established as $k_f \propto n^{1/2}$. If the rate is then independent on the agitation efficiency at a sufficiently high stirring speed (Fig. 3.28B), then it is usually assumed that mass transport effects are minimized.

However, it should be beared in mind that energy dissipation can have a complex behavior depending on the selection of solvent, stirring rate, presence of baffles, liquid volume and design of the reactor internals, including impellers. Experiments with an Autoclave Engineers EZE Seal 0.5 l reactor [25] demonstrated that the dissipation power at high stirring rates does not follow the square root dependence of mass transfer coefficient on the stirring speed. The energy of dissipation can also depend on the liquid load in an autoclave or a shaking reactor [26]. At the higher liquid load, increasing the shaking frequency does not necessarily lead to improved energy dissipation, therefore the rate of catalytic reactions could be independent on the agitation because the mass transfer coefficient is the same. As an illustration, Fig. 3.29 shows the dependence of the power number $Po = P/(\rho_{liquid}N^3D^5_{(stireer)})$ as a function of the Reynolds number Re. For laminar flow $Po \propto Re^{-1}$ while for turbulent flow it is constant. Such constant values of Po depend however on the impeller configuration, in particular the number of blades and the blade width, W.

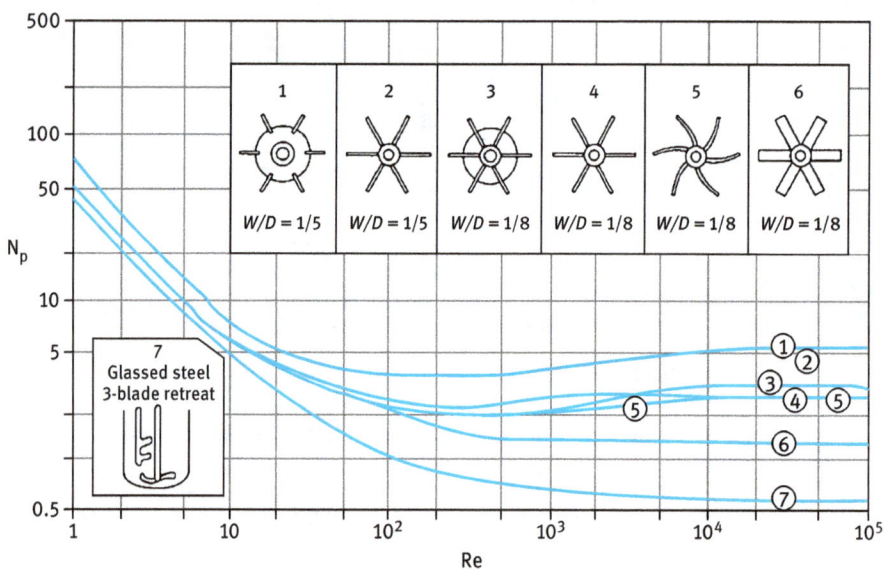

Fig. 3.29: Power curves for some typical impellers. (From [27]).

This shows the apparent danger of establishing the kinetic region based exclusively on the results of catalytic experiments at different stirring speeds, as the independence of catalytic activity on the agitation could be explained simply by the fact that higher stirring speed or shaking frequency does not necessarily influence the specific mixing power. Therefore, experimental verification of the absence of mass transfer should be combined with calculations.

Note that for lab reactors the stirring speed could be 1,000–2,000 rpm, while in plant conditions it is usually difficult to obtain sufficient agitator speeds.

When I was a PhD student in the 1980s I visited a caprolactam production plant and had a discussion with a plant manager. Experiments in the laboratory reactor for one of the process steps – hydrogenation of benzoic acid – were performed using a kinetic regime. Comparison with the plant data clearly demonstrated that liquid-solid mass transfer was present in industrial settings and it was proposed that they also increase the stirring speed in the industrial reactor. The response of the plant manager was negative because 120 rpm was the maximum technically feasible stirring speed for that plant reactor.

The absence of mass could be verified by applying several criteria, as discussed in more detail in Section 3.5.8. The criterion of external mass transport can be derived considering the mass transfer rate equal to the observed rate:

$$k_{LS}(c_b - c_s)\, 4\pi R^2 = \left(\frac{4}{3}\right) r_{obs}\pi R^3 \tag{3.197}$$

and subsequently:

$$c_b - c_s = r_{obs}\frac{R}{3}k_{LS} \tag{3.198}$$

It can be assumed that $c_b - c_s < 0.05c_b$, so the criterion for the nth order reaction takes the form:

$$\frac{r_{obs}R}{k_{LS}c_b} < \frac{0.15}{n} \tag{3.199}$$

3.5.3 Calculation of diffusion coefficients

Reliable calculation/estimation of diffusion coefficients should be done for correct evaluation of the influence of mass transfer. Different methods have been presented in the literature [28] to calculate the binary diffusion coefficients in the gas-phase. The most common is the Chapman-Enskog equation, which is based on kinetic gas theory:

$$D_{AB}(cm^2 s^{-1}) = 1.8829 \times 10^{-3} \frac{(T)^{1.5}\left(\frac{1}{M_{rA}} + \frac{1}{M_{rB}}\right)^{0.5}}{p(\sigma_{AB})^2\Omega_{AB}} \tag{3.200}$$

Here M_{rA} and M_{rB} are relative molecular masses (dimensionless), p is total pressure (kPa), σ_{AB} (nm) is the characteristic length (the Lennard-Jones parameter) for a pair of molecules, Ω_{AB} is a collision integral and is a function of $k_B T/\varepsilon_{AB}$, where ε_{AB} (J) is another Lennard-Jones parameter and k_B $(1.38 \times 10^{-23}$ J K$^{-1})$ is the Boltzmann constant. In the literature values of σ_{AB} are often given in Å $(10^{-10}$ m), then total pressure should be in 10^5 Pa, if the units of the diffusion coefficient are cm^2 s^{-1}. The Lennard-Jones parameter, σ_{AB}, is calculated from $\sigma_{AB} = 0.5 \, (\sigma_A + \sigma_B)$, while the collision integral depends on the constant ε_{AB}, which is calculated for a binary system according to $\varepsilon_{AB} = \sqrt{\varepsilon_A \varepsilon_B}$. The Lennard-Jones force constants for some molecules are given in Tab. 3.4 and the values for the collision integral are presented in Tab. 3.5.

Tab. 3.4: The Lennard-Jones force constants for some molecules.

Substrate g mol^{-1}	Molar mass	σ, Å	$\varepsilon/k_{B,k}$	Substrate g mol^{-1}	Molar mass	σ, Å	$\varepsilon/k_{B,k}$
H$_2$	2.016	2.915	38.0	CO	28.01	3.590	110.0
He	4.003	2.576	10.2	CO$_2$	44.01	3.996	190.0
Ne	20.183	2.789	35.7	NO	30.01	3.470	119.0
Ar	39.944	3.418	124.0	N$_2$O	44.02	3.879	220.0
Kr	83.80	3.61	190.0	SO$_2$	64.07	4.290	252.0
Xe	131.3	4.055	229.0	Cl$_2$	70.91	4.115	357.0
air	28.97	3.617	97.0	Br$_2$	159.83	4.268	520.0
N$_2$	28.02	3.681	91.5	I$_2$	253.82	4.982	550.0
O$_2$	32.0	3.433	113.0	CH$_4$	16.04	3.822	137.0
C$_2$H$_2$	26.04	4.221	185.0	CH$_2$Cl$_2$	84.94	4.759	406.0
C$_2$H$_4$	28.05	4.232	205.0	CHCl$_3$	119.39	5.430	327.0
C$_2$H$_6$	30.07	4.418	230.0	CCl$_4$	153.84	5.881	327.0
C$_3$H$_8$	44.09	5.061	254.0	C$_2$N$_2$	52.04	4.38	339.0
i-C$_4$H$_{10}$	58.12	5.341	313.0	COS	60.08	4.13	335.0
n-C$_5$H$_{12}$	72.15	5.769	345.0	CS$_2$	76.14	4.438	488.0
n-C$_6$H$_{14}$	86.17	5.909	413.0	C$_6$H$_6$	78.11	5.270	440.0
n-C$_8$H$_{18}$	114.22	7.451	320.0	CH$_3$Cl	50.49	3.375	855.0
Cyclohexane	84.16	6.093	324.0				

Typical values of gas-phase diffusion coefficients are $\sim D_A \sim 10^{-5}$m^{2}s^{-1}. Although values of diffusion coefficients calculated using the Chapman-Enskog equation generally agree quite well with experimental data, some modifications of this equation are applied as well as discussed in more specialized textbooks [2, 29] where values for binary diffusion coefficients are also presented.

For the calculation of binary diffusion coefficients in the liquid-phase D^0_{AB} semi-empirical equations such as Wilke-Chang equation are often used, giving accurate results for diffusion coefficients of gases in liquids:

Tab. 3.5: Values for the collision integral.

$k_B T / \varepsilon_{AB}$	Ω_{AB}	$k_B T / \varepsilon_{AB}$	Ω_{AB}
0.30	2.662	2.0	1.075
0.35	2.476	2.5	1.000
0.40	2.318	3.0	0.949
0.45	2.184	3.5	0.912
0.50	2.066	4.0	0.884
0.55	1.966	5.0	0.842
0.60	1.877	7.0	0.790
0.65	1.798	10.0	0.742
0.70	1.729	20.0	0.664
0.75	1.667	30.0	0.623
0.80	1.612	40.0	0.596
0.85	1.562	50.0	0.576
0.90	1.517	60.0	0.560
0.95	1.476	70.0	0.546
1.00	1.439	80.0	0.535
1.10	1.375	90.0	0.526
1.20	1.320	100.0	0.513
1.30	1.273	200.0	0.464
1.40	1.233	300.00	0.436
1.50	1.198	400.00	0.417
1.75	1.128		

$$D_{AB}^{o} = \frac{7.4 \times 10^{-8} (\Phi M_B)^{\frac{1}{2}} T}{\frac{\mu_b}{cP} V_{b(A)}^{0.6}} \; [\text{cm}^2 \text{s}^{-1}] \tag{3.201}$$

In eq. (3.201) the dimensionless association factor f is equal to 2.6 for water, 1.9 for methanol, 1.5 for ethanol and 1 for benzene, heptane, and other non-associated solvents, M_B is the molecular mass of solvent, μ_B is solvent viscosity in cP at temperature $T[\text{K}]$, $V_{b(A)}$ is the liquid molar volume at the solute's normal boiling point in cm^3/mol. Molar volumes of some molecules are given in Tab. 3.6.

The molar volume can be also estimated from the atomic increments given in Tab. 3.7.

Tab. 3.6: Molar volumes of some molecules.

Methane	37.7	Methanol	42.5	Acetonitrile	57.4
Propane	74.5	Dimethylether	63.8	Methyl chloride	50.6
Heptane	162	Acetone	77.5	Phosgene	69.5
Cyclohexane	117	Acetic acid	64.1	Ammonia	25
Ethylene	49.4	Methyl formate	62.8	Water	18.7
Benzene	96.5	Ethyl acetate	106	SO_2	43.8

Tab. 3.7: Atomic increments.

Atom	Increment, cm³ mol⁻¹	Atom	Increment, cm³ mol⁻¹
C	14.8	Br	27.0
H	3.7	Cl	24.6
O (except as noted below)	7.4	F	8.7
– in methyl esters and ethers	9.1	I	37
– in ethyl esters and ethers	9.9	S	25.6
– in higher esters and ethers	11.0	Ring	
– in acids	12.0	– 3-membered	−6
– joined to S,P,N	8.3	– 4-membered	−8.5
N		– 5-membered	−11.5
– double bonded	15.6	– 6-membered	−15
– in primary amines	10.5	– naphthalene	−30
– in secondary amines	12.0	– anthracene	−47.5

3.5.4 Size dependent internal (pore) diffusion

In a porous catalyst particle, the reacting molecules diffuse first through the fluid film surrounding the particle surface and then diffuse into the pores of the catalyst to the active sites. In a similar way the reaction products are diffusing out of the catalyst grains. As an outcome of pore diffusion for most common reaction kinetics the reaction rates inside the pores have lower values than expected with the concentration levels of the main bulk.

The mathematical treatment of pore diffusion is based on the general conservation laws. The amount of mass or heat transported per unit of time into and out of a differential volume element is balanced by accumulation in this volume and generation or consumption within the same volume

$$IN + GENERATION = OUT + ACCUMULATION$$

The complex diffusion process could be represented by Fick's first law, where for the species i the flux N_i (mol m² s⁻¹) is proportional to the concentration gradients of the components (dc_i /dr) and its effective diffusion coefficient (D_{ei}): $N_i = -D_e(dc/dx)$. In the calculation of the diffusion in a porous media the random pore model and the effective diffusion coefficient is applied. The latter is defined as $D_e = D(\varepsilon/\tau)$ which is smaller than the diffusion coefficient because the diffusional cross section is smaller than the geometric cross section (thus porosity ε is introduced) and the catalyst has irregular pore structure (expressed via tortuosity τ) as illustrated in Fig. 3.30. Typically $\varepsilon /\tau = 0.05 \div 0.2$

Fig. 3.30: Illustration of porosity and tortuosity.

Pore diffusion may occur by molecular diffusion, Knudsen diffusion and surface diffusion; the surface diffusion usually not seriously influencing catalytic kinetics.

In the gas-phase and within large pores, molecular diffusion prevails when the molecules collide with each other (Fig. 3.31A).

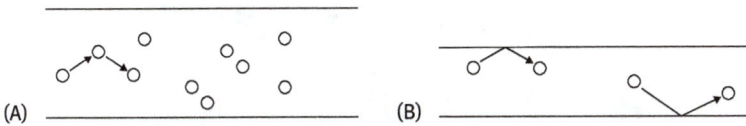

Fig. 3.31: (A) Molecular and (B) Knudsen diffusion.

In small pores or when the gas density is low, it is not only molecular but also Knudsen diffusion (Fig. 3.31B) that could play an important role. The mean free path becomes comparable to the size of the pore and molecules are colliding with the walls rather than with each other. Such diffusion is thus independent of pressure and is not observed in the liquid-solid catalytic reactions when the fluid density is much higher compared to the gas-solid catalysis.

The Knudsen diffusion coefficient is proportional to the pore radius r_e and the mean molecular velocity, giving proportionality of the Knudsen diffusion coefficient to the square root of temperature:

$$D_K = \frac{2}{3} r_e \sqrt{\frac{8\,RT}{\pi\,M}} \tag{3.202}$$

where R is the gas constant, T is the absolute temperature and M is the molecular mass.

As the pore radius in the case of cylindrical pores can be calculated from the total pore volume $V_p (cm^3\ g^{-1})$ and the total surface area $S\ (cm^2\ g^{-1})$, (i.e., $r_e = 2V_p\,/S$) the Knudsen diffusion coefficient for a porous solid becomes:

$$D_k = \frac{8\varepsilon_p}{3S\rho_p} \sqrt{\frac{2\,RT}{\pi\,M}} \tag{3.203}$$

where ε_p is porosity and ρ_p is the particle density.

A special case of diffusion is the so-called configurational diffusion in zeolite pores, where the pore size is comparable with the size of reacting molecules. Steric and diffusional constraints in zeolites are the origin of their shape-selective properties. Comparison of the values of diffusion coefficients for different cases (configurational, Knudsen and molecular diffusion) is given in Fig. 3.32. Such comparison should be taken with care because transition between different types of diffusion depends not only on the size of pores but also on the process conditions.

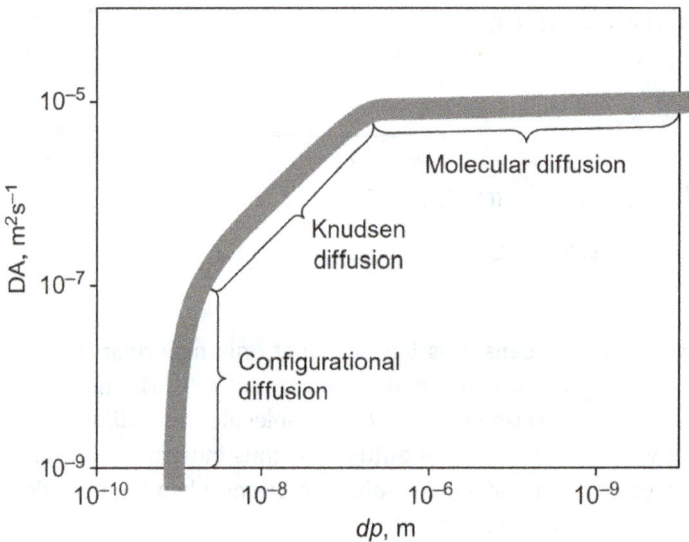

Fig. 3.32: Values of diffusion coefficients depending on the pore diameter [30].

If both molecular D_{AB} and Knudsen D_{KA} diffusion are present (for example, the catalyst contains broad distribution of pore radii), the Bosanquet approximation is used:

$$\frac{1}{D} = \frac{1}{D_{AB}} + \frac{1}{D_{KA}} \tag{3.204}$$

where the first term stems from molecular diffusion and the second from Knudsen diffusion.

Reaction and diffusion will be first considered for an isothermal case with only internal mass transfer control (Fig. 3.33).

The main aim of the mathematical treatment is to obtain an expression for the effectiveness factor $\eta = r_{obs}/r_{kinetics}$, which relates the observed reaction rate to the intrinsic chemical rate. If the value of the effectiveness factor is within $0.95 \div 1$ then internal diffusion could be neglected. As concentration changes with the position

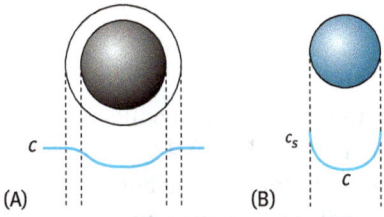

Fig. 3.33: Reactant concentration profiles (A) external and internal mass transfer control, (B) internal mass transfer control only.

inside the catalyst grain the reaction rate for non-zero order reactions is thus also a function of the position. Therefore, the concentration profiles should be obtained first.

Considering a pseudo-homogeneous model and the first law of Fick for a differential element of a spherical particle (Fig. 3.34) the following mass balance can be written:

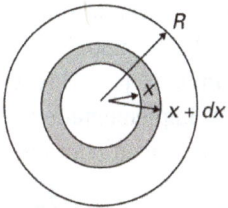

Fig. 3.34: Illustration of a differential element of a spherical particle.

$$A_{x+dx}D_e\frac{dc}{dx}\Big|_{x+dx} - A_xD_e\frac{dc}{dx}\Big|_x = rA_{dx}dx \tag{3.205}$$

where A is the geometrical surface area. As dx is very small, $x + dx = x$ and eq. (3.205) is rearranged to:

$$4\pi x^2 D_e\frac{dc}{dx}\Big|_{x+dx} - 4\pi x^2 D_e\frac{dc}{dx}\Big|_x = 4\pi x^2 r dx \tag{3.206}$$

A simple manipulation for the nth order kinetics gives:

$$\frac{d^2c}{dx^2} + \frac{2dc}{xdx} = \frac{kc^n}{D_e} \tag{3.207}$$

Introducing dimensionless concentration $y = c/c_0$ and the position inside the pore $x^* = x/L(R)$, eq. (3.207) is rewritten in a simple form:

$$\frac{d^2y}{dx^{*2}} + \frac{2}{x^*}\frac{dy}{dx^*} = \phi^2 y \tag{3.208}$$

where ϕ is the Thiele modulus:

$$\phi = R\sqrt{\frac{kc^{n-1}}{D_e}} \tag{3.209}$$

The analytic solution of differential eq. (3.209) is well-known. The boundary conditions reflect symmetry at the pellet center, giving $dy/dx* = y'(0) = 0$. Moreover at the external pellet surface the surface concentration is equal to the bulk substrate concentration $y(1) = 1$. The change of the dimensional concentration across the pellet radius is thus:

$$y = \frac{(e^{\phi x*} - e^{-\phi x*})}{e^{\phi} - e^{-\phi}} = \frac{\sinh(\phi x*)}{\sinh(\phi)} \tag{3.210}$$

Despite the simplicity of this case is it of great practical importance, because in many cases only small heat consumption or release is observed and the catalyst particle could be treated as an isothermal one. Normalized concentration profiles depending on the position inside the spherical particle are presented in Fig. 3.35 and it is apparently clear that for small values of Thiele modulus the reaction rate has a less steep decline of concentrations along the pore, while for large values of Thiele modulus the reactant concentration diminishes dramatically, indicating that in fact catalytic reactions occur only in the vicinity of the catalyst outer layer. Practical implementation of this conclusion for the design of catalyst particles used in methane steam reforming will be discussed in the corresponding section in Chapter 4.

During the stationary operation the amount of mass converted is equal to the flux across the external surface:

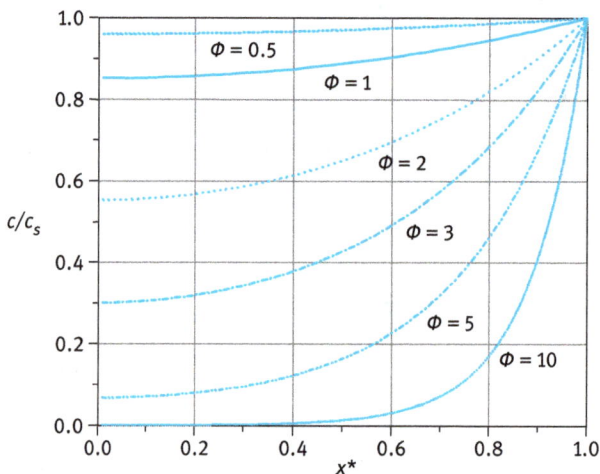

Fig. 3.35: The concentration profile dependence inside a spherical particle at different values of the Thiele modulus for isothermal, first-order irreversible reaction.

$$\frac{4}{3}\pi R^3 r_{ave} = 4\pi R^2 D_e \frac{dc}{dx}\bigg|_{x=R} \tag{3.211}$$

and the average rate is:

$$r_{ave} = \frac{3}{R} D_e \frac{dc}{dx}\bigg|_{x=R} \tag{3.212}$$

and subsequently the catalyst effectiveness factor takes the form:

$$\eta = \frac{3D_e \frac{dy}{dx^*}\big|_{x^*=1}}{R^2 k c_b^{n-1}} = \frac{3}{\phi^2} \frac{dy}{dx^*}\bigg|_{x^*=1} \tag{3.213}$$

By differentiating eq. (3.210) and substituting the result into eq. (3.213) the effectiveness factor η of a first-order irreversible reaction in a spherical pellet is obtained as a unique function of the Thiele modulus:

$$\eta = \frac{3}{\phi}\left(\frac{1}{\tanh(\phi)} - \frac{1}{\phi}\right) \tag{3.214}$$

Equation (3.214) clearly indicates that the Thiele modulus and the effectiveness factor depend strongly on the size of catalyst grains. At small values of the Thiele modulus (i.e., small particle size) the effectiveness factor is approaching unity. The flat dependence of effectiveness factor on the Thiele modulus occurs at small catalyst particles, low catalyst activity (small k), and large pore size and high porosity (large D_e). When $\varphi \gg 3$, the following dependence is valid $\eta \propto 1/\phi$ and the effectiveness factor is inversely proportional to the Thiele modulus and thus to the particle size. The asymptotic value for the effectiveness factor is $\eta = 3/\phi$ as $\lim(\varphi \to \infty)\tanh \varphi = 1$. For large values of Thiele modulus the overall rate is controlled by pore diffusion and for very active catalysts, or for catalysts with small pores, low porosity and/or large diameter the reactant concentration approaches zero in the center of a particle.

Obviously, in laboratory-scale reactors the size of catalyst particles can be rather small to diminish the impact of internal diffusion, while in fixed-bed pilot or industrial reactors the size of catalyst grains is unavoidably much higher because of increased pressure drop, resulting in a significant influence of internal diffusion. For slurry reactors, even at the pilot stage the size of catalyst powder could be still in the range of 50–100 μm, which in most cases (i.e., when catalytic reactions are not very fast) is sufficient to eliminate internal diffusion. However, external diffusion limitations can still play a role.

The analysis above was done for a spherical particle and the first-order kinetics. Analytical solutions are available for some isothermal steady-state systems and ideal geometric forms (slabs and spheres) and simple kinetics.

The treatment could be extended for arbitrary shapes typical for heterogeneous catalysis (Fig. 3.36) by introducing a generalized Thiele modulus, which is related to the ratio of the pellet volume to the external pellet surface:

Fig. 3.36: Typical shapes of industrial heterogeneous catalysts.

$$\phi_p = \frac{V_p}{A_p} \sqrt{\frac{kc^{n-1}}{D_e}} \tag{3.215}$$

where $V_p/A_p = L$, the characteristic length of diffusion. The effectiveness factor for different geometries for an isothermal, first-order irreversible reaction is presented in Fig. 3.37, demonstrating that the particle geometry is of less importance for the effectiveness factor.

Fig. 3.37 (presented as a double logarithmic plot) illustrates that for the values of Thiele modulus above four to five, dependences of the effectiveness factor on the Thiele modulus almost coincide independent to the pellet geometry. Therefore, for evaluation of the influence of internal mass transfer, even the simplest case [a porous catalyst slab (flat plate), for which the catalyst effectiveness factor is given by $\eta = \tanh(\phi)/\phi$], could be used.

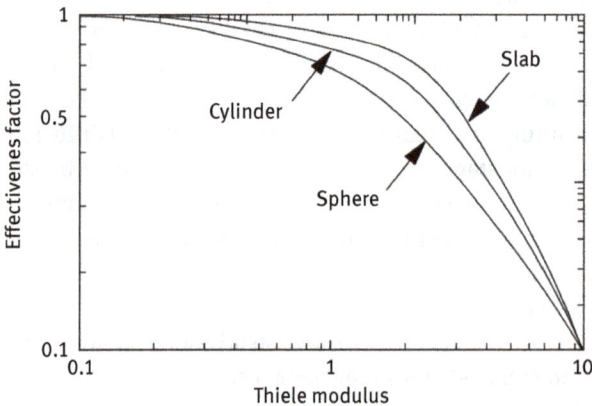

Fig. 3.37: Effectiveness factor η as a function of the generalized Thiele modulus φ_p for different pellet geometries.

For evaluation of the importance of internal diffusion in the literature it is mainly the expressions for first-order kinetics that are used. In general, for arbitrary kinetics a numerical solution of the balance equation that takes into account the boundary conditions is necessary.

Plots for a family of curves (irreversible Langmuir kinetics) are presented in Fig. 3.38, ranging from the zero order kinetics (denominator Kc_{As} is much larger than unity and approaches infinity) to first-order kinetics ($Kc_{As} \approx 0$).

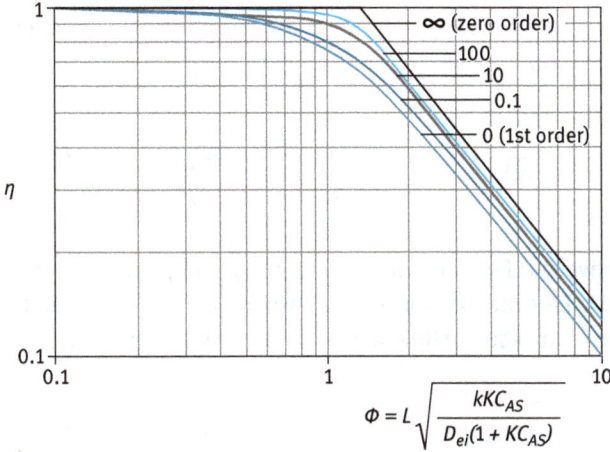

Fig. 3.38: Effectiveness factor as a function of Thiele modulus for different kinetic expressions [32].

As is clear from Fig. 3.38, the catalyst effectiveness factors can be obtained for Langmuir (Eley-Rideal) kinetics by interpolating between 0th and first-order kinetics. For some more complicated Langmuir-Hinshelwood rate equations negative order could be observed, because the numerator can be of a lower order than denominator. For such reaction kinetics, the effectiveness factor can even exceed unity, because the reaction rate increases along the pore length as the reactant concentration inside the particle becomes lower than in the bulk phase. Detailed analysis of the effectiveness factor for complex rate expressions is given in the review by Dittmeyer and Emig [31].

It is interesting to note that kinetic regularities and the values of activation energy are different in the regime, which is influenced by internal mass transfer. For the nth order reaction, when the catalyst effectiveness factor is inversely proportional to the Thiele modulus, an expression for the observed reaction rate is:

$$r_{obs} = \eta r = \frac{1}{\varphi_p} kC^n = kC^n \frac{1}{L} \frac{\sqrt{D_e}}{\sqrt{kC^{n-1}}} = \frac{1}{L}\sqrt{kD_e C^{n+1}} \tag{3.216}$$

Analysis of eq. (3.216) demonstrates that the observed reaction rate is inversely proportional to the characteristic length of diffusion and has the reaction order $(n + 1)/2$. Moreover observed activation energy has half the value of the corresponding to the true kinetics activation energy (Fig. 3.23) if neglecting the minor influence of temperature on the effective diffusion coefficient. Therefore, mechanistic conclusions based solely on kinetic observations obtained in the internal diffusion regime could be false, as the observed reaction orders for zero and second order reactions are 0.5 and 1.5, respectively. For the first-order kinetics the values of reaction orders are the same in kinetic and internal mass transfer regimes.

When internal diffusion is combined with the film diffusion the catalyst effectiveness factor can be extended to:

$$\eta = \eta_{int}\eta_{ext} = \frac{3}{\phi}\left(\frac{1}{\tanh(\phi)} - \frac{1}{\phi}\right) \frac{1}{1 + \frac{\phi}{Bi_M}\left(\frac{1}{\tanh(\phi)} - \frac{1}{\phi}\right)} \tag{3.217}$$

which gives the overall effectiveness factor for first-order reactions in a spherical catalyst particle. In eq. (3.217) the Biot number for mass transport Bi_M is defined as the ratio between the diffusion resistance in the fluid film and in the catalyst particle:

$$Bi_M = \frac{k_f R}{D_{ei}} \tag{3.218}$$

Usually $Bi_M \gg 1$ for porous catalyst particles and internal diffusion resistance prevails. The practical difficulties in using the eq. (3.217) are associated with the necessity to apply the Thiele modulus, which contains the rate constant, which might be unknown. The practical way to solve this problem is to use the Weisz modulus ψ, which only contains observable parameters and is defined as the ratio of effective (observed) reaction rate to the maximum effective rate of diffusion at the external surface:

$$\psi = \frac{r_{obs}R^2}{c_s D_e} = \phi^2\eta \tag{3.219}$$

The effectiveness factor as a function of the Weisz modulus for different Biot numbers is demonstrated in Fig. 3.39. The effectiveness factor is diminished when the Biot number is decreased, or in other words when external diffusion becomes more prominent. In practice the values of the Biot numbers are typically between 100 and 200 thus the influence of internal (pore) diffusion is more important than the impact of interphase (external) mass transport.

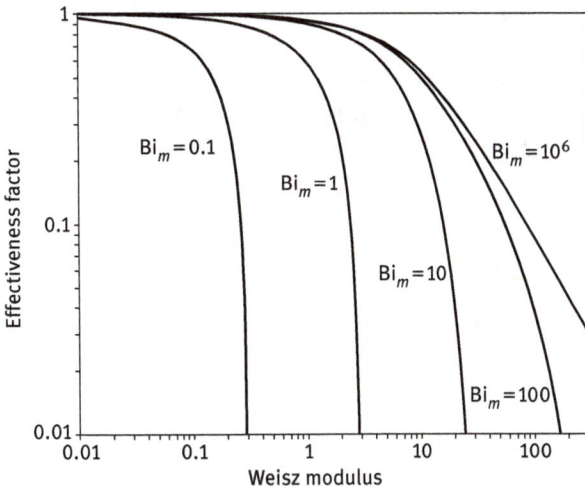

Fig. 3.39: Effectiveness factor η as a function of the Weisz modulus ψ for an isothermal, first-order irreversible reaction in a sphere.

3.5.5 Non-isothermal conditions

For some heterogeneous catalytic reactions (oxidations, hydrogenations, dehydrogenations) substantial consumption or release of heat results in non-isothermal temperature profiles inside the catalyst particle and in the film surrounding the particle (Fig. 3.40).

In general, the mass and the heat balances should be solved using the corresponding boundary conditions in order to get the concentration and temperatures profiles. The non-linear character of differential equations leads to a need for numerical solutions.

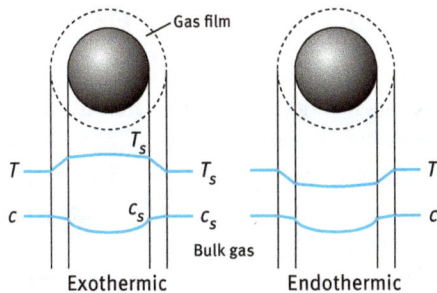

Fig. 3.40: Temperature and concentration gradients in the gas film and catalyst particle. (From [23].).

When interphase heat and mass transfer could be neglected the effectiveness factor depends on the Thiele modulus, the Prater and the Arrhenius numbers defined below. The Arrhenius numbers appear during the considerations of the heat and mass balances. The heat fluxes q are calculated according to the Fourier's law, relating the heat flux through a surface with the temperature difference along the path of heat flow:

$$q = -\lambda_e \frac{dT}{dx} \tag{3.220}$$

Another form of representing the Fourier law has been already presented (eq. 3.179) with just another notation also frequently applied in the literature. In eq. (3.220) the heat conductivity in the pores λ_e [k in eq. (3.179)] can be computed from:

$$\lambda_e = (1 - \varepsilon)\lambda_{solid} + \varepsilon\lambda_g \tag{3.221}$$

where λ_{solid}, λ_g are heat conductivity of the solid and gas in the pores, respectively, and ε is the porosity. As the heat conductivity of solid materials is several orders of magnitude higher than the conductivity of gases, the second term in eq. (3.221) can be neglected. The values of heat conductivities for industrial catalysts are $\sim 10^{-1} - 10^{-3}$ Watt $m^{-1} K^{-1}$.

When the interphase heat and mass transfer is neglected for a spherical particle, in addition to the mass balance equation in the particle (3.205), a heat balance equation for particles also needs to be considered, which defines the difference between the heat flows in and out of a differential shell as equal to the heat production within the shell because of a catalytic reaction of nth order:

$$- A_{x+dx}\lambda_e \frac{dT}{dx}\Big|_{x+dx} - \left(- A_x\lambda_e \frac{dT}{dx}\Big|_x \right) = kc^n(-\Delta H)4\pi x^2 \Delta x \tag{3.222}$$

The system of molar mass balances and the energy balance can be now written as:

$$\frac{d^2c}{dx^2} + \frac{2}{x}\frac{dc}{dx} = \frac{k(T)c^n}{D_e}, \frac{d^2T}{dx^2} + \frac{2}{x}\frac{dT}{dx} = \frac{k(T)c^n}{\lambda_e}(-\Delta H) \tag{3.223}$$

with the boundary conditions: $x = R$, $c = c_b$, $T = T_b$ (surface temperature is equal to the temperature in the bulk and the surface concentration is equal to the bulk concentration when external heat and mass transfer is absent) and $x = 0$, $dT/dx = dc/dx = 0$. This system is strongly coupled through concentrations in the reaction rate expressions and through the exponential temperature dependencies of the rate constants that require numerical solutions.

If introducing dimensionless concentration $y = c/c_b$; positions inside the pore $x^* = x/L(R)$, and temperature; $\Theta = T/T_b$ eqns. (3.223) are rewritten as:

$$\frac{d^2y}{dx^{*2}} + \frac{2}{x^*}\frac{dy}{dx^*} = \varphi^2 y^n \exp\left(\gamma\left(1 - \frac{1}{\Theta}\right)\right), \tag{3.224}$$

and

$$\frac{d^2\Theta}{dx^{*2}} + \frac{2}{x^*}\frac{d\Theta}{dx^*} = \beta_{Pr}\varphi^2 y^n \exp\left(\gamma\left(1 - \frac{1}{\Theta}\right)\right) \tag{3.225}$$

with the boundary conditions $x = 1$, $y = 1$, $\Theta = 1$ and $x = 0$, $d\Theta/dx^* = dy/dx^* = 0$ and the dimensionless Thiele, Prater (β_{Pr}) and Arrhenius (γ) numbers:

$$\phi = R\sqrt{\frac{k(T)c_b^{n-1}}{D_e}}, \beta_{Pr} = \frac{(-\Delta H)D_e c_b}{\lambda_e T_b}, \gamma = \frac{E_a}{RT_b} \tag{3.226}$$

The Arrhenius number is a measure for temperature sensitivity, while the Prater number is the normalized maximum temperature difference between the pellet center and bulk $\beta_{Pr} = \Delta T_{max}/T_b$. For exothermal reactions values of β_{Pr} are positive, rarely exceeding values of 0.1, which means that at reaction temperatures of 500–600 K the maximum temperature difference between the fluid bulk and the particle center is 50–60 K. Fig. 3.41 contains some numerically obtained typical curves for the effectiveness factor as a function of the Thiele modulus. As can be

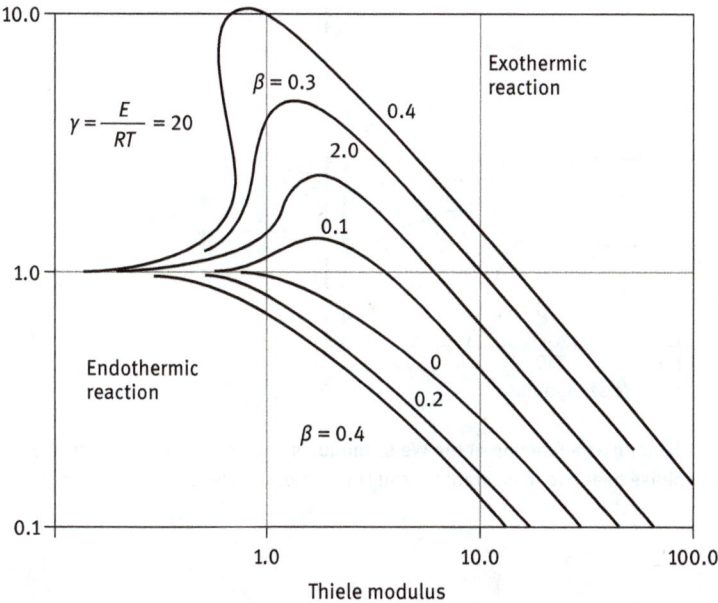

Fig. 3.41: The catalyst effectiveness factors as a function of Thiele modulus at different values of the Prater numbers.

seen from this figure, for exothermic processes the effectiveness factor can even exceed unity, because of a temperature rise inside the particle and increased values of the rate constants, which is not overcompensated by the lower concentrations inside the pellet because of the diffusion. This effect is particularly visible at small values of the Thiele modulus. Another interesting feature under non-isothermal conditions and strongly exothermic reactions is steady-state multiplicity, even for the simplest first-order reaction. It implies that there are several possible solutions for the effectiveness factor, with the middle one usually unstable. For the typical range of Arrhenius numbers (10–30) multiple steady states are possible only at large Prater numbers (highly exothermal reactions, low thermal conductivity). As mentioned above, for industrial catalysis the values of the Prater number are below 0.1 and multiplicities of the effectiveness factor are not important, although the values of the effectiveness factor in the real systems could be above unity.

The effective heat conduction with the catalyst particle is typically not that critical and the dominating part of the overall heat resistance is in the external boundary layer, but not inside the catalyst pellet. The mathematical treatment in this case is complicated and discussed in specialized monographs.

The effectiveness factor for this combined case is presented as a function of the observable Weisz modulus in Fig. 3.42 with a modified Prater number:

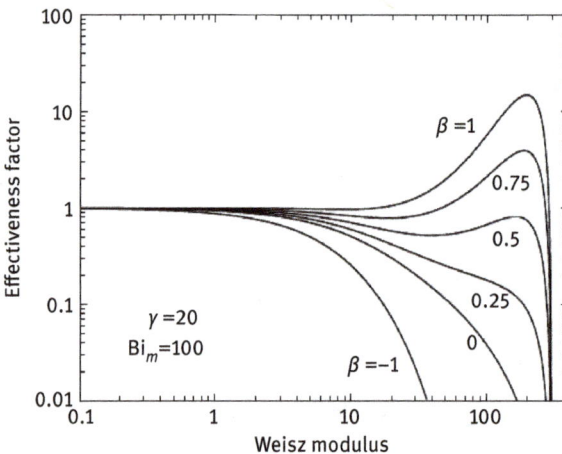

Fig. 3.42: Effectiveness factor η as a function of the Weisz modulus Ψ for combined intraparticle mass transfer and interphase heat and mass transfer and the first-order irreversible reaction in a sphere.

$$\beta^*_{Pr} = \frac{(-\Delta H)c_b Pr^{\frac{2}{3}}}{\rho c_p T_b Sc^{\frac{2}{3}}} \tag{3.227}$$

where ρ is density, c_p is heat capacity, Sc and P_r are, respectively, the Schmidt and the Prandtl numbers. In a similar way to the previous considerations for the large values of the Weisz modulus and exothermal reactions ($\beta*_{Pr} > 0$) the effectiveness factor can exceed unity, as the decline in concentrations is compensated by the temperature increase. For higher Arrhenius numbers the increase of the effectiveness factor above unity will be even more pronounced as the reaction rate will possess stronger temperature dependence.

3.5.6 Multiple reactions and diffusional limitations

It is not only catalytic activity but also selectivity that can be influenced by mass transfer phenomenon. Analytical treatment is available for simple reactions and as the complexity increases, numerical treatment becomes necessary. In the relevant literature it is typically independent reactions (less commonly observed in practice, although interesting from the theoretical viewpoint), consecutive and parallel reactions that are treated. The consecutive and parallel reactions will be considered here.

For a consecutive reaction sequence $A \Rightarrow B \Rightarrow C$ in a batch reactor when the rate equations are of the first-order, the differential selectivity is obtained from:

$$S = -\frac{dc_B}{dc_A} = \frac{k_A c_A - k_B c_B}{k_A c_A} = 1 - \frac{k_B c_B}{k_A c_A} \tag{3.228}$$

An analytical solution of (3.228) can be readily obtained that provides expressions for the concentration of the intermediate C_B and selectivity S_B:

$$c_B = \frac{1}{1-M}\left(\frac{c_A^M}{(c_A^0)^{M-1}} - c_A\right), S_B = 1 - \frac{M}{1-M}\left(\left[\frac{c_A}{c_A^0}\right]^{M-1} - 1\right) \tag{3.229}$$

where c^0_A is the initial concentration of the reactant at which $c_B^0 = 0$ and $M = k_B/k_A$.

The mass balances for pore diffusion with a slab geometry and a characteristic dimension L for the dimensionless concentrations y are:

$$\frac{d^2 y_A}{dx^2} = \phi^2 y_A, \frac{d^2 y_B}{dx^2} = -\phi^2 y_A + \gamma^2 \phi^2 y_B \tag{3.230}$$

Where:

$$\phi^2 = L^2 \frac{k_A}{D_{e,A}}, \gamma^2 = \frac{k_B D_{e,A}}{k_A D_{e,B}} \tag{3.231}$$

The solution of eq. (3.231) results in the concentration profiles, first making it possible to calculate the rates:

$$r_A^{bulk} = \sqrt{k_A D_{e,A}} c_A^{bulk} \tanh \phi \tag{3.232}$$

$$r_B^{bulk} = \sqrt{k_B D_{e,B}} c_B^{bulk} \tanh(\phi y) - \sqrt{k_A D_{e,A}} c_A^{bulk} \frac{\tanh \phi - y \tanh(\phi y)}{1 - y^2} \tag{3.233}$$

and then the selectivity towards the intermediate product

$$S = \frac{r_B^{bulk}}{r_A^{bulk}} = \frac{1 - y \frac{\tanh(\phi y)}{\tanh \phi}}{1 - y^2} - \frac{y \tanh(\phi y)}{\tanh \phi} \frac{D_{e,B} c_B^{bulk}}{D_{e,A} c_A^{bulk}} \tag{3.234}$$

Differential selectivity in consecutive reactions depends on the values of the Thiele modulus ϕ and parameter y. The higher is the value of $y = \sqrt{k_B D_{e,A}/k_A D_{e,B}}$, the more pronounced is the influence of diffusion resulting in lower selectivity towards the intermediate B (Fig. 3.43). This is an important conclusion – that internal diffusion limitations, prominent in industrial conditions because of the large size of catalyst pellets, lead to diminished selectivity towards the intermediate product in comparison with the intrinsic kinetic conditions.

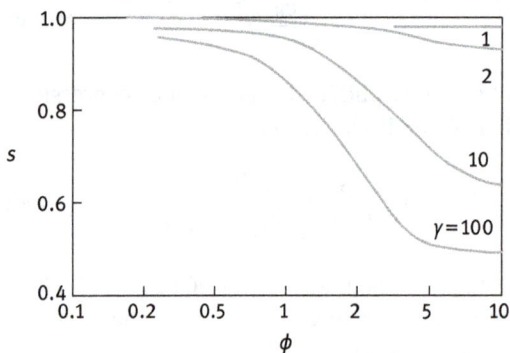

Fig. 3.43: Dependence of selectivity for consecutive reactions on Thiele modulus at different values of y. (Redrawn from [33].).

For parallel reactions $A \Rightarrow B_1$ and $A \Rightarrow B_2$ the differential equation describing the concentration profile of component A in the isothermal case is:

$$\frac{d^2 y_A}{dx^2} = \phi_1^2 (f_1(y_A) + \frac{1}{y} f_2(y_A)) \tag{3.235}$$

where:

$$\phi_1 = L\sqrt{\frac{k_1 r_1 \frac{(c_A^{bulk})}{c_A^{bulk}}}{D_{e,A}}}, y' = \frac{k_1 r_1 (c_A^{bulk})}{k_2 r_2 (c_A^{bulk})} \tag{3.236}$$

If the reaction orders are different, for instance the reactions are of the nth and the mth order, respectively $f_1(c'_A) = (c'_A)^n; f_2(c'_A) = (c'_A)^m$, then the concentration profile of component A is:

$$\frac{d^2 y_A}{dx^2} = \phi_1^2 \left((y_A)^n + \frac{(y_A)^m}{y'} \right) \tag{3.237}$$

with:

$$y' = \frac{k_1}{k_2}(c_A{}^{bulk})^{n-m}; \phi_1 = L\sqrt{\frac{k_1(c_A{}^{bulk})^{n-1}}{D_{e,A}}} \tag{3.238}$$

The ratio of the selectivity towards the product B_1 in the internal diffusion regime to the selectivity under kinetics when $y' \ll 1$ is:

$$\frac{S_{diffusion}}{S_{kinetics}} = \frac{m+1}{2n-m+1} \tag{3.239}$$

implying, that when the reactions are of the same order, the differential selectivity is independent on the presence of internal diffusion. If the desired reaction is of lower order (for example, $n < m$), then it is preferential to conduct the reaction in the diffusion region, as the highest penalty in the case of pore diffusion limitations is on the highest order reaction. Under non-isothermal conditions changes in apparent selectivity caused by temperature variations inside the pellet are expected when the activation energies are different irrespective of the reaction orders.

It is possible that one of the reactions occurring in the system is an undesired one, for example poisoning. For the slab geometry the diffusion model is:

$$\frac{d^2 y}{dx^2} = \phi^2 r(y) a \tag{3.240}$$

where a is the relative activity (or fraction of unpoisoned sites). For uniform poisoning and the first-order kinetics, eq. (3.240) takes the form:

$$\frac{d^2 c'}{dx^2} = (\phi^2 a) c' \tag{3.241}$$

Apparent activity is defined as the ratio of the effectiveness factors for the poisoned:

$$\eta^{poisoned} = \frac{\tanh(\phi\sqrt{a})}{\phi\sqrt{a}} \tag{3.242}$$

and unpoisoned catalyst under conditions of pore diffusion limitations:

$$X = \frac{\eta^{\text{poisoned}}}{\eta^{\text{unpoisoned}}} = \frac{\frac{\tanh(\phi\sqrt{a})}{\phi\sqrt{a}}}{\frac{\tanh(\phi)}{\phi}} = \frac{\sqrt{a}\tanh(\phi\sqrt{a})}{\tanh(\phi)} \tag{3.243}$$

For negligible diffusion limitations $\chi = a$ because $\tanh\phi\sqrt{a} \approx \phi\sqrt{a}$; $\tanh(\phi) \approx \phi$. Conversely, when intraparticle mass transport is significant, the Thiele modulus is large and $\chi = \sqrt{a}$ (Fig. 3.44), which means that poisoning is less prominent in the case of internal diffusional limitations.

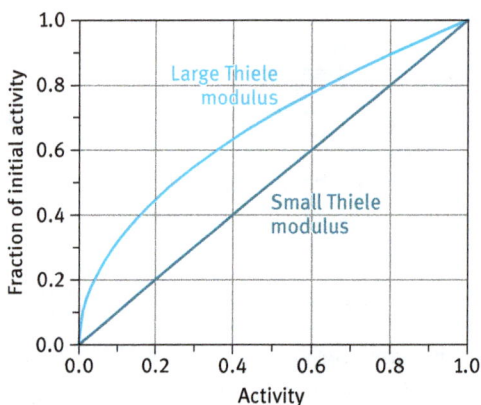

Fig. 3.44: Behavior of apparent activity for small and large Thiele moduli.

3.5.7 Diffusion in micropores

Mass transfer in micropores (configurational diffusion) has a complex behavior and is influenced by the interactions of the diffusing molecules with the walls of the pores as well as adsorbate-adsorbate interactions.

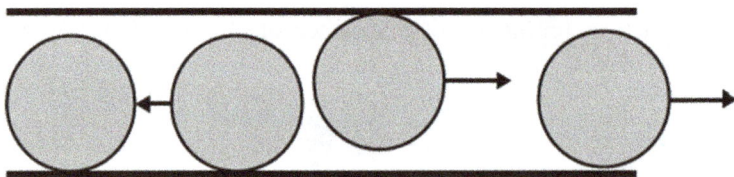

As a result, the diffusivity in this case (which could be anisotropic) is much lower than for molecular or Knudsen diffusion (Fig.3.32). There could be a great variation of several orders of magnitude in the diffusivity of molecules with not very different kinetic diameters. For mixtures caused by the similar size of the reacting species and the pores, diffusion of fast reacting species could be delayed.

In case of catalysis on zeolites, diffusion is sometimes described based on the adsorbed phase concentration, while the reaction rate is based on the bulk gas-phase concentration or pressure. Subsequently, an adsorption constant K is added to the Thiele modulus to maintain its dimensionless character

$$\varphi = L \sqrt{\frac{k}{KD_{eff}}} \tag{3.244}$$

The practical consequences of configurational diffusion are shape-selective properties of materials containing micropores, such as zeolites, as discussed in Chapter 1. In fact, reactant and product selectivity is a result of these mass transfer effects.

Modeling of shape selectivity effects is a complicated task and requires advanced numerical treatment. Molecular dynamics, stochastic and continuous models are applied. For molecular dynamics models a solution is needed for Newton's laws of motion for the molecules in the zeolites, which calls for the use of quantum mechanics. Stochastic or probabilistic models (such as Monte Carlo methods) use rather simple model reactions. Finally, continuous models are based on the conservation equations for mass and energy, which are certainly valid for macropores, while their applicability could be questioned for micropores when only few molecules are present within the pore cross section.

3.5.8 Criteria for the absence of diffusional limitations

In order to clarify the influence of mass transfer it is possible to calculate the mass transfer coefficients and catalyst effectiveness factors as explained above. Such analysis requires calculations of the diffusion coefficients and elucidation of several other physical and materials properties. Interested readers could consult the relevant literature [25, 34, 35] where details of calculations are provided.

For internal diffusion, analytical solutions can only be used directly for simple kinetics such as first or zero orders, thus approximations should be applied because many reactions are not of zero or first-order. Alternatively, it is possible to use the generalized diagrams by Aris [36] (Fig. 3.45). Such diagrams relate various kinetics effectiveness factors and the generalized Thiele modulus:

$$\phi^* = \frac{\phi}{(2 \int_0^1 r' dy)^{0.5}} \tag{3.245}$$

where y is the relative concentration and r' is the reaction rate for relative concentrations $r' = \dfrac{r}{r(c_b)}$.

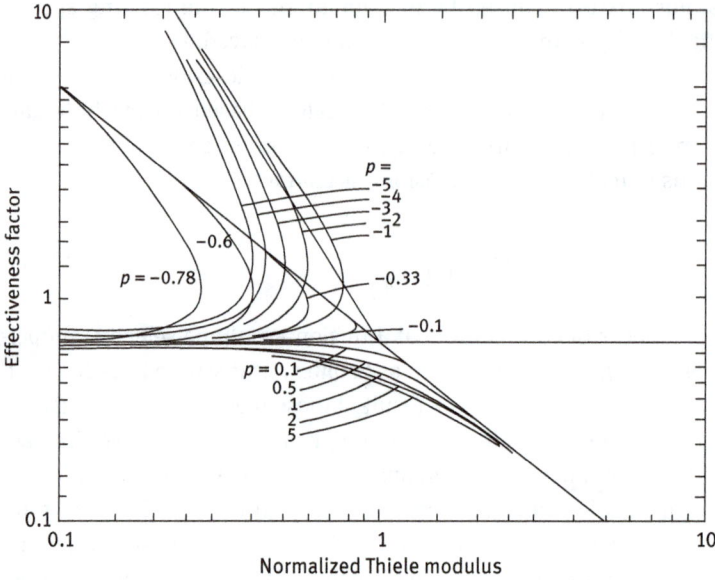

Fig. 3.45: Isothermal effectiveness factor for reactions of pth order.

For example, for a second order reaction:

$$\phi^* = \frac{\phi}{\left(2\int_0^1 y^2 dy\right)^{0.5}} = \sqrt{\frac{3}{2\phi}} = \sqrt{\frac{3}{2}}\sqrt{\frac{r_{obs}}{c_s D_e \eta}}R \qquad (3.246)$$

The intrinsic rate is evaluated from kinetic experiments for small catalyst particles (when the effectiveness factor is unity). The calculated value of the effective diffusion coefficient and the initial bulk phase concentration of the reactant (neglecting external diffusion and supposing that surface and bulk concentrations are equal) are inserted in eq. (3.246) and give a dependence of the generalized Thiele modulus on the effectiveness factor and the radius of the catalyst pellet. Then for a large catalyst particle size the generalized Thiele modulus is calculated by first setting the effectiveness factor equal to unity. The next step in the iteration loop is to check the value of effectiveness factor from the diagrams (Fig. 3.45) for the respective reaction order (2 for example).

This value of the effectiveness factor is further applied to recalculate the generalized Thiele modulus, followed by another estimation of the corresponding effectiveness factor from the plots. This iteration procedure is applied until convergence is reached.

Dittmeyer and Emig [31] proposed and reviewed some theoretical criteria for the absence of interphase (external) and intraparticle (internal) transport for simple reactions with arbitrary kinetics. Such theoretical criteria contain explicit expressions of the reaction rates and are more difficult to use than experimental criteria. Because of the complexity of using these criteria and a possibility of simultaneous concentration and temperature gradients in the catalyst grain, in order to make the theoretical approach more sound it could even be recommended that the level of complexity be increased and a numerical solution of the balance equation for the particle be generated.

Apparent complexity of the procedures described above led to the development of alternative approaches based on various experimental criteria, which became rather popular and are often applied to determine if heat and mass transfer are influencing a process under investigation. Such criteria are usually derived in a way that deviations of the observed rates from the ideal rates are not more than 5%.

$$\eta = \frac{r_{obs}}{r(T_b, c_b)} = 1 \pm 0.05 \tag{3.247}$$

The advantage of applying different criteria is that observed values of reaction rates are used, while the kinetics parameters (requiring calculation of, for example, Thiele modulus) could be still unknown, at least at the initial stage of a process investigation. The approximate nature of such criteria calls for a conservative approach in their use. A decision on the possible influence of mass transfer could be made when the calculated value is significantly different from the limiting value, otherwise an uncertainty remains. The criterion for external heat transfer was proposed by Mears [37]:

$$\chi = \frac{|\Delta H_r| r_{obs} R}{h T_b} < 0.15 \frac{R_g T_b}{E_A} = \frac{0.15}{\gamma} \tag{3.248}$$

where h is the heat transfer coefficient, T_b is the bulk temperature, ΔH_r is the reaction enthalpy, E_A is the activation energy and R_g is the universal gas constant. The criteria of external mass transport [see also eq. (3.197) for the derivation with $k_{LS} = k_f$] for a simple irreversible reaction of nth order ($n \neq 0$) is:

$$w = \frac{r_{obs} R}{k_f c_b} < \frac{0.15}{n} \tag{3.249}$$

and for a combined intra/interphase heat and mass transfer are also proposed by Mears [38]:

$$\frac{r_{obs} R^2}{D_e c_b} < \frac{1 + 0.33 \gamma \chi}{(n - \gamma \beta_{Pr})(1 + 0.33 n w)} \tag{3.250}$$

where y and β_{Pr} are the Arrhenius and the Prater numbers, χ and w are defined, respectively, by eqns. (3.248) and (3.249). It is apparent that the application of the latter criteria is not straightforward, as it requires knowledge of the Arrhenius number and thus the activation energy. However, activation energy is usually determined from the studies when the influence of mass and heat transfer is yet to be determined.

Internal heat transport limitations were considered by Anderson [39] leading to the following conditions when such limitations do not occur:

$$\frac{\Delta|H_r|r_{obs}R^2}{\lambda T_s} < 0.75\frac{R_gT_b}{E_A} = \frac{0.75}{\gamma} \tag{3.251}$$

where λ is the thermal conductivity.

The most widely applied criterion is probably the one for internal mass transfer limitations in an isothermal catalyst particle, that is, for pore diffusion. Following Weisz and Prater [40] no pore diffusion limitation occurs, if the Weisz modulus defined by eq. (3.219) is:

$$\psi = \frac{r_{obs}R^2}{c_sD_e} \begin{cases} <6, n=0 \\ <0.6, n=1 \\ <0.3, n=2 \end{cases} \tag{3.252}$$

Recently I reviewed a scientific paper where the calculation of the criterion for pore diffusion resulted in a value of 0.29, while for this particular reaction with kinetic order equal to 2 it should be below 0.3. Based on the approximate calculations of the criterion (which also required calculations of an effective diffusion coefficient and thus an estimation of porosity and tortuosity) the authors made a premature and possibly incorrect conclusion that their data are not influenced by mass transfer.

Strictly speaking, such criterion is applied for the spherical particles. For other geometries, the specific radius could be replaced by the pellet volume Vp, divided by the pellet surface area Ap. According to Villermaux [41] internal mass transfer is negligible when:

$$\frac{r_{obs}Vp^2}{c_sD_eA_p^2} < 0.1 \tag{3.253}$$

According to Weisz and Hicks [42], for a non-isothermal catalyst particle no internal heat and mass transport limitations occur if:

$$\psi \exp\left(\frac{\gamma\beta_{Pr}}{1+\beta_{Pr}}\right) < 1 \tag{3.254}$$

As the accuracy is questionable in the calculations of the various parameters in the criteria presented above, such as heat, mass transfer and diffusion coefficients, porosity, tortuosity, and particularly for bimodal pore size distribution, experimental tests are also conducted in addition to calculations of criteria for heat and mass transport.

As explained above an experimental way to determine whether external mass transfer limitations are present in flow reactors is to measure conversion dependence on the flow rate. At low Reynolds numbers the heat and mass transfer coefficients could be, however, insensitive to changes in the flow velocity. For the slurry systems experiments with different catalyst masses are used to determine the influence of film diffusion on the gas/liquid interface. At high catalyst mass, the rate becomes independent of the catalyst mass and is only limited by gas-liquid mass transfer, which is independent of the catalyst amount. The stirring speed is typically varied to check the influence of external (fluid/solid) mass transfer. As already emphasized the mass transfer coefficient depends on the energy of dissipation, rather than the agitation rate, thus special care should be taken to ensure that while varying the agitation rate the energy of dissipation (which depends also on the liquid load) is also changing.

An experimental approach to verify the impact of internal diffusion is to perform experiments with catalysts of different particle sizes. It should be kept in mind, however, that the flow field could be also changed while altering the pellet size. In addition, if there are no changes in activity for different pellet sizes for a catalyst with bimodal pore distribution, the influence of internal diffusion in the small pores may still be important, while the test may just indicate that there are no diffusional limitations in the larger set of pores. Another experimental criterion could be applied that was presented by Koros and Nowak [43] and elaborated on by Madon and Boudart [44]. This criterion is based on making rate measurements with catalysts in which the concentration of the catalytically active material has been changed. As the reaction rate in the kinetic regime is directly proportional to the concentration of active material, an increase in activity will be proportional to the amount of the active phase. The limitations of this method are associated with the practical difficulties of preparing catalysts samples with different loading without simultaneously altering some other important parameters, such as metal dispersion, diffusion characteristics or distribution of the active sites along the pore.

Exercise 3.1: The oxidation of methanol to formaldehyde is carried out over a Fe-Mo oxide catalyst. The following operating conditions are used: T_{bulk} = bulk gas-phase temperature = 536 K; P = total pressure = 0.12 MPa; C_A = feed composition = 6.0 mol% methanol in air; porosity = 0.4, tortuosity = 4; λ_e = effective thermal conductivity of catalyst = 0.17 (J m^{-1} s^{-1} K^{-1}); heats of formations for methanol, formaldehyde and water are correspondingly: 190, 112 and −239 kJ mol^{-1}. The reaction rate is first-order in methanol and zero order in oxygen. Activation energy is 80 kJ mol^{-1}.

1. Calculate the maximum temperature difference between the surface and center of catalyst pellet.
2. Determine whether temperature gradients in the catalyst pellets will have a significant influence on the reaction rate.
3. Is it possible for these reaction conditions that, depending on the catalyst particle size, the effectiveness factor will exceed 1?

Neglect temperature effects on the heat of the reaction

Exercise 3.2 (From [32])

A first-order catalyzed decomposition of N_2O in helium over spherical particles has been studied in a laboratory-scale reactor. Use the data below to answer the following questions:

1. What is the impact of internal diffusion?
2. Do temperature differences exist over the gas film or within the particle?

Data: Catalyst d_p = 2.4 mm, porosity = 0.2, tortuosity = 5, pore radius = 0.8 nm, λ_e = effective thermal conductivity of catalyst = 0.45 (J m^{-1} $s^{-1}K^{-1}$), Gas film k_f = 0.083 m s^{-1}, h = 46 (J m^{-1} s^{-1} K^{-1}), Reaction heat = −163 kJ mol^{-1}, c_b = 20 mol m^{-3} (at 1 bar, 609 K), rate$_{observed}$ = 27 mol s^{-1} m^{-3}.

3.6 Catalytic reactors

The choice of reactors is an important part of the engineering process and is dictated by the purpose of the investigation (the study of mechanism and kinetics; generation of data for process simulation and design; the investigation of catalyst performance in a certain range of parameters), process economics, deactivation, etc.

3.6.1 Laboratory reactors

Laboratory reactors are needed for the collection of representative kinetic data for activity and selectivity in the absence of mass transfer limitations and preferably deactivation. A high precision of the data is needed because large deviations in the values of the experimentally measured rates will have serious implications for quantitative considerations. Reproducibility of rate measurements over a broad range of parameters is also of importance. Another necessary feature is the possibility of obtaining the maximum amount of kinetic information in the minimum time. Analysis of products as well as reactor lay-out should preferably be as easy as possible.

The essential feature for catalytic reactions is the readiness in reduction/activation of heterogeneous catalysts and a possibility to utilize them in the needed geometrical form. Despite the strict definition of catalysis, which states that the catalyst does not change during the catalytic reactions, some activity deterioration takes place and therefore measurements of catalytic kinetics should always monitor the catalyst activity.

For example, tubular reactors with fixed (packed) beds of catalyst and plug flow (piston flow) (PFR) are often operated in a continuous mode (Fig. 3.46). In the set-up presented in Fig. 3.46 the check valves are needed to stop back diffusion and the tubing volume is minimized to avoid the "dead" volume.

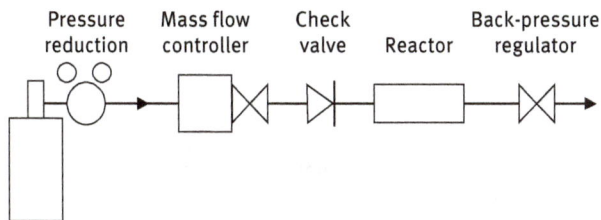

Fig. 3.46: Tubular reactor set-up.

This is a simple and inexpensive reactor type, which could be applied both as a differential or integral reactor. In the first case the main requirements is the ideal plug flow, low degree of conversion (typically below 10%) and high space velocities. Differential operation mode allows for direct measurement of the reaction rates with simple equipment, while obvious disadvantages are the difficulties in achieving steady-state rates because of transients and deactivation, and the long time to get just one experimental point, thus imposing serious limitations in studying multiroute processes. Conversely, the catalyst can be easily charged or discharged, and there is even an option to use the reactor tube with the catalyst only once and discharge the whole reactor. In addition, the catalyst deactivation could easily be seen with time-on-stream. The tube could be uniformly heated in a tubular furnace allowing isothermal conditions.

Dimensions of the reactor and the amount of catalyst depend on the scale (laboratory/bench) and the tasks. A typical tubular minireactor could contain 0.1–1 g of catalyst, while a bench scale reactor, operating in an integral mode might contain 50–200 g. Integral fixed-bed reactors of this size are suitable for measuring overall conversion and catalyst lifetime under realistic conditions, which is important in the scaling-up from a laboratory to a large-scale industrial reactor. Differential fixed-bed reactors, not suitable for simulation purposes, are rather efficient for measuring steady-state activity, selectivity and could be used for catalyst screening, especially when they are arranged in a parallel mode.

Quantitative treatment of plug flow reactors is cumbersome therefore several assumptions are usually made. The fluid compositions is considered to be uniform along the reactor cross section (i.e., there is no radial dispersion), which is valid only at certain conditions $d_{cat}/d_{reactor} < 1/30$, while for larger catalyst particles some gas can by-pass close to the reactor walls. When the reactants move along the reactor the composition of the mixture changes and axial diffusion could become prominent. To avoid possible impact of it in kinetic experiments the conversion level, as mentioned above, should be preferentially below 10–15%. It should also be mentioned that, in general, non-isothermicity of a tubular reactor should be considered (Fig. 3.47).

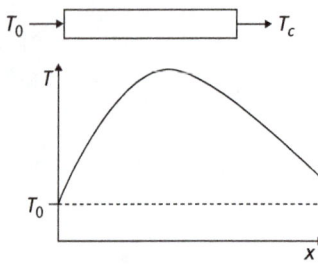

Fig. 3.47: Possible temperature profile in a single tube reactor.

Temkin and Kul´kova [45] developed a concept of single pellet string reactor (Fig. 3.48), where real-size catalyst particles can be tested by placing catalyst spheres and inert cylinders alternately in a tube of a diameter slightly larger than the catalyst spheres.

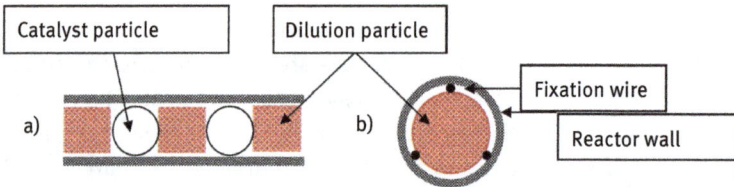

Fig. 3.48: Reactor concept of Temkin and Kul´kova [45]. (From [46]).

Absence of any contact between catalyst particles and a small space between the inert spheres and the reactor walls ensure efficient heat transport. The original design by Temkin and Kul´kova [45] was further optimized by Clariant [46] to minimize irregularities in the flow patterns and allow accelerated testing with the several so-called Temkin reactors in parallel. The current design comprises seven parallel tubular Temkin reactors, each filled with a series of reactor modules with eight spherical cavities that are connected by cross-shaped channels (Fig. 3.49).

Stirred tank reactors (STR) are very often applied for multiphase heterogeneous catalytic reactions and can be operated batchwise (batch reactor, BR, Fig. 3.50A), semi-batchwise (semi-batch reactor, SBR, Fig. 3.50B) or continuously (continuous stirred tank reactor, CSTR, Fig. 3.50C).

Batch mode of operation brings several advantages as it allows for monitoring the progress of the reaction over time and thus acquiring the whole kinetic curve in one experiment. High precision and a wide range of parameters afforded by this operation mode made batch reactors very popular for kinetic studies especially in the field of fine and pharmaceutical chemicals. The catalyst particles could be immersed in the slurry using a conventional autoclave (Fig. 3.51A), which could be arranged in series (6, 16 or 32 reactors) allowing

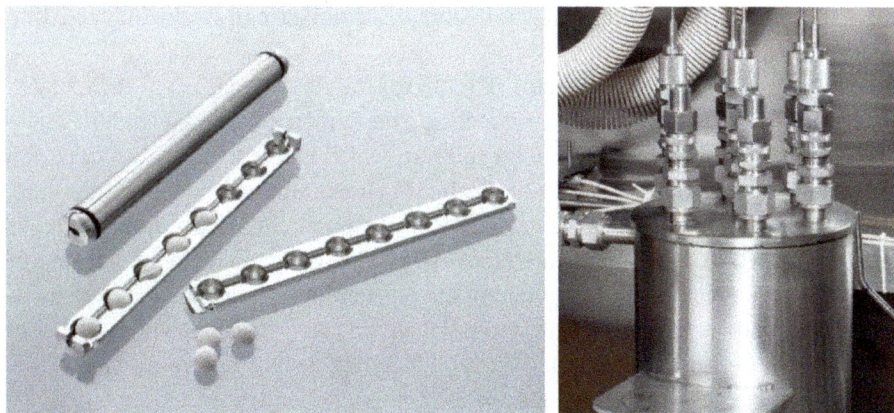

Fig. 3.49: Accelerated testing with the Temkin–Kul´kova reactor [47].

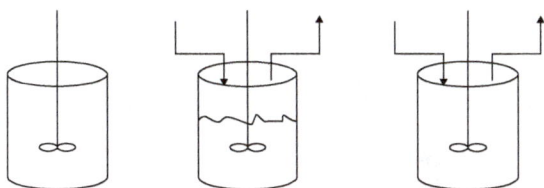

Fig. 3.50: Stirred tank reactor (STR) operating (A) batchwise, (B) semi-batchwise and (C) continuously.

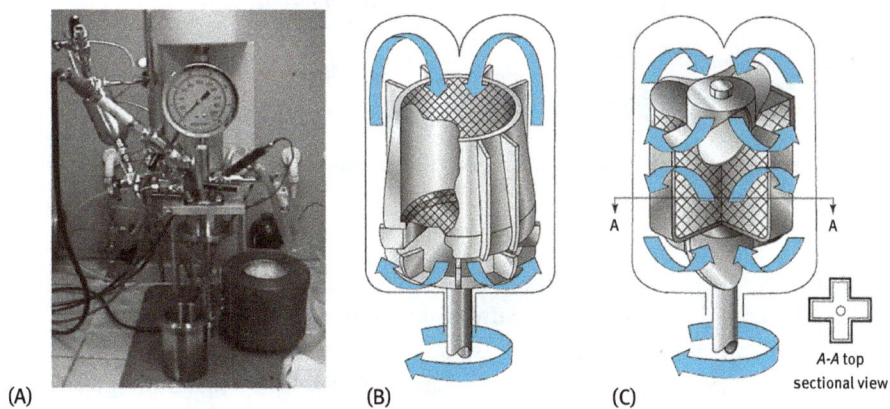

(A)   (B)   (C)

Fig. 3.51: (A) autoclave, (B) Berty and (C) Carberry reactors.

high throughput screening. Alternatively, the catalyst can be introduced into baskets (Fig. 3.51A,B).

For example, in a Berty reactor (Fig. 3.51B) the flowing gases are recirculated through the catalyst bed using an impeller. The Carberry reactor (Fig. 3.51C) contains a spinning basket with the solid catalyst, which is retained because of woven wire mesh basket, allowing efficient gas-liquid circulation with low pressure drop. While the advantage of such arrangements is the possibility to use heterogeneous catalysts of different geometrical shapes, large catalyst particles required for this operation mode often lead to conducting the reaction in the region of external diffusion. Such reactors or other similar types, when the catalyst is attached as a fiber or monolith to the impeller (Fig. 3.51), enable catalyst testing in the absence of the thermal and concentration gradients, representing gradientless reactors.

These gradientless reactors belong to the group of stirred flow reactors, characterized ideally by very small concentration and temperature gradients within the catalyst region.

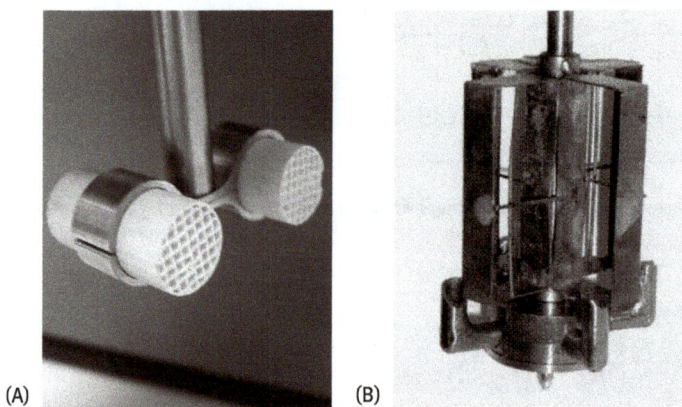

(A)   (B)

Fig. 3.52: Reactors when the catalyst is attached to the impeller: (A) monolithic catalyst, (B) holder for fiber catalysts.

Alternatively, recirculation reactors can be applied (Fig. 3.53), when there is external mixing, ensuring constant pressure in the system and no changes in the gas composition with time and in space as each single pass of reactants through the bed results in very small conversion because of the very high circulation rate.

In the flow circulation reactor developed by Temkin, Lukyanova and Kiperman (Fig. 3.53) the mixture of gases is fed through pipe (1) to the reactor (4) and then to the outlet (2) for further analysis, for example by GC. The pump (3) provides the necessary circulation of the mixture. The system is free from concentration gradients along the catalyst bed, concentration gradients caused by axial

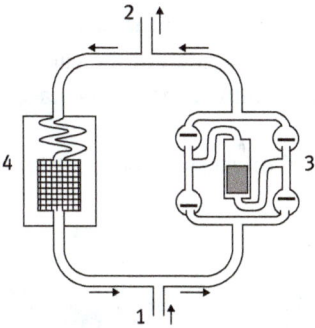

Fig. 3.53: Flow circulation reactor [48].

dispersion, and temperature gradients. The treatment of data is simplified as the rate is extracted directly in the differential form. Efficient circulation pumps are required and the inventory of reactants should be sufficient. This reactor can also operate in a batch mode, allowing measurements of the whole kinetic curve in one experiment with the disadvantage that the by-products are accumulated in the loop. The relaxation time (time to reach steady-state) could be significant. These disadvantages are, however, compensated by the easy mathematical treatment of experimental data.

The concept of circulation flow reactor was further developed on an industrial scale in the Buss reactor (Fig. 3.54), which can operate in batch, semi-continuous and continuous mode. Large quantities of the reaction gas are introduced via a mixer creating a well-dispersed mixture, which is rapidly circulated by a special

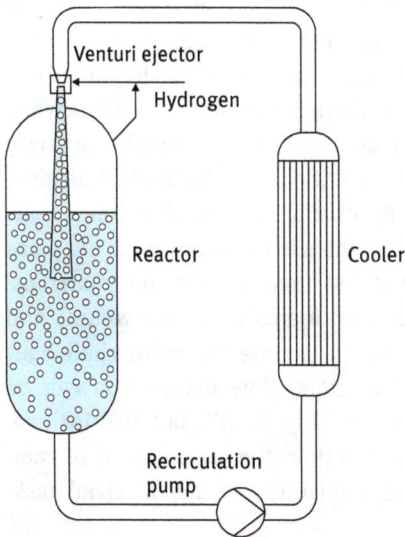

Fig. 3.54: Buss loop reactor. (From [49]).

pump at high gas/liquid ratios throughout the volume of the loop, allowing the maximum possible mass transfer rates. Independent optimization of the heat transfer is arranged by a heat exchanger in the external loop. Separation of the catalyst from the product is possible in a continuous operation, retaining the solid catalyst within the loop.

During recent years, research activities have focused on carrying out reactions on a small scale (Fig. 3.55).

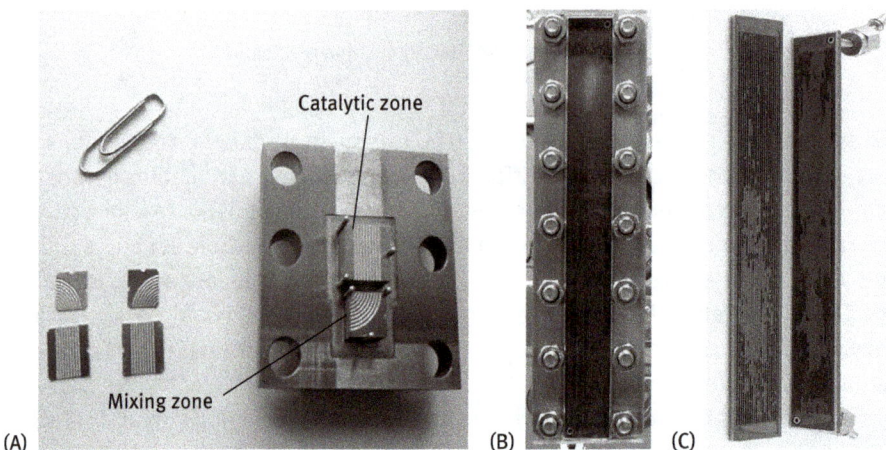

Fig. 3.55: Microchanneled reactor. (A) Two-phase reactor. (B) Flow pattern in a three-phase reactor. (C) Wash-coating stability.

Because of higher heat and mass transfer, high ratio of surface area to volume and short residence time, microreactors (or microstructured or microchannel reactors), with the dimensions of channels in microreactors typically between ≈10 μm ± 2 mm (Fig. 3.55A), provide efficient control over the exothermic reactions, reactions when low inventory is needed for toxic reactants or reactions carried out under runaway conditions. As a consequence, more aggressive conditions can be applied, leading to higher yields than in conventional reactors. Another advantage of the microreactors is a possibility of on-site production enabling the use of chemicals at the place of production and avoiding the transportation and storage of dangerous materials. Improvement of safety in microreactors is based on a small reaction volume and thus diminished inventory of dangerous chemicals. Among the main challenges related to the application of microchanneled reactors, flow distribution can be extremely uneven not only for three-phase systems (Fig. 3.55B), but also for two-phase (gas-solid and liquid-liquid) systems. In addition, the introduction of catalytic particles in microreactors, ensuring their robustness, is not a trivial task.

Fig. 3.55C illustrates the instability of alumina wash-coating used as a support for a ruthenium catalyst in the hydrogenation of sugars.

While an obvious use of microreactors is related to their application in generating kinetic data, somewhat larger scale production is also feasible. Options in the scaling-up of micro reactors could be external or internal numbering up. In the external numbering up many devices are connected in a parallel fashion, which implies a parallel connection of functional elements that provides compact architecture at reasonably high throughput. External numbering up preserves all the transport properties and hydrodynamics of a single device, requiring, however, sophisticated control systems, a large quantity of housing materials and thus high fabrication costs. Parallelization of reactors is also related to maldistribution of the reactants, as already mentioned.

According to Bartolomew and Farrauto [50], laboratory and bench scale reactors are selected to satisfy the intended application; avoid deactivation, heat and mass transfer limitations; minimize temperature and concentration gradients while maximizing the accuracy of measurements. In addition, minimization of the construction and operation costs is also taken into account. Among the laboratory reactors presented above for two-phase reactions differential fixed-bed reactors are usually selected to study intrinsic kinetics, with the preference given to gradientless reactors either with an external loop (flow circulation reactors) or to the stirred gas reactor (Berty or Carberry type).

For three-phase reactions kinetic experiments are typically done in autoclaves, although some researchers prefer CSTR. As batch operation mode is rarely used in industry for large-scale production, the preferred reactor types are either slurry reactors (bubble columns, CSTR and cascades of stirred tanks) or packed-bed reactors (up-flow packed-bed or down flow trickle bed reactors). Packed bed reactors are illustrated for up-flow (Fig. 3.56B) and down flow (Fig. 3.56A) modes. Trickle bed reactors can have multiple thermocouples (Fig. 3.57A) to measure temperature profiles and can be arranged in parallel fashion for throughput operation (Fig. 3.57B).

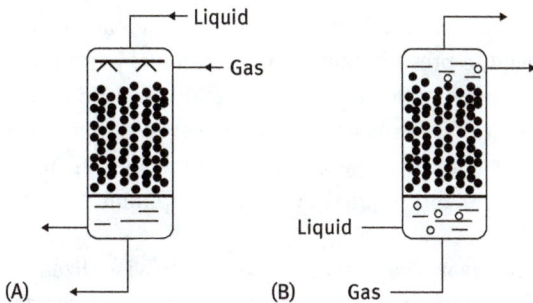

Fig. 3.56: Concurrent three-phase reactors: (A) down flow (trickle bed), (B) up-flow.

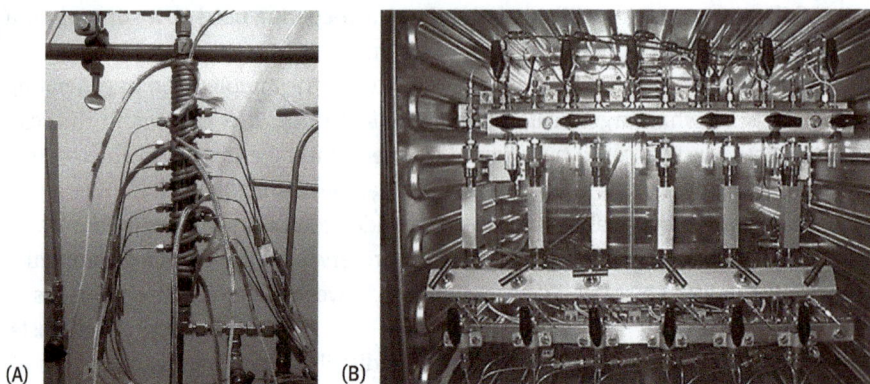

Fig. 3.57: Continuous reactors: (A) temperature control with multiple thermocouples (design by Dr. A. Tokarev), (B) parallel trickle bed reactors (design Dr. K. Eränen).

3.6.2 Industrial reactors

The discussion below will be focus separately on two (gas-solid or liquid-solid) and three-phase (gas-liquid-solid) reactors.

3.6.3 Two-phase reactors

Fixed-bed reactors with a packed catalyst bed are most commonly used in industrial practice (Fig. 3.58) for the production of petrochemicals and various basic chemicals. The catalyst is loaded in a vessel, through which the fluid is directed. The catalyst particles are in the form of pellets (extrudates, tablets) of a size that does not lead to too high-pressure drop, while increasing the diffusional length at the same time. Fig. 3.58 shows adiabatic fixed-bed reactors, when there is no temperature control in the bed and there is a monotonic temperature increase along the bed. For single-route reactions (such as, for example, water gas shift) the conversion can be calculated simply from the adiabatic temperature increase or decrease. When heat should be removed to prevent damaging the catalyst particles or because equilibrium composition is unfavorable at a high temperature, multibed reactors (Fig. 3.58B and 3.58C) can be used with heat removal achieved either using heat exchanges or cooling by cold reactants, respectively. Such procedures allow for adapting the temperature profile to the requirements of an optimal reaction pathway (Fig. 3.59).

Injection of hot or cold gas between the stages (or beds) as visualized in Fig. 3.58B is a simple option. The disadvantages for constant tube diameter are cross-sectional loading, which increases from stage to stage, and energetically

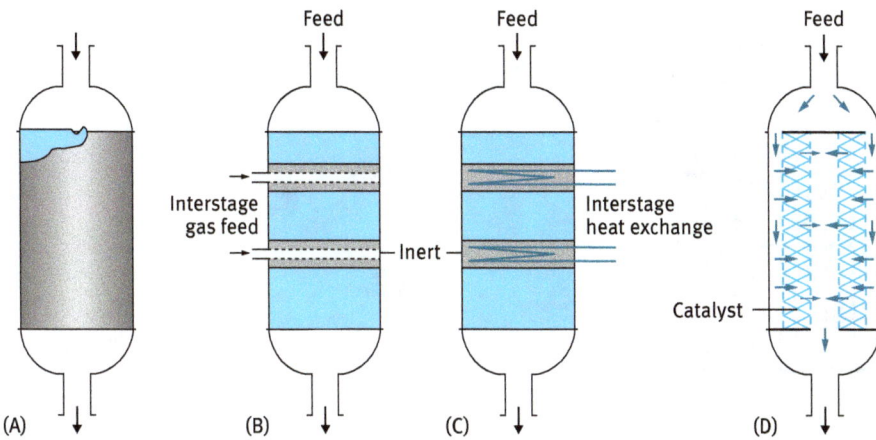

Fig. 3.58: (A) Single-bed adiabatic packed-bed reactor. (B) Adiabatic reactor with quenching. (C) Multibed adiabatic fixed-bed reactor with interstage heat exchange (from [51].) (D) Adiabatic reactor with radial flow.

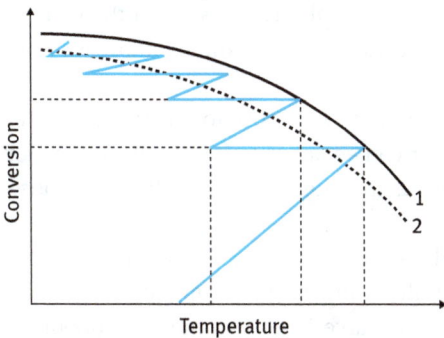

Fig. 3.59: Conversion vs temperature profiles in a multistage reactor compared with an optimum one reactor. 1, equilibrium; 2-optimum.

unfavorable mixing of hot and cold streams. As the composition is changing it can have an either positive or negative effect on the reaction kinetics and thermodynamics. Adiabatic multistage reactors with interstage heat transfer are typically used in reactions giving a single product, but limited by equilibrium (sulfur trioxide, ammonia, methanol syntheses) where intermediate cooling is used to displace the gas temperature in the direction of higher equilibrium conversion.

For strongly exothermic (partial oxidations and hydrogenations) or endothermic reactions (dehydrogenations) multitubular reactors with cooling or heating in between the tubes are applied. This allows for better temperature control (Fig. 3.60), although in this case the temperature profile is characterized by a maximum

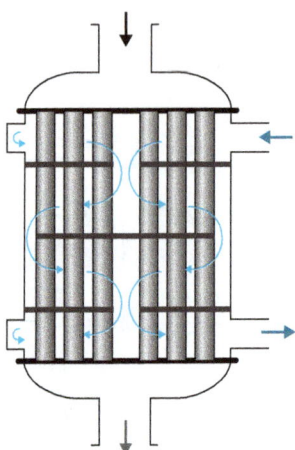

Fig. 3.60: A multitubular reactor. (From [51].).

(hot spot), which is related to enhanced heat release at the reactor inlet and a decline in heat release along the tube length caused by a diminished concentration of reagents. An important issue with a multitubular reactor is the uniform flow distribution among the tubes, which requires special measures when the catalyst is loaded in the reactor. The loading procedures will be separately addressed in a special section below in this Chapter.

The flow conditions in packed-bed reactors are very close to the plug flow (i.e., minor back-mixing), which implies a higher conversion rate for the most common reaction kinetics and better selectivity towards the intermediate products for consecutive and parallel-consecutive reactions.

The construction of the packed-bed reactors is simple, not requiring any moving parts, and therefore this is generally a low-cost, low-maintenance reactor. Among other advantages it is possible to have large variations in operating conditions (including high pressure) and contact times, little catalyst attrition and subsequent catalyst losses, high catalyst to reactant ratio and thus long residence time. In addition, principles of mathematical modeling of packed-bed reactors are well established. The most serious disadvantages are related to poor heat transfer in large fixed-bed reactors and thus the inability to control temperature properly. High-pressure drop is associated with limited productivity, while non-uniform flow patterns and broad-residence time-distribution may lead to poor selectivity. As mentioned above, the temperature control problems could be overcome by introducing internal and external heat exchanges, interstage cooling or heating and the application of diluents. The revamp of adiabatic axial flow operation to radial flow and the use of special geometrically shaped catalysts can improve pressure drop, simultaneously helping in removal pore diffusional problems.

In some cases, even making a hole in a catalyst pellet helps to improve the pressure drop. I was once involved in a case where the plant operator loaded the same amount of a tablet-ized catalyst as was used in a similar plant by the catalyst supplier, without taking into account different requirements for the maximum pressure allowed in these reactors (1.8 bar in this case and above 3 bar in the other plant). As the reactor at this plant was not de-signed to operate above 1.8 bar and the pressure drop was above 0.8 bar with the newly installed catalyst batch, the required productivity was not reached. The plant operator had to remove one-third of the catalyst load almost immediately and had to replace the catalyst charge after a while with the same size of tablets but with an additional hole, which solved the problem of the pressure drop.

Some examples of fixed-bed reactors will be discussed in Chapter 4. Regeneration and replacement of the catalyst are difficult and require shut-downs.

Uniform flow distribution could be achieved if the pressure drop is significantly high. It should be still ensured that the catalyst is packed uniformly in order to avoid bypassing. Such uniform flow distribution is more difficult to achieve if the pressure drop is low, i.e., in monolith reactors. The pressure drop in fixed-bed reac-tors depends very strongly on the void fraction, as follows from the Ergun equation, which is often applied for pressure drop calculations. For spherical particles the pressure drop Δp (Pa) per reactor length is

$$\Delta p/L = 150\eta \frac{(1-\varepsilon)^2}{\varepsilon^3 d_p^2} v_G + 1.75\rho \frac{(1-\varepsilon)}{\varepsilon^3 d_p} v_G^2 \qquad (3.255)$$

where v_G is the interparticle velocity (m/s), ε is the void fraction, m^3 of the free space/m^3 packing, η is dynamic viscosity, N s m^{-2}, and ρ is density, kg/m^3, L is the length, m, d_p the particle diameter, m. The first term in the Ergun equation corresponds to the laminar flow, while the second reflects the turbu-lent flow.

Recently, the author has received an e-mail, which had a following question: "We have bought your book "Engineering Catalysis" and teaching catalysis reaction engineering from it. I would like to know what is the maximum pressure drop that can be allowed while designing a fixed bed reactor for large scale industrial processes? Please excuse me for asking this very trivial question. I asked this because these issues are hardly mentioned in most of the undergraduate textbooks on reaction engineering." In my view, there is no general answer to this question as it depends on a particular case. Obviously, the pressure drop influences the reaction rates and the overall process economics, thus should be diminished still ensuring uniform flow distribu-tion. In some cases, the maximum allowed pressure drop could be say 0.5 bar, while for other cases it could be several bars. The reactor design including the pressure drop calculations should be viewed as a part of the overall (upstream and downstream) technology.

An important parameter during the reactor selection is the adiabatic temperature rise, which is defined as $\Delta T_{ad} = \Delta H_r C_o/c_p$ where ΔH_r is the heat of the reaction (kJ/mol), C_o

is the initial reactant concentration (mol m $^{-3}$), and c_p is the heat capacity (kJ/m^3 K). When $\Delta T_{ad} < 300$ K a multibed reactor can be used with heat exchanges or quenching in between the beds. If the adiabatic heat rise is $\Delta T_{ad} = 300 \div 700$ K a multitubular reactor is preferable. Sometimes two of these options are applied for the same reaction. Fig. 3.61 shows adiabatic and isothermal options for methanol synthesis reactor, offered by Uhde.

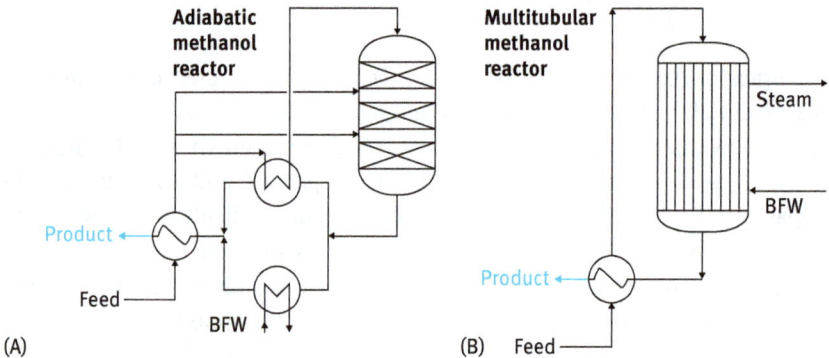

Fig. 3.61: (A) Adiabatic and (B) isothermal reactors for methanol synthesis. From [52].

The low-cost radial axial adiabatic multibed reactor with quenching is used in plants when extra steam is required in the synthesis unit. In the isothermal reactor the reaction heat is removed by partial evaporation of the boiler feed water (BFW) generating 1 metric ton of medium pressure steam per 1.4 metric ton of methanol. This isothermal reactor allows easier temperature control by regulating steam pressure, conditions close to isothermal, high heat recovery and low by-product formation.

An interesting option related to heat control is a reaction control with direct, regenerative heat exchange in the reactors with periodic flow reversal developed by Matros (Fig. 3.62). The catalyst packing acts simultaneously as the regenerative heat exchanger. The catalyst is initially pre-heated to the reaction temperature and then the cold reaction is put in contact with the hot catalyst packing. Thus, the reaction gas is heated by the catalyst. The front moves into the packing, however, before it reaches the outlet of the bed, the flow is reversed by special valves and then moves back again, heating the cooled part of the reactor. This operation ensures that a periodic steady- state is established. The hot zone is in the center of the packing, while both ends act as regenerative heat exchanger. Although such periodic flow operation was successfully tested in several medium-scale production plants for the synthesis of basic inorganic and organic chemicals, the main application

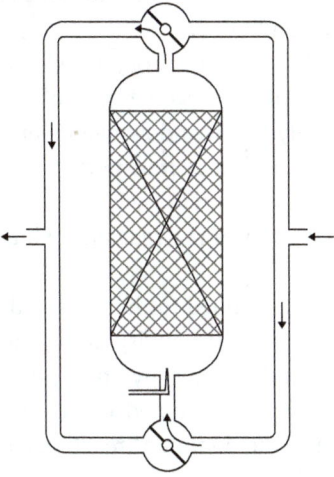

Fig. 3.62: Autothermal reaction control with direct (regenerative) heat exchange.

was found in catalytic off-gas purification, i.e., catalytic oxidation of volatile organic compounds (VOC) present in minute amounts in exhaust gases.

When the adiabatic temperature rise is above 700 K, fluidized bed reactors (Fig. 3.63A) are usually used. When heat exchanges are located directly in the fluidized bed of catalyst, it is possible to maintain the temperature at, for example, 300–400 °C, the order of magnitude is higher even in the adiabatic temperature

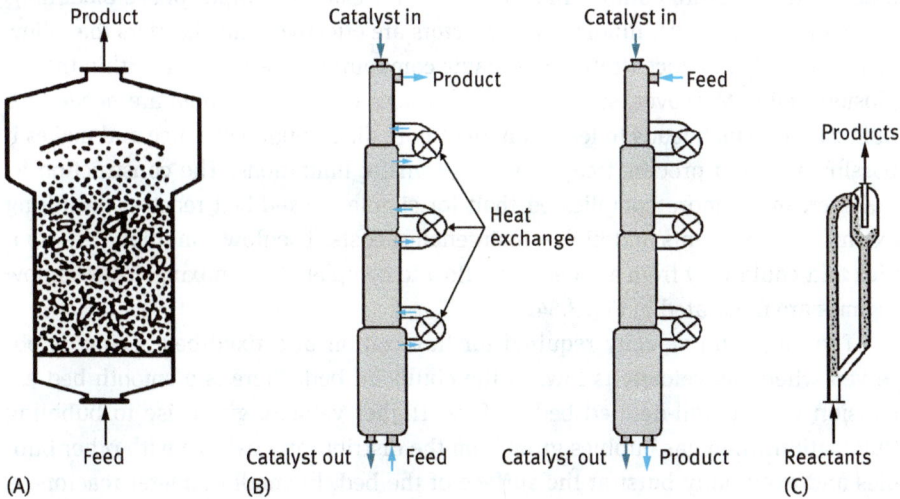

Fig. 3.63: Reactors with moving catalysts. (A) Fluidized bed reactor; (B) a moving-bed reactor (left, counter-current and right, concurrent flow of the feed and the catalyst); (C) entrained flow (riser) reactor.

rise. Because of intensive heat transfer the temperature in almost all parts of the fluidized bed reactor is the same, affording isothermal conditions.

When the catalyst undergoes strong deactivation the application of packed-bed reactors is cumbersome, as continuous addition, removal and regeneration of the catalyst is required. In the moving-bed reactor the catalyst moves downwards and the gas flows upwards (Fig. 3.63B, left). The removed catalyst is transported in a regenerator unit. For up-flow operation the catalyst, which is removed, is further deactivated because it is in contact with the feed most rich in the reactant, while less deactivated catalyst is in the reactor.

For the concurrent operation mode (Fig. 3.63B, right) applied for heavier feed-stock, as for example in catalytic reforming, uniform catalyst activity along the reactor is maintained. Typically, in such reactors the catalyst flows by gravity and the reaction mixture can be withdrawn from the reactor and cooled or heated. Heating is required in catalytic reforming, where one of the dominant reactions is dehydrogenation.

When deactivation is even more prominent and the catalyst life is just few seconds, an entrained flow or riser reactor is used when the catalyst powder is entrained by the catalyst in feed, thereafter being separated in a cyclone, regenerated (not shown in Fig. 3.63C) and sent back to the riser. A detailed discussion on the use of such reactors is given in a section devoted to fluidized catalytic cracking in Chapter 4.

The obvious advantages of the fluidized bed reactors are associated with the possibility to regenerate the catalyst continuously, high thermal efficiency and thus efficient temperature control. In some cases, for example, vapor phase oxidations of organic compounds, fluidized bed reactors are effective flame barriers that allow operation at high concentration of organic compound, possibly even within the explosion limits. Moreover, small catalyst particles (50–100 microns) are needed for fluidization, which leads to less prominent pore diffusional resistance and makes it possible to run a process free from mass transfer limitations. The construction is, however, much more complicated than for simple packed-bed reactors, requiring extensive investments and higher maintenance costs. The flow conditions are complex and could vary from an ideal plug flow to complete back-mixing. Various flow regimes are illustrated in Fig. 3.64.

The minimum velocity required for fluidization of a fixed-bed regime is observed when gas velocity is low. In the ebullated bed, there is a smooth bed expansion with a well-defined bed surface. Higher velocity gives rise to bubbling fluidization when gas bubbles grow from the distributor, coalesce with other bubbles and eventually burst at the surface of the bed. In small diameter reactors, a further size in fluidization velocity results in a slug flow regime with the bubble diameter approaching the reactor diameter. In a turbulent regime, the top surface of the bed cannot be distinguished. Transport of particles and their recycle back to the reactor is a feature of fast fluidization.

Fixed bed | Particulate fluidization | Bubbling regime | Slug flow | Turbulent regime | Fast fluidization

Aggregative   Fluidization

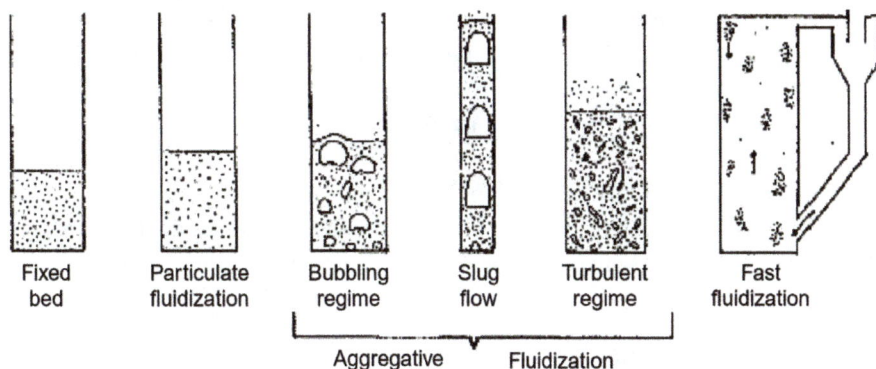

Fig. 3.64: Various flow regimes of fluidized bed reactors [53]. Reproduced with permission.

In three-phase fluidized beds, the solid volume fraction is much higher (10–50%) compared to slurry bubble columns (< 5%).

Scaling-up of fluidized bed reactors is challenging as the mass transfer characteristics of the fluidized bed are dependent on reactor dimensions. Moreover, some large gas bubbles are formed that could pass the reactor without being in contact with the catalyst, leading to lower conversion and requiring additional amounts of catalyst. The riser reactor (Fig. 3.63C), however, approaches the plug flow regime.

An example of a riser reactor usage will be discussed in detail in Chapter 4 in relation to an oil refining process – catalytic cracking. Some other examples of fluidized bed reactors usage in synthesis of basic organic chemicals could be mentioned, such as oxychlorination of ethylene (Section 4.8) and synthesis of acrylonitrile (Section 4.7.5.2). In the synthesis of acrylonitrile a single plant capacity can reach 100,000 tons per year. Oxidative ammonolysis of propene to acrylonitrile:

$$2CH_2 = CHCH_3 + 3O_2 + 2NH_3 \rightarrow 2CH_2 = CHCN + 6H_2O$$

occurs at 430–480 °C, pressures 1.3–1.5 bar, and contact times of a few seconds.

Serious disadvantages of the fluidized bed reactors are attrition, loss of catalyst, small variations in the residence times, which are typically low. Attrition of already small particles produces fines, which are expensive to separate from the product gas. It is, however, possible in principle to combine the catalyst with an attrition-resistant binder or coat the catalyst with some attrition-resistant material.

In order to achieve the fluidization regime for catalysts with the size d the velocity at which fluidization starts U_b is defined as:

$$U_b = Re_b \left(\frac{v}{d_p} \right) \tag{3.256}$$

where v is viscosity and d_p is the diameter of particles, while the Reynolds number $Re_b = Ar/(1400 + 5.22\sqrt{Ar})$ is related to the Archimedes number Ar:

$$Ar = \frac{gd_p}{v}\frac{\rho_p - \rho_s}{\rho_s} \tag{3.257}$$

Here ρ_p is the density of particles, ρ_p is the density of the reaction mixture, and $g = 9.8\ ms^{-2}$. The velocity at which carry over starts, $U_{c'}$ can be evaluated from

$$U_b = Re_b\left(\frac{v}{d_p}\right), Re_b = \frac{Ar}{(18 + 0.61\sqrt{Ar})} \tag{3.258}$$

The operational velocities are thus in between these values. For small catalyst particles of ~60 μm the ratio between U_c and U_b is ~50, while for some large particles (100 μm) it can be around 10. Because for real industrial catalysts there is a certain particle size distribution, the values of velocities at which the fluidized bed should operate are also confirmed experimentally.

As mentioned above, the formation of large gas bubbles (laminar fluidized bed) is undesirable because it decreases the interphase mass transfer. In order to achieve a desired turbulent regime the gas-flow rate should be sufficiently high, with Reynolds numbers of 100 ± 200. The catalyst particle must have sufficiently low density. Otherwise a transition from the bubble fluidized bed regime to the riser transport regime (carry over) will occur above Ar > 100 with an increase of the gas velocity.

3.6.4 Three-phase catalytic reactors

Catalytic three-phase (gas-liquid-solid) reactors are extensively used in industry in oil refining, synthesis of petrochemicals, and various basic, specialty and fine chemicals; these processes require different types of reactors. In addition to reactors with fixed catalyst beds using up-flow and downflow of liquids and well as co- and counter-current flows of gases and liquid, slurry reactors (bubble columns and stirred tanks reactors, reactors with external loops) and fluidized bed reactors are also applied (Fig. 3.65).

In some cases (for example hydrogenation of benzene or methanol synthesis) both gas-solid and gas-liquid-solid operations are possible. In general, in comparison with gas-solid reactors, three-phase reactors operate at low temperatures, which is translated into energy savings, also preventing potential damage to thermally unstable catalysts and support. In addition to better temperature control and more design options, low operation temperature usually also affords better

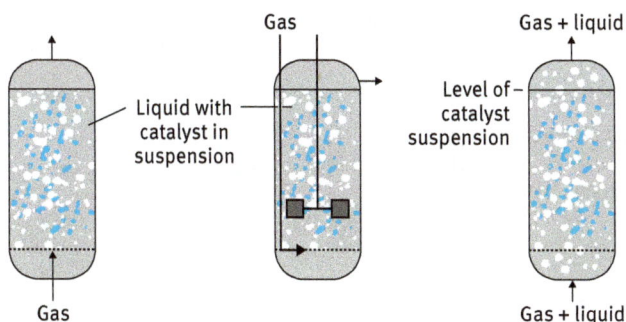

Fig. 3.65: Typical three-phase reactors (adapted from [29]). Left, bubble column; middle, stirred tank; right, fluidized bed.

selectivity and higher catalysts effectiveness. Conversely, the introduction of another phase boundary obviously results in an increase of mass transfer, while low temperature leads to a decrease in the chemical reaction rate. Moreover, batch slurry reactors are designed to operate in a limited pressure and (more importantly) a limited temperature range, which imposes some difficulties for catalyst reduction and in situ activation where they are needed. In addition, in industry the power input for impellers is limited and the same holds for the minimum catalyst particle size, which is defined by available separation systems, and thus the mass transfer cannot be neglected. In general, with slurry bed reactors the efficient separation of catalysts, which are either suspended in the slurry by stirring or bubbling gas up through the liquid, is difficult.

3.6.4.1 Fixed-bed multiphase reactors

Packed-bed reactors can operate in different modes with the downflow and upflow of the liquid and gas, and the gas can flow either co- or counter-currently. There are certain advantages for the liquid in operating either up or down flow. For example, the downflow option in fixed-bed multiphase reactors results in lower-energy input, lower pressure drop and an absence of fluidization, while upflow operation leads to better liquid distribution, mixing, heat dissipation, surface wetting, easier heat exchange, higher gas/ liquid mass transfer and effectiveness, as well as slow aging.

In the downflow operation mode the liquid moves in a laminar flow downwards, wetting the catalyst in a trickling flow, thus the name "trickle bed reactors" (Fig. 3.66). In these trickle bed reactors both the gas and liquid approach a plug flow condition.

The flow map of different flow regimes observed in trickle bed reactors is illustrated in Fig. 3.67.

Liquid
← Gas

Liquid
Gas →

Inlet distribution tray
Ceramic balls

Gas + liquid

Gas
Liquid

Solid catalyst

Catalyst support
Outlet collector

Gas

Gas + liquid
(A)

Liquid ←
(B)

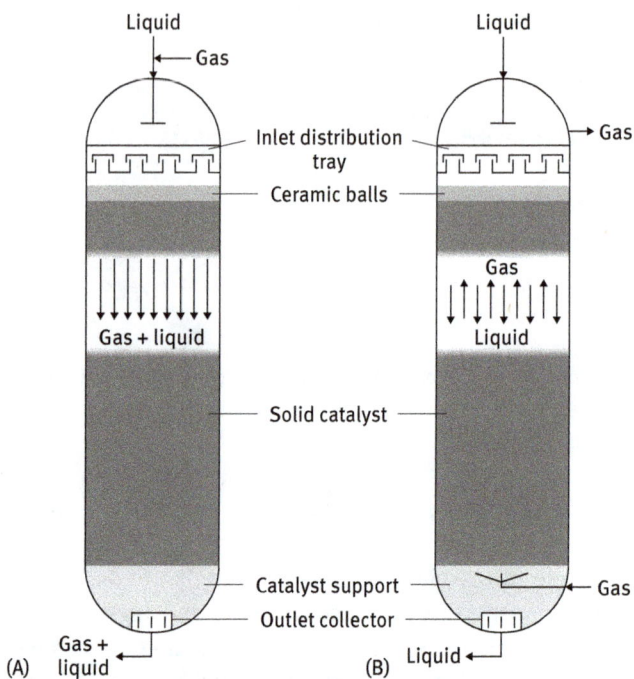

Fig. 3.66: Trickle bed reactors: (A) downward cocurrent, (B) counter-current gas flow. (Reproduced with permission from [54]).

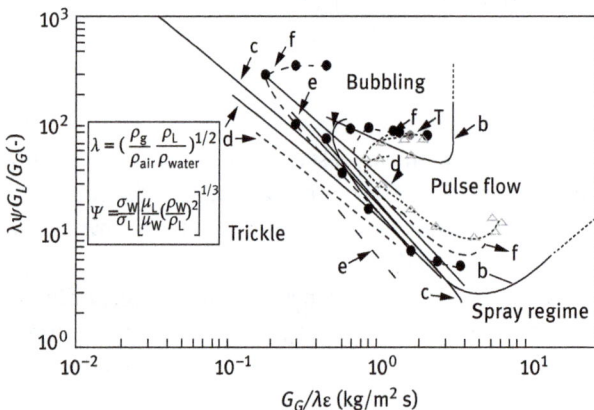

$$\lambda = \left(\frac{\rho_g}{\rho_{air}}\frac{\rho_L}{\rho_{water}}\right)^{1/2}$$

$$\Psi = \frac{\sigma_w}{\sigma_L}\left[\frac{\mu_L}{\mu_w}\left(\frac{\rho_w}{\rho_L}\right)^2\right]^{1/3}$$

Bubbling

Pulse flow

Trickle

Spray regime

$\lambda\psi G_L/G_G\,(\text{-})$

$G_G/\lambda\varepsilon\;(\text{kg/m}^2\,\text{s})$

Fig. 3.67: Flow regime map in fixed-bed reactors with co-current gas–liquid flow [55].

The trickle flow regime is observed at low gas and liquid flow rates, with such features as incomplete wetting, low pressure drop, low gas–liquid throughput, less catalyst attrition and suitability for foaming liquids. Heat and mass transfer rates are poor compared to other flow regimes in trickle bed reactors. Pulse flow regime is observed at moderate gas and liquid flow rates. The liquid phase occupies the entire flow cross section. Such regime has better utilization of the catalyst bed in terms of wetting, as well as higher heat and mass transfer rates, which makes the pulse flow regime potentially attractive for industrial application. In the bubble flow regime at a low gas flow rate and moderate/high liquid flow rates, the volume of the bed is occupied by the liquid with the gas phase flowing downward in the form of bubbles. In this regime, the pressure drip is higher. Moreover, higher liquid holdup leads to back-mixing, which depending on the reaction might not be suitable. When the liquid phase is a limiting component or the reaction is highly exothermic, complete wetting of the bed and high heat and mass transfer rates can be beneficial. The spray flow regime is observed at low liquid and high gas flow rates. In this regime, droplets of the liquid are formed, while the gas phase is a continuous phase. Subsequently this regime with a high gas–liquid mass transfer has a low liquid holdup and low foaming ability.

Trickle bed reactors usually operate in two regimes. At low gas and liquid flows [even as low as $0.1 \text{ kg m}^{-2} \text{ s}^{-1}$ in bench scale reactors and $0.6 \text{ kg m}^{-2} \text{ s}^{-1}$ in commercial scale reactors] a trickle flow dominates, while at low gas and high liquid flows the liquid-phase is continuous and gas bubbles flow through in a pulsed flow regime.

In such a trickling flow, maldistribution and incomplete wetting (Fig. 3.68) and losses caused by attrition counterbalance the advantages, while up-flow operation requires a larger energy input. In order to avoid stagnant regions and excessive catalyst loadings, more complicated geometrical shapes are preferable for trickle bed operations.

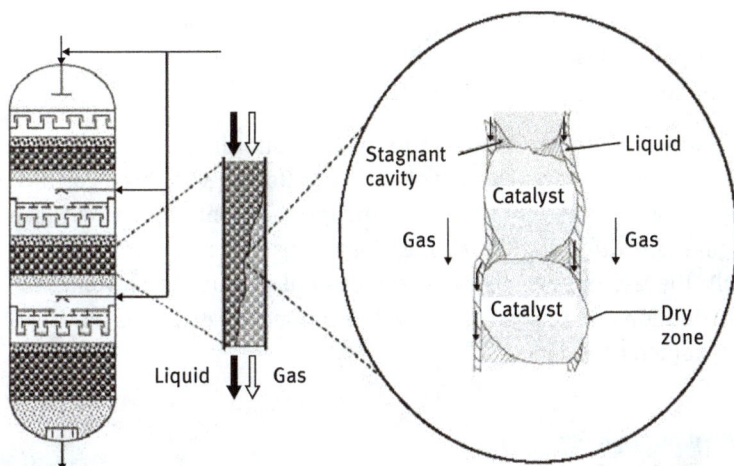

Fig. 3.68: Trickle bed reactors with maldistribution. (Reproduced with permission from [56]).

For up-flow reactors, in addition to the bubble flow (low gas and high liquid flows) and spray flow (higher gas and low liquid flows), a slug flow with uneven distribution of bubbles (high gas and low liquid flows) can also be developed, which gives in the latter case a plug flow for the gas-phase, but partially back-mixed liquid.

In summary, trickle bed reactors offer a simple structure, low investment costs and closeness to plug flow. It should be mentioned that the ideal plug flow is the one that is usually desired.

Catalytic monolith reactors, which were typically applied for two-phase systems (gas abatement, automotive exhaust cleaning) with high gas velocity (as discussed in Chapter 2), were proposed as an alternative to fixed-bed three-phase reactors (Fig. 3.69).

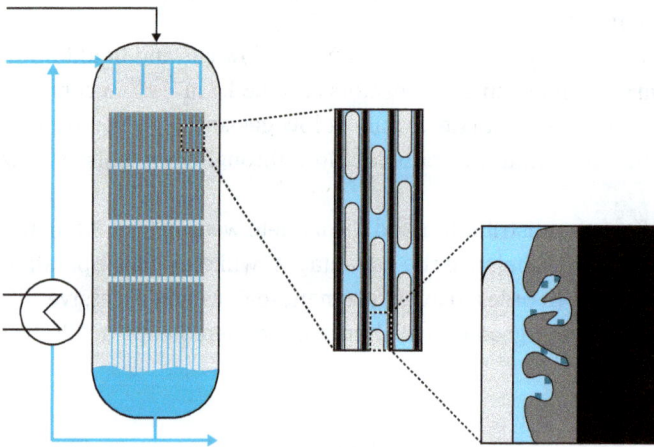

Fig. 3.69: Monolithic reactor for three-phase system. (From [23].).

At high gas-flow rates the bubbles fill the cross section, affording very good mass transfer. Special care should be taken to avoid annular flow at very high gas-flow rates when the gas flow becomes continuous. Although the monolithic reactor can allow very low pressure drop and negligible mass transfer as the wash-coating thickness is small, the use of such reactors in industrial practice is prevented by uneven liquid distribution and uneven metal profiling along the channels plus high catalyst costs and limited experience.

3.6.4.2 Moving-bed multiphase reactors

In stirred tanks, cascades of stirred tanks and bubble columns, the back-mixing is significant, broadening the residence time-distribution and making it more difficult to achieve high yields. There are, however, several advantages of the reactors with the moving catalysts beds, justifying their frequent application in industry.

Bubble columns (Fig. 3.70) often operating in a semibatch mode, i.e., continuously for gas and as batch for liquid, provide good mixing and heat transfer characteristics. Gas is distributed from the bottom of the reactor as bubbles, whose size depends on physical–chemical properties, sparger design and superficial gas velocity. The distribution of gas bubbles in radial direction is nonuniform, assisting in internal circulation within the bubble column driven by buoyancy (an upward force exerted by a fluid). Overall the back-mixing for the liquid and the solid catalyst particles is more intensive, thus the bubble columns operate as isothermal reactors.

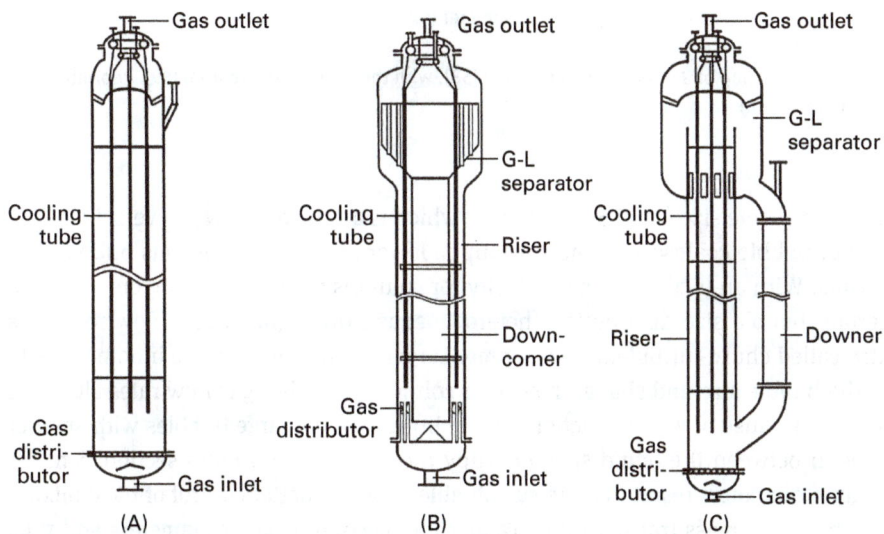

Fig. 3.70: Moving-bed multiphase reactors: (A) Bubble column, (B) Internal loop airlift reactor, (C) External loop airlift reactor. (Reprinted with permission from [57]).

Mixing can be enhanced by introducing gas (air) lift, ejectors and circulation pumps. The airlift reactor consists of risers and downcomers that provide the flowing channel for global circulation of the liquid-solid slurry phase (Fig. 3.70B,C).

Bubble columns exhibit very complex hydrodynamics due to spatiotemporal variations in interactions among gas, liquid and solid phases (Fig. 3.71).

Two major flow regimes exist in these reactors: the homogeneous and the heterogeneous ones. A homogeneous bubble flow regime is developed when bubbles

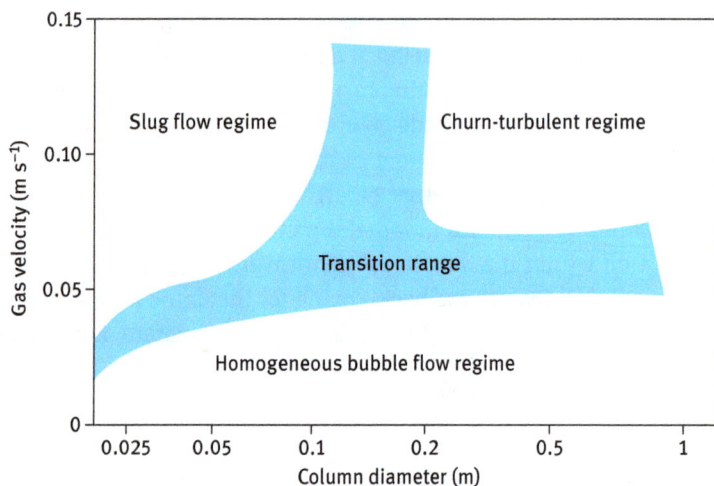

Fig. 3.71: Flow map of a bubble column. (From [58], with the original figure from [59]. Reprinted with permission).

are of the same size (ca. 0.5 to 2–3 mm), which is attained at low gas velocities. The rate of bubble coalescence and breakup is lower in the homogeneous bubble flow regime. With an increase of gas velocity for columns with large diameters there is a certain bubble size distribution (heterogeneous flow regime). This flow regime is also called churn–turbulent flow regime. Larger gas flow gives more inhomogeneity in the bubble size and shape. In smaller columns at higher gas flow rates slugs are formed because of wall restriction. The bubbles (mainly large bubbles with smaller ones in between the liquid slugs) occupy the entire reactor cross section. Smaller column diameters result in more sustainable slug and better control of back-mixing and heat and mass transfer rates. Taking into account smaller diameters and thus lower throughput, it is not surprising that these miniaturized bubble column reactors can have advantages compared to other reactor options in production of fine and specialty chemicals.

Mechanically agitated stirred tanks (with complete back-mixing profiles) can be used in batch and semi-batch (continuous for the gas) modes, as well as CSTR with a possibility of catalyst recycling. In CSTR (Fig. 3.72A) the mixing of the incoming reactant is instantaneous, thus the concentration in the reactor is the same as the concentration at the outlet. Such high degree of back-mixing is typically unfavorable for kinetics, resulting in a lower conversion of the reactants. Therefore, cascades of CSTR are used in industrial practice (Fig. 3.72B).

Various options for temperature control in slurry reactors will be discussed in Section 4.1.4. As already mentioned, the separation of fine catalyst particles can be

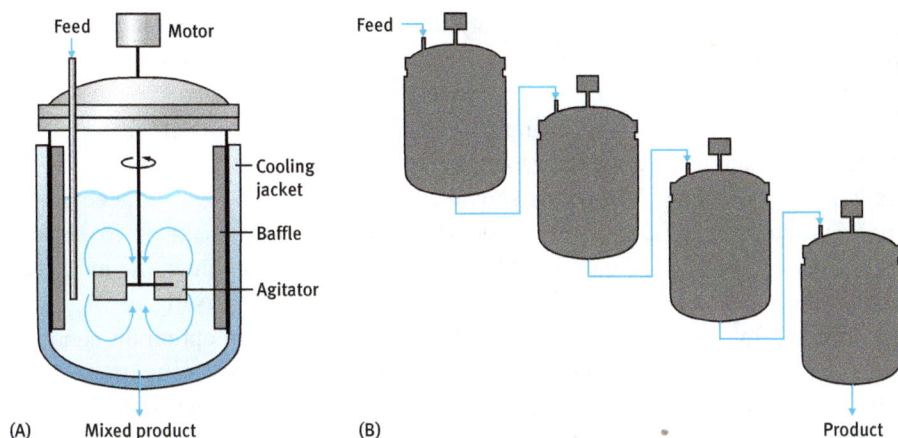

Fig. 3.72: Stirred tank reactor. (A) CSTR [60]. (B) Cascade of stirred tanks.

challenging, therefore somewhat larger particles are used in industrial practice with non-negligible diffusion resistance.

> Many years ago I had a discussion with a plant manager and suggested, based on experimental data in a laboratory reactor, that in order to avoid mass transfer limitations they should use catalyst particles below 50 µm for hydrogenation of an aromatic compound. Unfortunately, the filters available at the plant site could only handle catalyst particles that were several times larger, leading unavoidably to the presence of internal diffusion.

The fluid flow in stirred tanks is highly complex and requires computational fluid dynamics to get an exact solution of the partial differential equations associated with the Navier-Stokes momentum, energy and continuity equations. Tangential, axial and radial flows exist in stirred tanks depending on the impeller used.

Application of baffles (three or four flat plates attached to the walls), occupying up to 10% of the tank diameter, can stop the tangential flow. Such baffles cannot be recommended when the flow is laminar. Axial and radial flow impellers could be used to generate flow parallel or perpendicular to the axis (shaft), respectively. Pitched bladed turbines create a mixed flow. Axial flow impellers are recommended for turbulent mixing. Although variations in the size of catalyst particles mean that it is not possible to get suspensions that are completely uniform, the impellers are usually placed at one-quarter of the liquid level to achieve acceptable mixing. In order to increase the gas-liquid mass transfer, the gas bubbles should be dispersed by the impeller. It should be also remembered that because of the gas dispersed in the liquid, the volume of the liquid is increased (gas hold-up). Heat transfer is typically unaffected by mixing and could be improved by increasing heat transfer area (for example, installing coils in addition to a heat transfer jacket).

Industrial reactors cannot operate at the same level of mixing efficiency as the small scale reactors, thus simple geometrical similarities are insufficient in scaling-up, and location of the feed points should be carefully considered. Some engineering consultants even advise for process development to rather scale down and use laboratory reactors based on available large-scale reactors to mimic the industrial conditions [61].

Finally, fluidized bed reactors should be highlighted as an option for conducting three-phase reactions. Such reactors typically operate in a concurrent mode with up-flow of both gas and liquid. Conditions of high gas and low liquid flow (aggregative fluidization) resemble two-phase (gas-solid) fluidized beds, while at low gas flows the flow pattern (bubble flow) is obviously similar to fluidized beds with only liquid- and solid-phase. A slug domain with uneven distribution of particles is formed in between these two extremes.

Compared to bubble columns, larger particles are usually used in fluidized beds because of higher liquid flow velocities. The flow pattern is more of back-mixing type than plug flow.

3.6.4.3 Comparison between various reactor options

A general comparison between catalytic three-phase reactors was done by Trambouze et al. [62] and is presented in Tab. 3.8.

Tab. 3.8: Comparison between three-phase reactors.

Application criteria	Suspension catalysts	Three-phase fluidized beds	Fixed bed
Catalysts related			
Activity	Highly variable, but possible in many cases to avoid the diffusion limitations found in a fixed bed	Highly variable: intra- and extra-granular mass transfers may significantly reduce the activity, especially in a fixed bed Back-mixing unfavorable	Plug flow favorable
Selectivity	Selectivity generally unaffected by transfers	As for activity, transfers may decrease selectivity Back-mixing often unfavorable	Plug flow often favorable

Tab. 3.8 (continued)

Application criteria	Suspension catalysts	Three-phase fluidized beds	Fixed bed
Stability	Catalyst replacement between each batch operation helps to overcome problems of rapid poisoning in certain cases	Possibility of continuous catalyst renewal: the catalyst must nevertheless have good attrition resistance	This feature is essential for fixed-bed operation: a plug flow may sometimes be favorable because of the establishment of a poison adsorption front
Cost	Consumption usually depends on the impurities contained in the feed and acting as poisons		Necessarily low catalyst consumption
Technologies characteristics			
Heat exchange	Fairly easy to achieve heat exchange	Possibility of heat exchange in the reactor itself	Generally adiabatic operation
Design difficulties	Catalyst separation sometimes difficult: possible problems in pumps and exchangers because of the risks of deposit or erosion		Very simple technology for a downward concurrent adiabatic bed
Scaling-up	No difficulty: generally limited to batch systems and relatively small sizes	System still poorly known, should be scaled up in steps	Large reactors can be built if liquid distribution is carefully arranged

(From [62].)

3.6.3.4.1 Fischer–Tropsch synthesis

It is instructive to compare different reactors mentioned in Tab. 3.8 for a particular case of synthesis gas (mixture of hydrogen and CO with a stoichiometric ratio of 2:1) transformation through the Fischer–Tropsch (FT) synthesis (this section) and to methanol (section 3.6.3.4.2). In the former case the synthesis gas is converted to a wide range of hydrocarbons in a highly exothermic reaction:

$$nCO + 2nH_2 \rightarrow (-CH_2)_n + nH_2O \quad \Delta H = -165 \text{ kJ/mole (at } 227\,^\circ C)$$

The molar hydrogen to CO consumption ratio is 2:1. The inlet ratio of H_2 to CO in industrial reactors is different from the stoichiometry; thus, for example, in the fixed-bed FT reactor the H_2/CO ratio is changing along the reactor length.

WGS (Water-gas shift) reaction

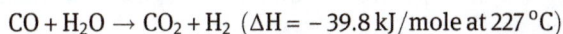

$$CO + H_2O \rightarrow CO_2 + H_2 \quad (\Delta H = -39.8 \text{ kJ/mole at } 227\,^\circ C)$$

is significant at low-temperature FT (LTFT) synthesis (ca. 230 °C) giving an overall stoichiometry

$$2CO + H_2 \rightarrow (-CH_2)_n + H_2 \ (\Delta H = -204\,kJ/mole \text{ at } 227\,°C).$$

Subsequently, the hydrogen to CO consumption ratio is only 0.5. For high-temperature FT (HTFT) synthesis the WGS reaction is less prominent.

Conditions of the FT process can be adjusted to the feedstock. When synthesis gas is produced by coal gasification at high (> 1,000 °C) temperature, the generated gas is rich in CO. To adjust the ratio between hydrogen and CO it is desirable to have WGS activity, which is provided by iron catalysts. When the feed, such as syngas from methane, is rich in hydrogen, no WGS activity is needed, and cobalt catalysts active at low temperature can be used with a danger of promoting an undesired, in this case, WGS reaction.

Utilization of iron catalysts at higher temperatures depending on the amount of alkaline promotes can cause formation of carbon deposits and subsequent deactivation through direct coke formation $H_2 + CO \rightarrow C + H_2O$ and the Boudouard reaction $2CO \rightarrow C + CO_2$. Another potentially significant reaction is methanation $CO+3H_2 \rightarrow CH_4 + H_2O$, which is thermodynamically more favorable than formation of higher hydrocarbons. Consequently, applied catalysts should be selective toward formation of less thermodynamically preferred products.

An overview of FT technologies and reactor types used is given in Fig. 3.73.

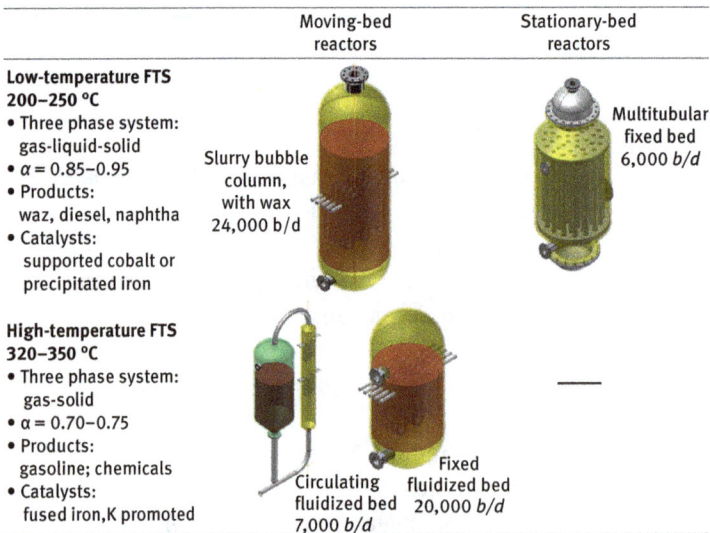

Fig. 3.73: Overview of Fischer–Tropsch technologies and reactor types used [63]. Capacities are in barrels per day (b/d).

More expensive cobalt-based catalysts that operate at lower temperature (200–250 °C) favor long-chain paraffins, are more robust and have low WGS activity contrary to a cheaper alternative, iron. Iron needs higher temperature (220–350 °C or higher for fluidized beds), possesses high selectivity to olefins and oxygenates, is a WGS catalyst and readily deactivates.

FT process was originally developed for the production of synthetic fuels from coal. The first FT plants began operation in Germany in 1938. There were nine low-temperature cobalt-based plants with a total annual product capacity of 660,000 tons eventually closed down after the Second World War. In fact, coal-to-liquid (CTL) had limited applications for many decades after WWII outside of the Sasol company in South Africa because of a number of political and technical reasons. A general overview of the process flow with coal as a feedstock is given in Fig. 3.74.

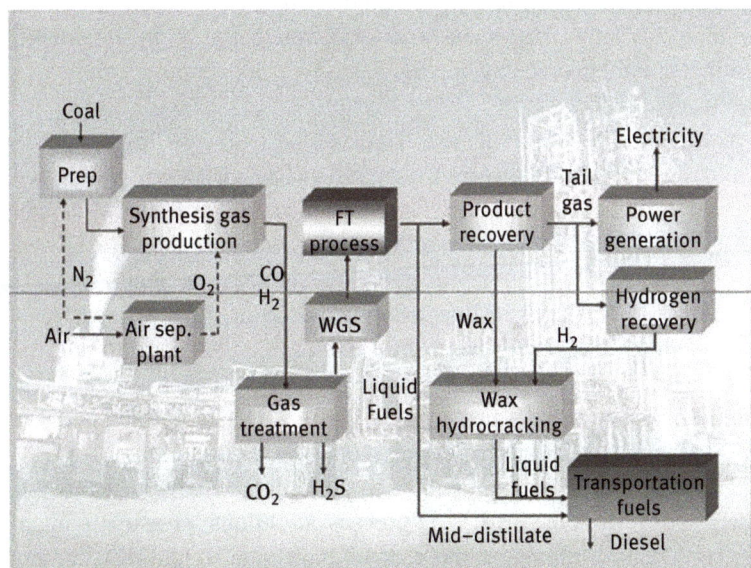

Fig. 3.74: A general overview of the process flow [64].

The first Sasol plant in Sasolburg, South Africa, started operation in 1953 and had annual production of million tons of FT products using coal as a feedstock and operating five tubular fixed-bed (ARGE) reactors for wax production and three circulating fluid bed reactors. A slurry reactor of the same production capacity replaced five ARGE reactors in 1993. In 2004 natural gas reforming was introduced instead of coal gasification at Sasol by transforming the plant technology from CTL to GTL producing waxes and chemicals.

Further expansions by Sasol in Secunda, South Africa, were done in 1980s, utilizing high-temperature Synthol reactors with improved heat exchange thereby boosting threefold the capacity compared to the first-generation circulating fluidized bed (CFB) reactors. The main focus of the production site of Sasol in Secunda is on motor gasoline and diesel, as well as some chemicals. High-pressure distillate hydrogenation section was also added to the tar refinery of Sasol III processing gasification derived coal pyrolysis liquids from Sasol II and Sasol III. Sasol Advanced Synthol reactors eventually replaced 16 second-generation CFB reactors with eight fixed fluid bed (FFB) reactors decreasing the operation costs at the same capacity.

In the early 1990s, Mossgas started up a natural gas based 1 million ton per year FT plant in South Africa using a high-temperature process with an iron catalyst for making motor gasoline, distillates, kerosene, alcohols and LPG, while Shell put onstream 500,000 tons/year natural gas-based FT plant using the Shell Middle Distillate Synthesis (SMDS) process for automotive fuels, specialty chemicals and waxes. The strategy of the LTFT synthesis is to produce heavier products with a cobalt catalyst, when formation of long-chain waxes is favored. The heavy alkanes are converted through mild hydrocracking to the desired carbon number range with subsequent product distillation.

Following this trend of renewed interest in using the FT process for the synthesis of gasoline and diesel from primarily natural gas, the Sasol Oryx GTL plant operating in Qatar using a cobalt catalyst at low temperature was commissioned in 2007 producing 34,000 bpd of diesel fuel mainly and naphtha as by-product. The FT syncrude is similar to SMDS process and is processed in a similar way. Shell Pearl GTL plant put onstream in 2011 in the same location in Qatar with a capacity of 140,000 bpd of petroleum liquids relies on the same LTFT technology to produce distillate and base oils.

Worth mentioning is the world's first coal to diesel conversion production line of Shenhua Group with the diesel production capacity of 860,000 tons in 2017.

As mentioned earlier, high exothermicity is a typical feature of the FT process. The choice of the catalyst (iron or cobalt) and process conditions (pressure, temperature, hydrogen to CO ratio) influences the product molecular weight distribution. The products in FT reactions are mainly n-paraffins, although terminal olefins and alcohols could also be formed.

A well-defined range of products, for example middle distillates, cannot be directly synthesized; thus, the strategy of newer and more economic plants is to generate high-molecular-weight linear waxes for further hydroprocessing.

The main reaction in FT synthesis follows a polymerization-like mechanism when a monomer CH_x species ($x = 1$–2) is added stepwise to a growing aliphatic chain (Fig. 3.75A).

Chain termination by desorption of unsaturated surface species and hydrogenation with subsequent desorption of saturated species relative to chain propagation determines the process selectivity. The weight fraction of a product with a carbon

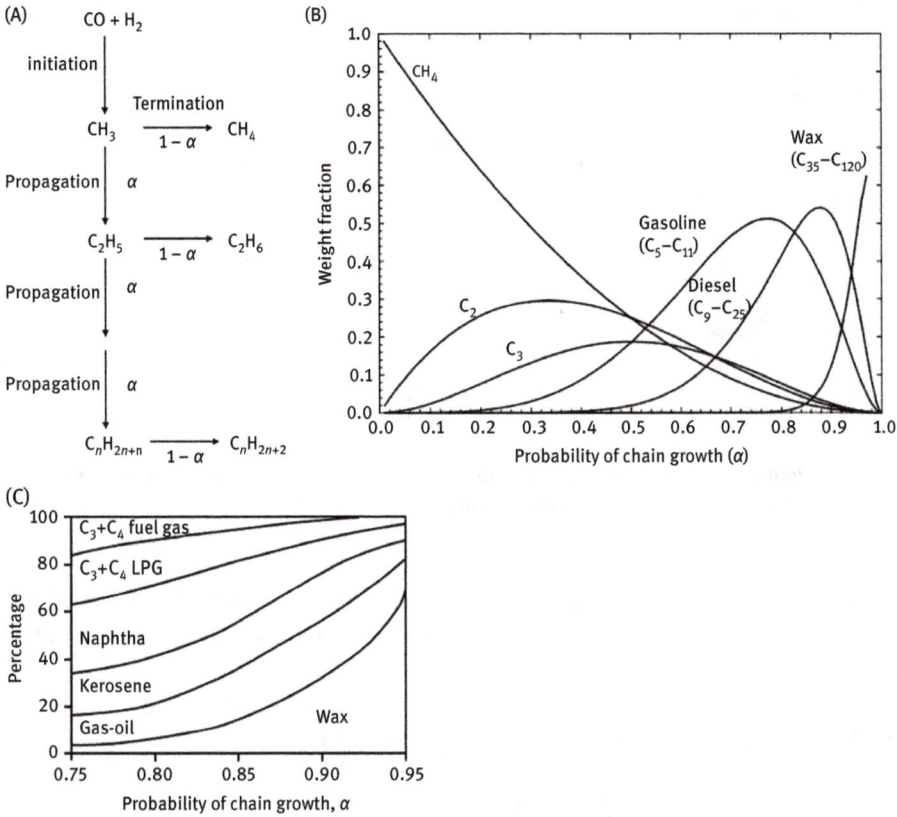

Fig. 3.75: Illustration of chain growth, weight fraction of hydrocarbons and percentage of different hydrocarbon products as a function of chain growth.

number n is defined through the Anderson–Schulz–Flory distribution $W_n = n\,\alpha^{n-1}$ $(1-\alpha)^{2n}$ (Fig. 3.75B), where parameter α is the chain growth probability being the ratio of the chain propagation to the sum of chain propagation and chain termination. This parameter is supposed to be independent on the carbon number.

As already mentioned, the product distribution depends on operating conditions (temperature, pressure and CO-to-hydrogen ratio) as well as the catalyst type. Cobalt gives more paraffins, while iron results in the product higher in olefins and oxygenates (Fig. 3.76).

Typically, the degree of polymerization α ranges from 0.7 to 0.95. Even with $\alpha = 0.95$, a range of different products is generated with predominant formation of high-molecular-weight linear waxes (Fig. 3.75C).

Low temperature (200–240 °C) and mid-pressure (2–3 MPa) are selected for FT process along with active catalysts based on iron and cobalt to get high selectivity to heavier products. Alternative utilization of nickel as a catalyst results in mainly

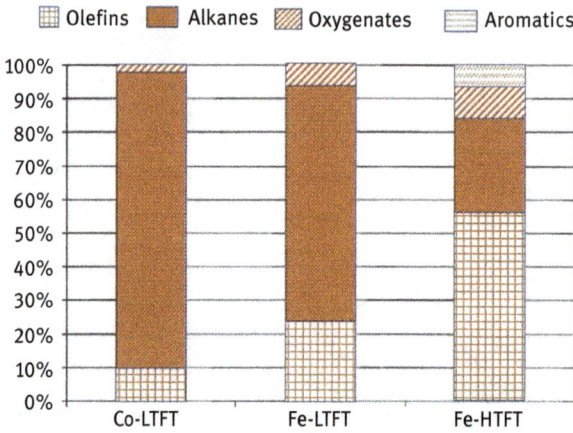

Fig. 3.76: Product distribution depending on the catalyst. Co-LTFT: cobalt low-temperature FT; Fe-LTFT: iron low-temperature FT; Fe-HTFT: iron high-temperature FT.

methanation. In FT synthesis, essentially no aromatic compounds are formed except for high-temperature processes (Fig. 3.76). The product is also free from sulfur and nitrogen compounds.

Cobalt-based catalysts are generally preferred for natural gas-based syngas giving FT stoichiometric H_2/CO ratio or close to it (Fig. 3.77).

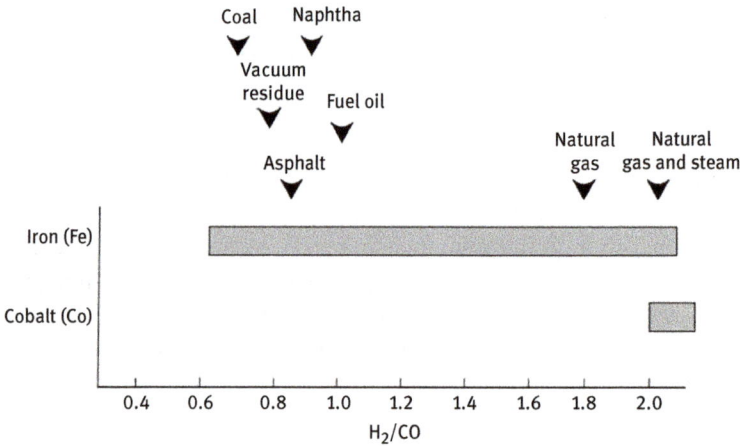

Fig. 3.77: Ratios of H_2/CO ratio for different feedstocks and catalysts [65].

Metallic cobalt which is considered to be the active phase in FT catalysts has low WGS activity. Earlier, Co catalysts for fixed-bed FT were prepared by co-precipitation, followed by shaping with extrusion, calcination and reduction. The novel generation applied in slurry-phase reactors is mainly synthesized by impregnation of oxides with aqueous or organic solutions of cobalt nitrates and other additives. Calcination of Co nitrates results in formation of Co oxide, which is reduced in a hydrogen containing gas.

During catalysis, Co metal crystallites are largely covered by active and inactive carbonaceous species. Co catalysts are more expensive than iron-based ones, at the same time possessing 10–20 times higher activity (calculated per weight for promoted Co vs promoted Fe catalysts), high selectivity to long-chain paraffins (C5+) and low selectivity to olefins and oxygenates, being also resistant to deactivation. The metal loading is typically 35 wt% with the metal dispersion of approximately 8–10%. A range of metal promoters (0.1–0.3 wt% Pt, Re or Ru) is used to increase reducibility and dispersion of Co, improve stability against carbon buildup and increase C_{5+} selectivity. Oxide promoters (i.e. 1–3% BaO or La_2O_3 or other additives) are used to stabilize cobalt crystallites and the support and promote hydrocarbon chain growth. As a support δ-alumina (ca. 150 m^2/g) stabilized with lantana is used. The support should be chemically and physically stable during catalyst preparation, activation, regeneration and reaction. The support can be stabilized with some other oxides. Besides alumina, supports such as silica and titania can be applied.

In comparison with cobalt, Fe-based catalysts are less expensive, display low selectivity to long-chain paraffins, generate more olefins and oxygenates, exhibit WGS activity and more prone to deactivation by coking.

Fe-based catalysts were used commercially in SASOL plants. They are generally preferred for coal-based plants with lower hydrogen to CO ratio. The active phase is Fe carbides (FexC, $x < 2.5$) covered by various carbonaceous species.

Fused iron oxide catalysts, for which alkali promotes are added during fusion at high temperature to improve activity and selectivity (Fig. 3.78), were applied in the HTFT synthesis using the fluidized bed (Sasol Advanced Synthol) reactors. Fusion is followed by casting into ingots and cooling, milling and activation in hydrogen.

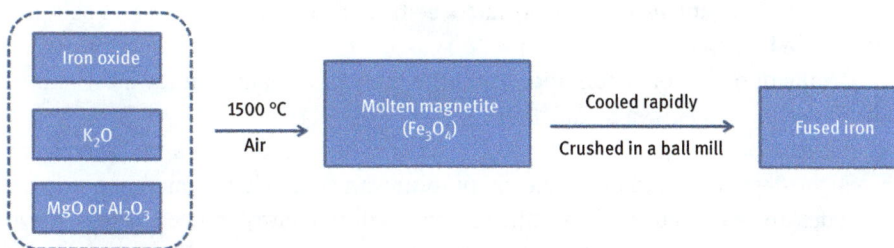

Fig. 3.78: Procedure for synthesis of fused iron oxide catalysts.

Precipitated iron oxide with addition of promoters (additives and modifiers) during precipitation is more expensive to prepare and has less structural strength. Such catalysts can be used in LTFT synthesis organized in trickle bed (ARGE) and slurry reactors. The catalysts are prone to deactivation and gradual loss of activity with a possibility for regeneration. Copper is added to the iron catalyst in order to increase the rate of iron reduction and the catalyst activity. Selectivity toward longer hydrocarbons (waxes) is improved through addition of alkali (K_2O). Silica is mainly used as a support for precipitated iron catalyst affording the highest activity and selectivity to waxes.

The catalyst is prepared by coprecipitation of iron and copper nitrates, followed by washing and impregnation with the potassium and structural promoters. Shaping of catalysts, done either by extrusion for fixed-bed or by spray drying for slurry-phase reactors, is followed by calcination. Prior to application, the oxidized iron catalysts are activated in hydrogen or synthesis gas.

Catalyst deactivation in FT synthesis occurs through poisoning of the active metal by sulfur or nitrogen compounds. This can be prevented by desulfurization of the syngas feed with formation of H_2S, which is captured by ZnO installed upstream an FT reactor. Prevention of fouling due to blockage of pores with hard waxes can be achieved by operating at lower value of parameter α, control of hydrogen to CO ratio, lowering temperature and in situ treatment of the deactivated catalyst in hydrogen at temperature 10–20 °C higher than the reaction temperature. Hydrothermal sintering of Fe happens at high pressure of steam. Keeping the latter below 0.5–0.6 MPa or operating below 50–60% of CO conversion prevents sintering as well as formation of iron oxides. Application of multiple reactors with intermediate removal of H_2O and stabilization of the supports with Ba, Zr or La oxides serve as a preventive measure against deactivation. The same approach is used to prevent formation of cobalt oxides. Generation of inactive cobalt carbides is minimized by keeping hydrogen to CO ratio above 2.1 in all reactor parts.

Loss of catalytic material due to abrasion and erosion in the case of fluidized or slurry reactors can be prevented by adequate preparation methods, including sol–gel granulation and application of binders.

Several different reactor types (fixed bed, fluidized bed or slurry phase) are used in the FT processes.

Details of operational conditions and the product composition for most important FT facilities are given in Tab. 3.9.

Because of the exothermicity of FT reactions, heat removal is a major concern in reactor design. In addition, selectivity to unwanted product – methane – is increased with temperature, thus efficient temperature control is needed to achieve desired selectivity. Several types of reactor systems presented in Fig. 3.79 (tubular fixed-bed, circulated and fixed fluidized beds, slurry bubble columns) were and are used in industrial practice.

Tab. 3.9: The most important FT plants [66].

Company	Plant, location, date	Feed	Technology/ reactor	Catalyst	Conditions	Products and capacity
Sasol	Sasol I, Sasolburg, South Africa, 1955	Natural gas	LT/slurry phase distillate + multitubular fixed bed	Precipitated Fe/K	220–250 °C	5,000 bbl/d Paraffin, waxes, oxygenates and fuel gas
Shell	Bintulu, Malaysia, 1993	Natural gas	LT/ multitubular fixed bed	Co/SiO_2	220 °C 25 bar	SMDS, 14,700 bbl/d LPG (0–5%), naphtha (30–40%), distillate (40–70%) and oils (0–30%)
Sasol	Oryx GTL Ras Laffan Industrial City, Qatar, 2007	Natural gas	LT/slurry phase distillate	$Co/Pt/Al_2O_3$	230 °C 25 bar	34,000 bbl/d LPG, naphtha and distillate (diesel blend)
Shell	Pearl GTL Qatar, 2011	Natural gas	LT/ multitubular fixed bed	Co/SiO_2	220 °C 25 bar	SMDS, 140,000 bbl/d LPG (0–5%), naphtha (30–40%), distillate (40–70%) and oils (0–30%)
Chevron	Escravos GTL, Nigeria, 2014	Natural gas	LT/ Similar to Oryx	$Co/Pt/Al_2O_3$	230 °C 25 bar	34,000 bbl/d LPG, naphtha and diesel blend
Sasol	Sasol 2 and 3 (Synfuels), South Africa, 1980	Coal	HT/fixed fluidized bed	Fused Fe/K	350 °C 24 bar	160,000 bbl/d. Fuel gas, oils, alpha-olefins, ammonia, gasoline, jet fuel, diesel

Tab. 3.9 (continued)

Company	Plant, location, date	Feed	Technology/ reactor	Catalyst	Conditions	Products and capacity
PetroSA	Mossgas, Mossel Bay, South Africa	Natural gas	HT/ circulating fluidized bed	Fused Fe/K	330–360 °C 25 bar	30,000 bbl/d. LPG, gasoline, diesel, fuel oil, kerosene, aromatics, alcohols

A tubular reactor was the first to be used commercially. Although this is a simple design, easy to scale-up, construction was rather expensive because of the large number of tubes needed for the industrial reactor. In the fixed-bed reactor, the catalyst tubes (ca. 2,000) with approximately 5 cm in diameter and 12 m in length are surrounded by boiling water, allowing efficient heat removal by evaporation.

As mentioned above, replacement of catalyst is an issue, as iron catalysts should be replaced periodically because of deactivation. Contrary to the lifespan of iron, cobalt-based catalysts are more robust, with a lifetime of several years and could be regenerated. The catalyst size is above 1 mm to avoid extensive pressure drop, thus the effectiveness factor is certainly below unity and mass transfer limitations are present. Possible temperature gradients in the tubes can lead to sintering and deactivation.

Alternatives to tubular reactors are CFB (Fig. 3.79) reactors (Sasol Synthol reactor) and fixed fluidized bed reactors (Sasol Advanced Synthol reactor). CFB reactors provide better heat removal and temperature control and experience less pressure drop problems than a tubular reactor. Moreover, catalyst removal and addition are possible on-line. Fluidized bed reactors operate with iron catalysts at high temperature (330 °C–360 °C) giving products in the gas phase with the chain growth probability of 0.7–0.75. Such reactors can be used for gasoline production, as well as olefins and oxygenates, which can be further processed to other types of fuels (e.g., oligomerization of olefins to diesel range hydrocarbons on HZSM-5 zeolites) and to chemical products.

The major disadvantage of fluidized beds for FT applications is that a low molecular weight product (gasoline) is obtained, while the concentration of diesel range products and waxes cannot be high because products must be volatile at the reaction conditions. If non-volatile hydrocarbons accumulate on the catalyst, the particle fluidization behavior is worsened as the particles stick to each other. Scaling-up of such reactors is more difficult in comparison to tubular reactors.

Fixed fluidized bed

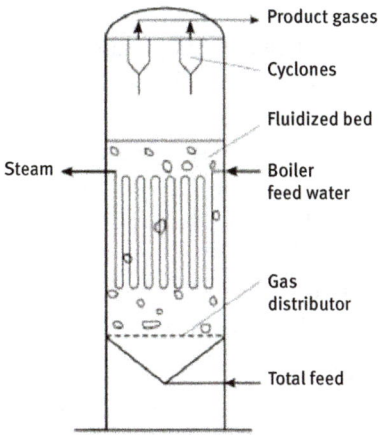

- Product gases
- Cyclones
- Fluidized bed
- Steam
- Boiler feed water
- Gas distributor
- Total feed

Tubular fixed reactor

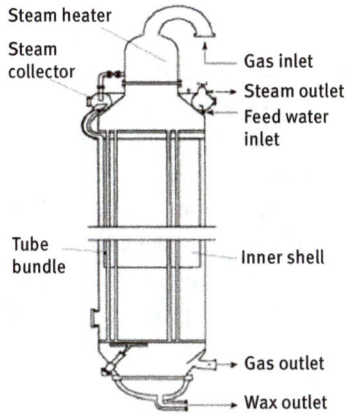

- Steam heater
- Steam collector
- Gas inlet
- Steam outlet
- Feed water inlet
- Tube bundle
- Inner shell
- Gas outlet
- Wax outlet

Slurry phase reactor

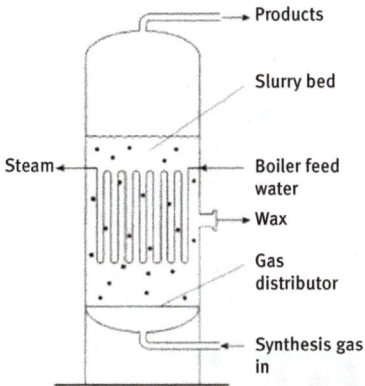

- Products
- Slurry bed
- Steam
- Boiler feed water
- Wax
- Gas distributor
- Synthesis gas in

Circulating fluidized bed

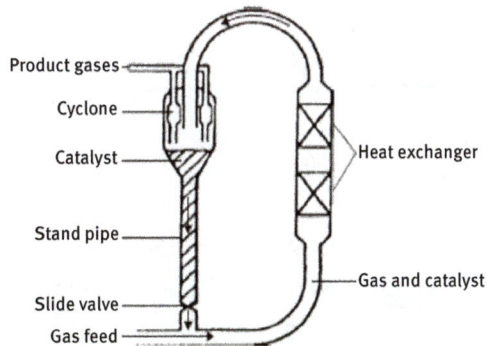

- Product gases
- Cyclone
- Catalyst
- Stand pipe
- Slide valve
- Gas feed
- Heat exchanger
- Gas and catalyst

Fig. 3.79: Reactors for FT synthesis.

Slurry phase bubble columns (Fig. 3.79) are considered to be the best choice for newer FT reactors with more active cobalt catalyst, which is suspended in a slurry. The synthesis gas is bubbled through this slurry that contains hydrocarbon waxes, liquid at reaction conditions, and catalyst particles of 50–80 mm in size. Such reactor could weigh ~2,200 tons and be up to 60 m high with an outer diameter of 10 m. Simpler heat removal and easier scaling-up are advantages of the slurry phase bubble columns. For LTFT synthesis the chain growth probability is much higher than for HTFT producing also liquid waxes. In the liquid slurry phase reactor, the catalyst is suspended in the liquid product wax.

Such reactors provide good heat transfer and temperature control, low pressure drop and are suited for the synthesis of higher boiling products. This could be an advantage, as it gives more overall flexibility if there is a downstream unit for hydrocracking of the long-chain waxes, thus generating high-quality middle distillates (diesel and jet fuel). The product of LTFT synthesis contains naphtha, which is a feedstock for chemicals. The design of slurry phase bubble columns is rather simple, allowing easy addition and removal of catalysts. The catalyst, however, should be highly mechanically stable and be resistant against attrition. The gaseous products are removed from the top, while there is a need to separate waxes from the catalyst, which is send back to the reactor.

Table 3.10 presents a comparison between different three-phase reactors illustrating clearly that product composition and operation parameters are reactor dependent.

Tab. 3.10: Comparison between different three-phase reactors for Fischer–Tropsch synthesis (From [67]).

Application criteria	Slurry bubble column	Riser	Fixed bed
Conditions			
Inlet/Outlet T (K)	533/538	593/598	496/509
Pressure (bar)	15	23	25
H_2/CO ratio	0.68	2.5	1.7
Conversion (%)	87	85	60–66
Products (wt%)			
CH_4	6.8	10.0	2.0
C_2	4.4	8.0	1.9
C_3	9.3	13.7	4.4
C_4	8.0	11.3	4.5
C_5–C_{11} (gasoline)	18.6	40.0	18.0
C_{12}–C_{18} (diesel)	14.3	7.0	14.0
C_{19+}waxes	37.6	4.0	52.0
Oxygenates	1.0	6.0	3.2

As already mentioned the strategy of the LTFT synthesis illustrated in Fig. 3.80 is to produce heavier products with a cobalt catalyst, when formation of long-chain waxes is favored (α-value of 0.9 and higher). The heavy alkanes are converted through mild hydrocracking to the desired carbon number range with subsequent product distillation.

HTFT syncrude conducted in CFB or FFB reactors with Co catalyst has obviously a higher naphtha yield, while application of LTFT using slurry (Co catalyst) or tubular (Fe catalyst) results in higher boiling point hydrocarbons. LTFT refineries are less complex and typically have hydroprocessing (wax hydrocracking) steps and

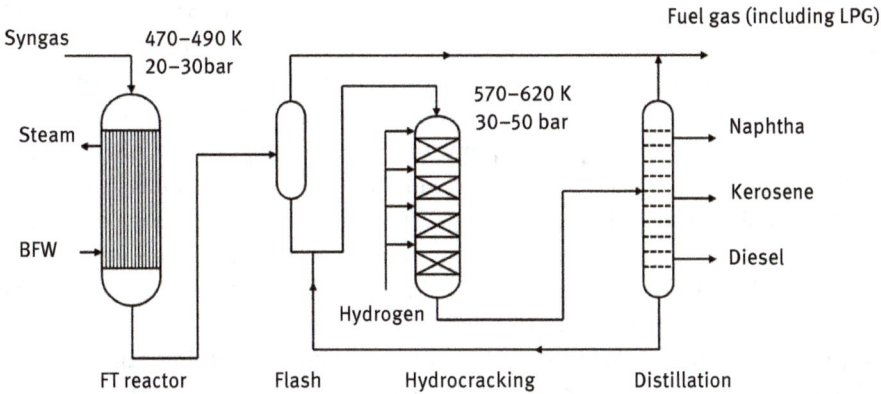

Fig. 3.80: Simplified flow scheme of the Shell Middle Distillate synthesis plant.

fractionation to produce naphtha and middle distillates. At the same time, LTFT refineries are making fuel blending stocks rather than final fuels. On the contrary, diesel production can be achieved in HTFT process as the syncrude contains aromatics, naphthenes giving a diesel density closer to the required specification.

Hydrocracking and hydroisomerization of the heavy product fraction are different compared to similar processing in oil refineries (Chapter 4). More specifically, the feed after an LTFT consists of only paraffins without sulfur compounds and aromatics. This allows using a highly active hydrogenation metal (e.g., platinum) on an acidic support (silica–alumina or proton forms of zeolites) with a danger of secondary cracking.

3.6.3.4.2 Methanol synthesis

Methanol is synthesized from CO and hydrogen according to the following reversible exothermal reaction:

$$CO + 2H_2 \leftrightarrow CH_3OH, \quad \Delta H = -90.8\,kJ/mol$$

Because the reaction is exothermal, equilibrium constant decreases with the increase in temperature. Elevation of pressure results in shifting equilibrium toward the product side.

Another reaction leading to methanol is related to hydrogenation of carbon dioxide

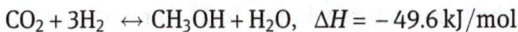

$$CO_2 + 3H_2 \leftrightarrow CH_3OH + H_2O, \quad \Delta H = -49.6\,kJ/mol$$

These two reactions are coupled by the WGS reaction

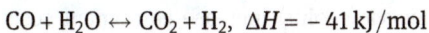

$$CO + H_2O \leftrightarrow CO_2 + H_2, \quad \Delta H = -41\,kJ/mol$$

By-products in this process are higher alcohols and hydrocarbons. Formation of dimethylether is also possible due to methanol dehydration. Application of active catalysts based on copper ($CuZn/Al_2O_3$) allowed to decrease operation pressure of 25–35 MPa used in the classical gas phase processes with $ZnO–Cr_2O_3$ catalysts to approximately 5–10 MPa. Selectivity toward the desired product in low-pressure plants is above 99%. It should be kept in mind that the modern catalysts allow to obtain such high selectivity toward the product, which is not the most thermodynamically preferred, namely, methane is a more thermodynamically favored product than methanol. Conversion of CO and CO_2 to methanol is limited by chemical equilibrium, thus a temperature rise being in principle beneficial from the viewpoint of kinetics negatively influences thermodynamic equilibrium. In addition, highly active catalysts are sensitive to temperature rise, which leads to catalyst deactivation through irreversible sintering. Although initial activity declines substantially during operation as illustrated for different commercial catalysts in Fig. 3.81, a careful catalyst design allows to extend the lifetime to 4–6 years and even more.

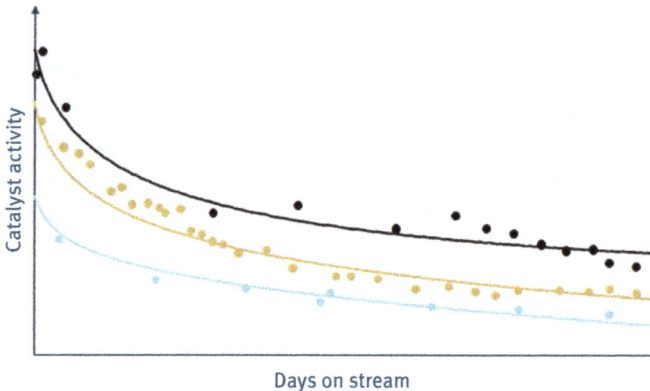

Fig. 3.81: Dependence of catalytic activity in methanol synthesis with time-on-stream for different catalysts.

A typical measure for counterbalancing deactivation in various catalytic processes is to increase temperature restoring activity but often compromising selectivity. In the case of methanol synthesis such approach cannot be easily applied as the temperature should not exceed approximately 270 °C. Because copper is marginally active below 230 °C, the temperature window for the process is rather low.

Increase in pressure is an alternative way of compensating for activity loss due to sintering. At the same time, too high pressures of CO and CO_2 (favoring conversion from the thermodynamic viewpoint) increase equipment costs in the synthesis loop and for syngas compressor. The synthesis loop is thus required as pressure in the modern plants of 5–10 MPa gives only moderate conversion levels (15–30% in adiabatic reactors). Unreacted gas is recycled back acting as a syngas quench cooler. The ratio between the recycle gas and the fresh feed ranges from 3:1 to 7:1, which along with the purge allows to prevent buildup of impurities (methane and argon) in the loop.

Raw methanol containing water and impurities is condensed and sent to the distillation unit, whose design depends on the desired product purity. Typically, from one to three distillation columns are used with the first one (so-called topping column) acting as a stabilizer for removal of dissolved gases (CO, CO_2, H_2, N_2 and CH_4) and some of the light by-products (aldehydes, ketones and dimethyl ether). In the downstream columns, raw methanol, containing besides water also minor amounts of higher alcohols, is fractionated (Fig. 3.82). The heat input is optimized in the three-column system.

An important theoretical and practical issue is related to a question which reactant is leading to methanol. A long controversy surrounded this topic, and either CO or CO_2 or both were proposed as the true reactants. Methanol can be produced from both H_2-CO and H_2-CO_2 mixtures, while a mixture containing H_2, CO and CO_2 gives much higher yields of methanol. Isotopic labeling studies suggest that the source of carbon in methanol is CO_2, while CO is mainly converted to CO_2 via a WGS reaction. CO_2 is also influencing the properties of the catalyst keeping an intermediate oxidation state of copper (Cu^0/Cu^+) and preventing reduction of ZnO. High concentrations of CO_2 are, however, inhibiting methanol synthesis, in which rate drops slightly up to 12 vol% of CO_2 and thereafter more steeply.

A typical gas composition (can be different depending on the synthesis gas generation procedure) could be thus: 67.5% H_2, 21.5% CO, 8% CO_2 and 3% CH_4 for high capacity plants and 69% H_2, 18% CO, 10% CO_2 and 3% CH_4 for lower and medium capacity.

Modern commercial catalysts for methanol synthesis are supplied as tablets and contain above 55 wt% CuO, 20–25% ZnO with 8–10% Al_2O_3 and also catalyst promoters, as well as catalyst binders (e.g., graphite). For such structure-insensitive reaction as methanol synthesis the activity is dependent only on the total exposed copper area and is not affected by the structure of the crystallites. This means that large loading of copper (reaching 64% in some commercial formulations) and small cluster sizes are in fact need for efficient catalysts. High metal dispersion as such is not enough for successful industrial operation, as the catalyst should be stabilized against sintering. It was mentioned earlier that thermal sintering is a key mechanism for deactivation with temperature approaching 315 °C depending on the reactor type. Moreover, sulfur and, in some cases, iron and nickel carbonyls introduced into the loop with fresh syngas contribute to catalyst deactivation. ZnO used in the commercial formulations for many decades is a textural and chemical promoter being introduced as small crystallites (2–10 nm). It helps to stabilize copper against sintering, facilitating formation of small copper clusters and also scavenging sulfur. Alumina needed in the catalyst to stabilize both ZnO and copper oxide might in principle lead to formation of dimethylether;

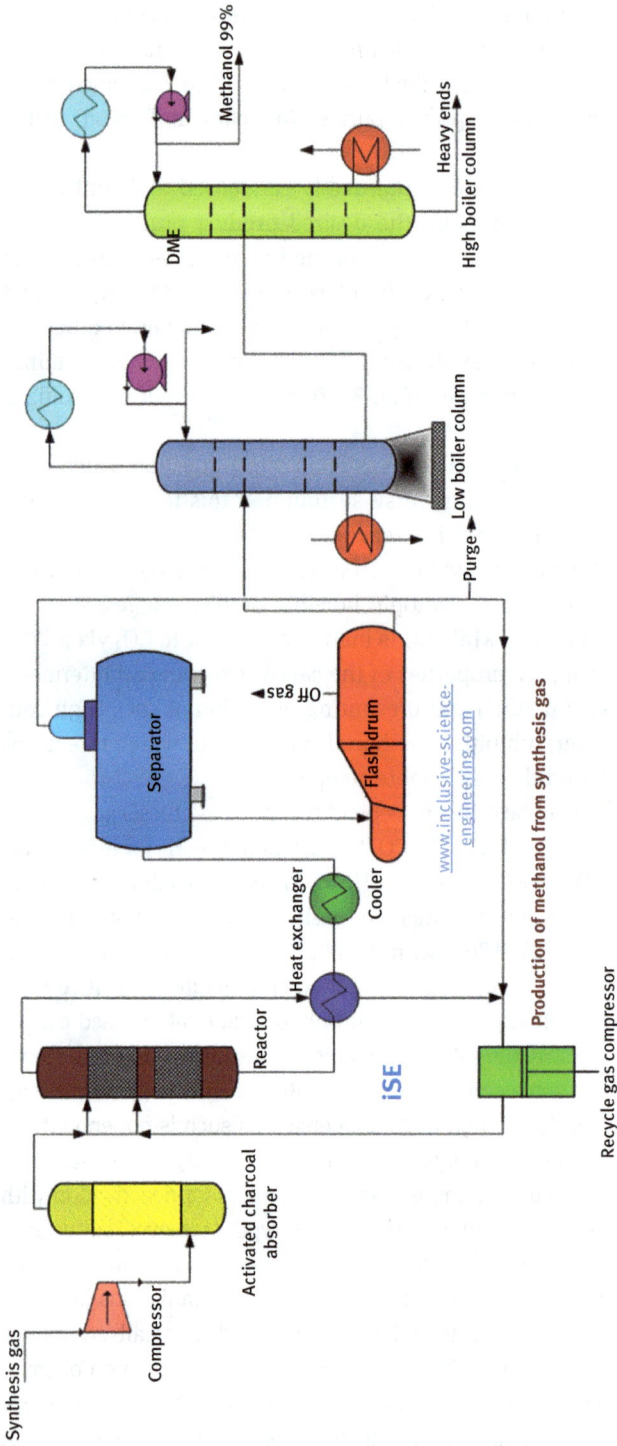

Fig. 3.82: Methanol production flow scheme with purification section [68].

however, presence of ZnO oxide neutralizes acidic sites of alumina. Other promoters (such as MgO) were also introduced in commercial formulations.

Utilization of the commercial catalysts in the form of cylindrical pellets of 5–12 mm implies that diffusional limitations can be significant.

Several reactor designs and synthesis flowsheet arrangements for methanol production can be utilized. Reactor choice depends on a plant size and the syngas generation method. In reactor selection, conversion temperature profiles should be optimized being close to the equilibrium one and affording lower peak temperature. Moreover, in addition to optimized temperature profile, proper mixing and reactant distribution allow higher selectivity, thus diminishing amounts of by-products with substantial savings in product purification, as well as lower deactivation and longer catalyst life.

Mainly multibed (three to four) adiabatic fixed-bed reactors are applied in low-pressure methanol processes with heat removal either by quenching with the cold feed (quenching) or using heat exchangers. The temperature profiles are far from the maximum rate curve as illustrated in Fig. 3.83 for a four-bed adiabatic reactor with heat exchangers. Despite this apparent inefficiency, multibed reactors represent an attractive low-cost reactor concept when there is no need for steam generation.

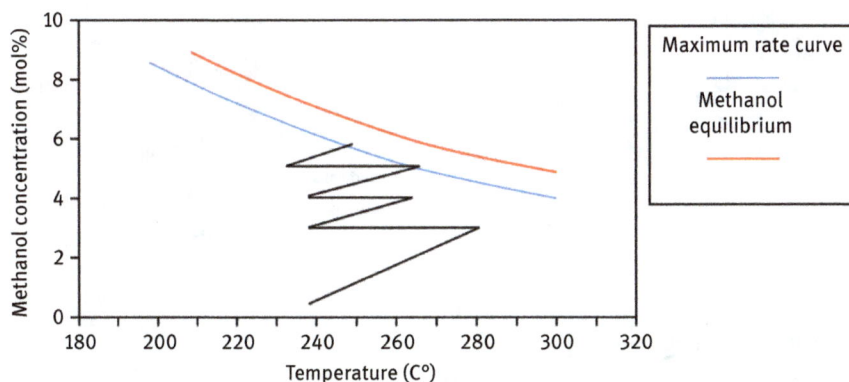

Fig. 3.83: Temperature profile for multibed adiabatic methanol synthesis reactor with heat exchangers. Adapted from [69].

Not only cylindrical but also spherical adiabatic reactors (Fig. 3.84A) are applied when the catalyst is located between two perforated spherical shells. Such reactor type allows a decrease in the vessel wall thickness at a given pressure and thus affords lower reactor costs. The flow in such reactors is organized from the outside of the catalyst layer to the center of the spherical core. Pressure drop is minimized as a relatively thin catalyst layer is used.

In a tube-cooled converter (Fig. 3.84B) the feed enters the reactor at the bottom and flowing upward through the tubes with minimum thickness is preheated

(A)

Feed gas

BFW → Steam

BFW → Steam

Product gas

(B)

Tube-cooled converter (TCC)

Syngas from compressor

Circulator

Heat recovery

Condenser

Purge

Crude methanol to distillation

(C)

Axial steam-raising converter (A-SRC)

Steam drum

Syngas from compressor

Circulator

Interchanger

Condenser

Purge

Crude methanol to distillation

(D)

Gas inlet

Catalyst loading

Cooling tube

Catalyst

Central pipe

Inert balls

BFW inlet

Steam outlet

Gas outlet and catalyst unloading

(E)

Manway   Manway

Steam outlet

Outer tube

BFW inlet

Gas out

Flexible hose

One touch coupling

Catalyst

Support grid

Manway

Manway

Diaphragm

Feed gas inlet

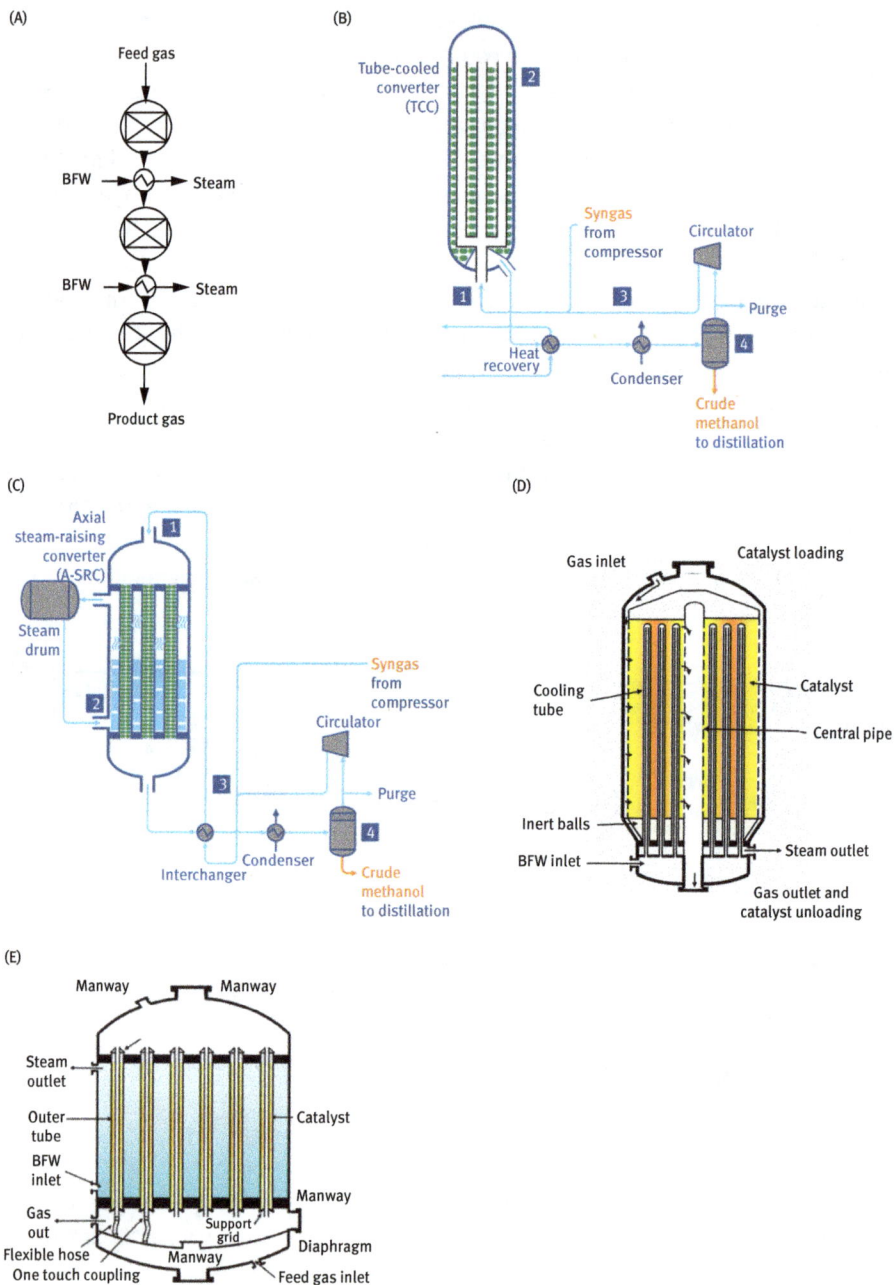

Fig. 3.84: Special reactor types used for methanol synthesis: (a) spherical adiabatic reactors, (b) tube-cooled converter, (c) converter with axial steam rising [69], (d) Toyo's MRF-Z reactor adapted from [70] and (e) Mitsubishi superconverter, adapted from [71] .

by the product gas flowing downward through the catalyst bed. This way of arranging the heat exchange gives a temperature profile (Fig. 3.85) much closer to the maximum rate curve than the case of a multibed reactor with heat exchangers. The catalyst amount in such a tube-cooled reactor with the axial flow is limited by pressure drop considerations. For large capacity plants, several reactors might be needed.

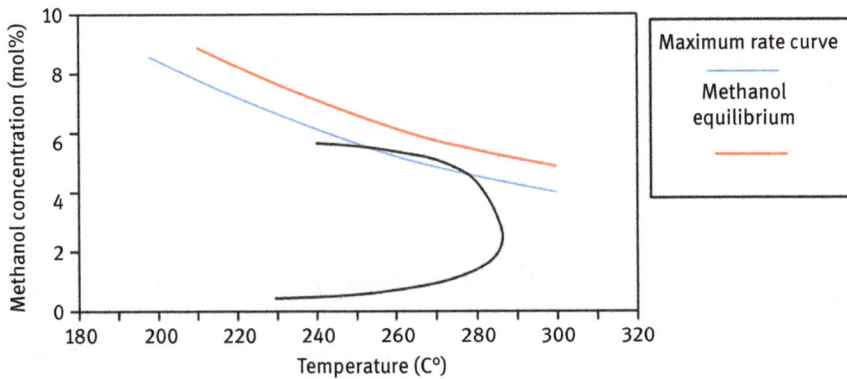

Fig. 3.85: Methanol concentration profile in tube-cooled converter. Adapted from [69].

Near isothermal operation (Fig. 3.86) is provided in a tubular boiling water reactor with axial flow (Fig. 3.84C), where the catalyst is located on the tube side. Temperature is controlled by pressure of water, which is circulated on the shell side generating steam at the maximum possible pressure without overheating the catalyst. The temperature profile is close to the maximum rate curve and allows somewhat low temperatures than in tube cooler converters, still requiring a significant

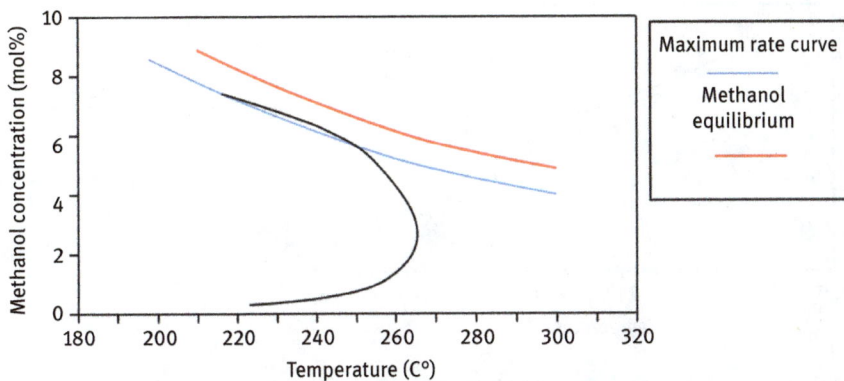

Fig. 3.86: Concentration profile in a steam-generating multitubular reactor. Adapted from [69].

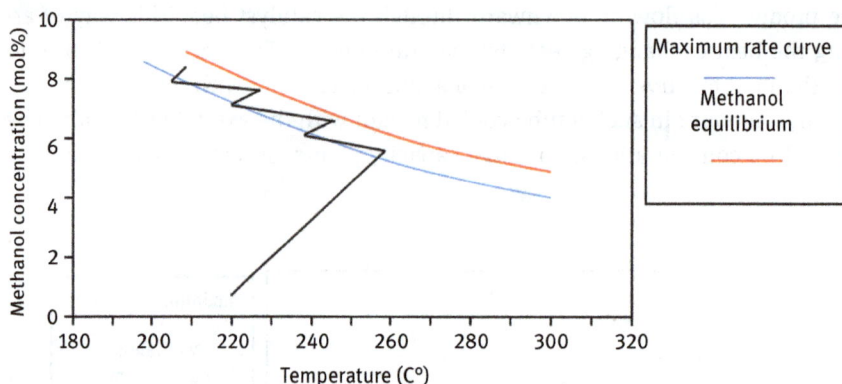

Fig. 3.87: Concentration profile in Toyo's MRF-Z reactor. Adapted from [69].

recycle ratio. High investment costs for this reactor concept limit the maximum plant size to approximately 1,500 tpd and require several reactors in series for larger capacity.

A specific feature of Toyo's MRF-Z reactor (Fig. 3.84D) is multistage indirect cooling and a radial flow facilitating the capacity increase of methanol plants. This reactor type generates steam of approximately 3 MPa, has a good approach to equilibrium (Fig. 3.87), a small number of tubes and a low pressure drop (0.05–0.075 MPa) while it can be in the range of 0.3–1 MPa for fixed-bed adiabatic reactors.

Mitsubishi reactor (Fig. 3.84E) for methanol synthesis can be viewed as an integration of interchange and steam rising. The design is rather complex consisting of a large number of tubes, a manifold, two tube sheets. It generates approximately 4 MPa of steam, closely following the maximum rate line (Fig. 3.88) and thus allowing high conversion per pass and a lower recycling rate.

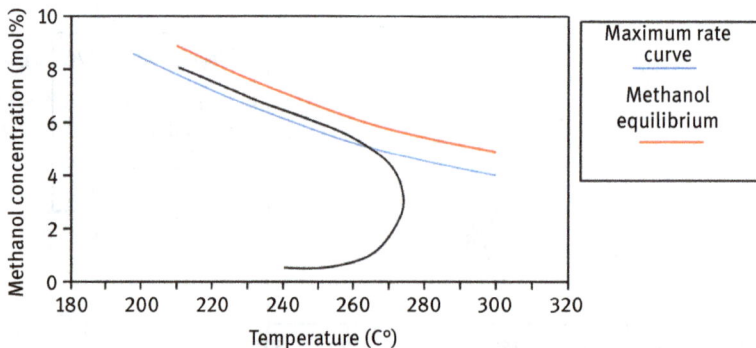

Fig. 3.88: Concentration profile in Mitsubishi superconverter. Adapted from [69].

The reactors presented above correspond to gas-phase methanol synthesis. An alternative liquid-phase process for methanol production was developed by Air Products and Chemicals.

The technology relies on a bubble slurry reactor (Fig. 3.89), in which an inert hydrocarbon acts a reaction medium and a heat sink. As the feed gas bubbles through the catalyst slurry forming MeOH, the mineral oil transfers the reaction heat to an internal tubular boiler where the heat is removed by generating steam. The reactor operates at isothermal conditions being able to handle CO-rich (in excess of 50%) syngas with wide compositional variations. Such operation mode allows to reach much higher concentration of methanol (ca. 15%) than in the gas-phase process increasing conversion from 15% to ca. 35%. This technology was proven at the demonstration plant level [73] but has not yet been commercialized. Few key scale-up-related issues were identified based on the experience with the demonstration plant. A particular important in the context of this textbook was related to catalysis. Iron from residual construction debris caused severe poisoning of the catalyst, requiring replacement of the full catalyst charge.

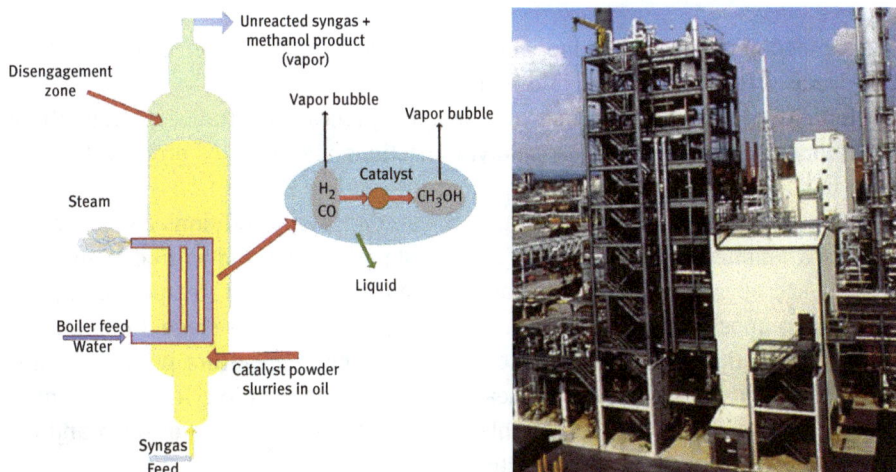

Fig. 3.89: Bubble slurry reactor for liquid-phase methanol synthesis: (a) schematics [72] and (b) column at the demonstration plant [73].

3.6.5 Reactor modeling

Nowadays many industrial companies realize that simulation and optimization of a catalytic process require sound knowledge on reaction kinetics, transport phenomena and hydrodynamics. The principles objectives of reactor modeling and design include determination of the most technically and economically relevant reactor type; its sizing by establishing the reactor and catalyst volume for grass root reactors or estimation of throughput for available reactors; and determination of the optimum conditions and operation policy.

3.6.5.1 Reactor modeling and process development strategy

The chemical development, process development and process engineering phases proceed both consecutively and in parallel. A typical interrelated workflow for the development of chemical manufacturing processes is presented in Fig. 3.90.

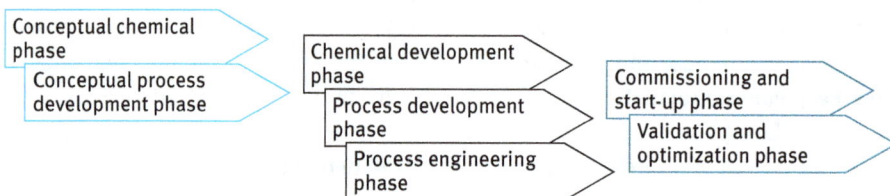

Fig. 3.90: Workflow for the development of a chemical manufacturing process. (From [74].).

The methodology of reactor modeling, on which the discussion below partly is based, was presented by Dutta and Gualy [75].

Initially the reactions and separations of a process are studied thoroughly in laboratory and bench scale. For catalytic reactions two types of process development units are usually applied. In the first, the catalyst is used in the form of powders to generate reliable kinetic data; in the second, step conditions correspond to those of a commercial plant in terms of catalyst particle size and hydrodynamics.

The key operating variables (space velocity or throughput, temperature, pressure and composition) are usually carefully investigated at this stage. In particular, a special attention should be paid to monitoring the temperature profiles, as this parameter is the most dominant in reactor design and operation. It is important to cover various combinations of variables, explore the broad range of them and get information on each component reaction.

One of the most critical tasks is to establish reaction kinetics in a reliable fashion. Otherwise the kinetic models are applicable only to a narrow range of conditions. It is advisable in mechanistic studies to change one parameter at a time, at least initially, and then cover several possible and diverse combinations of variables. An extremely important aspect is catalytic kinetics under non-steady-state conditions, with transient experiments providing additional information necessary in dynamic simulations. Non-stationary kinetics is also needed when a process in organized in a non-stationary model (for example, reaction-regeneration cycles).

Experimental data generated in process development units should cover not only the domain of pure kinetics, but also the region of significant mass transfer resistance. For three-phase systems, after experimentation with the catalysts in powder form (micron level), a catalyst basket containing pellets of mm size could be placed inside an autoclave to determine pore diffusion.

An important question in the generation of catalytic data is related to the purity of the feedstock and use of model compounds in test units. Preliminary laboratory studies, aimed at initial screening of potential catalysts, evaluation of intrinsic kinetics and mass transfer, could probably be performed with a model feedstock. In scaling-up batch operations the real feedstock can in principle be used at an early stage, but this masks, however, intrinsic kinetics caused by the presence of various impurities, which influence for instance catalyst deactivation.

The safety aspects should be studied carefully. This is important because laboratory-scale continuous reactors are made of small diameter tubes that provide efficient heat transfer to the environment. For large commercial reactors, adiabatic conditions could be more prominent, leading to heat build-up.

The operating conditions are typically selected to be outside of explosion limits, however, for fluidized bed reactors no hot spots are generated because of vigorous stirring, thus operation within the explosion limits is possible, provided special care is taken about the way the reactants are introduced and removed from the reactor.

In many cases an extensive pilot plant campaign is undertaken after laboratory tests. In fact nowadays, extrapolation of laboratory results to industrial conditions can be rigorously performed through adequate mathematical modeling.

A number of years ago at one of the conferences on chemical reactors I was approached by a chemical engineer who worked at a petrochemical company in South America, and who very enthusiastically acknowledged the efforts made by my colleague Prof. T. Salmi. It turned out that the rigorous mathematical model for hydrogenation of aromatics, which included detailed kinetics and transport phenomena, was directly used for building a commercial plant with annual capacity of 300,000 tons.

Thus, from the modeling perspective, pilot plant experiments should be carried out only when it is necessary to improve and validate models and when the passage from the lab to production scale is too complex because of sophisticated reactor types. However, in practice, piloting is still widely applied. The reasons for this are as follows. The management, responsible for making decisions about building a large-scale plant, needs to have more confidence in the process. The same holds for the potential customers if licensing is an issue. Representative product samples might be needed for marketing purposes and various application tests. Moreover, pilot units allow investigating such parameters as the feed type, impurities, build-up of by-products, catalyst deactivation, corrosion, etc. Stability tests are usually performed at a pilot scale with the real feedstock. It is even possible to utilize side reactors (for example, in parallel with industrial ones) for piloting as practiced by some large chemical manufacturers.

The complexity of the tasks, which a chemical engineer faces, justifies the necessity and the need for reliable kinetic and transport models, which are based on physicochemical understanding of catalytic processes. The reactor models cover several levels: an active site (mechanism and kinetics), a catalyst grain (mass and heat transfer and thus knowledge of catalyst effectiveness factor, interphase mass transfer, phase holdups) and the reactor itself (hydrodynamics, bed voidage profile, axial and radial dispersion, bubble properties, wall heat transfer, etc.).

The degree of complexity of the reactor model is still a matter of debate. A very complex model with a very large number of adjustable parameters might be found impractical by some industrial companies.

Through my own industrial experience and collaboration with various companies I have seen utilization of reactor models of different complexity levels ranging from very simple to extremely sophisticated. On average, over simplified or over complicated models should be avoided, as the former are applicable only to a narrow range of parameters, while the latter could be difficult to handle during parameter estimation.

A few reactors-related issues should be specifically highlighted. Changing geometrical dimensions for fixed-bed reactors and trying to keep the same contact time means that the ratio between the length of the catalyst bed and the linear velocity should be kept constant. At the same time, it will certainly change the Reynolds and Peclet numbers, which determine hydrodynamic conditions.

The importance of the regeneration model in systems such as fluidized bed reactors should not be underestimated. In particular, the circulation lines should be properly sized, which is important for cases when the catalyst itself acts as an oxygen carrier and demands high solids circulation rates. Maleic anhydride synthesis by n-butane oxidation is an industrial example of such a riser-regenerator reactor arrangement. In this case, hydrocarbons and oxidants are not in contact, which provides inherent safety to this technology. For alternative partial oxidation technologies that use multi-tubular reactors, there are risks of explosions and thus special care should be taken regarding the mixing zone and injection procedures.

Design of the entrance zone and fluid (gas or liquid) distributors is, in general, important as conversion is significant close to the rector inlet. In CFBs the entrance and exit zone are also critical in influencing the overall reactor hydrodynamics.

On a more general level the reaction model should be clarified first before properly addressing reactor hydrodynamics. Although the actual hydrodynamics should be modeled adequately, the impact of an inadequate reaction model is more serious than an inadequate hydrodynamic model.

It is advisable during the model building to check the model consistency against simple kinetic and reactor models where analytic solutions are available, as the more complex model should be able to account for known simple systems.

Nowadays process simulators are commercially available with dynamic reactor models, which are able to predict not only normal but also transient operations (start-up and shut-downs). These simulators contain various reactor models and make it possible to determine the best reactor configuration. Fig. 3.91 illustrates such a multiple reactor model package.

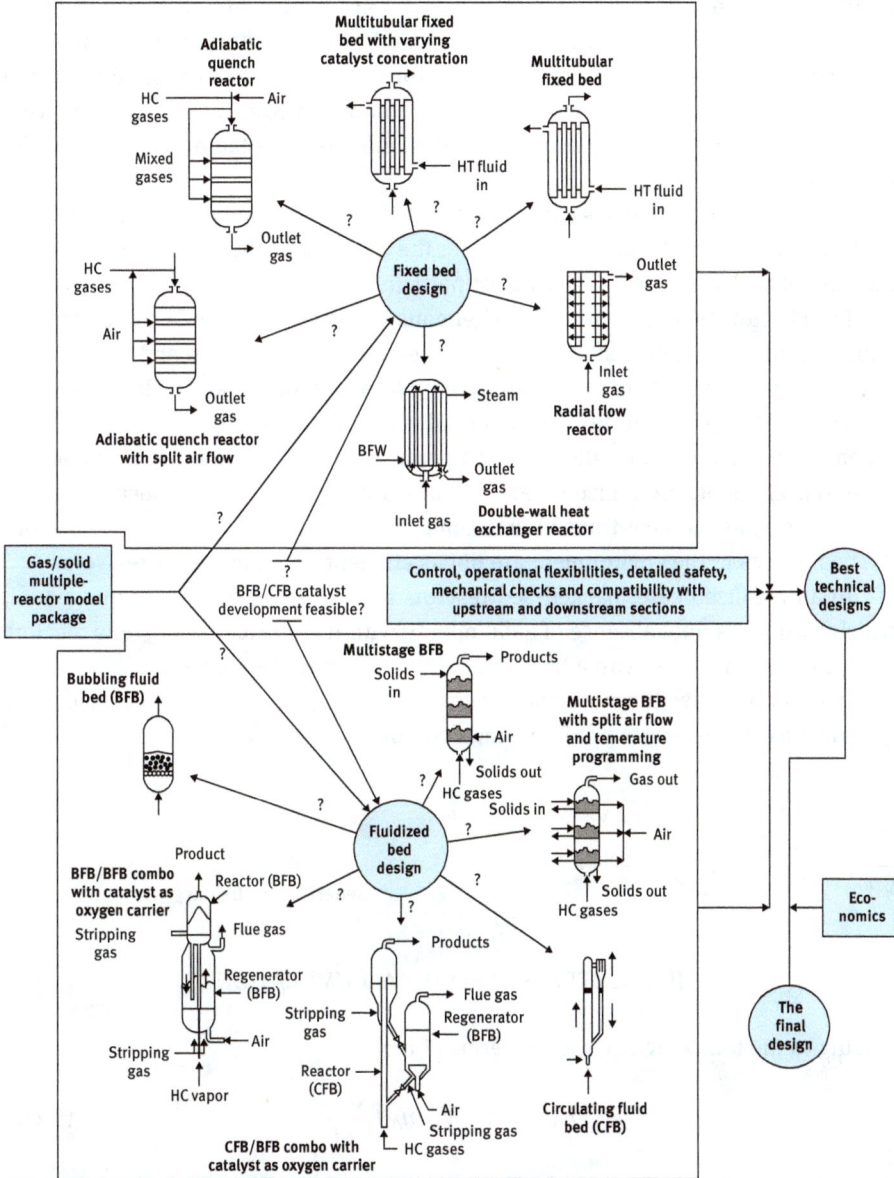

Fig. 3.91: Decision making with a multiple reactor model package. (From [75].).

3.6.5.2 Reactor models: mass and heat balances

The reactor type is defined by the physical configuration, the flow mode and hydrodynamics.

The first model for discussion is catalytic packed-bed reactors. They can be classified in two main categories: pseudo-homogeneous and heterogeneous models. In pseudo-homogeneous models the diffusional limitations (concentration and temperature gradients) inside the catalyst are neglected and the fluid phase (gas or liquid) and the solid catalyst are considered as a continuous phase. Mass balance equations are thus only required for the fluid (bulk) phase. Even when internal diffusion is prominent the pseudo-homogeneous model can be used through use of a correction term, the catalyst effectiveness factor, extensively discussed in Section 3.5.

Heterogeneous models account for diffusion effects inside catalyst particles by introducing separate balance equations for the fluid inside the catalyst pores and in the bulk phase surrounding the catalyst particles.

Pseudo-homogeneous and heterogeneous models can be either one- or two-dimensional. For one-dimensional models the radial concentration and temperature gradients in the reactor tube are neglected and a chemical reaction is assumed to proceed at the same rate in various radial positions in the reactor cross section. When such radial concentration and temperature gradients are present, radial dispersion coefficients are utilized. Severe temperature gradients can appear in industrial reactors, as the heat transfer area diminishes with an increase in the size and the heat losses to the environment are not as efficient as in laboratory reactors.

In an idealized case, packed-bed reactors are described by a plug flow model, which assumes a total absence of axial mixing within the reactor. Plug flow reactors are typically long tubes with a high ratio of the length to the diameter.

In order to derive the mass balance for such reactors, we must first consider the general mass balance for an infinitesimal volume element ΔV (Fig. 3.92):

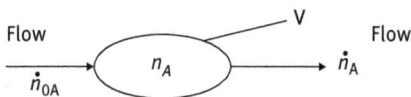

Fig. 3.92: General mass balance.

$$\text{IN} + \text{GENERATION} = \text{OUT} + \text{ACCUMULATION}$$

leading to the following equation in terms of moles:

$$\dot{n}_{0A} + \eta_A r_A \Delta V = \dot{n}_A + \frac{dn_A}{dt} \tag{3.259}$$

where $\dot{n}_{0A}$ and $\dot{n}_A$ are the mole fluxes, η_A is the catalyst effectiveness factor, which takes into account mass transfer.

Equation (3.259) can be rewritten:

$$\eta_A r_A \Delta V = \Delta \dot{n}_A + \frac{dn_A}{dt} \tag{3.260}$$

where $n_{A_{in}} - n_{A_{out}} = \Delta \dot{n}_A$. Replacing the amount in moles through concentration and volume, considering an infinitely small volume $n_A = c_A \Delta V$, $\Delta V \to 0$ one arrives at:

$$\eta_A r_A = \frac{d n_A}{dV} + \frac{dc_A}{dt} \tag{3.261}$$

When the volumetric flow $\dot{V}(m^3/s)$ is constant $\dot{n}_A = c_A \dot{V}$. Defining the residence time in the reactor (τ, s) as $\tau = V/\dot{V}_0$ the dynamic mass balance equation is obtained:

$$\frac{dc_A}{dt} = -\frac{dc_A}{d\tau} + \eta_A r_A \tag{3.262}$$

At steady-state the concentration is independent on time $dc_A/dt = 0$, thus:

$$\frac{dc_A}{d\tau} = \eta_A r_A \tag{3.263}$$

and the design equation is given by:

$$\tau = \int_{c_{0A}}^{c_A} \frac{dc_A}{\eta_A r_A} = c_{0A} \int_0^{\alpha} \frac{d\alpha}{-\eta_A r_A} \tag{3.264}$$

where α is conversion. For packed-bed reactors the balance is exactly the same as for homogeneous plug flow reactors, except that the rate in the balance equation is replaced by $r_A \rho_B$, where ρ_B is the catalyst bulk density, leading to the steady-state design equation:

$$\tau = \int_{c_{0A}}^{c_A} \frac{dc_A}{\rho_B \eta_A r_A} \tag{3.265}$$

As an example, let us consider an Eley-Rideal type rate equation:

$$r = \frac{kK c_A}{1 + K c_A} \tag{3.266}$$

When introducing this rate expression into eq. (2.265), and assuming that the catalyst effectiveness factor is close to unity, the following design equation is obtained:

$$\tau = \frac{1}{\rho_B} \left[\frac{1}{kK} \ln \frac{c_{0A}}{c_A} + \frac{1}{k}(c_{0A} - c_A) \right] \tag{3.267}$$

The CSTR is the opposite of plug flow hydrodynamics, which assumes that the fluid stream is completely mixed and the concentration in the reactor is the same as in the reactor outlet. For a perfectly mixed CSTR at steady-state it holds that there is no accumulation $dn_A/dt = 0$ and thus:

$$\dot{n}_{0A} - \dot{n}_A + \eta_A r_A V = 0 \tag{3.268}$$

Molar flows in eq. (3.268) can be substituted by concentration and volumetric flows: $\dot{n}_{0A} = c_{0A} c_A \dot{V}_0, \dot{n}_A = c_A \dot{V}$. The volumetric flows in and out are assumed to be equal (for example, the density is constant) $\dot{V}_0 = \dot{V}$. When introducing residence time as the ratio of reactor volume to volumetric flow rate in the same way as for the plug flow reactor, a design equation is then obtained:

$$\tau = \frac{c_A - c_{0A}}{\eta_A r_A} \tag{3.269}$$

and for heterogeneous catalytic reactions the design equation becomes:

$$\tau = \frac{c_A - c_{0A}}{\alpha \eta_A r_A} \tag{3.270}$$

with $\alpha = m_{cat}/V_L = \rho_B$ as the catalyst bulk density. For simple kinetics, analytical solutions are possible, while numerical solutions provide a general approach to model simulations and parameter estimation. For the Eley-Rideal reaction, and assuming $\eta_A = 1$, the expression for the residence time becomes:

$$\tau = \frac{V}{\dot{V}} = \frac{1}{k\rho_B}\left[K^{-1}\frac{(c_{0A} - c_A)}{c_A} + c_{0A}x\right] \tag{3.271}$$

Industrial reactors cannot be represented by these idealized plug flow and CSTR models and axial dispersion concept is applied in all mentioned variations (one- or and two-dimensional pseudo-homogeneous and heterogeneous cases).

The axial dispersion models add a second order spatial derivative to the mass balances. For a steady-state, neglecting internal diffusion, the mass balance for a compound i becomes:

$$D_a\frac{d^2c_i}{dl^2} - \frac{d(c_i w)}{dl} + r_i\rho_B = 0 \tag{3.272}$$

where l is the length, w is the fluid flow rate, D_a is the axial dispersion coefficient. When there is no axial dispersion $D_a \approx 0$ and a reactor approaches the plug flow regime. For complete mixing $D_a \approx \infty$ and an axially dispersed reactor displays CSTR behavior.

The values of the axial dispersion coefficient are difficult to predict theoretically. They are usually determined by experiments with tracers or through correlations relating the Peclet number $\text{Pe} = wL/D_a$ (where L is the reactor length) with the Reynolds number. When $\text{Pe} > 100$ the reactor behaves like a PFR, while values below unity indicate a CSTR-like behavior. It should be noted that during scaling-up hydrodynamic conditions are changed and thus the values for axial dispersion coefficients cannot be kept constant.

Limitations in applying one single axial dispersion coefficient are related to much more complicated flow patterns in industrial reactors. These patterns could be of complex geometry, unusual shape or the reactors can have some complicated internals influencing the flow pattern. For example, the presence of a stagnant zone leads to long tailing, which cannot be predicted by the axial dispersion model.

Details on complex reactor models for different cases are presented in specialized textbooks on chemical reactors, for example, Salmi et al. [29], and will not be treated here. As one example, a mass balance is given below for the liquid component in a one-dimensional dynamic three-phase fixed-bed with axial dispersion:

$$\frac{dc_{Li}}{dt} = -(\varepsilon_L \tau_L)^{-1}\frac{dc_{Li}}{dz} + (Pe_L \varepsilon_L \tau_L)^{-1}\frac{d^2 c_{Li}}{dz^2} + r_i \rho'_B \tag{3.273}$$

where $\rho'_B = \rho_B/\varepsilon_L$, ε_L is the liquid hold-up, Pe_L is the liquid-phase Peclet number defined by $Pe_L = w_L L/D_L$, D_L is the liquid axial dispersion coefficient, w_L is the liquid velocity, L is the total reactor length, z is the dimensionless length coordinate ($z = l/L$), τ_L is the space-time defined as $\tau_L = L/w_L$.

In the derivation of eq. (3.273) several assumptions and simplifications were made, including, constant fluid velocity, applicability of the Fick's law for gas-liquid and liquid-solid fluxes, steady-state for the outer surface of the catalyst particle, absence of internal diffusion resistance in the pores, large values of liquid-solid mass transfer coefficient and thus kinetic control, as well as non-volatility of the organic liquid-phase component.

The axial dispersion model in eq. (3.273) is a dynamic one-dimensional model with respect to the spatial coordinate, containing just two adjustable hydrodynamic parameters, $\tau_L \varepsilon_L$ and Pe_L. This model can be used for an isothermal case without the need to consider radial concentration gradients. For non-isothermal cases with strong heat effects a two-dimensional model is required, consisting of a set of coupled second order partial differential equations (PDEs). The strategy for solving the problem is to discretize the reactor coordinate (z), i.e., to divide the area between the boundaries into grid points, where the values of derivatives are approximated, and to integrate the ODEs in time (t). The number of ODEs is the number of components times the number of discretization points. Two methods are mainly applied for discretization – lines or orthogonal collocation. In the method of lines the derivatives, which originate from the plug flow, should be described with backward differences, whereas central differences should be used for the first and second derivatives originated from diffusion and dispersion. In the orthogonal collocation method the areas between collocation points are approximated by orthogonal polynomials.

For gas-phase reactions in catalytic microreactors there are several differences between the regular size reactors and microdevices. The flow in microchannels is usually laminar, while in macrochannels it is turbulent. Because of the geometry of microchannels and high surface-to-volume ratios, the diffusion paths for heat and mass transfer are very small, surface effects dominate over volume effects, and the surface heat transfer is important.

The Knudsen number (Kn), which represents the ratio between the main free path of the gas molecules and the characteristic length of the flow domain (diameter of the channel in this case), is, however, quite small in microchannels, as it is several orders of magnitude below unity. When Kn > 1, which is realized in the pores of zeolites (*nm* size compared to tens of microns in microchannels) the molecules are most likely to collide with the walls rather than with each other. From the modeling viewpoint Kn < 1 implies that the typical Navier-Stokes equations, which are valid for macroflows, can be also applied for microreactors.

Depending on the thickness of the catalytic layer and on the reaction conditions, diffusion limitation inside the microchannels might play a role. Moreover, mass transfer limitation from the bulk phase to the surface of the coating could appear, mainly via molecular diffusion.

For fluidized bed reactors the ideal models of the plug flow and CSTR are very crude approximations and the reactor design should be based on a hydrodynamic model coupled with pilot scale experimentation.

Rather complex models for fluidized beds, such as the Kunii-Levenspiel model, assume negligible gas flow in the emulsion phase and movement of the bubble phase in the reactor following the characteristics of a plug flow. The cloud and wake phases are also present, which are presumed to have similar chemical contents. Fig. 3.93 illustrates the transport in these phases.

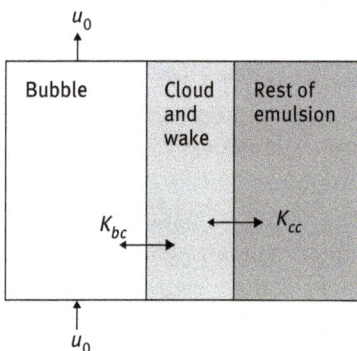

Fig. 3.93: The schematic structure of a fluidized bed according to the Kunii-Levenspiel, where b, *c* and *e* refer to the bubble, cloud and emulsion phases respectively.

Details on this model are available in textbooks on chemical reactors [29].

For manufacturing of organic fine chemicals multipurpose semi-batch and batch reactors are mainly used, equipped with necessary auxiliary equipment for mixing, heating and cooling as well as for distillation or evaporation of solvents or light reaction products. The reactions typically take place in the liquid-phase.

The simplified mass balance for the batch reaction does not contain any terms related to in- or out-flow of compounds, thus:

$$\eta_A \rho_B r_A V = \frac{dn_A}{dt} \tag{3.274}$$

If the volume is constant $dn_A = d(c_A V) = V dc_A$ after some rearrangements, taking into account that concentration is related to conversion α through $c_A = c_{A_0}(1 - \alpha)$ and that at $t = 0$, $\alpha = 0$ a design equation is obtained:

$$t = c_{A_0} \int_0^\alpha \frac{d\alpha}{-\rho_B \eta_A r_A} \tag{3.275}$$

which is the same as for a packed-bed reactor with the only difference that in eq. (3.275) time is used instead of the residence time.

More complicated reactor models for three-phase systems include internal diffusion:

$$\frac{dc_{Li}}{dt} = \varepsilon_p^{-1} \left(D_{ei} \frac{d\left(\frac{dc_{Li}}{dr} r^s\right)}{r^s dr} + r_i \rho_p \right) = 0 \tag{3.276}$$

where D_{ei} is the effective diffusion coefficient of a species (i) in the catalyst particle, r is the coordinate of the catalyst particle, s is a shape factor: $s = 2$ for spheres, $s = 1$ for long cylinders, $s = 0$ for slabs and ε_p^{-1} is the particle porosity.

Considerations above were related only to mass balances. However, design of catalytic reactors should involve also energy balances. In general, it is enough for catalytic reactor engineering to include only heat balances

$$\left\{ \begin{array}{c} \text{energy} \\ \text{accumulation} \end{array} \right\} = \left\{ \begin{array}{c} \text{energy} \\ \text{added} \\ \text{by} \\ \text{convection} \end{array} \right\} - \left\{ \begin{array}{c} \text{energy} \\ \text{leaving} \\ \text{by} \\ \text{convection} \end{array} \right\} + \left\{ \begin{array}{c} \text{energy} \\ \text{exchanged} \\ \text{with} \\ \text{surroundings} \end{array} \right\} + \left\{ \begin{array}{c} \text{energy} \\ \text{generation} \end{array} \right\} \tag{3.277}$$

or more specifically

$$\frac{dQ}{dt} = (\dot{n} c_p T)_0 - \dot{n} c_p T + UA(T_c - T) + V \sum_j r_j(-\Delta H_{r,j}) \tag{3.278}$$

where Q is the heat (J), T the temperature of the reaction mixture (K), c_p the molar heat capacity (J/mol/K), $\dot{n}$ the molar flow (mol/s), U the overall heat transfer coefficient (W/(m²/K) or J/s/m²/K), A the heat exchange surface area in the reactor (m²), r the reaction rate (mol/s/m³), V the volume (m³), ΔH the enthalpy (J/mol), t the time (s).

In batchwise-operated stirred tank reactors, the reaction mixture is homogeneous without any concentration and temperature gradients. The general heat balance is then

$$(n\bar{c}_p + \bar{C}_W)\frac{dT}{dt} = UA(T_c - T) + V\sum_j r_j(-\Delta H_{r,j}) \tag{3.279}$$

The total heat capacity of the reactor $\bar{C}_W$ is considered to be temperature independent. The same is assumed for the average specific heat $\bar{C}_p$, which is additionally assumed to be constant throughout the reactions. This heat balance equation together with the mass balance equation describes the batch slurry reactor behavior and should be solved simultaneously. The time-dependent temperature profile is

$$\frac{dT}{dt} = \frac{UA}{n\bar{c}_p + \bar{C}_W}(T_c - T) + \Delta T_{ad}\frac{da}{dt} \tag{3.280}$$

With the adiabatic temperature rise defined as

$$\Delta T_{ad} = \frac{V(-\Delta H_r)c_{A,0}}{(n\bar{c}_p + \bar{C}_W)} \tag{3.281}$$

where $C_{A,0}$ is the initial concentration of the key component. The temperature profiles are thus determined by the heat of the reaction, dependence of the reaction on temperature, the adiabatic temperature rise and the heat removal rate. The latter is influenced by the average temperature difference between the cooling(heating) medium and reaction mixture as well as the heat exchange area A and coefficient U.

A constant coolant temperature implies a linear removal of heat with the temperature in the reactor, while the exponential increase of heat generation follows from the Arrhenius equation. Subsequently, extremely high-temperature peaks (so-called hot sports) can be generated.

For an ideal CSTR without any concentration gradients the heat balance at steady state is

$$\bar{V}_0\rho_0 c_{p0}T_0 - \bar{V}\rho c_p T = UA(T_c - T) + V\sum_j r_j(-\Delta H_{r,j}) = 0 \tag{3.282}$$

where the subscript "0" corresponds to the reactor inlet.

For the ideal plug flow reactor at steady state, there are concentration and temperature gradients along the reactor length without any gradients over the cross section in the radial direction. Moreover, any axial mass and heat dispersion is neglected.

The heat balance for a volume element in an ideal plug flow reactor is

$$\dot{n}\,\bar{c}_p \frac{dT}{dV} = \sum_j r_j(-\Delta H_{r,j}) + U(T_c - T)\frac{dA}{dV} \tag{3.283}$$

The heat balance equation for a tube with a constant diameter d_t takes the form

$$\frac{dT}{dZ} = \frac{U\tau}{\rho_0\,\bar{c}_p}\frac{4}{d_t}(T_c - T) + \Delta T_{ad}\frac{r\tau}{c_{A,0}} \tag{3.284}$$

where the adiabatic temperature rise is defined in the following way:

$$\Delta T_{ad} = \frac{(-\Delta H_r)c_{A,0}}{\rho_0\,\bar{c}_p} \tag{3.285}$$

Equation (3.284) should be solved simultaneously with the mass balance to determine the axial temperature and concentration profiles. It contains two terms, the first one determined by the heat exchange rate and heat capacity of the mixture and the second one is influenced by temperature dependence of the heat generation. Exponential dependence of the reaction rate in temperature implies potentially prominent hot spots in case of exothermic reactions.

3.6.6 Catalyst handing in a plant

3.6.6.1 Delivery to a plant

Catalysts are usually supplied to the plant sites in steel drums and should be handled as gently as possible to avoid additional attrition. In some cases, additional sieving could be done prior to loading the catalyst into a reactor, using a simple inclined screen without any vibrations, as such vibration screens might additionally damage the catalyst and lead to losses of the material.

> Sieving of catalysts sometimes is done after discharging them from a reactor. I was once involved in techno-economical negotiations about the purchase of industrial sieves for the separation of catalysts used in oxidation of SO_2. Such screens are needed when the spent catalyst from the top bed of a five-bed converter will be discharged after (for example) 1 or 2 years of operation, then screened and utilized in a lower bed. The sieves can be of vibrating nature and should be compact as they are placed at various levels of a converter. Such converters could have catalyst loading of 250–300 m^3 and be as large as a five-story building with a diameter of 10–13 m.

When the catalyst is sensitive to air moisture, the drums are kept closed until the reactor is filled, which will be discussed in the following section.

3.6.6.2 Catalyst charging

Catalyst charging in adiabatic fixed beds (Fig. 3.94) and multitubular reactors is an extremely important part of catalyst handling in the plant. Correct packing and even distribution ensure a satisfactory gas flow through the bed, while poor packing can lead to an unsatisfactory performance because of uneven gas flow.

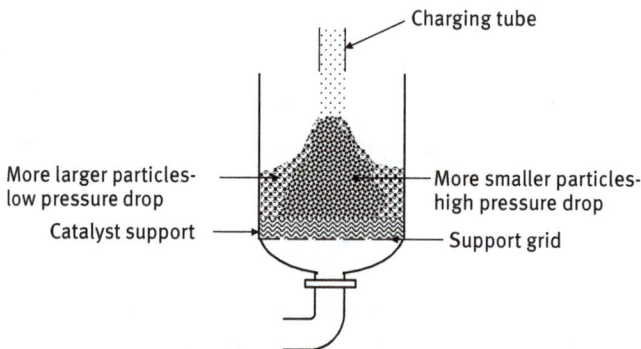

Fig. 3.94: Segregation of particles with one-spot charging.

Generally, when the catalyst is charged into a vessel it should not have a free fall of more than 0.5–1 m. Some catalyst forms (spheres) withstand a drop better than extrudates or soft pellets. Pouring a catalyst from a charging tube into a vessel results in a particle hill in one spot (Fig. 3.94), which should be eventually smoothed by raking or by a wooden plate. The charging tube is raised slowly to allow a controlled flow of catalyst in the vessel.

The heap is segregated, with smaller particles and dust staying in the center. This leads to a higher pressure drop and low flow velocity, while close to the edge the gas velocity is higher because of larger particles. The variations in the bed voidage could be ~10%. It should be noted that the uneven loading results not only in the distribution of the gas flow through the bed and thus pressure drop variations, but also in radial temperature gradients at the bed outlet. For example, in methanol synthesis with a temperature of 220 °C at the inlet and the average outlet temperature of 290 °C, a very pronounced hot spot corresponding to the hill center is observed, with the maximum outlet temperature of 315 °C.

In a dense loading method (Fig. 3.95), a special rotating device is used to spread the catalyst more evenly and diminish the void spaces and bridging. Higher catalyst bed density allows to load more catalyst into the same reactor volume and decrease channeling at the expense of a higher pressure drop.

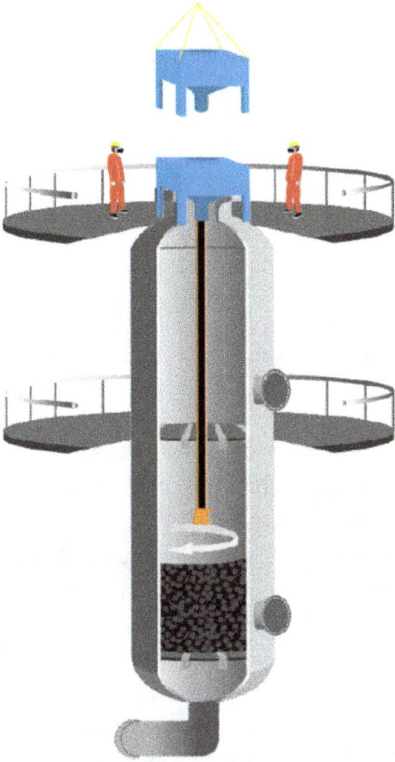

Fig. 3.95: Dense loading method principle [76].

Obviously, deep narrow beds are easier to charge, averaging of irregularities, while such even distribution might be challenging for wide beds (where charging is done manually with a conveyor, Fig. 3.96) represented, for example, by some sulfur dioxide oxidation reactors (Fig. 3.97) or radial flow beds.

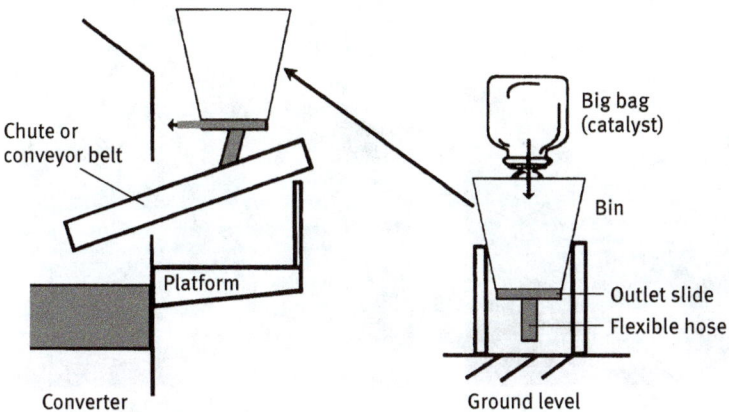

Fig. 3.96: Charging of a converter with a conveyor.

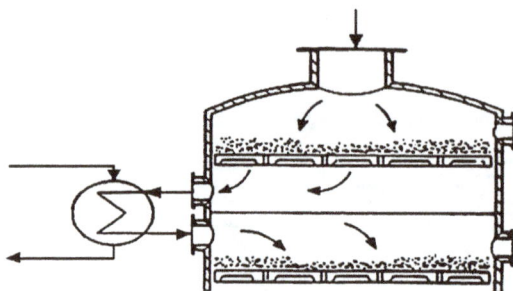

Fig. 3.97: The upper two beds of an SO$_2$ oxidation reactor. Access points are indicated on the right.

For a wide and narrow reactor (as shown in Fig. 3.97), operators have to enter the vessel through the side manholes during charging/discharging or when examining the bed at regular shut-downs. As walking on catalyst beds is not permitted, planks should be used. Although, in principle, raking can be used to evenly distribute the catalyst, this creates fines, particularly close to the manhole. Therefore, for such wide and shallow beds a small conveyor could be used, if it can fit into the side manhole and is long enough to be used across the vessel.

The size of access points on Fig. 3.97 might be a bit exaggerated. I had to take samples of the spent sulfuric acid catalyst by entering a reactor for SO$_2$ oxidation through a side manhole 1 m^2 while wearing overall protection and a gas mask, which was not very comfortable. The non-uniformity in distribution and also the "waves" of catalysts, similar to those in Fig. 3.97, were clearly visible as the bed had a diameter of 13 m and a catalyst bed height of ~0.5 m.

Special care should be also taken when charging catalysts in multitubular reactors [Fig. 3.98, such as those described in detail in Chapter 4 for steam reforming of methane and o-xylene oxidation)].

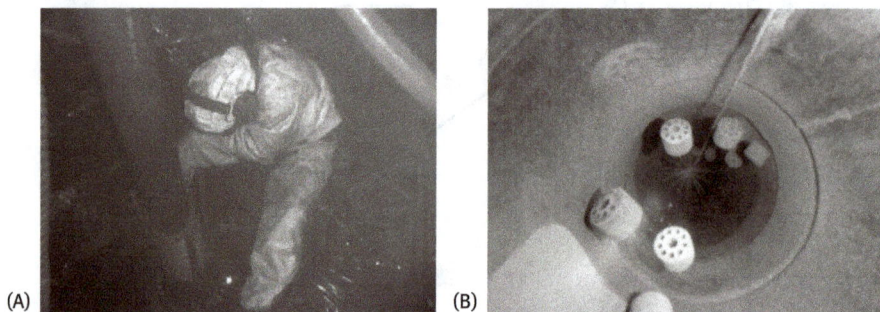

Fig. 3.98: Loading of multitubular reactors: (A) sock loading (from [77]); (B) loading using a brush (from [78]).

A sock method can be used for steam reforming, where the catalyst is first put in socks of 1.5–2 m in height and with an outside diameter that should be 15 mm less than the tube diameter, otherwise sliding will be troublesome (larger diameter) or unfolding of the lower portion of the sock might be incorrect (smaller diameter). The loose end of the sock is folded 20 cm from the bottom. The sock is introduced into the reactor and a strong cord, which is hooked onto the top, is then pulled, unfolding the loose end and releasing the catalyst. This is followed by topping up the tubes with the catalyst to an exact level, careful checking of the catalyst level and measuring the pressure drop in each tube. The pressure drop across each tube should be within 5% deviation of the mean value. Higher values of pressure drop indicate the presence of broken catalyst particles and require the tube to be discharged, the broken extrudates removed, and the tube to be recharged. When the pressure drops, low gaps in the packing are present and vibration of the tube is required. In fact, the tubes are vibrated periodically during charging either by electric vibrators or by gently hammering the top flange with a soft-faced hammer. A more detailed description with illustrations is available in [79].

Tab. 3.11 contains data for temperature and conversion variations (slippage of methane) obtained for a case of equal pressure drops, and for tubes showing pressure drop (or flow rate) variations. As shown in Tab. 3.11, when there are no variations in the flow rate, the methane slippage is 3.64%, while the methane slippage is higher when deviations are significant. In the latter case approach to equilibrium, which is a good indication of the catalyst activity, displays a high value of 9 °C, sometimes incorrectly ascribed to poor catalyst activity rather than to unsuccessful catalyst loading in the tubes. The catalyst performance is worsened further with operation time, as activity of the catalyst in different tubes will decline in a different way.

Tab. 3.11: Temperature and conversion variations methane steam reforming with differently loaded catalysts.

	Tubes with gas-flow deviations						No deviations
Flow rate							
% of tubes	10	20	40	20	10	all tubes	100
% of mean	80	90	100	110	120		100
T at outlet (°C)	922	894	870	848	843	872	870
CH₄ slippage [% (dry gas)]	1.64	2.55	3.64	4.9	6.54	3.86	3.64
Approach to equilibrium (°C)	1	1	2	3	3	9	2

(From [80]).

For multitubular reactors, other methods of catalyst loading are used besides the socks method. The SpiraLoad method, developed by Haldor Topsoe A/S, uses loading equipment that consists of a number of equally sized tube sections with spiral-shaped guide elements placed along the inner walls. The tube sections can be assembled and inserted in the reformer tubes to form one long loading tube (Fig. 3.99).

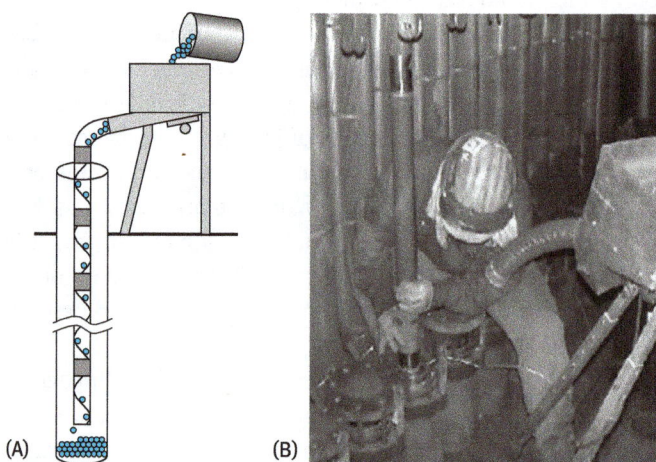

Fig. 3.99: The SpiraLoad method: (A) general scheme, (B) industrial practice. (From [81].).

Among other methods that were used in the past for loading steam reformer tubes (and that are still applied for other reactors), the use of ropes that contain knots along the rope length after ~1 m is of note. The catalyst can be loaded directly from drums. When the catalyst is experiencing free fall, the knots stop the fall and prevent catalyst damage and the formation of fines. Another option is to use metal brushes (Fig. 3.98B) attached to either a rope or a metal stick. Obviously, during loading the rope should be continuously pulled.

The procedure of using a rope with knots might sound very primitive, however I was once involved in the charging of a dehydrogenation multitubular reactor with 600 tubes, which was successfully loaded with catalyst tablets using this method.

For loading of other multitubular reactors, such as those used in o-xylene oxidation, special loading devices are available and are provided by the catalyst manufacturers. When complicated catalyst profiling along the tubes is required, as will be discussed in detail in Chapter 4, thousands of reactor tubes are first loaded with the reactor bottom type of catalysts, pressure drop is carefully monitored, the level of this catalyst layer is adjusted and the procedure is repeated with all other types of catalysts.

The justification for such a tedious procedure is that correct loading is very important for successful catalyst performance. Catalyst loading can be therefore very time-consuming and take up to several weeks depending on the catalyst and reactor types.

3.6.6.3 Catalyst activation, shutdown, restarts and discharging

The catalyst performance in a plant and its lifetime are influenced not only by the way the catalyst is charged into a reactor, but also by pre-treatment, activation and operation.

An activation procedure might involve, for example, sulfidation or reduction of metal oxide. When hydrodesulfurization catalysts are charged directly as supported oxides and put in contact with a feed containing sulfur compounds, the catalyst is progressively transformed into sulfides. Sulfidation of the precursor oxide can be performed either outside of the plant or specifically at the plant prior to regular operation.

Similarly, reduction of metal oxides can be done on-site and should be carefully controlled to achieve the best performance. For example, the reduction of palladium catalysts used in selective hydrogenation of acetylenic compounds is done after the first start-up and each regeneration. After regeneration by burning coke, the reactor is blanketed from air by nitrogen. The heating is done under inert gas flow up to the required temperature (~120 °C) with a specific heating gradient, not exceeding 50 °C per hour. When the reduction temperature is reached, hydrogen is slowly added to the inert gas. The temperature increase should be marginal to avoid overheating. The reduction proceeds for 10–12 h. The catalyst is cooled down to a normal operation temperature with the subsequent replacement of hydrogen with nitrogen, which is then used when the reactor is in idle mode.

Another more detailed reduction procedure of iron oxide an ammonia synthesis catalyst, is described in Chapter 4.

When I was working for a catalyst supplier almost two decades ago, I took part in the commissioning of the revamp of an ammonia synthesis reactor from axial to radial flow. The long-term performance of the catalyst was of major concern, not only for the plant operator, but also for the catalyst supplier, as it was part of the warranty. Thus, catalyst reduction with controlled temperature increase while monitoring the concentration of water exiting the reactor was done. As uneven reduction could take place, which is especially pronounced in radial flow converters because of shallow beds compared to axial flow reactors, special care was taken with respect to the gas flow. During the reduction procedure the mantra of the catalyst supplier was: "more flow".

During shut-downs and restarts, damage to the catalyst should be prevented, especially when the catalyst is in the reduced state and is pyrophoric.

I had once to fly by private jet to a remote plant site, where there was an emergency situation resulting in overheating of a WGS catalyst. Unfortunately, the catalyst had been damaged and despite all the efforts of the plant personnel to make a careful shutdown by slow cooling, the catalyst charge (which in that reactor was exceeding a hundred tons) had to be replaced.

In that case, the local overheating, which apparently exceeded 900 °C, was a result of an accident.

Two other examples not from my personal experience, but also related to WGS, illustrate more negligence during catalyst start-up and reduction. After loading the newly developed high-temperature WGS catalyst based on iron and lead (the story is from 1947 and happened in Novomoskovsk) in the reactor, there was a profound runaway when the catalyst was exposed to a mixture of 50% CO and 50% H_2. Apparently, a possibility of such runaway with a more active than the previous generation Fe–Mg catalyst was unnoticed in lab-scale experiments. The whole reactor was damaged, as not only the catalyst melted, but the same happened with the catalyst grid and thermocouple sockets.

In another example of a profound overheating, now from ammonia plant in Rovno in 1974, the low-temperature WGS catalyst based on copper experienced temperature of 1,350 °C (instead of 250 °C). During a start-up of the reactor with 80 tons of catalyst, after reaching 200 ° C under an inert gas (nitrogen), two plant operators decided to bring some innovation into the catalyst reduction procedure. Against all instructions instead of adding 0.3% of hydrogen, they decided that 15% of hydrogen would be more efficient to shorten the reduction time. Obviously, after the accident the whole catalyst charge had to be replaced and the damaged walls of the reactor had to be repaired.

For some catalysts that gradually deactivate, regeneration is required from time to time. As already mentioned, a palladium catalyst that is active in selective hydrogenation of triple bonds needs to be regenerated after ~6 months of operation. Regeneration can be undertaken with a steam/air mixture at temperatures between 400 and 450 °C. After blocking the process gas flow and depressurizing the reactor, it is heated under an inert gas (nitrogen) at a certain temperature ramp (~40 °C h^{-1}). At ~150 °C the inert gas is continuously replaced by overheated steam while heating to 400 °C. At this point, air is added to the steam and the temperature might increase slightly. Regeneration is carried out while controlling the outlet concentration of CO_2. At a certain value of CO_2 slip (~500 ppm), which indicates almost complete burn-off of carbon deposits, the air is switched off and cooling is carried out under steam to ~200 °C. Below this temperature cooling continues with nitrogen to the required reduction temperature (~120 °C, as mentioned above). Thereafter the catalyst is reduced. Reduction procedures, available from catalyst suppliers, typically include detailed information about exact values of temperatures and flow rates of steam, nitrogen, air, hydrogen, etc.

As air should not be in contact with the catalyst, an inert atmosphere (nitrogen or process gas) is kept in the plant where it is off-line. Nitrogen can be considered a suitable gas, as the oxygen content in it is below 10 ppm. Blanketing of the reduced catalyst is typically done under slight pressure with static nitrogen. When the reactor must be opened under a nitrogen blanket, the rule is to have one opening at a time. In some cases natural gas can be also applied for blanketing, however, it represents a fire hazard.

Discharging a catalyst that is not going to be used further is usually a simple process, where the discharge is done by gravity through the bottom of a vessel. Alternatively, suitable chutes, conveyors or vacuum equipment could be used to carry the catalyst out into drums or trucks.

Special care should be taken when a pyrophoric catalyst is discharged, as it should be done under a nitrogen blanket to prevent oxidation, overheating and potential damage to the reactor vessel. When the catalyst does not react with water, resulting in a build-up of hydrogen, filling of the vessel with water and draining it several times can be done prior to discharge as an alternative to discharge under blanketing.

Top discharge for a vessel with a large top opening could even be done by shoveling the catalyst into buckets attached to ropes and pulled out of the reactor. Obviously, the proper ventilation should be done prior to discharge and the personnel should have appropriate protective clothing.

Another possible method is for a top discharge is to use industrial vacuum cleaners.

> Sometimes catalyst discharge with vacuum cleaners is complicated by the required precision regarding the exact amount of catalyst to be removed. In one industrial example that I was involved in, one-third of the catalyst charge had to be withdrawn from a multitubular reactor because of the high-pressure drop, with a target of minimizing deviations in pressure drop in each tube after partial catalyst discharge.

3.6.6.4 Metal recovery

Catalysts containing nickel, copper, zinc and precious metals (PMs) may be sent for the recovery of these metals. The economics of such refining depends on the amount of metal, the plant location, as well as current metal prices. Thus the value of precious metals in catalysts easily exceeds the price of the catalyst and significantly exceeds the recovery cost. Moreover, PM recovery provides a safe and environmentally responsible disposal of contaminated catalysts. Recovery and recycle of PMs also minimizes the effect of metal market price fluctuations.

One of the most important economical issues is an accurate, transparent and reliable determination of PM content in the spent catalyst. A 2% deviation from this during the catalyst handling, weighing and/or sampling corresponds to a 2% deviation in PM credit, which for a catalyst with Pt content of about 0.3% and a price of $2100/ troz is translated into a US $400,000 discrepancy for 10 tons of spent catalyst.

The refining process can be either pyro-metallurgical (i.e., smelting) or a hydrometallurgical recovery (i.e., a wet chemical process). The modern smelters offer a broad range of applications as, for example, refining Pt, which is not dissolved in sodium hydroxide or sulfuric acid, the same can hold for such insoluble supports as carbon or zeolites.

A flow diagram of the supercritical water oxidation process, based on the Chematur Engineering supercritical water oxidation process, is presented in Fig. 3.100. Water is pressurized to ~240 bar and then heated to ~385 °C to ensure formation of the supercritical phase. Heterogeneous catalysts are fed in a slurry form from the feed tank through an economizer prior to oxygen treatment. Homogeneous catalysts are injected into the reactor after the oxygen points to avoid the pyrolysis of elements such as phosphorus during preheating.

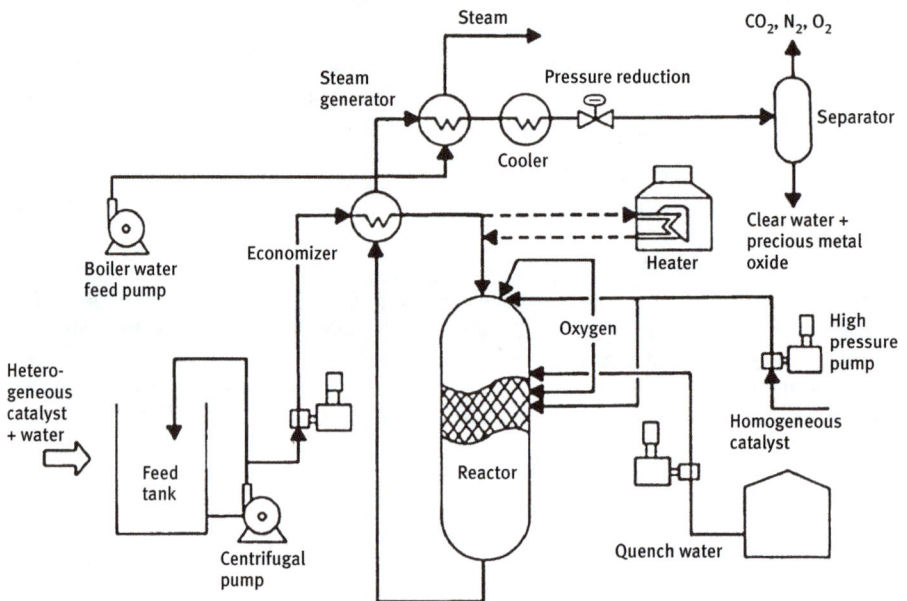

Fig. 3.100: Johnson Matthey-Chematur Engineering process flow scheme. (From [82]). Copyright Johnson Matthey.].

References

[1]   Kusema, B.T., Mikkola, J.-P., Murzin, D.Yu. (2012) Kinetics of L-arabinose oxidation over supported gold catalyst with in-situ catalyst electric potential measurements. Catal. Sci. Technol. 2: 423–431.

[2]   Murzin, D.Yu., Salmi, T. (2016) Catalytic Kinetics. Science and Engineering, Amsterdam, The Netherlands: Elsevier.

[3]   Murzin, D.Yu. (1995) Modeling of adsorption and kinetics in catalysis over induced nonuniform surfaces: surface electronic gas model. Ind. Eng. Chem. Res. 34: 1208.

[4] Kiss, V., Osz, K. (2017) Double exponential evaluation under non-pseudo –first order conditions. A mixed second-order process followed by a first-order reaction, Int. J. Chem. Kinet., 49: 602–610.

[5] Lente, G., (2015), Kinetics of irreversible consecutive processes with first order second steps: analytical solutions, J. Mathem. Chem., 53: 1172–1183.

[6] Murzin, D.Yu., Simakova, I.L., Wärnå, J. (2019) Selectivity analysis for networks comprising consecutive reactions of second and first order, Int. J. Chem. Reactor Eng, 17: https://doi.org/10.1515/ijcre-2018-0161.

[7] Boudart, M. (1969) Catalysis by supported metals. Adv. Catal. 20: 153–166.

[8] Murzin, D.Yu. (2010) Size dependent heterogeneous catalytic kinetics. J. Mol. Catal. A. Chem. 315: 226.

[9] Murzin, D.Yu. (2010) Kinetic analysis of cluster size dependent activity and selectivity. J. Catal. 276: 85.

[10] van Hardeveld, R., Hartog, F. (1969) The statistics of surface atoms and surface sites on metal catalysts. Surface Sci. 15: 189.

[11] Lazman, M.Z., Yablonskii, G.S. (2008) Overall reaction rate equation of single-route complex catalytic reaction in terms of hypergeometric series. Adv. Chem. Eng. 34: 47.

[12] Glasstone, S., Laidler, K., Eyring, H. (1941) The Theory of Rate Processes. New York: McGraw Hill.

[13] Eyring, H. (1935) The activated complex in chemical reactions. J. Chem. Phys. 3:107.

[14] Temkin, M. (1938) Transition state in surface reactions. Acta Physicochimica. URSS, 8:141

[15] Campbell, C. T., Arnadottir, L., Sellers, J.R.V. (2013), Kinetic prefactors of reactions on solid surfaces, Z. Phys. Chem. 227: 1.

[16] Dumesic, J.A., Rudd, D.F., Aparicio, L.M., Rekoske, J.E., and Trevino, A.A. (1993) The Micro-Kinetics of Heterogeneous Catalysis. ACS Professional Reference Book, American Chemical Society, Washington, DC.

[17] Kiselev, V. D., Miller, J. G. (1975) Experimental proof that the Diels-Alder reaction of tetracyanoethylene with 9,10-dimethylanthracene passes through formation of a complex between the reactants, J. Am. Chem. Soc., 97: 4036.

[18] Kiselev, V. D., Konovalov, A.I. (2009) Internal and external factors influencing the Diels–Alder reaction, J. Phys.Org. Chem. 22: 466.

[19] Han, X., Lee, R., Chen, T., Luo, J., Lu, Y., Huang, K.-W. (2013) Kinetic evidence of an apparent negative activation enthalpy in an organocatalytic process, Sci. Rep., 3: 2557.

[20] Wei, J. (1996), Adsorption and cracking of n-alkanes over ZSM-5: Negative activation energy of reaction. Chem. Eng. Sci., 51: 2995.

[21] Moulijn, J., Diepen, van A.E., Kapteijn, F. (2001) Catalyst deactivation: is it predictable? What to do? Appl. Catal. A-Gen. 212:3.

[22] Cornish-Bowden, A. (2001) Detection of errors of interpretation in experiments in enzyme kinetics. Methods 24: 181.

[23] Kapteijn, F., Berger, R.J., Moulijn, J.A. (2008) Rate procurement and kinetic modeling. In: Ertl, G., Knözinger, H., Schüth, F., Weitkamp, J., editors. Handbook of Heterogeneous Catalysis. Weinheim, Germany: Wiley-VCH, p.1693.

[24] Gianetto, A., Silveston, P.L. (1986) Multiphase Chemical reactors. Theory, Design and Scale-Up. Hemisphere Publishing Corporation, Washington.

[25] Hájek, J., Murzin, D.Yu. (2004) Liquid-phase hydrogenation of cinnamaldehyde over Ru-Sn sol-gel catalyst. Part I. Evaluation of mass transfer via combined experimental/theoretical approach. Ind. Eng. Chem. Res. 43: 2030–2038.

[26] Murzin, D.Yu., Konyukhov, V.Yu., Kul'kova, N.V., Temkin, M.I. (1992) Diffusion from surfaces of suspended particles and specific mixing power in shaker reactors. Kinet. Catal. 33:728.

[27] Hemrajani, R., Tatterson, G.B. Chapter 6. Mechanically Stirred Vessels. In: Paul, E.L., Atiemo-Obeng, V.A., Kresta, S.M. Handbook of Industrial Mixing: Science and Practice, DOI: 10.1002/0471451452.ch6. John Wiley and Sons. Hoboken, NJ, USA.

[28] Sherwood, T.K., Pigford, R.L., Wilke, C.R. (1975) Mass Transfer. New York: McGraw-Hill.

[29] Salmi, T.O., Mikkola, J.-P., Wärnå, J.P. (2019) Chemical Reaction Engineering and Reactor Technology. Second Edition. Boca Raton, FL: CRC Press.

[30] Kolodziej, A., Lojewska, J. (2013), Engineering aspects of catalytic converter designs for cleaning of exhaust gases. In: S.L. Suib, A.V. Sapre, R.J. Katzer (Eds.), New and Future Developments in Catalysis: Catalysis for Remediation and Environmental Concerns. Amsterdam, p.257.

[31] Dittmeyer, R., Emig, G. (2008) Simultaneous heat and mass transfer and chemical reaction. In: Ertl, G., Knözinger, H., Schüth, F., Weitkamp, J., Editors. Handbook of Heterogeneous Catalysis. Weinheim, Germany: Wiley-VCH, pp. 1727–1784.

[32] Moulijn, J.A. (2003) NIOK course on catalysis, Lecture notes.

[33] Ioffe, I.I., Reshetov, V.A., Dobrotvorskii, A.M. (1985) Heterogeneous catalysis. Leningrad, Chimia.

[34] Sandelin, F., Oinas, P., Salmi, T., Paloniemi, J., Haario, H. (2006) Dynamic modelling of catalytic liquid-phase reactions in fixed beds-Kinetics and catalyst deactivation in the recovery of anthraquinones. Chem. Eng. Sci. 61: 4528–4539.

[35] Leveneur, S., Wärnå, J., Eränen, K., Salmi, T. (2011) Green process technology for peroxycarboxylic acids: Estimation of kinetic and dispersion parameters aided by RTD measurements: Green synthesis of peroxycarboxylic acids. Chem. Eng. Sci. 66: 1038–1050.

[36] Aris, R. (1976) The Mathematical Theory of Diffusion and Reaction in Permeable Catalysis, vol.1. Oxford: Clarendon Press.

[37] Mears, D.E. (1970) Diagnostic criteria for heat transport limitations in fixed bed reactors. J. Catal. 20: 127.

[38] Mears, D.E. (1971) Tests for transport limitations in experimental catalytic reactors. Ind. Eng. Chem. Proc. Dd. 10: 541.

[39] Anderson, J.B. (1963) A criterion for isothermal behavior of a catalyst pellet. Chem. Eng. Sci. 18: 147.

[40] Weisz, P.B., Prater, C.D. (1954) Interpretation of measurement in experimental catalysis. Adv. Catal. 6: 143.

[41] Villermaux, J. (1993) Génie de la réaction chimique – Conception et fonctionnement des réacteurs. Tec&Doc Lavoisier. Paris.

[42] Weisz, P.B., Hicks, J.S. (1962) The behaviour of porous catalyst particles in view of internal mass and heat diffusion effects. Chem. Eng. Sci. 17: 265.

[43] Koros, R.M., Nowak, E.W. (1967) A diagnostic test of the kinetic regime in a packed bed reactor. Chem. Eng. Sci. 22: 470.

[44] Madon, R.J, Boudart. (1982) Experimental criterion for the absence of artifacts in the measurement of rates of heterogeneous catalytic reactions. Indust. Eng. Chem. Fund. 21: 438.

[45] Temkin, M., Kul'kova, N. (1969) Ideal-displacement laboratory reactor. Kinet. Katal., 10: 461.

[46] Mestl, G. (2012). High throughput development of selective oxidation catalysts at Sued-Chemie, Combinat. Chem High Throughput Synth. 15: 144.

[47] http://www.wraconferences.com/wp-content/uploads/2016/11/ILS-ERTC-Presentation-V3b-2.pdf

[48] Temkin, M. I., Kiperman, S. L., Luk'yanova, L. I. (1950) Flow-circulation method of investigation of the kinetics of heterogeneous catalytic reactions. Dokl. Akad. Nauk SSSR 74: 763.

[49] Stitt, E.H. (2002) Alternative multiphase reactors for fine chemicals: a world beyond stirred tanks? Chem. Eng. J. 90: 47.

[50] Bartolomew, C.H., Farrauto, R.J. (2006) Fundamentals of Industrial Catalytic Processes. Wiley. Hoboken, NJ.

[51] Eigenberger, G., Ruppel, W. (2012) Catalytic Fixed-Bed Reactors, Ullmann's Encyclopedia of Industrial Chemistry. 10.1002/14356007.b04_199.pub2.

[52] http://www.uhde.eu/en/competence/technologies/hydrogen/129/133/methanol-reactors.html.

[53] Gunjal, P.R., Ranade, V.V. (2016) Catalytic Reaction Engineering, in Industrial Catalytic Processes for Fine and Specialty Chemicals, http://dx.doi.org/10.1016/B978-0-12-801457-8.00007-0, Elsevier.

[54] Mederos, F.S., Ancheyta, J. (2007) Mathematical modeling and simulation of hydrotreating reactors: Cocurrent versus countercurrent operations. Appl. Catal. A: Gen. 332: 8.

[55] Sato, Y., Hirose, T., Takahashi, F., Toda, M. (1972), Performance of fixed-bed catalytic reactor with cocurrent gas liquid flow, Pac. Chem. Eng. Cong, Sess. 8, Pap. 8–3: 187–196.

[56] Mederos, F.S., Ancheyta, J., Chen, J. (2009) Review on criteria to ensure ideal behaviors in trickle-bed reactors. Appl. Catal. A: G. 355: 1.

[57] Wang, T., Wang, J., Jin, Y. (2007) Slurry reactors for gas-to-liquid processes: a review. Indust. Eng. Chem. Res. 46: 5824–5847.

[58] Kantarci, N., Borak, F., Ulgen, K.O. (2005) Bubble column reactors. Proc. Biochem. 40: 2263.

[59] Deckwer, W.-D., Louisi, Y., Zaidi, A., Ralek, M. (1980) Hydrodynamic Properties of the Fischer-Tropsch Slurry Process. Ind. Eng. Chem. Proc. DD. 19: 699.

[60] Post, T. (2010) Understanding the real world of mixing. Chem. Eng. Proc. 25–32.

[61] Daniel A. Hickman, D.A., Holbrook, M.T., Mistretta, S., Rozeveld, S.J. (2013), Successful scale-up of an industrial trickle bed hydrogenation using laboratory reactor data, Ind. Eng. Chem. Res. 52: 15287–15292

[62] Trambouze, P., van Landeghem, H., Wauquier, J.P. (1988) Chemical Reactors – Design/Engineering/Operation. Paris: Editions Technip.

[63] Loosdrecht, van de, J., & Niemantsverdriet, J. W. (2012). Synthesis gas to hydrogen, methanol, and synthetic fuels. In R. Schlogl (Ed.), Chemical Energy Storage. Walter de Gruyter GmbH, pp. 443–458. DOI: 10.1515/9783110266320.443.

[64] http://what-when-how.com/energy-engineering/coal-to-liquid-fuels-energy-engineering/

[65] van der Laan, G.P. (1999) Kinetics, Selectivity and Scale Up of the Fischer-Tropsch Synthesis, PhD Thesis, University of Groningen.

[66] Advanced liquid biofuels synthesis. Adding value to biomass gasification. ECN-E–17-057 – February 2018, www.ecn.nl

[67] Moulijn, J. A., Makkee, M.I., van Diepen, A.E. (2013) Chemical Process Technology, 2nd Edition. Chichester, Wiley.

[68] http://www.inclusive-science-engineering.com/wp-content/uploads/2012/01/Production-of-Methanol-from-Synthesis-Gas.png

[69] http://www.gbhenterprises.com

[70] http://www.gbhenterprises.com/methanol%20converter%20types%20wsv.pdf

[71] http://www.slideshare.net/GerardBHawkins/methanol-flowsheets-a-competitive-review

[72] Salehi, K., Jokar, S.M., Shariati, J., Bahmani, M., Sedghamiz, M.A., Rahimpour, M.R. (2014) Enhancement of CO conversion in a novel slurry bubble column reactor for methanol synthesis, J. Natur Gas Sci. Eng. 21: 170.

[73] https://www.osti.gov/servlets/purl/823132

[74] Tirronen, E., Salmi, T. (2003) Process development in the fine chemical industry. Chem. Eng. J. 91: 103.

[75] Dutta, S., Gualy, R. (2000) Reactor models. Chem. Eng. Prog. N10, 37–51.

[76] https://www.catalystseurope.eu/images/Documents/ECMA1004q_-_Catalyst_handling_best_practice_guide.pdf

[77] http://www.crealyst.fr/lang/en-us/activities-2.

[78] http://www.catalysthandling.com/content/services/catalyst-loading-services.

[79] Twygg, M. (1997) Handbook of Industrial Catalysis. Manson Publishing.

[80] Farnell, P.W. Methods of charging and discharging of steam reforming catalysts, ICI Katalco, 410W/037/0/RUS.

[81] http://www.topsoe.com/business_areas/hydrogen/~/media/PDF%20files/Steam_reforming/ Topsoe_spiraload_technology.ashx.

[82] Grumett, P. (2003) Precious metal recovery from spent catalysts. Platin. Met. Rev. 47: 163.

4 Engineering technology

4.1 General structures of chemical processes

This chapter will discuss the practical implementation of heterogeneous catalytic processes. The general structure of chemical processes is presented in Fig. 4.1.

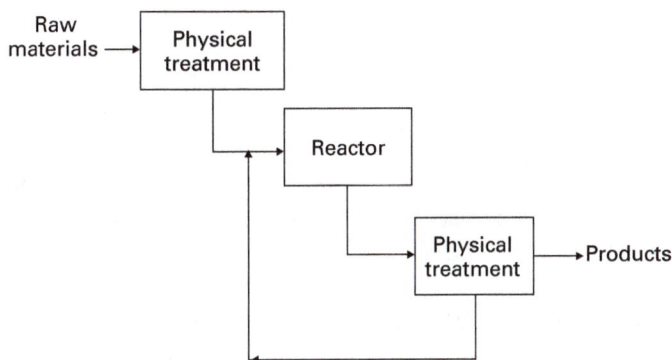

Fig. 4.1: General structure of chemical processes.

As can be seen in Fig. 4.1 the main features of the chemical processes could be identified as: feed purification, reaction, separation and product purification.

Chemical technology is the science of those operations, which convert raw materials into desired products on an industrial scale, applying one or more chemical conversions. A new technology can succeed only if it is robust, reliable, safe, clean, cheap, easy to control, and if it provides significant gains over existing processes. Moreover, technology should be based on sound economical considerations, safety requirements and labor conditions.

It is important to note that in chemical technology the process should be viewed in its whole complexity, rather than a combination of individual steps. For example, the performance of a reactor unit can depend on the performance not only of the units located upstream to reactor vessel, but also downstream. Obviously the upstream units would influence the inlet composition or the purity of the feedstock, and thus have an impact on the reactor. Selectivity changes made by adjusting the catalyst performance would influence separation. A loss of pressure downstream from the reactor could lead to an increase of pressure in the reactor and might damage the catalyst support grid. Improvement of one unit (reactor) usually improves the overall performance. Thus, a slight improvement in catalyst selectivity would sometimes result in very high savings in separation.

https://doi.org/10.1515/9783110614435-004

It should be also mentioned that optimum conditions for a system element are not necessarily optimum for the system as a whole. Thus the optimization of a particular chemical technology should include not only the reactor unit, but other units as well.

Process technology design includes a number of aspects to be considered: the reactions involved, thermodynamics, operating parameters such as temperature and pressure, kinetics, conversion levels and thus a need for recycle, separation of products, types of catalysts (homogeneous vs heterogeneous), catalyst stability, deactivation and regeneration if needed, safety aspects, environmental issues, corrosion, etc.

Such process design involves quite complicated flow charts and is not a straightforward application of the disciplines on which chemical technology is based (chemistry, physical transport, unit operations, reactor design), but rather the integration of this knowledge.

Complications also arise from having to choose from many possibilities, taking into account product markets, geographical location, social situation, legal regulations, etc. and that the final result must be economically attractive.

As an example, we can consider the synthesis of caprolactam from benzene (Fig. 4.2). In fact, various options are available: hydrogenation of benzene in the liquid or gas phases to cyclohexane; partial hydrogenation of benzene to

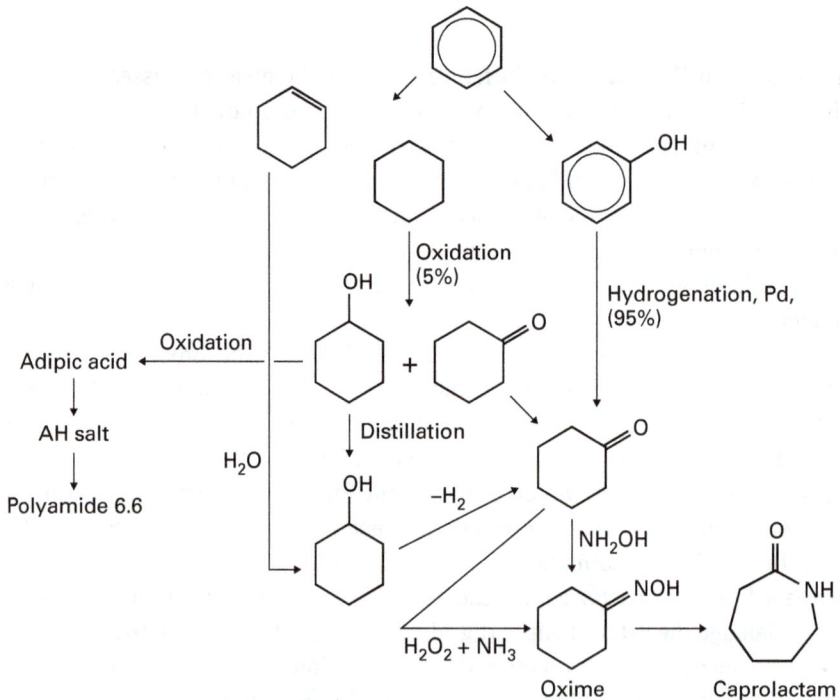

Fig. 4.2: Several routes for caprolactam synthesis.

cyclohexene; formation of phenol from benzene first with subsequent hydrogenation to cyclohexanone or cyclohexanol. Several options are available for Beckmann rearrangement, synthesis of cyclohexanone oxime and even the production of hydroxylamine. For example, Beckmann rearrangement of cyclohexanone oxime to caprolactam can be done in the liquid phase using sulfuric acid as a catalyst or alternatively in the gas phase over a zeolitic catalyst.

In addition to the routes sketched in Fig. 4.2 other options are also possible, thus for instance BASF-Du Pont technology relies on synthesis of adiponitrile (ADN) from butadiene by two-step hydrocyanation, partial hydrogenation of ADN to hexamethylenediamine and 6-aminocapronitrile, with this being converted to caprolactam through a cyclization reaction.

Many of the routes presented in Fig. 4.2 are used commercially, emphasizing the fact that there is not only one way to produce caprolactam, and many variants for a particular technology are possible.

Regarding the requirements for catalysts that were mentioned in Chapter 1 – they should be active, selective and stable. For reactors, a requirement of high selectivity means low capacities and thus savings in capital costs. High selectivity also implies few separation steps for product purification. Catalyst stability is an important issue and will be discussed below in greater detail.

4.1.1 Safety in design

The need for adequate design of a particular process, including safety issues, should not underestimated. It is helpful to consider the worst industrial disaster of the twentieth century, which occurred on December 2nd 1984 at the Union Carbide Corporation, in the city of Bhopal, in India, which had a population of about one million people. Over 40 tons of methyl isocyanate (MIC), as well as other lethal gases including HCN leaked from the plant side to the city. There are different numbers available in the literature about the casualties, but according to the Bhopal Peoples Health and Documentation Clinic, 8,000 people were killed in immediate aftermath of the leak, and over 500,000 people suffered from injuries.

On the night of the disaster, water that was used for washing the lines entered the tank containing methyl isocyanate through leaking valves. The refrigeration unit designed to keep the methyl isocyanate close to 0 °C had been shut off in order to save on electricity costs. The entrance of water to the tank, which was full of methyl isocyanate at an ambient temperature, initiated an exothermic runaway process and the subsequent release of the gases. The safety systems, which were not properly designed to handle such runaway situations, were non-functioning and under repair.

Unfortunately workers ignored early signs of disaster, as gauges measuring temperature and pressure in the various parts of the unit, including the methyl isocyanate storage tanks, were known to be unreliable.

It was assumed that the methyl isocyanate would be kept at low temperatures by the refrigeration unit; however, this unit was shut off. In addition the gas scrubber, meant to neutralize methyl isocyanate if released, had been shut off for maintenance.

In any case, the design was inappropriate, as the maximum designed pressure was only 25% of that actually reached during the disaster.

Moreover, the flare tower, which was installed to burn off escaping methyl isocyanate, was not in operation awaiting the replacement of a corroded piece of pipe. Even if it had been in operation it would have only been able to process a fraction of the gas released. There were some other reasons for the disaster, such as a too short water curtain, a lack of effective warning systems and a failure of the alarm on the storage tank to signal a temperature increase. Overfilling of the storage tank beyond the recommended capacity and methyl isocyanate filling a reserve tank, which was supposed to be empty, are added to the overall picture.

I remember the day in 1984 that the news about the disaster broke, as I was doing some experiments in the lab, synthesizing substituted carbamates using alkyl isocyanates.

In the case of the Bhopal disaster, there are many reasons for such unfortunate events, including poor maintenance, design and inventory excess.

It would have been possible to prevent the release of methyl isocyanate if the technology had been organized in a different way. As illustrated in Fig. 4.3, methyl isocyanate was formed by a reaction of methylamine with phosgene, with a subsequent reaction with 1-naphtol. If the process had been designed in another way, for example, phosgenation of naphthol, the release of methyl isocyanate could have been avoided.

4.1.2 Conceptual process design: examples

Before discussing conceptual process design, a few words should be mentioned about flow diagrams in general. These diagrams should provide a clear simple outline of the steps involved in the process and must cover all the major steps, including the steps before and after the main chemical processing (Fig. 4.4).

Fig. 4.4, for example, demonstrates cyclohexanol dehydrogenation in a multitubular reactor. An interesting feature is a heat exchanger upstream reactor, which is used by the products to warm-up the reactants. The scheme also comprises a distillation column. A stream of unreacted cyclohexanol from this column back to the reactor points out incomplete conversion during the reaction.

Another example of heat integration is shown in Fig. 4.5 for catalytic incineration of volatile organic compounds, comprising an efficient heat exchanger in addition to a burner for preheating the gases and a catalyst chamber with monolithic or pelletized catalyst.

$$CH_3NH_2 \; + \; COCl_2 \; \longrightarrow \; CH_3N{=}C{=}O \; + \; 2\,HCl$$

Methylamine   Phosgene   Methylisocyanate

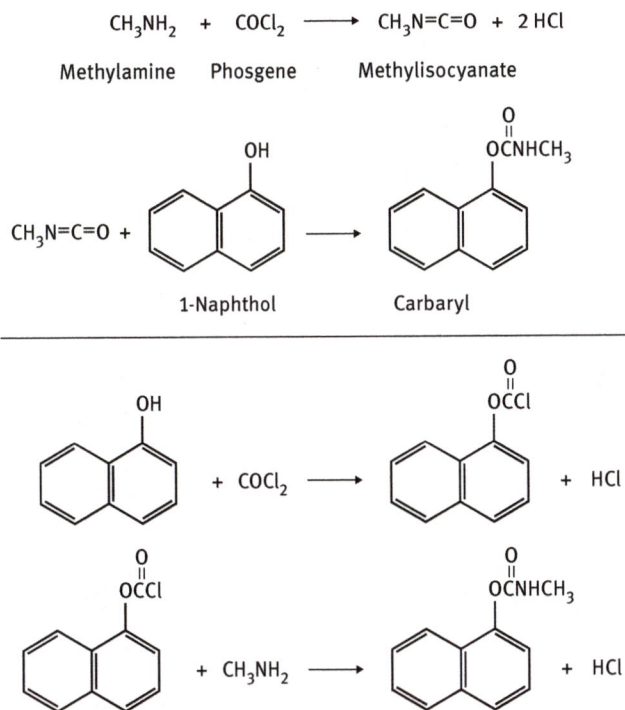

Fig. 4.3: Alternative pathways for synthesis of carbaryl. Upper scheme: process at UC Bhopal plant, lower scheme: alternative design.

A block type diagram is usually sufficient for understanding and evaluating the process flow and complex engineering drawings (Fig. 4.6) are not required.

Let us consider the treatment of V.S. Beskov and V.S. Safronov in [2] as an example of conceptual process design synthesis of nitric acid. The process consists of several steps. Initially, ammonia is combusted to form nitric oxide:

$$4NH_3 + 5O_2 \; \rightarrow \; 4NO + 6H_2O \qquad (4.1)$$

This reaction is highly exothermic and occurs at high temperatures of 850–900 °C over platinum catalysts. The residence time should be minimized (using high flow rates) to prevent side reactions, such as:

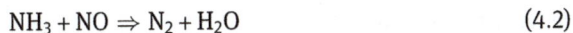

$$NH_3 + NO \; \Rightarrow \; N_2 + H_2O \qquad (4.2)$$

In order to achieve this, rather atypical Pt gauzes are utilized. As the reaction rate is fast, the influence of external diffusion at such conditions can be prominent. The influence of reaction parameters on this reaction is not straightforward. With an increase of temperature the reaction rate increases at the expense of selectivity. An increase of pressure similar to temperature enhances the reaction rate, but leads to

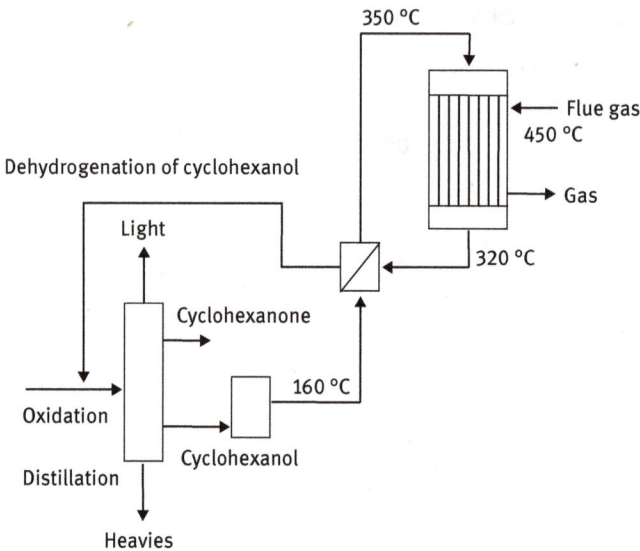

Fig. 4.4: A diagram of cyclohexanol dehydrogenation.

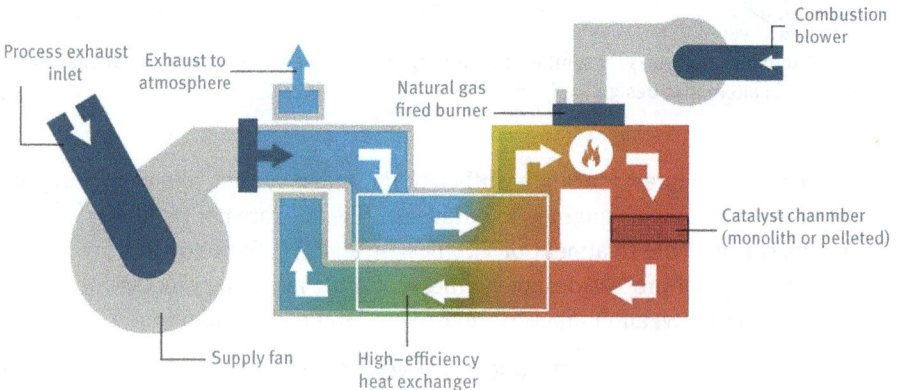

Fig. 4.5: A diagram of heat integration in oxidation of VOC [1].

higher metal losses. Because of severe conditions the catalyst lifetime is limited usually to one year.

The gauzes gradually deactivate because of platinum volatilization in the form of platinum oxide (Fig. 4.7a).

The fresh gauze is initially activated for several hours, leading to formation of crystallites with more active crystallographic planes and increase in the surface area of palladium. Thereafter, gradual deactivation of the gauzes occurs because of platinum oxide volatilization. Elevation of the severity of operation from 800°C and

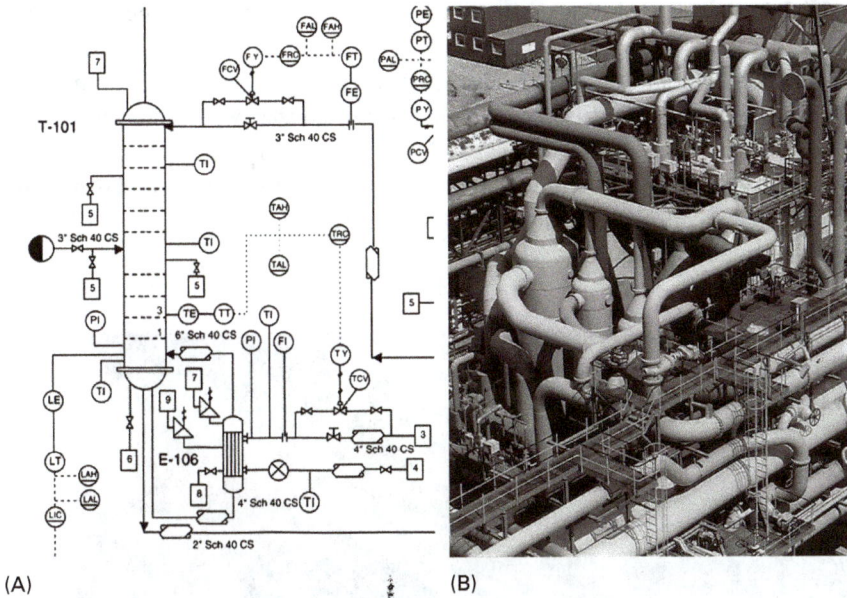

(A) (B)

Fig. 4.6: Examples of (A) engineering drawings and (B) complex piping networks.

(A) (B)

Fig. 4.7: Catalysts for ammonia oxidation: (A) comparison between the fresh and deactivated 90% Pt–10% Rh commercial catalyst [3], and (B) an assembled pack of platinum–rhodium gauzes [4].

atmospheric pressure to 900°C and 8 bars leads to four- to eightfold increase of the lost palladium amount. Recovery of palladium was first realized by placing a woven Pd-rich alloy (platinum–palladium) gauze immediately below the oxidation gauze, giving 70% recovery. Further improvement was introduction of 90% platinum/10% rhodium alloy not only reducing platinum losses and increasing lifetime but improving as well selectivity to more than 94%. Gauzes are changed when the amount of Rh is increasing to 12% with a subsequent decrease of platinum content. As there is a certain activation of the catalyst after the start-up, the fresh low-activity gauze is always packed below the used gauze. Apart from Pt/Rh gauzes also 90% Pt + 5%

Rh + 5% Pd alloy is used. Such gauzes are typically woven or knitted using wire diameters between 0.06 and 0.12 mm.

A pack installed in the reactor (Fig. 4.7b) can contain up to 40 platinum–rhodium gauzes with the catalysis occurring in just few gauzes, while the rest helps to mitigate deactivation and improve heat and flow distribution.

Monolith nonplatinum catalysts for oxidation of ammonia have been introduced in commercial nitric acid plants (Fig. 4.8) for partial substitution of Pt-Rh gauzes to decrease platinum losses.

Fig. 4.8: Monolith nonplatinum catalysts for oxidation of ammonia in the production of nitric acid [5].

After the catalytic reaction in the subsequent step, oxidation of nitric oxide to nitrogen dioxide is done:

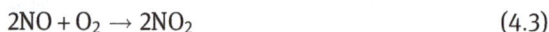

$$2NO + O_2 \rightarrow 2NO_2 \tag{4.3}$$

This is a gas-phase, reversible exothermic reaction, which does not require the presence of any catalysts. Thereafter nitrogen dioxide is absorbed in water to form nitric acid in a heterogeneous gas-liquid process:

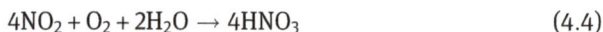

$$4NO_2 + O_2 + 2H_2O \rightarrow 4HNO_3 \tag{4.4}$$

This reaction is more complicated as other components can react (NO, N_2O_3, N_2O_4, etc.) in addition to NO_2. In fact the main reactions happening in an absorption tower could be written in the following way:

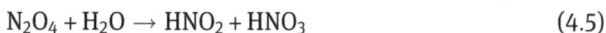

$$N_2O_4 + H_2O \rightarrow HNO_2 + HNO_3 \tag{4.5}$$

$$3HNO_2 \rightarrow HNO_3 + H_2O + 2NO \tag{4.6}$$

A preliminary flow diagram for nitric acid synthesis is given in Fig. 4.9.

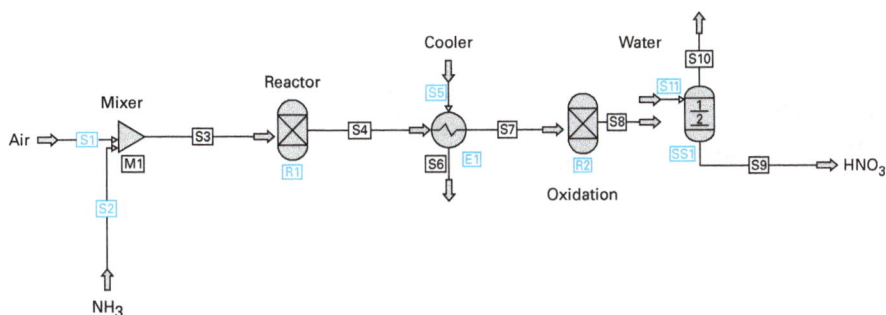

Fig. 4.9: Flow diagram of nitric acid synthesis (first iteration).

The flow diagram in Fig. 4.9 consists of a mixer, where air and ammonia are mixed, followed by an adiabatic reactor, and then a cooler to make the oxidation reaction as complete as possible. This is important because reaction 4.4 has an unusual temperature dependence and is more active at low temperatures. This reaction represents an example of tri-molecular reactions, whose temperature dependence can be explained within the framework of the transition state theory, taking into account temperature dependence of the partition functions for the transition state and reactants.

Several improvements should be made however in Fig. 4.9 as it suffers from obvious deficiencies. For example, air will not flow to a mixer without a blower or compressor, which should be added to the flow scheme. Fig. 4.9 shows that ammonia enters the mixer as a liquid. Dosing of low-boiling liquids is difficult and thus ammonia evaporation should also be included. As mentioned above, the oxidation of ammonia is an exothermic reaction. The reaction temperature is 850–900 °C and the adiabatic heat release is equivalent to 720 °C, thus the reactor inlet temperature should be ~130–180 °C. This implies that a heat exchanger should be installed upstream to the catalytic reactor.

Furthermore, NO oxidation is also exothermic, and thus gases are heated during this reaction. As a result, the absorption of NO_2 is worsened. In order to circumvent this a cooler upstream absorber would be required.

In addition, the heat integration in Fig. 4.9 is far from optimized. At least high-energy gases after the reactor could be used for steam generation, and thus a steam reclaimer downstream to the reactor could be applied.

Finally, an optimum ammonia/oxygen ratio in the reactor should be 1:1.8, but two volumes of O_2 per one volume of NH_3 are required for the acid production, and therefore extra air could be added to the absorber. With these changes, the next iteration of the nitric acid flow scheme is illustrated in Fig. 4.10.

This scheme can be still improved by considering the following issues. The catalyst is sensitive to impurities; however, no cleaning of air is used in Fig. 4.10. Air

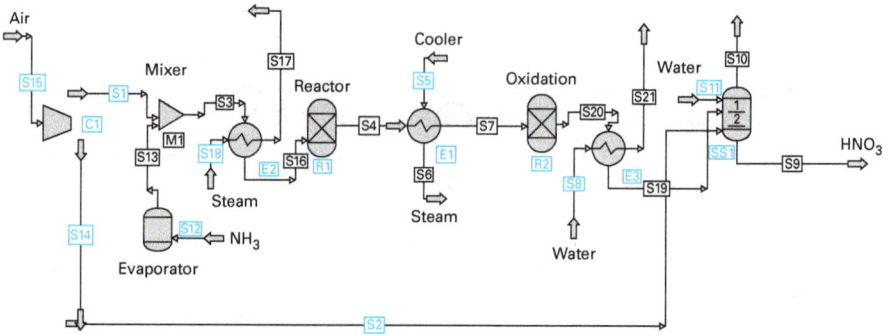

Fig. 4.10: Flow diagram of nitric acid synthesis (second iteration).

filtration should be then installed in the third iteration scheme (Fig. 4.11). The heat exchanger network in Fig. 4.10 is not rational and could be significantly improved, for example the gases coming from the reactor could be used to heat the reactants and a heat exchanger upstream to the reactor could be arranged in a similar way to that in Fig. 4.4.

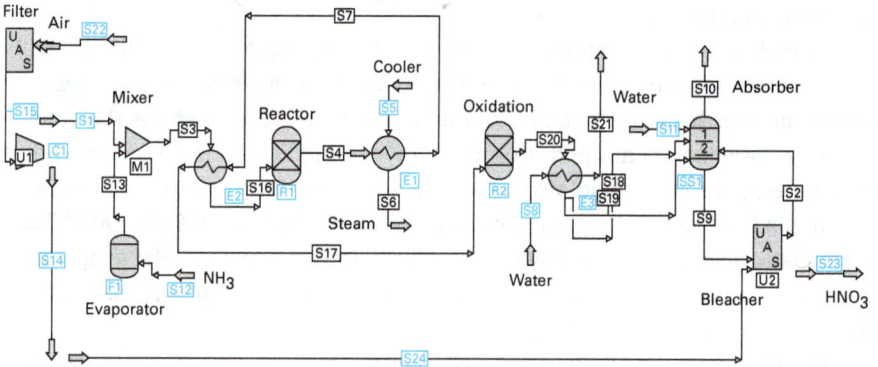

Fig. 4.11: Flow diagram of nitric acid synthesis (third iteration).

As water is produced in ammonia oxidation, low concentration nitric acid already in the heat exchanger upstream absorber could be produced by water condensation. This in turn means that an extra stream could be introduced to the absorber.

Nitric acid will contain some NOx resulting in a low quality product with a yellowish color, thus a bleach (stripper) should be introduced to treat unwanted emissions of NOx through contacting the product acid with air.

An essential question to be addressed regarding the scheme in Fig. 4.11 is at which pressure the nitric acid process should be operated, as absorption of NO_x

should preferably be performed at high pressure, while atmospheric pressure is beneficial for oxidation. In fact, several options exist (Tab. 4.1), including carrying out the whole process at atmospheric pressure or combining oxidation at low pressure and absorption at high pressure. In the latter case a compressor should be added between the ammonia conversion stage and the absorption stage.

Tab. 4.1: Modern nitric acid plants [6]*.

	Atmospheric pressure (AOP)	Intermediate pressure (IOP)	High pressure (POP)
NH_3 concentration (vol%)	12	12	11
Gauze temperature (°C)	810–850	810–850	870–890
Burner Pressure (atm)	1	1	4
Absorber pressure (atm)	1	3	6
Number of gauzes	3–6	10–20	35–45
Gauze diameter (m)	3–5	3–5	3–5
Life of gauze (months)	8–12	8–12	4–6
Pt loss (g/ton HNO_3)	0.05	0.05	0.1
Acid strength (%)	49–52	55–69	60–62
Conversion efficiency (%)	97–98	97–98	96–96.5

*Ammonia concentration below 12% is used to avoid explosive limits.

The technological scheme developed by Krupp Udhe GmbH (Fig. 4.12) is obviously more complicated than Fig. 4.11, but still contains the same essential features as in a simplified diagram.

Another example worth considering when discussing conceptual process design is the synthesis of methyl tert-butyl ether (MTBE). It can be achieved by reacting isobutene with methanol.

As a raw material, isobutylene from a C4 fraction could be used either from steam cracking (20–30%) or fluid catalytic cracking (10–20%).

As the tertiary carbon atom in i-C_4H_8 is much more active than primary and secondary carbon atoms in other C_4 hydrocarbons, only small quantities of other esters will be obtained as by-products. The reaction in Fig. 4.13 is in equilibrium, thus in order to obtain high product yields it should be shifted to the right side. The etherification reaction in Fig. 4.13 is an exothermic one, therefore it is thermodynamically favored at low temperatures.

One of the questions that arises when devising a flow scheme is in which phase the reaction should be carried out, for example, the liquid- or the gas-phase. Non-

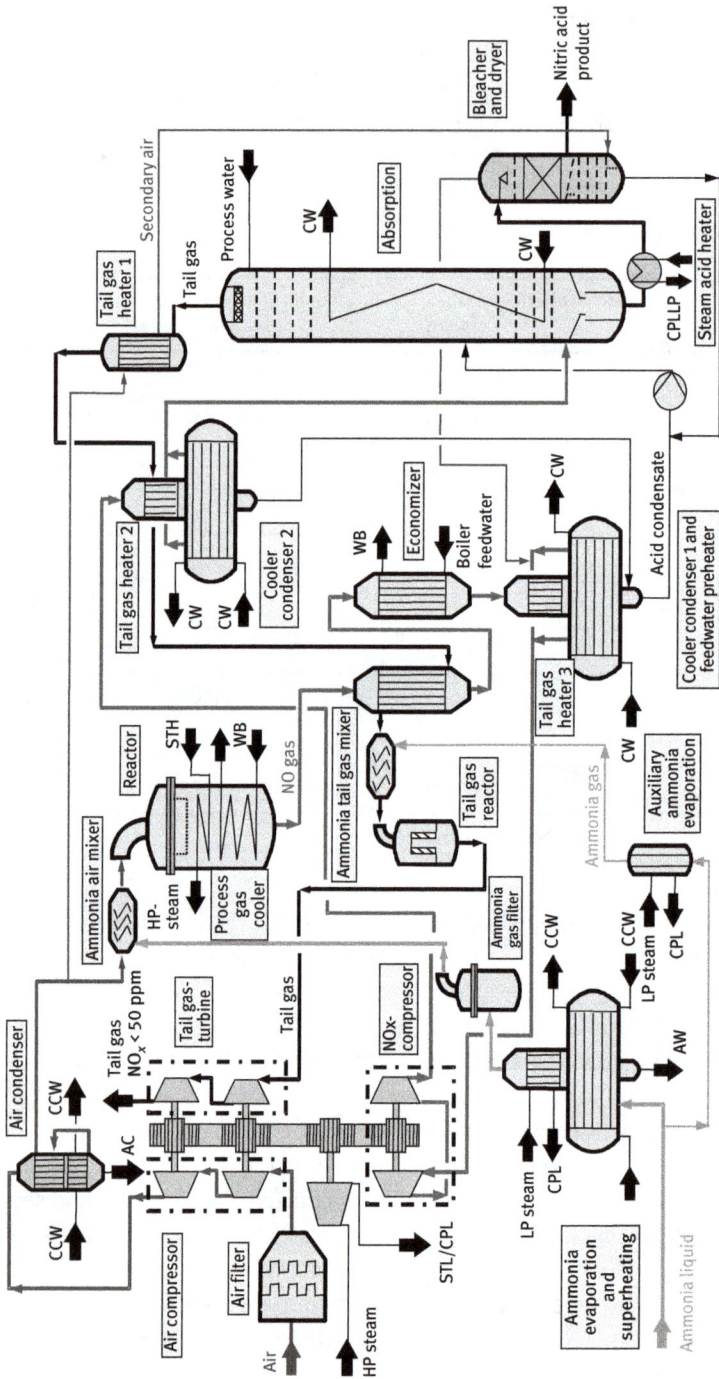

Fig. 4.12: Krupp Uhde dual pressure plant. (From [7].).

$$\underset{\text{CH}_3}{\overset{\text{CH}_3}{\text{CH}_3\text{-C}=\text{CH}_2}} + \text{CH}_3\text{OH} \;\rightleftharpoons\; \text{CH}_3\text{-}\underset{\text{CH}_3}{\overset{\text{CH}_3}{\text{C}}}\text{-O-CH}_3$$

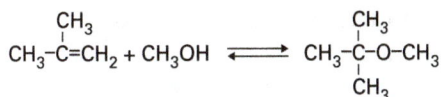

Fig. 4.13: Synthesis of MTBE.

ideality of the mixture between methanol and hydrocarbons increases the activity coefficient of methanol with respect to the vapor mixture. This leads to a much higher equilibrium concentration of the product in the liquid-phase at the same T.

The process should thus be preferentially performed in the liquid-phase. This type of etherification reactions can be catalyzed by sulfonic acid resins, which are sufficiently active at rather low temperatures (40–50 °C). Requirements of low temperatures inevitably mean that pressure is needed (around a few bars) to keep the reactants in the liquid-phase. This results in a rather high equilibrium concentration (90%).

This high concentration (along with other parameters) determines the input-output structure (Fig. 4.14). Two reactants (C_4 and methanol) will result in the product. Unreacted C_4 hydrocarbons other than isobutene (remember that the feedstock is coming from steam cracking or FCC and contains a mixture of hydrocarbons) and unreacted isobutylene will be a part of the outlet stream.

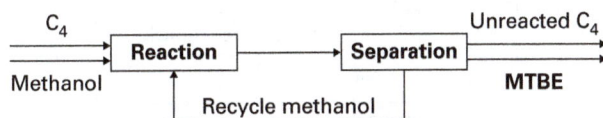

Fig. 4.14: Input-output structure in MTBE production.

Recycling of this unreacted substrate is not worthwhile at such high conversion levels. In fact this will be a recycle of normal butenes, leading to their build-up. Therefore no recycling is done and unreacted C_4 is used in other refinery processes. The equilibrium could, however, be shifted by applying an over than stoichiometric ratio between methanol and isobutene, and thus only methanol is recycled (Fig. 4.14). A high excess of methanol is not usually used at it is slightly over stoichiometric.

Fig. 4.14 is somewhat simplified as it does not contain by-products, which are in fact formed because of the dehydration of methanol to dimethylether $CH_3\text{-O-}CH_3$ and the dimerization of C_4 to C_8 hydrocarbons. The first reaction is a non-equilibrium one and is favored at high T, high methanol concentration and low space velocity. The product is a poison in downstream alkylation of C_4 hydrocarbons. The second reaction is also a non-equilibrium one, being favored at low methanol concentration and must be limited because C_8 hydrocarbons leave with MTBE and decrease its purity.

In addition, the hydration of isobutene to tert-butyl alcohol can also occur, while its equilibrium is limited by low water content. As mentioned above the reaction of methanol with normal butenes is another side process which is not that important if MTBE is used for gasoline blending. Several options could be considered for the reactor. A plug flow reactor should afford the highest reactant concentrations. As the reaction is an exothermic one, being favored at low T, thus a low inlet T is beneficial. In order to avoid unnecessary overheating a multitubular reactor with external water cooling could be applied, which is obviously not a purely isothermal reactor. A hot spot exists, as heat generated close to the inlet by a high reactant concentration cannot be totally transferred by water. With operation time the catalysts gradually deactivates, which is associated with the fact that the inlet streams contain impurities with deactivated catalyst. As purification is considered to be not economical, gradual deactivation means that the hot spot also moves gradually towards the reactor exit.

A Huels design of the MTBE process is illustrated in Fig. 4.15.

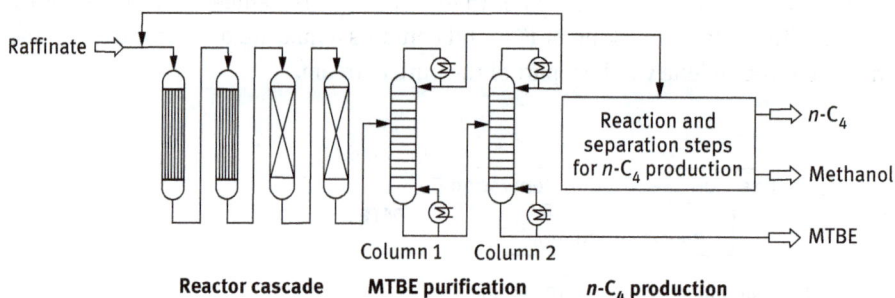

Fig. 4.15: Huels design of MTBE synthesis. ([8]). Reproduced with permission.

In the reactor cascade, i-butene is completely converted with methanol to MTBE. The effluent from the reactors contains MTBE, linear butenes and methanol, which was in the excess. As methanol and C4 form an azeotrope the bottom product in column 1, operating at 6 bar, consists of ~97 wt% MTBE and 3 wt% methanol. In column 2, MTBE with over 99 wt% is obtained. The MTBE/methanol azeotrope at the top of column 2 is recycled into the reaction zone.

4.1.3 Conceptual process design: general comments

As illustrated in these two examples, quite a number of issues should be considered in conceptual process design. In addition to the chemistry of the main reaction including its stoichiometry, stoichiometry of the parallel or series reactions should be also determined. Temperature, pressure and concentration domains should be carefully considered and a decision should be made on which phases are present in the reaction system.

Analysis of kinetics and thermodynamics makes it possible to address composition at chemical equilibrium as well as to calculate how feasible the reaction rates are, and which type of catalysts might be needed.

The application of a catalyst eventually raises a question of its stability, possible deactivation and whether it is necessary to perform regeneration.

The desired reactant and product purity at the available quality of feedstock would also impose requirements on the feed purification and separation of products.

Obviously, processing constraints such as explosivity, corrosivity, toxicity and overall safety should be carefully evaluated.

Thus, conceptual process design should answer a number of questions, some of which are presented below:
- Is continuous or discontinuous processing to be preferred?
- What are the optimal regions of process conditions?
- Which process conditions are dangerous?
- How is the reaction T reached?
- Which type of reactor is preferred?
- Is pre-treatment of the reactor feed necessary?
- How is the reaction mixture processed?
- Are there any special measures in relation to the co-products and waste required?

For example, if conversion is not complete there is an option of recycling (Fig. 4.16) that should be carefully considered, as recycling at rather high conversions is not economical. Moreover, a simple recycling could lead to the build-up of impurities, which might be present in the feedstock, thus the introduction of a purge stream is necessary.

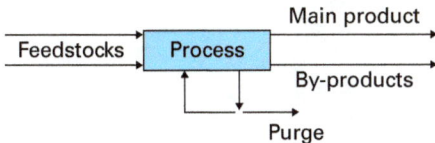

Fig. 4.16: Input-output structure with purge.

Common sense rules should be used in the phase of conceptual design, such as those for separations: avoid unnecessary separations; do not separate fuel and waste stream products any further; do not separate and then remix.

Basic engineering design based on the conceptual one will then answer the question is the production process technically feasible in principle. More detailed calculations should address the issue of economic attractiveness and the level of the risk in economic and technological terms.

Technical risks are associated, for example, with a need to use technically non-established equipment or equipment exceeding the conventional limits such as very high distillation columns. A particular company might be not familiar with a certain technology requiring for example the application of high pressure or fluidized bed technology. Pharmaceutical companies might not be willing simply to switch to continuous processes from conventional batch ones. Use of units, which are difficult in scaling-up, such as in general solid processing, might also be a risk. With solids even apparently simple operations as bunkering and transportation can be difficult.

> A couple of decades ago during a visit to an ammonia plant, a plant manager, who was previously involved for many years in production of fertilizers, told me that it was so nice to deal with only gases and never in his life he would like to return to production of any solids.
>
> Once upon a time, my father-in-law was working at a plant producing phthalic anhydride from naphthalene. Later the feed was changed to o-xylene. Naphthalene is a solid with a melting point of approximately 80°C. Once a railroad car was delivered to the plant site and naphthalene had to be melted and pumped to the storage facilities during the winter time when the outside temperature was approximately –30°C. Switching the feed to o-xylene (liquid at room temperature) was considered a blessing.

There could be risks of operational failure of a single-train plant when there is a strong coupling between steps when some streams are fed back to earlier stages. Decoupling by inserting buffer tanks for intermediates can be an option for risk mitigation, although this approach is not practical for larger product quantities of product or gaseous intermediates.

When a new technology is introduced, it can be anticipated that the operational safety is lower. It is possible to diminish the risks by increasing expenditure on research and development and developing failure scenarios.

4.1.4 Reactor selection

Reactor selection is an important issue in process design. An economic option is to use fixed adiabatic beds when the temperature rise corresponds to the conversion.

> When I was working in industry, one of the experienced ammonia plant managers mentioned to me that for some reactions happening in ammonia production (to be discussed later), such as a water-gas shift reaction, he did not even need to wait until the analysis of the product mixture, as the temperature difference between the inlet and outlet is an excellent indication of conversion.

As already mentioned in Section 3.6, for adiabatic fixed-bed reactors when too high temperatures should be avoided, several beds are applied with interbed cooling either using heat exchanges or quenching with cold reactants or an inert gas. The synthesis of ammonia or hydrotreating of various streams in oil refining are typical examples of this approach. In the case of ammonia synthesis, driven by equilibrium, quenching is a viable method to shift equilibrium. In other cases, the downside of quenching, resulting in an increase of the gas mixture volume, decrease of the reactants partial pressure and lower reaction rates, can overweight the benefits of interbed cooling. For various hydrogenation or hydrotreating reactions, quenching is done by hydrogen.

For adiabatic reactions, a particular care should be taken of the first bed. Higher concentrations of reactants imply higher rates, and thus larger temperature gradients than for other beds. Subsequently, the height of the bed has to be decreased, which in turn negatively influences gas flow distribution in the radial direction across the bed. An option would be to dilute the first bed with an inert or introduce a less active catalyst.

For very strongly exothermic or endothermic reactions too many beds would be required in order to control temperature rise, thus hundreds (for benzene hydrogenation, or methane steam reforming) or thousands of catalysts tubes filled with solid (ethylene oxide or phthalic anhydride synthesis) are arranged parallel to each other, with cooling or heating in between the tubes (Fig. 4.17). Because of better control temperature, such approach might either prevent excessive deactivation and/or is needed to improve selectivity.

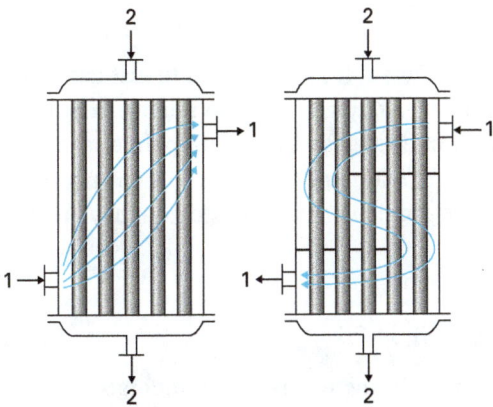

Fig. 4.17: Different arrangements of heat media circulation in between the tubes. 1, Heat medium; 2, reactants. (From [9]).

Deactivation is in fact an important issue that often determines the type of reactors used in industry. For example, for gas-solid catalytic reactions if deactivation is not that profound (months to years) then packed beds of catalysts can be used.

Poisons present in the feed can be removed if necessary by installing guard beds, which can be done either by use of a separate adsorbent, or by over sizing the catalyst where this additional volume of catalyst is used to adsorb impurities.

Usually if the catalyst lifetime is sufficiently long (several years) then no regeneration is done and the catalyst is simply removed when it is considered uneconomical to continue with it. For example, in the synthesis of ammonia the lifetime of catalysts is usually 14–15 years. In fact, as impurities do not influence the catalyst performance in this case and no carbon deposition occurs, the lifetime could be even longer. However, ammonia synthesis requires the use of high pressures, therefore according to legislation the reactor vessels should be regularly inspected. For this reason, after 14–15 years the catalyst charges are unloaded from the reactor. Other catalysts in the same ammonia train, such as catalysts for natural gas primary and secondary reforming, high temperature and low-temperature shift, can operate for several (2–6) years without any regeneration. An interesting example in the same process is hydrodesulfurization, containing NiMo or Co-Mo catalysts. The lifetime of such catalysts depends heavily on the presence of sulfur in the natural gas.

In 1999, I visited an ammonia production plants in Russia that utilized natural gas from Siberia, and I asked about the situation with the hydrosulfurization (HDS) catalysts. It turned out that the plant was put on stream in 1972 and since that time there were no changes in this catalyst, which was obviously connected to low sulfur content in the feed. Similar plants, for example in the Persian Gulf, change HDS after ~5 years.

The example of HDS demonstrates a case where it is not sufficient to just use a bed with adsorbent or install a bigger volume of catalyst in order to remove impurities in the feed (for example, mercaptanes in natural gas). In fact a separate reactor is used upstream to the main one, where steam reforming of natural gas on supported nickel catalysts is performed. Mercaptanes should be removed because they are poisons for nickel.

In an HDS reactor the following reaction occurs:

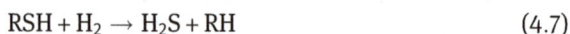

$$RSH + H_2 \rightarrow H_2S + RH \tag{4.7}$$

Additional hydrocarbons, which are formed during this reaction undergo transformation to hydrogen along with methane that is exposed to steam over nickel catalysts. H_2S should be, however, removed upstream of steam reforming and this is done by contacting it in a separate reactor with zinc oxide, which reacts (noncatalytically) to zinc sulfide, which is then is discharged when all zinc oxide is consumed.

After several years in operation the activity of catalysts in adiabatic or isothermal fixed-bed reactors could decline. To compensate activity losses, an engineering practice during industrial operation is to increase temperature (Fig. 4.18).

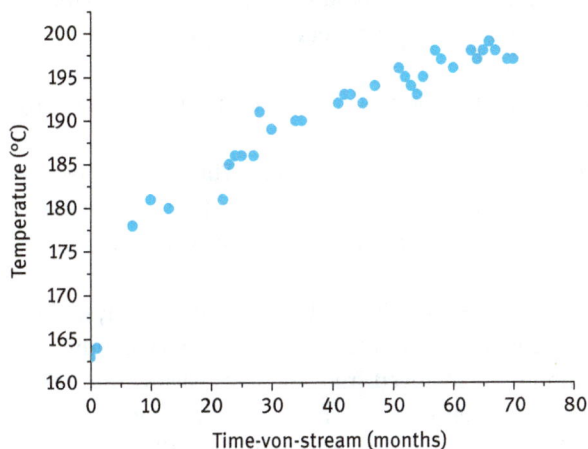

Fig. 4.18: Compensation of activity losses by temperature increase.

Although the policy of temperature increase can allow the constant level of production output, the obvious drawback with this operation policy is that with a temperature increase side reactions are becoming more prominent, thus further deteriorating the catalyst performance.

It is, however, possible that deactivation increases too strongly above a certain T. For example, copper catalysts are very sensitive to sintering, which prevents a policy of temperature increase in methanol synthesis over copper-containing catalysts. In such cases the pressure could be gradually increased to compensate the activity decline with operation time.

When deactivation is more profound because of coking and carbon deposition it becomes necessary to regenerate the catalysts after several months. Fixed-bed reactors could be applied and regeneration (by coke burning) is done while the reactor is offline. As regeneration should be done very carefully in order to prevent, for example, catalyst sintering, this regeneration process could be rather time-consuming. Let us consider, as an example, a hydrogenation process where the catalyst deactivates. As direct exposure of the catalyst to air (or oxygen) is very dangerous and can lead to explosions, because the catalyst could still contain some hydrogen, the reactor should be first purged with an inert gas, and thereafter the coke from the catalyst should be carefully oxidized thus controlling the amount of oxygen in the feed. If it is not properly done, heat released during a highly exothermic coke oxidation can promote catalyst sintering and irreversible losses of catalyst activity. After burning of coke and subsequent purging with an inert gas, the catalyst should be once again activated.

These lengthy procedures can significantly influence production output, thus an additional reactor is usually installed in parallel. This allows the first reactor to be isolated and regenerated as the feedstock is rerouted to the second reactor, allowing the plant to operate continuously. After regeneration the first reactor remains in a stand-by mode.

If the catalyst is active for several days or weeks, an option is to use a moving-bed reactor with continuous catalyst regeneration.

Catalytic reforming should be mentioned in this context. This process helps to transform heavy naphtha feedstock into high octane reformate for gasoline blending, as well as to generate high purity hydrogen for use in hydrotreating and hydro-cracking (HC) applications.

In the CCR reforming process developed by UOP (Fig. 4.19), catalyst flows through the reactors in series instead of remaining static in fixed beds of individual reactors. Spent catalyst is continuously removed from the last reactor, transferred to the regeneration section, where it is regenerated in a controlled way and transferred back to the first reactor. Such frequent regeneration allows CCR reforming to operate in more severe conditions than for fixed-bed reforming.

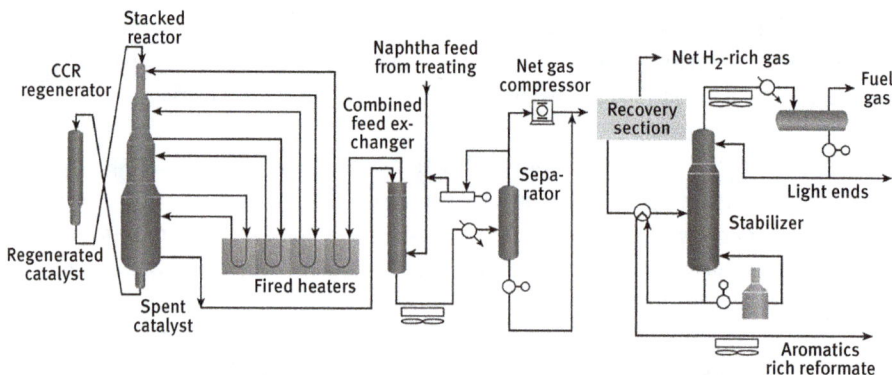

Fig. 4.19: CCR platforming from [10].

Deactivation could be even more severe, as will be discussed later on for fluid catalytic cracking. A fluidized catalyst bed could be used when very small catalyst particles are fluidized in the flow of feed gas and there is continuous flow of the deactivated catalyst to a regenerator and back to the reactor. It was already mentioned in Chapter 3 that fluidized beds of catalyst allow operation at a uniform temperature.

A similar concept of separating the reactor and the regenerator can be applied when a riser is used instead of a fluidized bed reactor.

It became apparently clear from the discussion in Chapter 3 and also in this chapter, that catalyst handling, including installation, activation (reduction, oxidation, sulfidation) and regeneration, is extremely important as modern plants use hundreds of tons of different catalysts, representing a significant cost.

I was once involved in a pre-scheduled supply of 100 tons of catalyst to a plant operator as the result of the catalyst deactivation because of a maloperation. The catalyst was substantially overheated by more than 200 °C, which resulted in an activity loss after just 1 year of operation, while the expected lifetime was 5 years. The plant operator suffered significant financial losses caused by an unscheduled purchase of catalyst, as well as a need to discontinue production for a substantial time. Closing down such a large plant usually is very expensive.

Example of catalyst deactivation, time scales and the consequences for processes are given in Tab. 4.2. As could be seen from the table, a range of reactions involving organic compounds are prone to deactivation by coking, while the time scale of deactivation depends on a particular reaction and can vary from seconds to days, weeks and even years. Regeneration by coke combustion, as mentioned above, could be made either continuously when the catalyst is recycled between a reactor and regenerator or discontinuously.

Another option to prevent deactivation by the deposition of carbon is to operate under conditions that will minimize sintering, for example, using excess of steam in steam reforming of natural gas, which will be discussed below.

Other types of catalyst deactivation (discussed in Chapter 3), such as poisoning, could be irreversible and thus catalyst regeneration is not an option. In those cases the technological scheme should include a purification section, as exemplified above for the hydrodesulfurization of natural gas prior to steam reforming.

One interesting historical example in this context is the nickel-based catalyst for synthesis of so-called protective atmosphere for different metallurgical processes. In the mid-1960s, a truck producer ZIL headquartered in Moscow with the annual capacity of approximately 200,000 trucks per year and approximately 100,000 employees working in approximately 20 locations across the former Soviet Union (20,000 only in metallurgical plants) had to temporary shut down production. The reason was severe deactivation of nickel catalysts, which were used for generation of protective atmospheres containing hydrogen from reforming of natural gas with a mixture of steam and CO_2. For a long time, the factory operated with a relatively sulfur-free natural gas originating from Samara and the Soviet Central Asia. Due to expansion in the needs of natural gas in rapidly growing city of Moscow, an alternative supply was found. The natural gas from Orenburg had a much higher content of sulfur. Obviously, absence of any natural gas hydrodesulfurization resulted in severe deactivation of nickel catalysts.

Another issues that should be considered while selecting a reactor are injection and dispersion strategies, which will be discussed below, following the approach introduced by Krishna and Sie [12].

Tab. 4.2: Deactivation in different industrial processes. (From [11]).

Process	Catalyst	Main deactivation mechanism	Time scale of deactivation	Consequences for catalyst	Regeneration	Consequences for process
FCC	Zeolite	Coke	Seconds	Regeneration on s scale	Coke combustion	Recirculation catalyst between reactor and regenerator
Oxidative dehydrogenation	Various oxides	Coke	Seconds	idem		Similar schemes as for FCC
Catalytic reforming	Pt/γ-Al$_2$O$_3$	Coke, Cl loss	Months Days	Alloying	Coke combustion Cl supply redispersion	Fixed-bed, swing operation, moving-bed
Hydrotreating	Co/Mo/S/Al$_2$O$_3$	Coke Metal sulfides	Months Days	Once-through catalyst Adapted porosity	Coke combustion	Fixed-bed, slurry, moving-bed
Methanol	Cu/ZnO/Al$_2$O$_3$	Sintering (Cl)	Years	Stabilization		Feed purification
Water-gas shift	Cu/ZnO/Al$_2$O$_3$	Poisoning (S, Cl)	Years	Stabilizers (ZnO)		Feed purification
Three-way catalyst	Pt, Pd	Sintering, loss of active components. deposits (Zn, P from lubricants)	Years	Noble metals. Stabilized alumina (La, Ba)	Rejuvenation by leaching	
Steam reforming	Ni/Al$_2$O$_3$	Coke, whiskers		K. Mg Gasification catalysts	Coke combustion	Excess steam
Dry reforming	Ni	Coke		S-doping	Coke combustion	Excess steam

Diesel soot	Cu-Cl	Evaporation	Minutes Hours	Select other catalyst	Add catalytic additives to fuel (Ce)
DeNO$_x$	V$_2$O$_5$/Al$_2$O$_3$	Formation surface salts	Months	Select other carrier	
Wacker oxidation	Pd, Cu	Catalyst deposit			Low pH
Xylene oxidation	Co, Mo, Br	Mo, Co deposits		Add new catalyst	Deposits in reactor and downstream
Styrene	Iron oxide	Coke, sintering movement promoters		Structural promoters	Coke gasification in steam

For example, reactants can be introduced in a one-shot mode as for batch reactors or in a step function mode as for continuous reactors. Staged injection is an intermediate case and is applied in semi-batch reactors. Another option is to apply pulsed feed, as for flow reversal types of reactors or semi-batch ones. Energy supply could be also envisaged in many ways. In adiabatic reactors it can de done through quenching by the cold reactant or by introducing intermediate heat exchangers. The application of fluidized bed reactors can be an option for exothermal reactions, for example, selective (or partial) oxidation of alkanes, i.e., n-butane to maleic anhydride (Fig. 4.20).

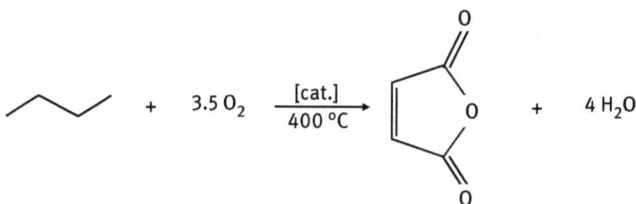

Fig. 4.20: Butane to maleic anhydride oxidation.

This reaction is the complex one, requiring extraction of 8H atoms and insertion of three oxygen atoms to butane along with a ring closure. A high concentration of butane is advantageous, but also dangerous because butane and oxygen could be within explosion limits, thus the concentration of butane in air is ~2.5%. The application of fluidized bed reactors allows the concentration of butane in the feedstock to be increased by 50–100% compared to that in fixed-bed operations.

In this particular case of butane oxidation there is also a possibility to utilize a transport reactor [13]. The advantage of such system, when a metal oxide oxidation (in this case V_2O_5) is changing during the reaction (from V^{+5} to V^{+4}), is that donation of oxygen from the catalyst lattice to the substrate with subsequent reduction of V^{+5} to V^{+4} is separated from oxidation of V^{+4} to V^{+5}. The latter process is conducted in a separate reactor, which prevents butane from being in contact with air, making the process thus much safer (Fig. 4.21). Operating temperature is similar to the conventional processes (Tab. 4.3), while the pressure is somewhat higher. The throughput can be increased by 10% with 20% higher yield [6]. The catalyst microspheres of approximately 40–150 μm are made by spray drying of $(VO)_2P_2O_7$ with 5% of polysilicic acid at pH 3 resulting in a hard porous shell of silica over the surface of catalyst particles.

A number of oxidation reactions are conducted in a batch mode in slurry reactors. For such reactors, several options for energy removal could be considered, including evaporative cooling (Fig. 4.22).

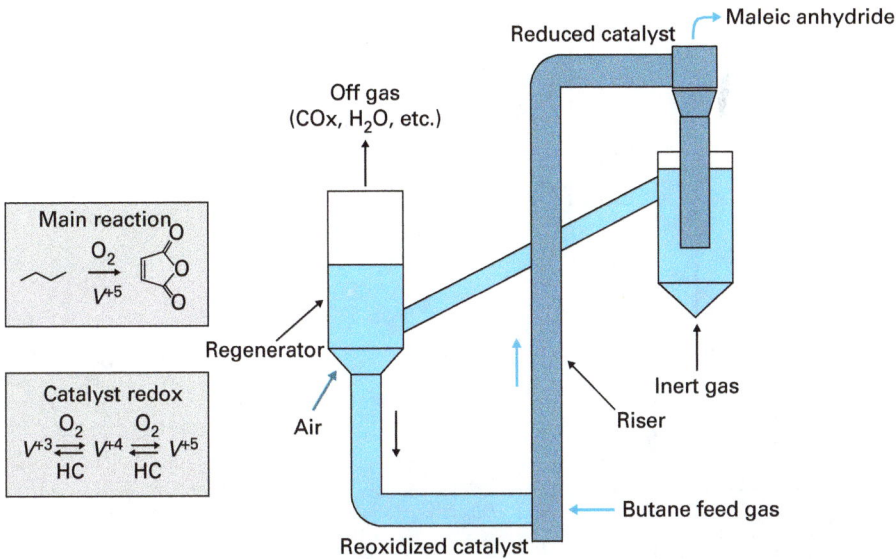

Fig. 4.21: Maleic anhydride synthesis in a transport reactor-regenerator tandem.

Tab. 4.3: Oxidation of butane to maleic anhydrate in fluidized beds [6].

	Alusuisse	DuPont
Design	Conventional fluidized bed with separate feed and air injection	Circulating fluidized bed reactor with no oxygen supply and a separate regenerator with air addition
Operating temperature, °C	Ca. 350	Ca. 350
Butane concentration in the reactor (vol.%)	4	100
Conversion	Ca. 60	Ca. 75

Other options for slurry reactors could be to have an external heat exchanger (Fig. 4.23) or arrange heating through a double jacket or internal coils (Fig. 4.24).

The methods of energy input could be also different. In addition to those mentioned above, energy removal could be arranged through programmed temperature cooling.

From the point of view of concentration profiles, it could be more attractive to have a continuous plug flow reactor with no mixing of reactants. Conversely, such types of arrangements for exothermal reactions lead to hot spots (Fig. 4.25), which

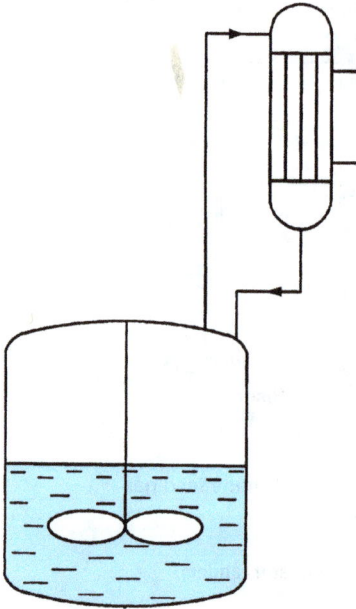

Fig. 4.22: Slurry reactor with evaporative cooling.

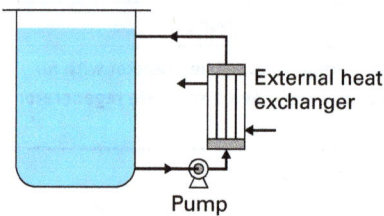

External heat exchanger

Pump

Fig. 4.23: Slurry reactor with a heat exchanger.

not only determine conversion and selectivity, but also catalyst lifetime, reactor materials and safety of the whole process.

A specific method of energy removal could be to load catalysts with different activity along the tube. In this way the amount of the active phase on a support can be profiled along the tube length. This counterbalances the excessive temperature increase and hot spots through deliberately minimizing the activity of the catalyst layer close to the reactor inlet. Another possibility is to keep the amount of metal or metal oxide on a support the same, but dilute the catalyst with the support at a different ratio changing along the reactor length (Fig. 4.26).

During reactor selection for a particular technology a decision should be also made on whether in situ product removal will be beneficial for the process.

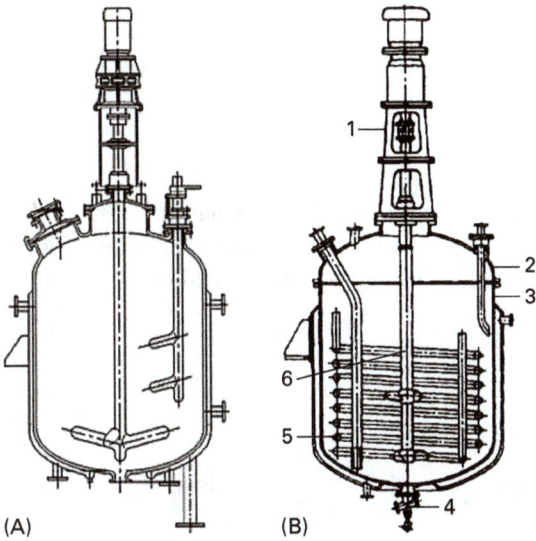

Fig. 4.24: Slurry reactors with (A) a double jacket and (B) a double jacket and internal coil. 1, motor; 2, hood; 3, reactor walls; 4, discharge valve; 5, internal coil; 6, shaft with impeller.

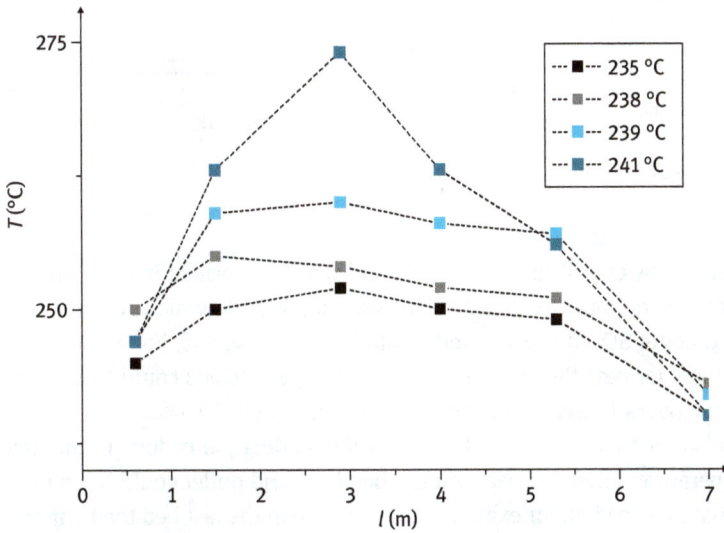

Fig. 4.25: Temperature profiles in a multitubular reactor along the length for different heating media temperatures (ethylene oxidation to ethylene oxide, from [9]).

Fig. 4.26: Catalyst profiling along the bed.

There are examples of reactions driven by equilibrium when in order to shift equilibrium, the products should be withdrawn. A typical example is the production of sulfuric acid by oxidation of sulfur dioxide to sulfur trioxide over vanadium pentoxide catalysts.

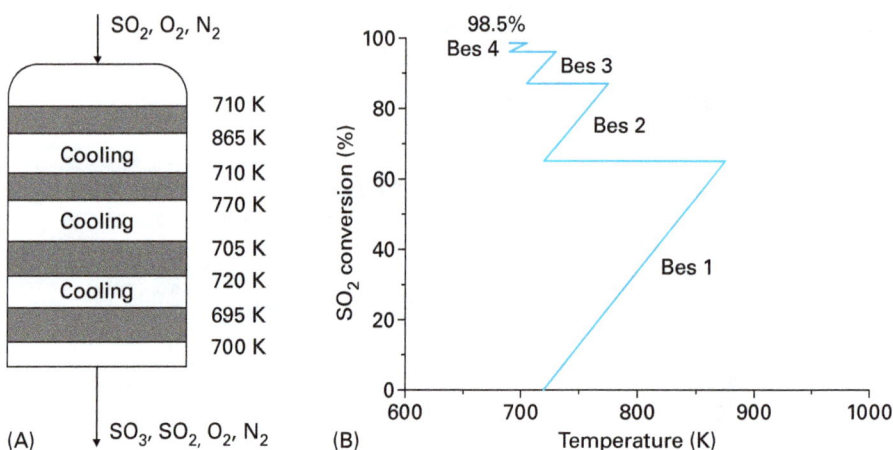

(A)

(B)

Fig. 4.27: Adiabatic fixed-bed reactor for SO_2 oxidation.

The reaction is limited by equilibrium (Fig. 4.27) and when performed in an adiabatic fixed-bed reactor after the fourth bed the conversion is 98–99%, which still must be improved to afford better SO_2 utilization and diminish SO_2 emissions, For a plant producing 1 million tons per year the annual decrease of SO_2 emissions could be as high as 14,000 tons if the overall conversion is increased from 97.5 to 99.5%.

It should be also mentioned here that the most demanding job is done in the first bed, and the temperature difference between the bed inlet and outlet could be as high as 200 K with the catalyst bed of, for example, 0.5 m, while in the last bed the temperature difference could be just be 1 K. Note that the bed height of the last bed is usually higher, than the first bed. The lifetime of catalysts in different beds thus varies dramatically from 1 year for the top layer of the first bed to 10–15 years in the last one.

The intermediate removal of SO_3 from the gas stream, usually after the third bed and a quench, which can be done upstream of bed 4 (Fig. 4.28) or bed 5, allows a shift of equilibrium leading to conversion above 99.5%.

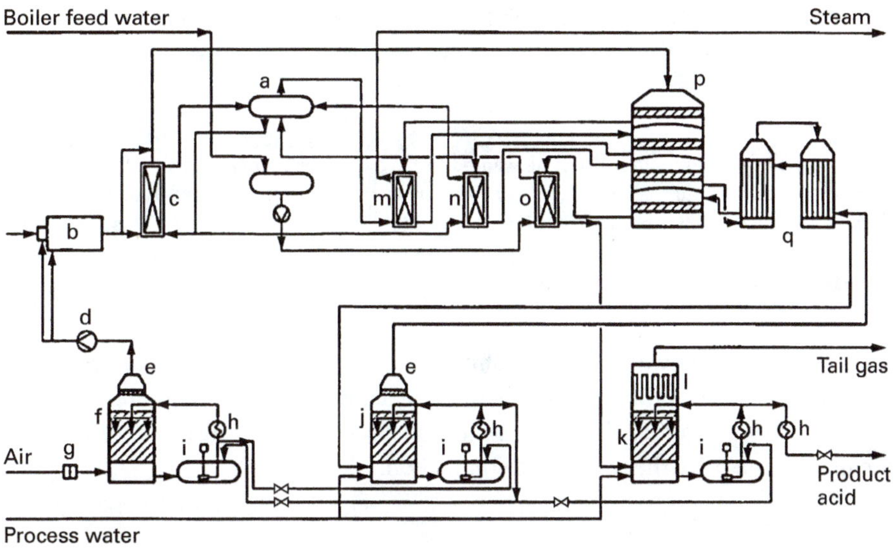

Fig. 4.28: Sulfur-burning double-absorption sulfuric acid process (Lurgi). (a) Steam drum; (b) Sulfur furnace; (c) Waste heat boiler; (d) Main blower; (e) Mist eliminator; (f) Drying tower; (g) Air filter; (h) Cooler; (i) Acid pump tank; (j) Intermediate absorber; (k) Final absorber; (l) Candle filters; (m) Steam superheater; (n) Boiler; (o) Economizer; (p) Converter; (q) Intermediate heat exchanger (From [14]).

In Fig. 4.28 SO_2 gas from the blower passes through heat exchangers before entering the converter. After the first bed the partially converted gas is cooled in the tube side of the hot heat exchanger and then flows to the second bed. The same is done with the gases leaving the second bed. Hot gases leaving the third converter bed and still containing unreacted SO_2 are cooled and routed to the first absorbing tower, where SO_3 is efficiently combined with sulfuric acid. Unreacted gas is directed to the fourth and subsequently to the final bed (if there are five beds). Thereafter the gas leaving the converter passes the second absorption tower. This double-absorption technology affords high overall SO_2 conversion and lowers SO_2 emissions.

In fact, process efficiency can even be improved if the catalyst with a cesium promoter is employed instead of potassium. The main components of the conventional catalyst include: SiO_2 as a support, vanadium (V), potassium (K) and various other additives. The reaction occurs within a molten salt consisting of potassium/cesium sulfates and vanadium sulfates, coated on the solid silica support. Vanadium is present as a complex sulfated salt mixture and not as vanadium pentoxide (V_2O_5).

Cesium-promoted catalysts were developed specifically for lower temperature operations that can lead to greater SO_2 conversions and hence lower SO_2 emissions to the atmosphere. The cesium salt (Cs) promoter reduces the required operating temperature for the sulfuric acid catalyst by as much as 40 °C. The cesium/vanadium catalyst can be used in the first bed to reduce the bed inlet temperature (saving energy and start-up time). The Cs-catalyst can be also used in the final cata-lyst bed (at a low inlet temperature) to maximize the SO_2 conversion and reduce emissions.

> Once a plant manager told me that he wanted to have Cs promoted catalysts in the final bed, because in addition to the interests of his company to save on payments for SO_2 emissions, he wanted to have a better environment for his daughter living nearby the plant. The same plant, even if expressing an interest in applying Cs promoted catalyst in the first bed, in fact had prob-lems with cooling gases to the required temperature after sulfur burner.

The example above was related to shifting equilibrium by removal of product. A similar approach could be applied for other reactions, limited by equilibrium. As an example, in an esterification reaction the removal of water (Fig. 4.29) through a membrane can drive the reaction to completion.

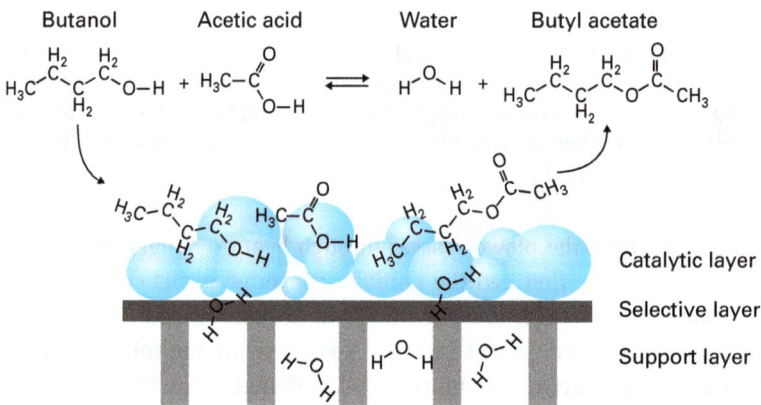

Fig. 4.29: Esterification reaction with water removal. (From [15] copyright Elsevier.).

Another example of a reaction/separation combination using a membrane reactor is worth mentioning even if it is related not with equilibrium, but rather with safety. The synthesis of ethylene oxide from ethylene over silver catalysts (Fig. 4.30) has some limitations associated with the fact that at certain concentrations ethylene and oxygen could be within explosivity limits, thus an excess of ethylene is applied in industry.

An interesting concept would be to utilize a catalytic membrane reactor (Fig. 4.31).

Reactive distillation along with a membrane reactor could be also applied to sepa-rate products from the reaction mixture for equilibrium-limited reactions such as esteri-fication mentioned above. Conversion can be increased far beyond the equilibrium because of the continuous removal of reaction products from the reactive zone.

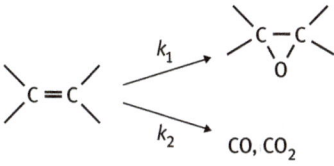

Fig. 4.30: Oxidation of ethylene.

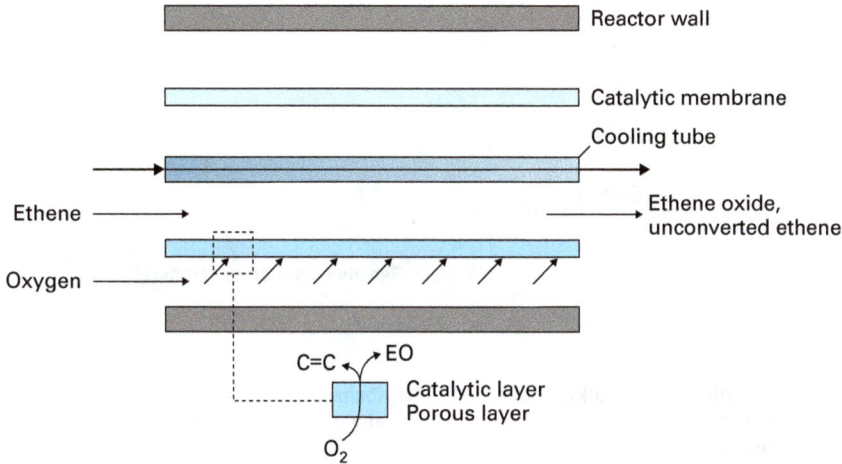

Fig. 4.31: Catalytic membrane reactor for ethylene oxide synthesis.

Heterogeneous reactive distillations could be performed in distillation columns illustrated in Fig. 4.32: the reaction section is the middle section containing a solid catalyst, while above and below there are rectifying and stripping sections.

A clear advantage of combined separation and reaction is that a single piece of equipment is used, making a considerable cost saving.

A similar strategy is applied in synthesis of ethylbenzene by alkylation of benzene with ethylene (Fig. 4.33).

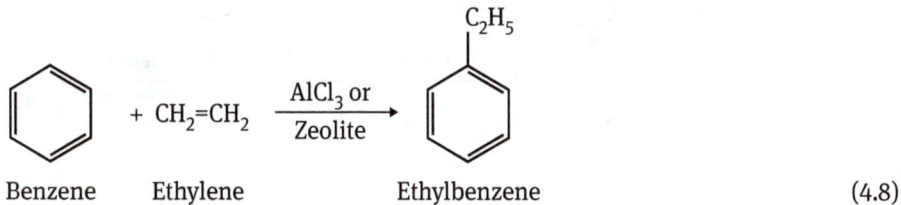

$$\text{Benzene} \qquad \text{Ethylene} \qquad \qquad \text{Ethylbenzene} \qquad\qquad (4.8)$$

Benzene is fed to the top of the alkylation reactor, while ethylene is fed as a vapor below the catalytic distillation section, making a counter-current flow of the

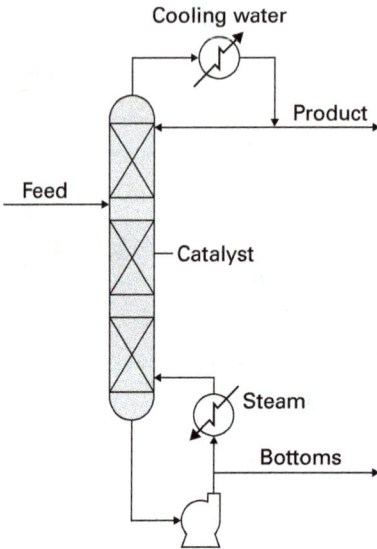

Fig. 4.32: Reactive distillation. (From [16]).

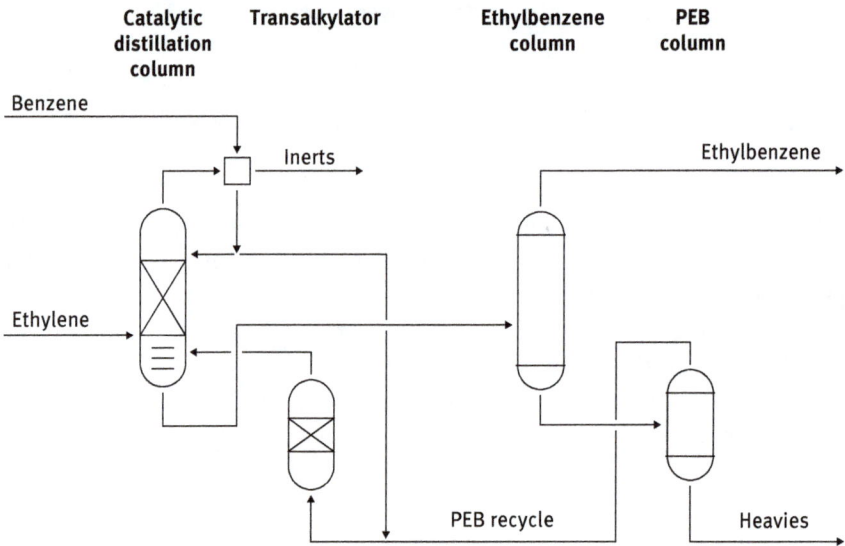

Fig. 4.33: Flow scheme of CDTECH EB technology. (From [17]).

alkylation reactants. In the catalytic distillation section vapor-liquid equilibrium (VLE) is established, with ethylene being mainly in the gas-phase. The reaction heat provides the necessary vaporization to influence distillation. The bottom of the reactor/separator operates as conventional distillation columns.

An important issue in the reactor selection is the injection strategy, for example directions of flow, as discussed in detail for downflow and up-flow options in fixed-bed multiphase reactors in Chapter 3.

> Selection of an operation mode could be far from straightforward. Several years ago, at a conference after a presentation about hydrotreating reactors, I posed a question about which operation mode is preferential. A very famous Dutch chemical engineering professor replied "Wise men do downflow". A manager from one of the leading European chemical companies answering the same question mentioned that all pilot plant reactors at that company operate in up-flow mode.

4.2 (Petro)chemical industry

The chemical industry comprises the companies that produce 70,000 different chemical products starting from basic raw materials, such as oil, natural gas, air, water, metals, and minerals.

It is certainly beyond the scope of this book to describe various processes that are currently applied in industry. There exists several excellent textbooks, handbooks and encyclopedias that cover various aspects of chemical process technology in general and processes where catalysts are applied.

Among the major processes that require application of heterogeneous catalysts, fluid catalytic cracking (FCC), hydrocracking, stream reforming of natural gas and synthesis of ammonia as well as a range of oxidation reactions including oxychlorination were selected for more detailed description in this textbook.

FCC catalysts account for a significant part of all refinery catalysts, creating a demand of approximately 650,000–750,000 Mt annually. Worldwide annual HC capacity was approximately 200 million tons some years ago. The synthesis gas catalysts (including ammonia and methanol synthesis) market is also very large.

These examples and others, which will be discussed in detail below, are not only related to important chemicals and fuels generated in these processes, but also illustrate different types of reactors and catalysts, which were addressed in the previous sections.

4.3 Fluid catalytic cracking

The thermal cracking process was patented in 1891 and substantially modified later. Thermal cracking is still used to upgrade very heavy fractions or to produce light fractions or distillates, burner fuel and/or petroleum coke.

Catalytic cracking as introduced by E. Houdry in 1928 is a flexible process to reduce the molecular weight of hydrocarbons. The first full-scale commercial fixed-bed catalytic cracking unit began production in 1937. In the first plants cyclical fixed-bed operations were used, which were replaced by a more efficient operation with continuous catalyst regeneration in 1941.

The reasons for such modifications were associated with the fact that during the catalytic cracking process the catalysts deactivated after a short time because of coke deposition. Although coke can be removed and regenerated by burning, the regeneration time is relatively long compared to the reaction time. Application of a fixed-bed reactor with very frequent shut-downs for regeneration is not an economically viable option, while use of two reactors, one for hydrocarbon cracking and another for catalyst regeneration along with moving of catalyst between these two, is much more efficient. The first continuous circulating catalyst process used a bucket elevator.

Although this technology of the moving-bed solved the problem of moving the catalyst between efficient contact zones, the catalyst beads used still were too large. Large catalyst particles result in large temperature gradients within catalyst particles, which restrict regenerator temperatures and result in a large regenerator and catalyst hold-up.

The next step in the development of catalytic cracking was introduction of the fluid catalytic cracking. This process uses fine powdered catalysts that can be fluidized. The first commercial circulating fluid bed process was put on stream in 1942 and by the 1970s FCC units replaced most of the fixed and moving-bed units.

Initial process implementations were based on synthetic low activity silica-alumina catalysts and a reactor where the catalyst particles were suspended in a rising flow of feed hydrocarbons in a fluidized bed.

A dramatic increase of catalyst activity and gasoline selectivity was achieved with the introduction of zeolite cracking catalysts modified with rare earths in 1962 by Mobil. Thus, conversion and gasoline yield could be increased from 56% and 40% respectively for silica-alumina gels to 68–75% for conversion and 52–58% for gasoline yield depending on the catalyst.

A significant development in the 1980s was the introduction of ZSM-5 zeolite into the catalyst matrix as an octane enhancer.

Even today, FCC remains the dominant conversion process in petroleum refineries, producing a high yield of gasoline and liquefied petroleum gas (LPG). Approximately 1,500 tons per day were consumed worldwide in 722 refineries according to Bartolomew and Farrauto [11].

4.3.1 Feedstock

In the 1950s kerosene gasoil fractions were mainly applied as the feedstock, while later vacuum gas oil and even heavier feedstocks started to be used. The boiling range of the feedstock could be 540–550 °C with a molecular mass 1.5 times larger than for a lighter feedstock. It also means that vaporization of the feedstock is not complete.

A switch to heavier feedstock also required efforts aimed at a decrease of the coke yield, which for vacuum gas oil can reach 5–6%. The trend of using heavier metal-containing feedstock, such as atmospheric residues and synthetic crude, is continuing and will only accelerate. The atmospheric residue can contain Ni and V in the amounts of up to 10 ppm and even higher, while the content of these metals is below 1 ppm in vacuum gas oil. Undesirable effects of nickel and vanadium are associated with increased gas and coke selectivity for Ni and loss of activity for V. In addition, this heavier feedstock can contain sulfur. One of the options of handling it on a catalyst level is to incorporate an additive into the catalyst, which will capture SO_x formed during regeneration. Other possibilities might include a hydrotreating unit upstream FCC.

Typical FCC unit feeds and operating conditions are presented in Tab. 4.4.

Tab. 4.4: Typical FCC unit feeds and operating conditions.

Operation		
Feedstock	**Vacuum gas oil**	**Atmospheric residue**
API gravity	25.5	22.4
Sulfur (wt%)	0.7	0.8
Nickel (ppm)	0.4	3
Vanadium (ppm)	0.6	3.5
Conradson carbon (wt%)	0.2	4.0
Conversion (vol%)	86	76
Fuel gas (wt%)	4.4	4.6
Total C_3 (vol%)	13.9	10.4
Total C_4 (vol%)	18.6	13.2
C_5 + gasoline (vol%)	62.6	57.4
LCO (vol%)	6.8	11.6
Slurry (vol%)	6.9	11.9
Coke (wt%)	5.6	7.5
Reactor temperature (°C)	535	535
Regenerator temperature (°C)	720	720

(From [6].)

The amount and the quality of FCC products depends on the feedstock characteristics as well as the process parameters leading to fuel gas, gasoline and light cycle oil (LCO). LCO, when used as a blending component in heating oil, can have a greater value than that of gasoline. An example of the product distribution is given in Tab. 4.5. Under such circumstances, many refineries adjust their FCC unit operation to increase LCO yield at the expense of gasoline.

Tab. 4.5: Product distribution of FCC unit [18].

Product	Yield, vol%	Yield, wt%
Light gases (C2−)		4.93
Propane	2.56	1.43
Propylene	5.80	3.34
Isobutane	5.59	3.48
n-butane	1.96	1.27
Butenes	7.61	5.06
Gasoline (C5+)	57.05	47.15
Light cycle oil (LCO)	21.20	20.60
Heavy cycle oil (HCO)	6.80	7.84
Coke		4.90
Total	108.57	100
Cycle oils	28.00	28.44
Total LPG	23.52	14.59

Capacity 25,000 bbl/d or 65.61 m^3/h. Conversion of gas oil = 72 vol%.

LCO compared to diesel fractions has lower cetane number and higher sulfur content. The cetane number depends on the feedstock and operation temperature. Obviously an increase of LCO yield can be achieved by diminishing FCC unit cracking severity, eventually leading to higher yields of heavy products (light cycle oil, heavy cycle oil, and clarified oil) and lower yields of light products (gasoline, LPG, and gas) as well as coke. Reducing catalyst activity lowering reactor temperature, and reducing the catalyst/oil ratio can lead to increase LCO yield.

4.3.2 Reactions/mechanism

Many reactions happen during catalytic cracking, such as isomerization, cracking, proton transfer, alkylation, polymerization, cyclization, dehydrogenation, condensation, coking, etc. Some of them are primary reactions, while the majority are secondary.

A somewhat simplified overview of the reactions is given in Fig. 4.34.

During the cracking of normal paraffins the cracking reactions are the dominant ones. The products contain mainly paraffins of lower molecular weight and olefins:

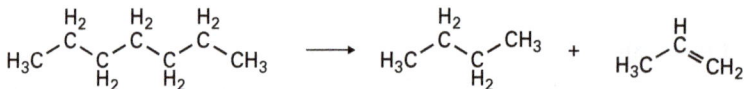

The yield of the latter increases with the increase of the feedstock molecular mass. Heavier fractions are less stable and can be cracked more easily than light fractions. Isoparaffins crack easier than paraffins.

Long-chain alkanes

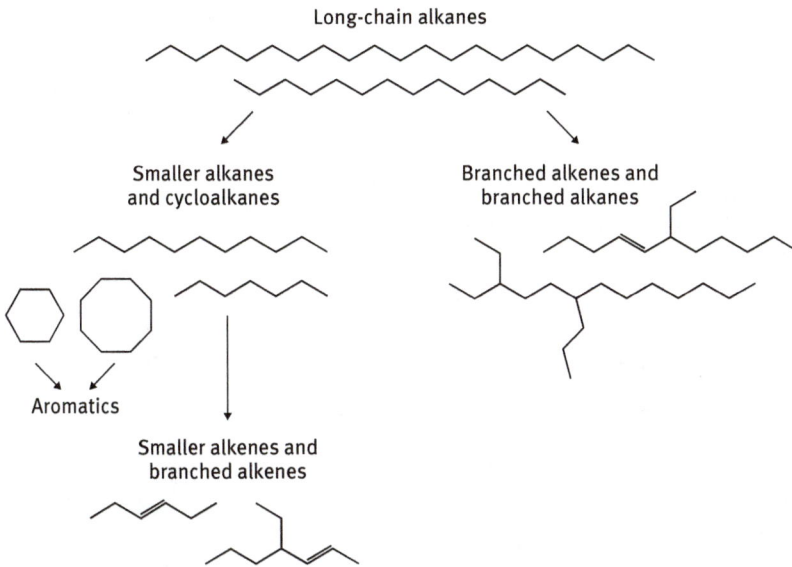

Smaller alkanes
and cycloalkanes

Branched alkenes and
branched alkanes

Aromatics

Smaller alkenes and
branched alkenes

Fig. 4.34: Diagrammatic example of the catalytic cracking of petroleum hydrocarbons. (From [19]).

Naphthenes are cracked at the side chain. Alkylnapthenes or aromatics with shorted side chains are formed first as they are stable, especially with methyl and ethyl substituent. Cracking of olefins, which is a secondary reaction, leads to smaller olefins:

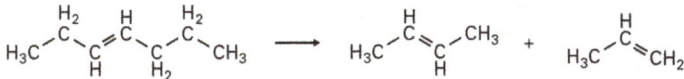

Initiation of the cracking process of olefins occurs (as for many other reactions of olefins) as a result of formation of carbocations, catalyzed by Brønsted and Lewis acids. A carbenium ion is a positively charged tricoordinated carbon atom, while a carbonium ion is a positively charged pentacordinated carbon atom. If a carbenium ion is large enough (C6+) it can crack, forming alkene and another carbenium ion, which isomerizes if possible into a secondary or tertiary ion. When the carbenium ion is not large (C3–C5), it will be terminated into an olefin through proton abstraction to the catalyst or another olefin, or through termination to a paraffin via addition of a hydride.

The mechanistic scheme for primary catalytic cracking is presented in Fig. 4.35. Branched polycyclic aromatics are often dealkylated

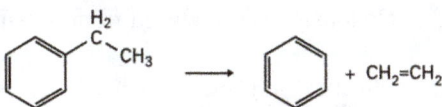

$H_3C-CH_2-CH_2-CH_2-CH_2-CH_2-CH_3$ *n*-**Alkane**

$\downarrow$ Initation

$H_3C-\overset{+}{C}H-CH_2-CH_2-CH_2-CH_2-CH_3$ **Classical carbenium ion**

$\updownarrow$

$H_3C-CH-CH-CH_2-CH_2-CH_3$ **Protonated cyclopropane**

$H^+\cdots\diagdown\diagup$
    CH_2 $\downarrow$ Hydride shifts +   Isomerization
          C–C bond breaking

$H_3C-\overset{+}{C}-CH_3 + H_2C=CH-CH_3$ $H_3C-CH-\overset{+}{C}H-CH_2-CH_2-CH_3$
     CH_3 *n*-**Alkene** CH_3

                                                      etc.

$\downarrow$ Hydride transfer

$H_3C-CH-CH_3$
     CH_3

iso-Alkane

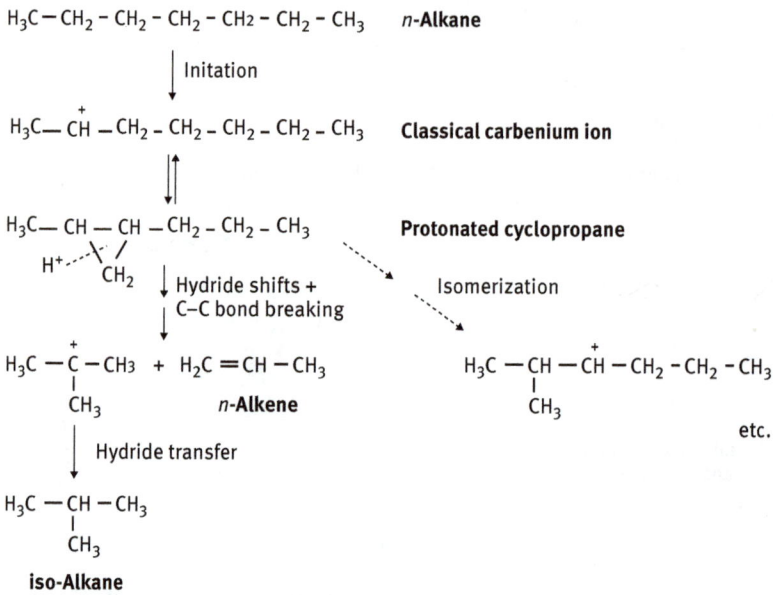

Fig. 4.35: Mechanistic scheme for catalytic cracking [20].

or also can be cracked in the side chain:

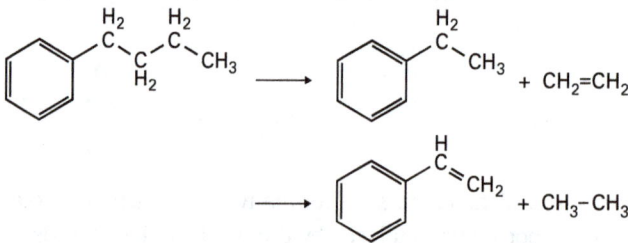

Carbon-carbon bond rapture occurs at the ring and is not very profound when an alkyl chain contains less than three carbon atoms. Polycyclic naphthenes are transformed to monocyclic naphthenes and alkenes. Naphthenes can react with alkenes through hydrogen transfer, leading to aromatics and alkanes:

A large amount of aromatic compound is obviously also produced from paraffins, whose structures allow cyclization.

Straight-chain alkenes can undergo acid-catalyzed skeletal isomerization reactions, leading to branched alkenes:

$$H_3C-\underset{H}{\overset{H_2}{C}}-\underset{H}{\overset{H}{C}}=\overset{H}{\underset{}{C}}-CH_3 \longrightarrow H_3C-\underset{H}{\overset{CH_3}{C}}=\underset{}{\overset{}{C}}-CH_3$$

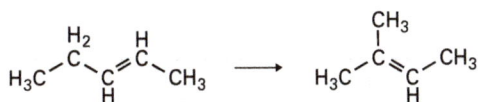

This is an important reaction because it increased the octane rating of gasoline range hydrocarbons. In addition to skeletal isomerization, double-bond migration can also occur. In addition, dehydrogenation can take also place, which together with cyclization gives polyaromatics, which are adsorbed on the catalyst surface, causing catalyst deactivation through coke formation.

The secondary reactions also produce unwanted light gases in addition to the formation of coke, which could be very substantial (as shown in Tab. 4.4). In a conventional operation around 2–5% coke is typically deposited on the catalyst.

Among secondary reactions, ring opening of cycloalkanes could be highlighted, as well as alkyl groups transfer:

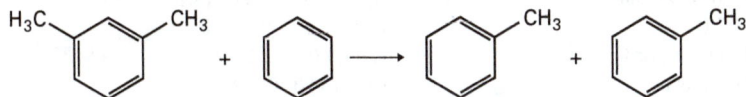

(reaction scheme of aromatic ring opening/alkyl transfer)

and dehydrocyclization:

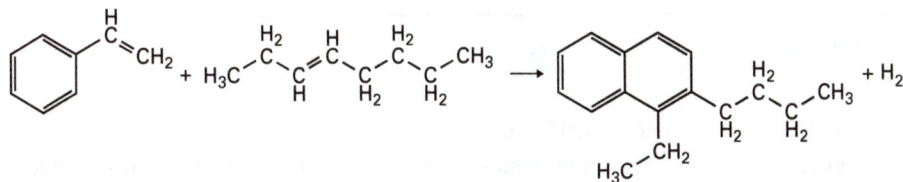

(reaction scheme of dehydrocyclization)

An overview of cracking reactions and the products is provided in Tab. 4.6.

The handbook by Lloyd [6] gives a very thorough description of FCC as well as of a number of other important industrial reactions and it is strongly recommended to those interested to learn more about these processes.

Dehydrogenation reactions does not usually play a significant role during the cracking of high molecular-mass paraffins, while for low molecular weight paraffins these reactions could be important as they result in the formation of valuable components – olefins and from low value ones – paraffins.

Coke, which usually has an atomic ratio between H and C from 0.3 to 1 and spectral characteristics typical for polycyclic aromatic compounds, can be classified into several groups:

Tab. 4.6: Catalytic cracking reactions.

Hydrocarbon	Initial products	Further products
Paraffins	Branched paraffins and olefins mainly in the C_3–C_{10} range.	Olefins crack and isomerizes and are also saturated by hydrogen transfer to give paraffins. Olefins also cyclize to naphthenes.
Naphthenes	Crack to olefins. De-hydrogenate to cyclic olefins. Isomerize to smaller rings.	Further dehydrogenation to aromatics, by hydrogen transfer.
Aromatics	Alkyl groups crack at the ring to form olefins.	Further dehydrogenation and condensation forms coke.
	Dehydrogenation and condensation to polyaromatics.	
Typical products (approximate)	Light gas (3%) LPG (17%) Naphtha (52%) LCO (16%) HCO (5%) Coke (5%)	H_2, CH_4, C_2H_6, C_2H_4, C_3H_6, C_3H_8 C_4H_8, C_4H_{10} Light, 40 °C–110 °C, Heavy 110 °C–220 °C Jet fuel 220 °C–340 °C Kerosene, diesel, heating oil Recycle. Higher than 340 °C

(From [6].)

- catalytic coke formed on acid sites
- dehydrogenated coke, which results from dehydrogenation reactions over metals coming with the feedstock and deposited on the catalyst
- chemisorbed coke, which is a result of irreversible chemisorption of high boiling point polycyclic arenes
- reversible coke, resulting from incomplete desorption after exposure to steam

Catalytic coke is directly related to cyclization of olefins, condensation, alkylation and hydrogen transfer. The highest yield of dehydrogenated coke occurs over such metals as Co, Ni, Cu and to a lesser extent V, Mo, Cr and Fe.

One of the criteria for determining the significance of the secondary reactions is the ratio between gasoline and coke. If the ratio is high it shows the dominance of the desired reactions, provided that the octane number of gasoline is high enough. Isomerization, hydrogenation, cyclization and mild aromatization of olefins could be considered as desired reactions. They lead to high yields of paraffins, including branched ones, as well as aromatics with a boiling range close to that of gasoline and to a favorable ratio between iso- and normal paraffins. Excessive cracking, dehydrogenation and polymerization of olefins, as well as condensation of aromatic hydrocarbons results in high yields of hydrogen and coke, low yields of olefins, giving a heavier product with lower yields of gasoline and a lower octane number. The amount of coke is dependent, both on properties of the catalyst and the feedstock but also on the kinetic parameters of the process, which will be discussed below.

4.3.3 Kinetics/process variables

Because of the high complexity of cracking reactions and hundreds of them occurring at the same time, it is not realistic to develop a detailed model for cracking, therefore for modeling purposes empirical models are usually applied. Simplified kinetic models are applied when several compounds are lumped together. For example, a simple three lump model given in Fig. 4.36 consists of a) one feedstock lump (gas oil, VGO or any other heavy feed) and two product lumps: b) gasoline and c) coke + light gases. The gasoline lump contains the fractions between C_5 up to the hydrocarbons with a 220 °C boiling temperature. The coke + light gases lump contains (in addition to coke) C_4 and lighter than C_4 hydrocarbons.

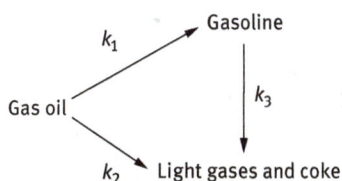

Fig. 4.36: The three lump model.

As example of this reaction mechanism is gasoline formation from gas oil and it is expressed in the following way:

$$r_{\text{go to gasoline}} = k_1 \varphi_1 C_{\text{go}}{}^n \tag{4.9}$$

where φ_1 is the deactivation function and n is the reaction order in gas oil. Although the first-order reactions are mainly considered in the literature, sometimes second order is also assumed. In order to explain the effect of coking on the catalyst decay two main approaches have been used. The first one relates deactivation with the catalyst coke content, while the second approach links catalyst deactivation to time-on-stream.

Another approach to kinetic modeling is to lump different products based on their chemical structure, for example, separately lump paraffins, olefins, naphthenes, and aromatics. This approach also gives an option to include the reaction type (cracking, hydrogen transfer, isomerization, etc.) and stoichiometry.

Although a detailed kinetic analysis is challenging, it is however worth considering various kinetic parameters and their influence on the catalytic cracking. Catalytic cracking is conducted in the absence of any heat supply and is thus considered adiabatic. The main process parameters are temperature, pressure, the ratio between the feedstock and the catalyst, as well as the circulation ratio.

In catalytic cracking the catalyst is circulated continuously between the two reactors and acts as a vehicle to transport heat from the regenerator to reactor. The spent catalysts, which are partially covered with oil, are sent to a steam stripping unit to remove the adsorbed oil before entering the regenerator. The stripped oil vapors are sent to a fractionation tower for separation into streams with desired boiling range.

Catalytic cracking is usually conducted at 450–550 °C decreasing from bottom to top in a riser as the cracking reactions are endothermic. Typical temperature profiles for catalysts with different amounts of rare earth metals are illustrated in Fig. 4.37.

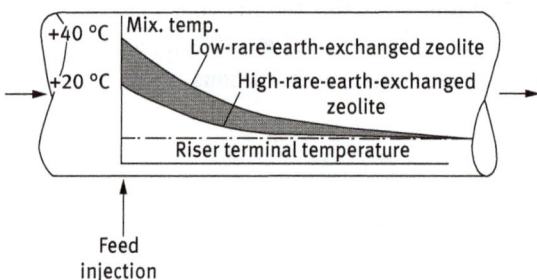

Fig. 4.37: Typical riser reactor temperature profile [21].

After stripping, the catalyst, which still contains typically 0.8–1.3 wt% coke, is fed to the regenerator where the coke is burnt off. The temperature in the regenerator is 540–680 °C being carefully controlled as high temperatures can deactivate catalysts. The residence time is in the order of minutes. Pressure is 1.6–2.4 bar in the reactor and 1.3–3.1 bar in the regenerator. Because steam generated by combustion has an undesired effect of dealuminating the framework of the zeolite and degrading zeolite crystallinity, the fresh catalyst is fed to the unit continuously to partially replace a spent catalyst. Thus, in FCC the catalyst (coined as "equilibrium catalyst") is a physical mixture of the fresh catalyst and regenerated catalyst with a broad age distribution circulating within the FCC column.

Temperature in the riser has a following influence on the reactor performance. With temperature increasing to 470–480 °C the gasoline yield is increasing, passing through a maximum at 490 °C as secondary cracking of formed hydrocarbons is becoming more prominent and the yield of gaseous products as well as coke is increasing. Moreover, with a temperature increase the octane number is increasing, along with the C1 to C3 in the gases, while C4 is decreasing. With the pressure increase there is a decrease in the gasoline and C1–C3 yields, as well as the total amount of olefins and aromatics. The coke yield is almost independent of pressure. More catalysts are introduced to the riser than the feed: the ratio of the catalyst to oil is 5–30, typically approximately 10. Increase of this ratio increases conversion and diminishes the coke content on the catalyst surface, while the yield of the desired products (gasoline) is increasing.

Another important parameter is the flow rate (16–20 h^{-1} in a riser) or the contact time (few seconds). The process usually operates at 75–80% conversion, which also determines the contact time. Longer contact times will result in secondary cracking and undesired low molecular weight gases and coke.

In summary, the yield of the main products can be influenced by the following parameters. Dry gas (an undesirable product) can be diminished by decrease in temperature and residence time; metal content and aromatic content of the feed. The yield of the desired product is increased by: increasing the catalyst to oil ratio; operation at the maximum possible temperature (without over cracking) and increasing the catalyst activity. Production of LCO can be maximized by decreasing reaction temperature, decreasing catalyst to oil ratio and a decrease in catalyst activity.

4.3.4 Catalysts

Cracking catalysts are solid acids, therefore amorphous silica-aluminas were initially used for this process. Acidic strength is usually higher for crystalline zeolites compared to amorphous ones, also displaying higher activity, which is essential for the yield of the desired product. Because of such features, compact riser reactors with zeolites replaced rather large fluidized bed reactors.

The highest acid strength was found in zeolites containing the lowest concentrations of AlO_4 tetrahedra, for example, ZSM-5 and Y. Some early zeolite catalysts contained X-zeolite, probably because it was cheaper to manufacture, however, later on it was demonstrated that Y zeolite was more stable under typical operating conditions. One of the reasons could be that zeolite X is hydrophilic (contrary to Y zeolite, which is hydrophobic).

It should be also noted that if a catalyst is too active this might lead to higher yields of gas and coke, and during regeneration substantial amounts of heat would be released leading to the collapse of the zeolite framework. Undiluted zeolites are too active for use in existing units and would be immediately deactivated as coke is deposited on the surface.

Thermal stability increases with increasing Si/Al ratio, thus ultrastable zeolite Y (prepared by ion exchange with rare earth ions, La^{3+} and Ce^{3+} or by dealumination through treatment with steam), found widespread application in fluid catalytic cracking. Note that the regeneration of zeolites includes treatment with steam, thus hydrothermal stability is essential.

In addition to the modification of acidity and stability as described above in order to achieve better heat removal (~10%) the catalyst also contains a matrix in addition to zeolite. Initially the matrix was simply a diluent and binder to form porous particles strong enough to resist attrition while circulating continuously between the reactor and the regenerator. As this matrix is also used to transport reactants, large pore materials are applied. Nowadays, aluminosilicate is applied as an amorphous matrix, also providing the required mechanical stability. Typical commercial cracking catalysts are thus a mixture of Y zeolite and SiO_2–Al_2O_3. Alumina solutions made by dissolving pseudoboehmite in a monobasic acid, such as formic acid, had previously been used to bind silica/alumina catalysts. Other binders now include silica sols and aluminum chlorhydroxide. A fresh FCC catalyst consists of spray-dried spherical, fluidizable, attrition-resistant particles, typically containing 20–40% ultrastable Y (USY) zeolite, a binder, a catalytically active acidic matrix and various additives.

The use of cracking catalysts containing zeolite increased rapidly following their introduction in 1962, and by 1972 they were being used in at least 90% of the FCC units in the United States. The most stable and successful catalysts for producing high yields of gasoline were made from rare earth zeolites. Various manufactures prepared FCC catalysts with different amounts of rare earths depending on operational severity and product objectives.

The flow scheme of the rare earth zeolite Y preparation is given in Fig. 2.121 and the synthesis of microspherical catalysts is presented in Fig. 2.122. Spatial distribution of the zeolite Y in the matrix is addressed in Section 2.2.15.6.

The most important achievement of the application of zeolites compared to aluminosilicates was not the activity increase, but rather an increase of selectivity towards some products. For example, the amount of olefins is much less over zeolites, which is associated with a higher rate of hydrogen transfer on zeolites than on aluminosilicates, and stronger adsorption of olefins compared to paraffins. Because olefins have a longer residence time on the active sites, they are able to be saturated with hydrogen, which is generated because of naphthene dehydrogenation to aromatics. Yields of gasoline are higher on zeolites, while the yields of coke and gas (<C4) are not less than on aluminosilicates.

In my first discussion with the leader of the research group in a company where I used to work in the 1990s, it was mentioned that selectivity improvements by just 1% are extremely important for a chemical company dealing with large production volumes, and that a substantial amount of work in the company was just about that.

A similar quote from Robert S. Langer's Priestley Medal address (2012), when he was describing his experience in 1970s, is: "One job interview made quite an impression on me. I went to this interview at Exxon in Baton Rouge, La., and one of the engineers there said to me that if I could increase the yield of a particular petrochemical by about 0.1%, wouldn't that be wonderful? He said that would be worth billions of dollars".

In very many cases selectivity issues are the main focus of industrial research and development work, although activity and stability to deactivation should not be undermined.

During catalyst preparation, the initial exchange with rare earth and ammonium chloride removes sodium ions from the supercages. When calcined the rare earth oxides and hydroxides decompose and migrate into the sodalite cages, where they can exchange for more sodium ions. In the calcination process, some of the rare earth ions are converted to cationic polynuclear hydroxy complexes that provide additional acid sites.

Because the demand for gasoline grew with time, the level of rare earths in the catalyst formulation refiners tended to increase, being on average ~3%. A recent spike in the prices of rare earth metals has put an additional pressure on catalyst manufacturers as well as refiners to diminish the role of rare earths for FCC.

Although better conversion and an increased yield of gasoline could be achieved with rare earths Y zeolite catalysts, it was found that octane levels were lower because fewer olefins were produced. As gasoline quality could be improved by either increasing cracking severity or using antiknock lead that contained compounds such as tetraethyl lead, this lower octane level was not a problem before new constraints were enforced.

In the mid 1970s, the application of small pore ZSM-5 (0.55×0.5 nm channel) as a co-catalyst with rare earths Y zeolite increased the octane number of gasoline. Straight-chain C6–C10 olefins produced by normal cracking were further processed in a shape-selective cracking over ZSM-5, resulting in an increase in the gasoline octane number with the same dry gas, heavy oil, or coke production at the expense of decreasing gasoline yield by up to 2%. This loss could be compensated because propylene and n-butylene are used in an alkylation unit. ZSM-5 in an inert matrix is added along with Y zeolite on a daily basis to FCC units. ZSM-5 framework is extremely stable, affording less deactivation than typical Y zeolites.

Low hydrothermal stability of ZSM-5 leads to relatively fast dealumination and the relative activity loss if no special measures are taken. Modification with phosphorus is used to limit this relative activity loss.

The feedstock contains small amounts of metals even after hydrotreating (in particular Ni and V), which deactivate zeolites, acting as dehydrogenation catalysts and giving more coke, thus passivators as Sb, P, Sn and B compounds are added because they react with the impurities. Nickel, which is a concern when its concentration on the equilibrium catalysts exceeds 800 ppm, may be passivated by the addition of low levels of antimony or bismuth to the feed. A negative side of this method is an increase of NO_x emissions and potential bottoms fouling.

A more permanent solution is to use a material within the catalyst that is able to deposit Ni. In the current residue FCC catalysts a special alumina is integrated in the catalyst throughout the catalyst microsphere in order to trap nickel forming nickel aluminate. More efficient use of alumina could be achieved if it is located only at the outer layer of the catalyst. Such catalysts were recently introduced into the market and the production technology is presented in Fig. 4.38.

Further development of more stable catalysts was related to introduction of boron, which migrates within the catalyst by solid-state diffusion to passivate Ni.

Step 1: Inner stage microsphere (MS)   Step 2: Outer stage microsphere

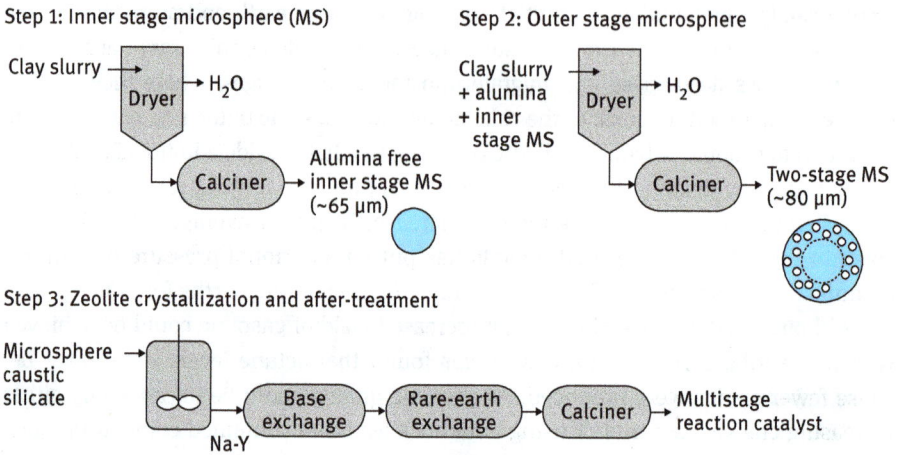

Fig. 4.38: Steps in preparation of multi-stage reaction catalyst. (From [22]).

This innovation helps to increase the liquid product yields with heavy residue feeds by better metal passivation and lower hydrogen and coke production (Fig. 4.39).

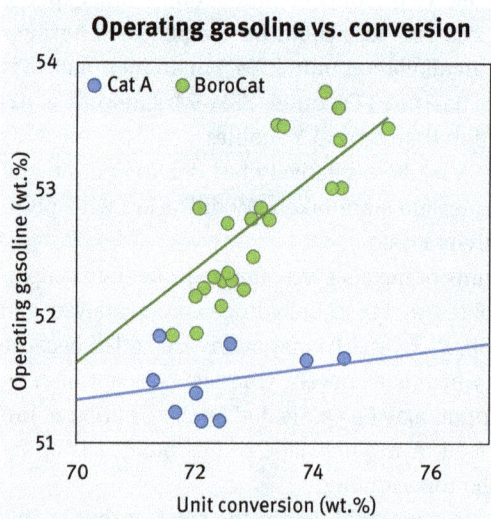

Fig. 4.39: Comparison of boron containing catalyst with a reference commercial catalyst (From [23]).

A decrease of the amount of carbon on the regenerated catalyst and an improvement in burning CO in the regenerator can be achieved by adding a CO combustion promoter, such as Pt, Pd, Ir, Ru, etc, in small amounts (0.01 to 50 ppm) to the catalyst of the regenerator.

Another key property of FCC catalysts is attrition resistance, as it determines the catalyst addition rate (typically less than 1% per day) and the catalyst fluidity. This property increases with increasing bulk density and decreasing pore volume. Conversely, catalyst activity and resistance to deactivation by the plugging of pores increases with a decrease in bulk density and increased pore volume. The application of microspherical catalysts (Fig. 4.40) in aluminosilicate amorphous matrix (10–150 μm) improved abrasion resistance and diminished the formation of fines.

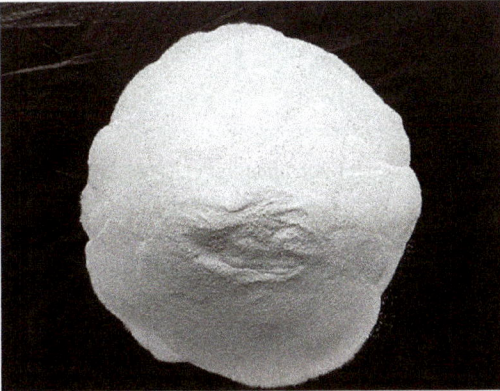

Fig. 4.40: Microspherical catalyst for FCC.

4.3.5 Technology

In its simplest form, the FCC unit (Fig. 4.41) consists of three sections: reactor section, fractionation and gas concentration (Fig. 4.42).

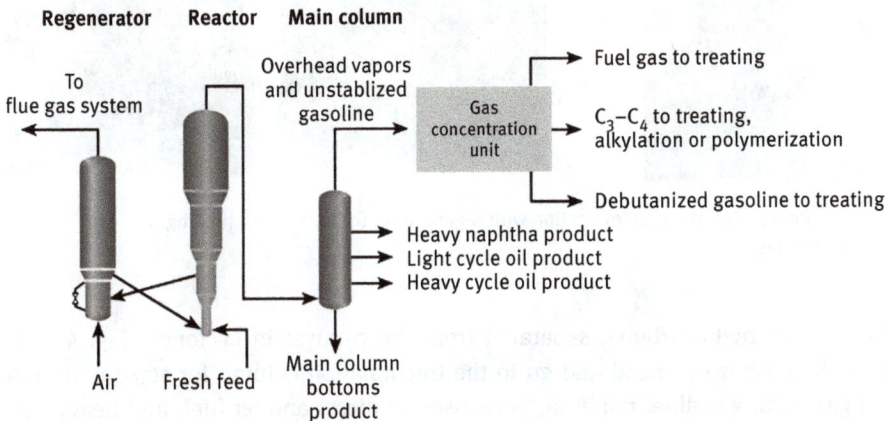

Fig. 4.41: Flow scheme of a FCC unit [24].

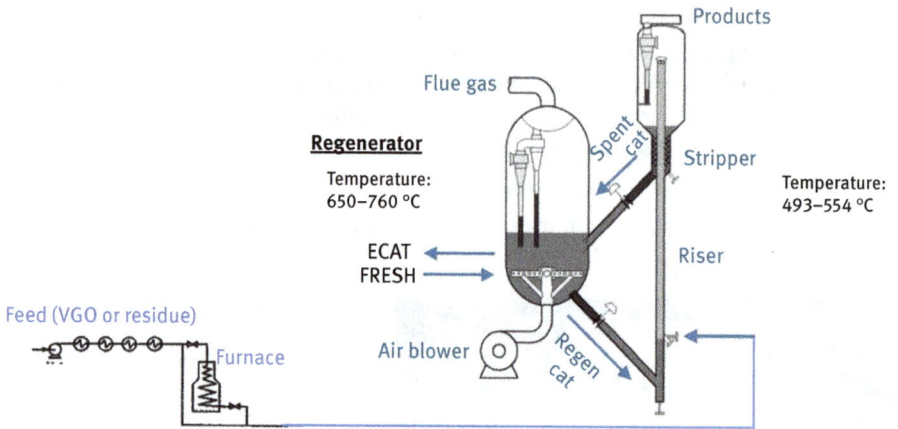

Fig. 4.42: Fluid catalytic cracking with reactor and regenerator (modified from [25]).

In the riser reactor design introduced by UOP in 1971 the pre-heated raw oil feed, together with hot catalyst from the regenerator, enters at the lower part of the reactor (Fig. 4.42). Vaporization of oil (vaporizing temperature depends on the feed, usually between 350 °C and 450 °C) occurs because of heat provided by the catalyst. In fact, the catalytic cracking unit (Fig. 4.43) also has a furnace (Fig. 4.43b), which is required during the start-up or insufficient heat supply from the regenerator.

Fig. 4.43: Photo of (a) the reactor section with regenerator, (b) furnace for heating the feed and (c) cyclones.

The cracked hydrocarbons, separated from the catalyst in cyclones (Fig. 4.43c), leave the reactor overhead and go to the fractionation column for separation into fuel gas, LPG, gasoline, naphtha, LCOs used in diesel and jet fuel, and heavy fuel oil. A large part of cracking occurs within a few seconds (2–5 s). Minimum gas velocity should be at least 10 times higher than the minimum fluidization velocity of the catalyst particles. Because of the increase in volume inside the reactor, the gas velocity increases axially and the highest gas velocity at the outlet of the riser can be as high as 25 m s^{-1}.

The spent catalyst first falls down into the stripping section within the reactor vessel. Steam (3–5 kg of steam per 1,000 kg of the circulating catalyst) removes most of the hydrocarbon vapor and the catalyst then flows down a standpipe to the regenerator. If these hydrocarbons are not removed, temperature in the regenerator will be increased beyond that required for smooth operation limits. The residence time of the catalysts in the stripper is 0.5–1 min.

The spent catalyst mixes with air and clean catalyst at the base of the regenerator, where the coke deposited during cracking is burned off, thus regenerating the catalyst and additionally providing heat for the endothermic cracking reactions. The temperature in the regenerator can rise from 500 °C to 650 °C because of the exothermicity of the coke burning. Gases leaving the regenerator contain substantial amounts of CO and have significant heating value, which is used to generate steam, thus the flue gas is treated in a CO boiler where CO is transformed to CO_2.

Catalysts are partially deactivated in each cycle. Moreover, because of catalyst transport in the reactor and the regenerator and catalyst attrition, losses of catalyst occur. To compensate for this, make-up catalyst is added to the process continuously, with a replacement rate of about 1–3% of the total inventory every day. Catalyst attrition in the unit is directly related to the catalyst circulation rate. Increasing superficial gas velocity (with increase of feed rate, reactor temperature, quench steam or decrease of operation pressure) increases catalyst entrainment to the cyclones and results in increased catalyst losses.

More than 2,000 tons of catalysts are replaced every year in a FCC unit with an inventory of 200 tons of catalyst and a replacement rate of 3% a day.

Because of the constant addition of catalysts FCC units operate continuously and are very seldom closed down to replace the catalyst inventory completely. Such shutdowns would obviously lead to significant decreases in production capacity.

Several types of FCC units are currently in operation, using various designs, not only the one presented in Fig. 4.42. For example, the R2R process (Fig. 4.44) was initially developed by Total to process feedstock with a high residue content and is now licensed by Axens/IFP and Stone and Webster. In addition to a riser, a stripper, a disengager and two standpipes; in this process there is a regeneration system composed of two regenerators linked by a lift.

The first regenerator acts as a mild pre-combustion zone at a temperature not higher than 700 °C to achieve 40–70% of the coke combustion. The partially regenerated catalyst with less than 0.5 wt% coke is transported to the elevated second regenerator where complete regeneration is achieved at almost 900 °C with slight air excess and under a low steam partial pressure. Such conditions allow almost complete removal of coke.

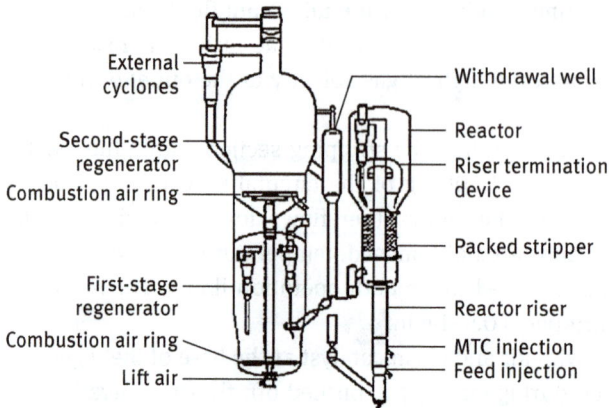

Fig. 4.44: R2R FCC process. (From [26] copyright Elsevier.).

Rather recently, UOP has started to market a new concept of millisecond catalytic cracking (MSCC), which is designed to operate at very short contact times.

Short contact times could have advantages, as product selectivity is higher over freshly regenerated catalysts than with the catalyst that was in contact with feed. In particular, short contact times favor the formation of desirable liquid product and minimize the formation of gas and coke. The process (Fig. 4.45) is organized in a way that instead of widespread and more costly riser reactors, the feed is injected perpendicular to a falling curtain of the regenerated catalyst. The feed, which is vaporized, and the catalyst move horizontally across the reaction zone and immediately enter the separation section.

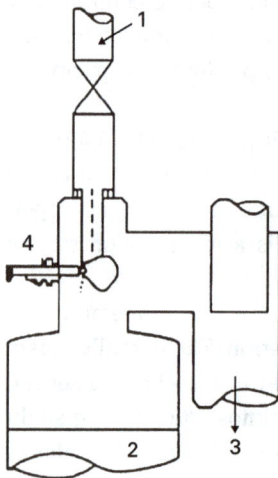

Fig. 4.45: Scheme of millisecond catalytic cracking: 1, regenerated catalyst; 2, steaming section; 3, to separation; 4, feed.

4.4 Hydrocracking

4.4.1 Overview of hydrocracking

Hydrocracking (HC) process was developed to produce high yields of distillates with better qualities than can be obtained by FCC (fluid-bed cracking cracking), which is described in the previous section. FCC is performed under more severe conditions than HC and is mainly aimed for gasoline production. Milder conditions of HC allow also to produce middle distillates, thus making the process more flexible.

An example of how HC can be incorporated in an oil refinery with some other catalytic processes is presented in Fig. 4.46.

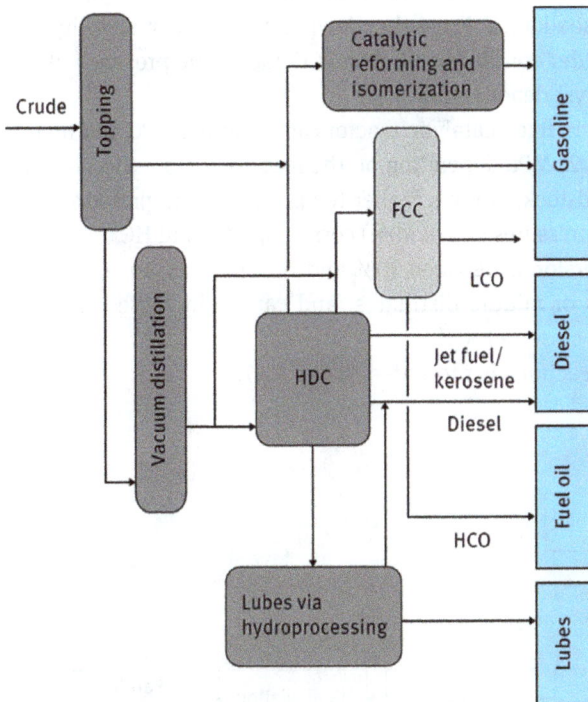

Fig. 4.46: An example of utilization of hydrocracking (HDC) in oil refining for fuel production [27].

Importance of HC, especially in Europe and the USA, is illustrated by the fact that currently HC of vacuum gas oil represents approximately 7–8% of processes in oil refining in terms of volume. Capacity of hydrocrackers can be up to 3–4 million tons per year. In the USA, hydrocrackers were traditionally used to produce naphtha from low-value aromatic streams including LCO product from fluid catalytic cracking; however, more units are shifting to production of middle distillates.

HC feedstocks are typically flashed distillates from vacuum distillation, from catalytic and thermal cracking, or deasphalted oils. Operating conditions in the reactor section of hydrocrackers are usually about 400 °C and 8–15 MPa.

Major advantages of the process include the full-scale production of high-quality products, that is, no low-grade fuel oil remains as residue, and the high flexibility toward product yields, that is, possible gasoline maximization, or gasoline–kerosene, or kerosene–gas oil maximization.

The reactions in hydrocrackers take place on metal sulfide catalysts in the presence of hydrogen resulting in hydrotreating, in which impurities such as nitrogen, sulfur, oxygen and metals are removed from the feedstock, and HC per se when carbon–carbon bonds are cleaved with hydrogen addition over bifunctional catalysts. Besides these reactions, there are several others such as hydrogenation of olefins generated during cracking and skeletal isomerization of alkanes giving equilibrium composition of normal and isoalkanes. From the viewpoint of catalyst stability, HC catalysts operating under milder conditions than in FCC and in the presence of hydrogen are less prone to catalyst deactivation.

The process is carried out in fixed catalytic reactors at 280 and 475 °C and hydrogen pressure between 3.5 and 22 MPa depending on the feedstock properties and the products desired. Heavier feedstock requires higher temperatures and pressures and longer residence time. Lower pressures (3.5–8 MPa) correspond to mild HC.

Typically, cracking conversion is between 30% and 70% per pass to achieve required selectivity to naphtha or middle distillates, and can be increased by recycling of the unconverted feed close to 100%.

A schematic view of HC reactions is presented in Fig. 4.47.

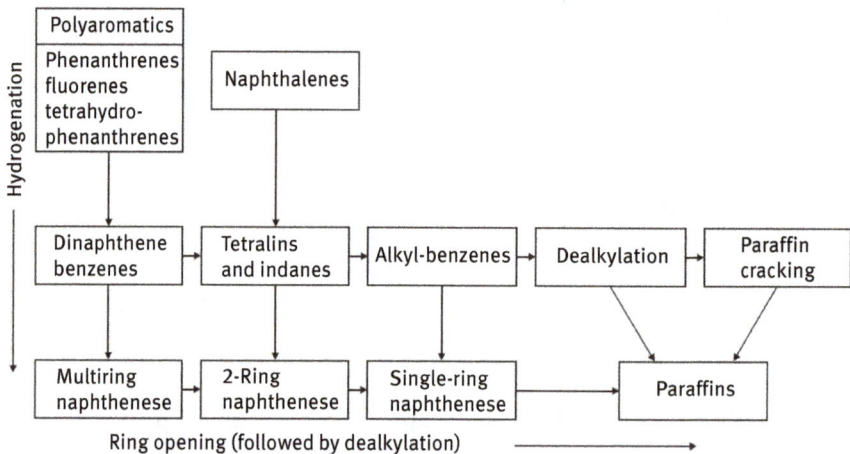

Fig. 4.47: Schematic hydrocracking chemistry.

The first step in the mechanism of HC is dehydrogenation of normal alkanes to alkene intermediates at very low near-equilibrium concentrations, which react on Brønsted-acid sites giving alkoxides. Subsequent acid-catalyzed isomerization and beta-scission cracking is followed by desorption from the acid sites and shuttling to metal sites, where isomerized and cracked species are hydrogenated. In parallel to this route, alkenes can form coke. Keeping low concentrations of alkanes allows to maintain activity of HC catalysts for several years, while FCC catalysts as presented in Section 4.3 deactivate in seconds.

One of the design parameters for HC units is hydrogen pressure, whose elevation diminishes coking and extends time between regeneration. Obviously, in bifunctional catalysis with consecutive reactions the rates of dehydrogenation/hydrogenation on metal sites and reactions on acidic sites should be harmonized. The same holds for location of these functions, which, in an ideal HC catalyst for a long time, were supposed to be in a relatively close proximity after the initial hypothesis of P. Weisz. Such proximity ensures that the equilibrium composition of alkenes is maintained at the acid sites.

An example of the product distribution of vacuum gas oil at different temperatures is given in Tab. 4.7, while compositions of the product depending on the main target of HC is illustrated in Tab. 4.8,

Tab. 4.7: Yield of vacuum gas oil hydrocracking products at different reaction temperatures.

Product	360–400 °C	400–420 °C
H_2S	2.3	2.3
C1–C2	5.7	6.5
C3–C4	4.3	10.6
C5–C6	2.6	17.6
Gasoline (b.p. up to 180 °C)	12.7	33.4
Diesel (180–350 °C)	66.9	8.3
Recycle (above 350 °C)	7.9	8.3
Hydrogen consumption, vol %	2.4	4.1

Occurrence of two main types of reactions, hydrotreating and HC, allows to split processes in modern hydrocrackers into units with two reaction stages – a hydrogenation step or desulfurization, denitrogenation and deoxygenation using cobalt–molybdenum catalysts and an HC step using nickel–tungsten catalysts. Hydrotreating catalysts are used upstream of conventional HC resulting in higher yields of gasoline. There is also a possibility to use milder conditions for HC will lower catalytic activity and higher yields of middle distillates.

Tab. 4.8: Yield of vacuum gas oil hydrocracking depending on the desired main target.

Yield, vol%	Gasoline	Jet	Diesel
Butane	16.0	6.3	3.8
Light naphtha	33.0	12.9	7.9
Heavy naphtha	75.0	11.0	9.4
Jet fuel	–	89.0	–
Diesel		–	94.1
Hydrogen consumption, m³/m³ of feed	361	312	260

4.4.2 Hydrocracking catalysts

HC catalysts (more than 200 available) combine acid and hydrogenation components; therefore, during HC, acid-catalyzed isomerization and cracking reactions as well as metal-catalyzed hydrogenation reactions occur resulting in the products with lower aromatic content, more naphthenes and highly branched paraffins. As mentioned earlier, olefins are completely hydrogenated.

Hydrogenation function is provided by noble metals and combinations of certain base metals. Either platinum and palladium or sulfided forms of molybdenum and tungsten-promoted nickel or cobalt are applied. The noble metal HC catalysts are commonly used in very-low-sulfur or sulfur-free environments with the metal loading typically below 1 wt%. The base metal content is substantially higher, with 2–8 wt% cobalt or nickel, and 10–25 wt% molybdenum or tungsten. The cracking function is provided by one or a combination (Fig. 4.48) of zeolites (e.g., zeolite Y) and amorphous silica–aluminas selected to suit the desired operating and product objectives.

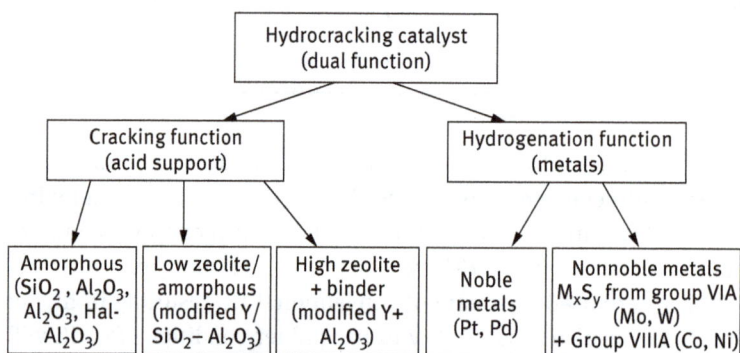

Fig. 4.48: An overview of hydrocracking catalysts.

Acidity of amorphous silica–alumina HC catalysts is moderate, requiring higher temperature and longer residence time. At the same time, such catalyst can give a higher yield of diesel fraction and prevent extensive HC. As illustrated in Fig. 4.49, there is also a possibility to have a complex catalyst loading.

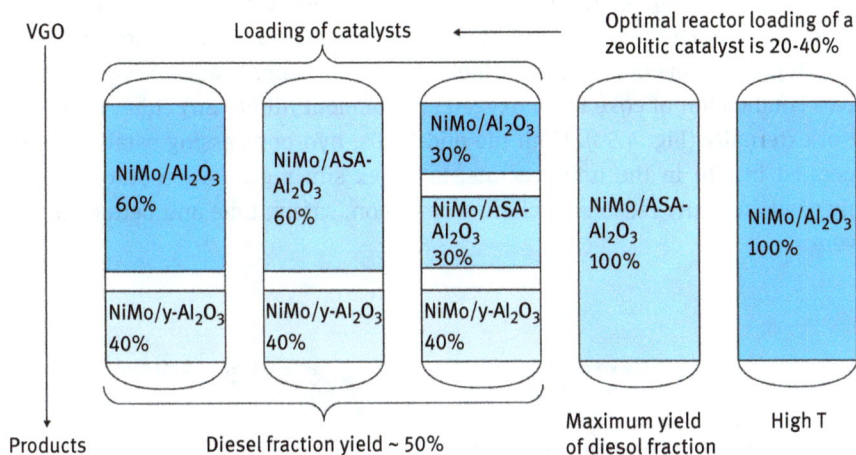

Fig. 4.49: Options for using complex catalyst loading with different acidic catalysts (ASA stands for aluminosilicates).

Unlike the amorphous-based catalysts, the zeolite-containing materials are usually more acidic and thus selective to lighter products. Microporous character of zeolites with narrow channels restricts accessibility of larger organic molecules to the active sites. Y-type zeolites are thus modified by removing aluminum through hydrothermal treatment (steaming at temperatures above 540 °C) or first acid washing with subsequent steam calcination. An alternative approach is to improve acidity of amorphous aluminosilicates.

An HC catalyst support can therefore be tailor-made depending on how the catalyst is to be operated and the product type needed. To obtain maximum conversion to gasoline at the lowest operating temperature, up to 80% Y zeolite and about 20% of a peptized alumina binder is used as support. High middle-distillate conversion is obtained at higher operating temperatures with a support containing about 10% dealuminated Y-zeolite, 70% alumina or silica/alumina in varying proportions and 20% of the binder. The appropriate Y-zeolite and the proportion to be used in the catalyst can be selected for the products, as required.

For a long time, the concept of ideal HC was used as a guideline for development of bifunctional HC catalysts, as well as skeletal isomerization of hydrocarbons. It implies careful optimization of the cracking activity of the support and the hydrogenation activity of the metal component, allowing maximum conversion with different operating conditions and feeds. Namely, an ideal HC catalyst was supposed to have a close proximity between the metal and acid sites; thus, an optimal location of the metal function in bifunctional catalysts was thought to be in the micropores of the zeolite. More recently, it was shown that the metal particles can be located on the binder or on the surface/in the mesopores of the zeolite [28].

HC catalysts during operation gradually lose activity, which is compensated by an increase in temperature. HC catalysts typically operate for cycles of several years between regenerations depending on the process conditions. Catalyst regeneration involves combustion of coke in an oxygen environment (dilute air) either in the reactor or externally (Fig. 4.50). Until the mid-1970s, hydroprocessing catalysts were regenerated in situ in the unit reactors, lately ex situ regeneration started to be mainly used for many reasons, including corrosion, safety, time and better activity recovery.

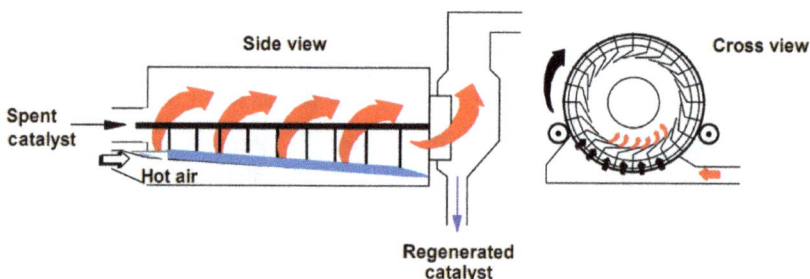

Fig. 4.50: Ex situ regeneration of spent hydrocracking catalyst by combustion in air.

Such regeneration procedure used for nickel-molybdate or nickel-tungstate catalysts is followed by resulfidation. The situation with deactivated palladium supported on zeolite is more complicated because of metal sintering. Treatment with an excess of ammonium hydroxide solution dissolves agglomerated palladium giving a tetramine complex, which is decomposed after calcination in air. In this way, the original palladium distribution and activity are restored.

Activation of nickel catalysts by sulfidation is done either ex situ or in situ using the same type of organic sulfur compounds and procedures as implemented for hydrotreating catalysts. Palladium catalysts should be carefully reduced in hydrogen at approximately 350°C, avoiding overheating and thus sintering.

Development of commercial HC catalysts is aimed not only on optimizing the chemical composition, but also on the catalyst shape. Some of the shapes are presented in Fig. 4.51. In Criterion's Advance Trilobe Technology, a replacement of a cylindrical catalyst to a trilobe can elevate the diesel yield by 1.5%, which for a hydrocracker capacity of 50,000 bbd capacity gives daily additionally 750 bbl.

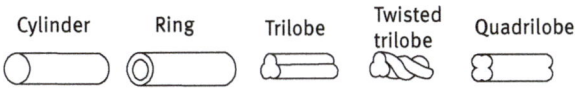

Fig. 4.51: Some shapes of commercial hydrocracking catalysts.

Sulfided Ni–Mo catalysts supported on a suitable γ-alumina with the nickel molybdate content higher than in typical hydrotreating catalysts are used for first-stage HC reactions. Sulfided Ni–W or palladium supported on zeolite or silica/alumina are applied in the second HC stage. Palladium catalysts are active if the feed contains residual sulfur compounds but are not active for the hydrogenation of benzene rings.

A scheme for production of sulfided Ni–W catalyst is shown in Fig. 4.52.

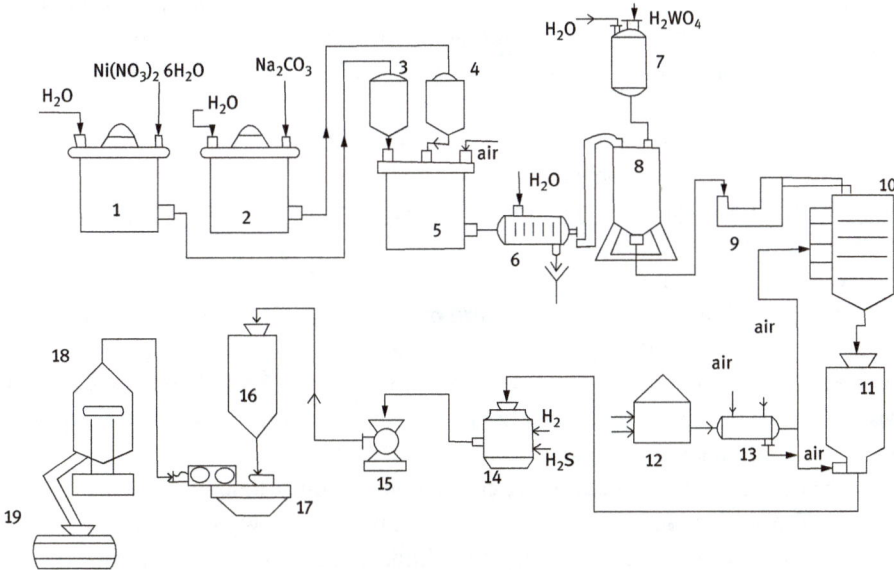

Fig. 4.52: A scheme for production of sulfided Ni–W catalyst. 1,2,7, tanks for aqueous solutions of $Ni(NO_3)_2$, Na_2CO_3 and H_2WO_4; 3, 4, measuring tanks; 5, precipitation vessel; 6, filter press; 8, mixer; 9, tabletting; 10, 11, drying ovens; 12, furnace; 13, heat exchanger; 14, sulfidation oven; 15, milling; 16, bunker for milled catalyst; 17, mixing sliders; 18, tabletting; 19, packaging [29].

The catalyst is prepared in the following way. The aqueous solutions for nickel nitrate and sodium carbonate (pos.1) and (2) through the metering tanks (pos. 3, 4) are introduced in vessel 5 (pos.5), where mixed nickel carbonate–nickel hydroxide is precipitated according to

$$5Ni\,(NO_3)_2 6H_2O + 5Na_2CO_3 + (n+3)H_2O = 2NiCO_3 3Ni(OH)_2 + nH_2O + 10NaNO_3$$
$$+ 3CO_2 + 30H_2O$$

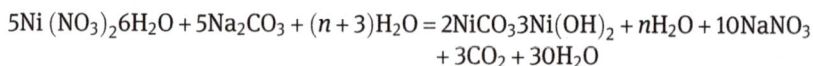

The suspension is filtered (pos. 6), washed and milled (pos. 8) in the presence of tungstic acid H_2WO_4. From the obtained suspension, the tablets are formed (pos. 9), which are then dried for one day in pos. 10. In the downstream dryer (pos.11), the temperature is 378–388 K. Sulfidation with a mixture of hydrogen and H_2S is performed in oven pos. 14. Milling (pos. 15) gives particles of the size 0.15 mm, which after additional milling (pos. 17) are tablettized (pos. 18) to the size of 8–10 mm with the bulk density of 2,000 kg/m^3. Nitric acid is added during forming as a peptizing agent.

4.4.3 Hydrocracking technology

The following factors affect product quality and quantity and the overall economics: catalyst type, process configuration (e.g., one or two stages) and operating conditions (e.g., conversion, hydrogen pressure, liquid hourly space velocity, feed/hydrogen recycle ratio).

The role of catalyst was described in Section 4.4.2. Here other parameters will be analyzed. One of the parameters in HC which prolongs catalyst lifetime by diminishing deactivation is hydrogen pressure. At the same time, an increase of pressure enhances hydrogenation of aromatic compounds diminishing the octane number.

Another important operation parameter is catalyst average temperature (CAT), which is the volume average temperature of catalyst bed. Higher CAT allows lower residence time and thus higher throughput for the same product quality or improved quality at the same feed rate. Moreover, more difficult feedstock with, for example, higher sulfur content can be processed. The downside is increased deactivation and thus shorter cycle time.

Cracking of large multiaromatic compounds requires first saturation of rings. Since hydrogenation is an exothermic reaction, from the viewpoint of equilibrium it should be conducted at lower temperature. Thus, higher temperature makes it more difficult to saturate and subsequently crack such aromatic compounds. An increase of T also enhances production of naphtha and light gases.

Gas-to-oil (hydrogen-to-oil) ratio in most modern hydrocrackers is designed to be 4 to 5 times larger than hydrogen consumption in various reactions. Increased hydrogen recycle improves catalyst stability, minimizes overcracking and acts as a heat sink, limiting the temperature rise in the catalyst beds. At the same time, too high recycle rate is not economical.

Several flow schemes of HC are possible. The once-through arrangement is the most cost-effective option and processes very heavy, high boiling feed. There is also a possibility to have a separate hydrotreating unit, which is installed when there are special feed considerations, such as high content of product in the feed or high nitrogen content.

Otherwise, either single- or two-state arrangements are commonly applied. The first option is cost-effective for moderate capacity giving a moderate product quality, while the two-stage configuration is more effective for large capacity units affording high product quality and can be applied for difficult feedstock.

Flow schemes for HC unit comprise such sections as a high-pressure reaction loop, a low-pressure vapor–liquid separation section, a product fractionation section and a make-up hydrogen compression section. As mentioned earlier, the gas-to-oil ratio is higher than hydrogen consumption; thus, the excess of hydrogen is recovered. A high-pressure loop can account for 70–85% of the installed cost of the hydrocracker. The reactors are typically trickle-bed downflow reactors with multiple beds.

Such multibed arrangement is needed as exothermic cracking and saturation reactions result in a large heat release; therefore, a cold recycle gas is introduced between the beds to quench the reacting fluids and to control the extent of temperature rise and the reaction rate. An alternative for heat management is to use heat exchangers between the beds. Depending on the conversion level and the type of feed processed, the number of beds can vary from two to eight.

Important for trickle-bed reactors is uniform reactant distribution across the catalyst; otherwise, maldistribution leads to hot spots that deteriorate catalyst performance and life.

A typical example of a two-stage hydrocracker is shown in Fig. 4.53. In the two-stage flow scheme, feedstock is treated and partially converted in a first reactor section. Products from this section are then separated by fractionation. The bottoms from the fractionation step are sent to a second reactor stage for complete conversion. This flow scheme is most widely used for large units as mentioned earlier.

The partially cooled reactor effluent is often flashed in a hot high-pressure separator (HHPS). This allows withdrawal of the heavy unconverted oil at a high temperature minimizing heat input to the product recovery section. The vapor from HHPS is first cooled in a heat exchanger. Addition of water is needed to prevent formation of solid ammonium hydrogen sulfide (NH_4HS) from ammonia and H_2S, which can otherwise deposit on the air-cooler tubes, reduce heat transfer and eventually plug the tubes.

After cooling the HHPS effluent vapor in an air cooler in the downstream, cold high-pressure separator (CHPS) hydrogen-rich vapor is separated from water and hydrocarbon liquid phases. The latter is sent to the cold low-pressure separator (CLPS), while hydrogen-rich gas is either directly recycled to the reactor as feed or quench gas with gas compressor or is first purified from H_2S using absorption with MDEA (methyl diethanol amine) or diethanol amine.

After expansion (pressure reduction), the liquid from the HHPS is sent to the hot low-pressure separator (HLPS). Hydrocarbon liquid from the CHPS is combined with the vapor from the HLPS and cooled in an additional air cooler before entering

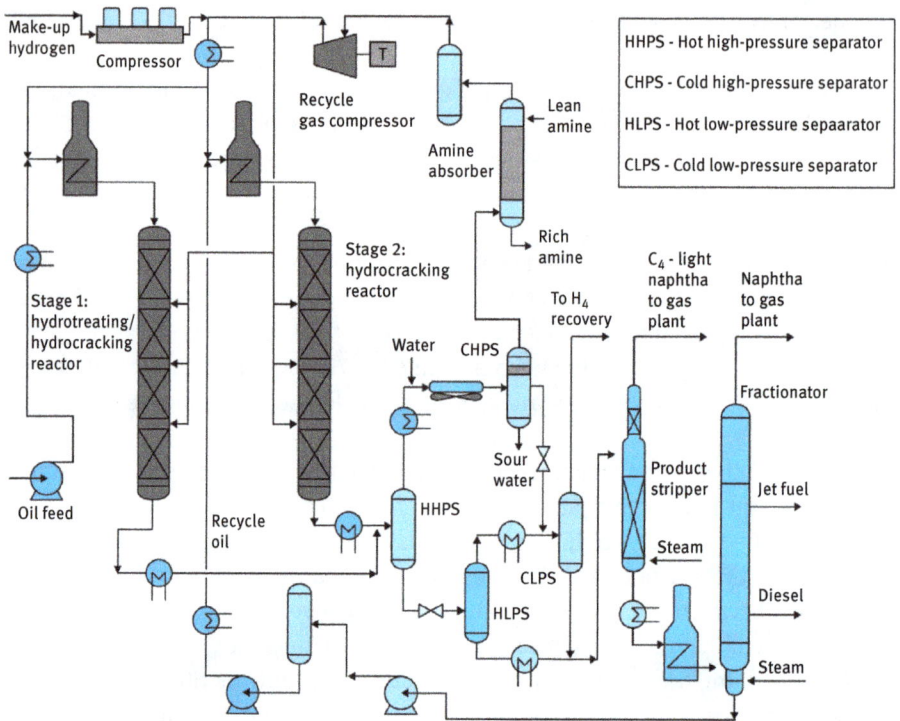

Fig. 4.53: Two-stage hydrocracking [30].

the CLPS. Hydrogen still present in the vapor from CLPS is recovered, for example, by pressure swing adsorption.

The fractionation section consists of a product stripper and an atmospheric column, separating reaction products into light ends, heavy naphtha, kerosene, light diesel, heavy diesel and unconverted oil.

When a refiner wishes to convert all the feedstock to lighter products, the fractionator bottoms stream can be recycled back to the reactor and coprocessed with fresh feed.

The single-stage flow scheme with recycle is illustrated in Fig. 4.54. Typically, there are two reactors with the first one for hydrotreating and the second for HC catalyst. In most modern units, recycle oil is blended in with fresh feed before processing in the first reactor.

Hydroconversion of heavy oil feeds (with up 4% sulfur and more than 400 ppm of metals) – gas oils, petroleum atmospheric and vacuum residues, coal liquids, asphalt, bitumen and shale oil is done using fixed, moving and ebullated bed reactors. This process results in a wide spectrum of lighter products such as naphtha, light and middle distillates, and atmospheric and vacuum gas oils. Such feedstock

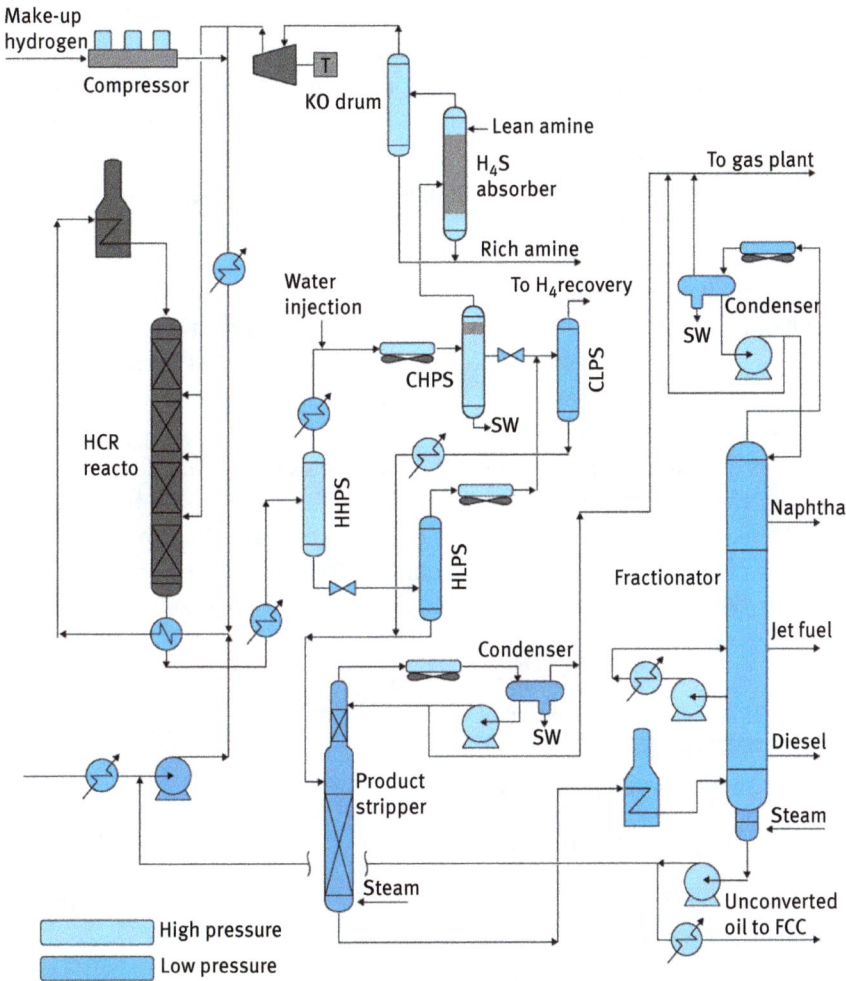

Fig. 4.54: One-stage hydrocracking [30].

contains high-molecular-weight polynuclear aromatic, coke-forming and organo-metallic compounds (with nickel and vanadium).

Residium HC requires temperatures between 385 and 450 °C, hydrogen partial pressures from 7.5 to 15 MPa and space velocities from 0.1 to 0.8 vol h^{-1}. During HC besides cracking of large molecules, heteroatoms (S, N, O) and metals (vanadium, nickel) are removed. Removal of nitrogen is more difficult than sulfur removal, thus some nitrogen compounds are merely reduced only to lower boiling range nitrogen compounds rather than to ammonia.

Metal sulfides formed during HC are absorbed on the catalyst, plugging the pores and resulting in catalyst deactivation. Minimization of coking can be done by elevation of hydrogen pressure, introduction of aromatic diluents and continuous physical removal of coke precursors from the reactor loop.

As catalysts in residuum fixed-bed hydroprocessing, extrudates (8–16 mm) are used with the active metals (Co, Ni, Mo, W, etc.) impregnated on alumina. The pores should be large enough to allow transport of large asphaltene molecules.

Different catalysts can be used in different reactors, for example, a demetalliza-tion catalyst in the first reactor, and highly active Ni–Mo desulfurization catalysts in the subsequent two reactors.

Catalyst addition to optimize catalyst usage is done in a countercurrent fashion with installation of the fresh catalyst into the third reactor. After a certain time, it is withdrawn and reused in the second reactor. Finally, the catalyst from the second reactor is reused in the first reactor. Accumulation of carbon can be as high as 40% in the case of vacuum bottoms.

Few types of hydroprocessing reactors rely on moving or ebullating catalyst beds with catalyst grain size less than 1 mm in size to facilitate suspension by the liquid phase in the reactor. Ebullated bed reactors operating also with NiMo or CoMo cata-lysts with the catalyst bed expansion of about 30–50% have up to 30 m height and 5 m diameter (Fig. 4.55). The feed and hydrogen are added from the bottom. In LC-

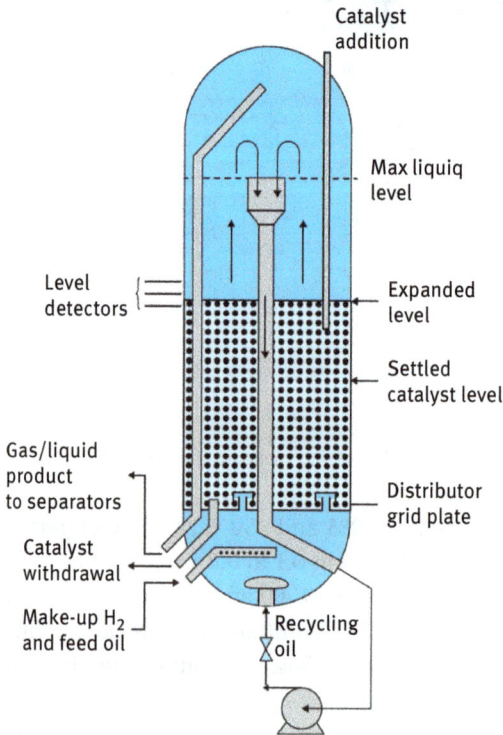

Fig. 4.55: Ebullated bed reactor for LC fining [31].

fining technology, several reactors are arranged in a series with the final product going to a separator. The vapor stream containing hydrogen is treated with amine in an absorber, purified with pressure swing adsorption, recompressed and recycled back to LC-fining reactors. A decrease of pressure before heat exchange, removal of condensates and purification gives considerable savings in investment compared to a conventional high-pressure recycle gas purification system. The liquid stream after a separator is sent to the hydrotreated distillate fractionator.

In the ebullated bed reactor, a part of the liquid product is recycled through a large pan at the reactor top and the central downcomer, by means of a pump mounted in the bottom head of the reactor. Such flow is needed to keep the bed in an expanded (ebullated) state and assure near-isothermal reactor temperature. Catalyst is added and withdrawn from the reactor to maintain a stable activity without a need for unit shutdown.

Fixed-bed hydrotreater/hydrocracker can be also put downstream of the ebullated bed reactors operating at the same pressure level. The feed for such fixed-bed reactor is the vapor stream from the ebullated bed reactors and the distillate recovered from the heavy oil stripper overhead and the straight run atmospheric and vacuum gas oils. This design makes use of excess hydrogen in the effluent vapors to hydrotreat the distillate fractions. Additional hydrogen, equivalent only to the chemical hydrogen consumed in the fixed bed reactor, is introduced as quench to the second and third catalyst beds.

In LC-fining technology with ebullated bed reactors, HC is performed at approximately 410–440 °C, pressure 110–180 bar, giving HDS efficiency of 60–85% and the following product distribution (% w/w): C4 2.35; C5 177 °C – 12.6; 177–371 °C – 30.6; 371–550 °C – 21.5; above 550 °C – 32.9 [32].

Eni slurry technology (Fig. 4.55) with the capacity of 2 million tons annually is operating at Eni's Sannazzaro refinery at 15–16 MPa and 410–450 °C with molybdenite MoS_2-nanosized catalyst generated in situ from oil-soluble precursors. An attractive feature of such technology is the absence of aging and plugging of pores, avoiding, therefore, catalyst substitution with subsequent shutdown of HC plants typical for other catalytic hydrotreating processes.

Small size of catalysts leads to the absence of mass transfer limitations, increasing the overall catalyst activity and allowing to keep a very low (few thousand ppm) catalyst concentration.

High degree of back-mixing in a bubble column reactor (height 58 m, diameter 5.4 m, weight 2,000 t) operating in slurry phase ensured almost flat axial and radial temperature profiles making the reactor intrinsically safe against temperature runaway.

The separation system in this technology is similar to other HC units with the main difference that the unconverted bottom material is recycled back to the reactor with the dispersed catalyst. A small purge (<3%) needed to limit the buildup of metals (Ni and V) fed when the heavy feed is processed.

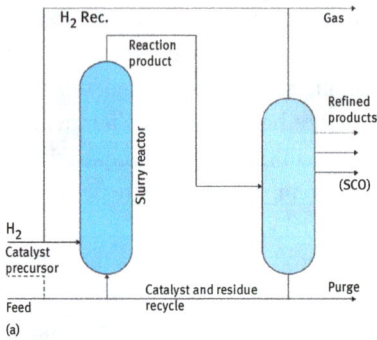

Fig. 4.56: Eni slurry moving bed hydrocracking technology [33]: (a) scheme and (b) reactor.

4.5 Steam reforming of natural gas

4.5.1 General

Steam reforming of natural gas or methane is the most common method of producing commercial hydrogen as well as the hydrogen used in the industrial synthesis of ammonia and methanol (Fig. 4.57). In addition to natural gas, other hydrocarbons

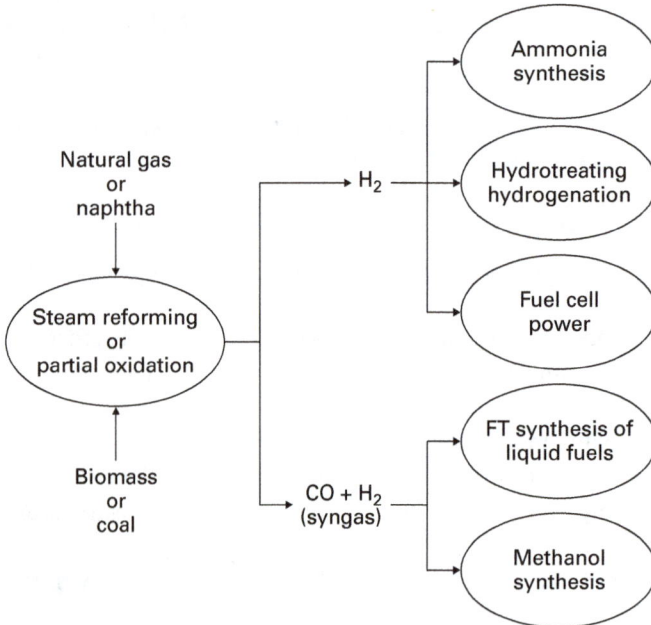

Fig. 4.57: Routes to hydrogen and syngas.

containing streams such as associated gas, liquid petroleum gas, and naphtha boiling up to 220 °C, can be fed to ammonia plants. The application of higher hydrocarbons, however, can lead to excessive coke formation on the catalysts.

At high temperatures (700–1,100 °C) steam yields carbon monoxide and hydrogen over a nickel-based catalyst in a reversible endothermic reaction with methane:

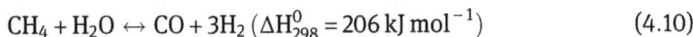

$$CH_4 + H_2O \leftrightarrow CO + 3H_2 \ (\Delta H^0_{298} = 206 \ kJ \ mol^{-1}) \tag{4.10}$$

Although natural gas consists of some other compounds this reaction is the main one occurring during steam reforming and is accompanied by extensive coke formation, which leads to significant catalyst deactivation by methane decomposition:

$$CH_4 \leftrightarrow C + 2H_2 \ (\Delta H^0_{298} = 75 \ kJ \ mol^{-1}) \tag{4.11}$$

or by Boudouard reaction

$$2CO \leftrightarrow C + CO_2 \ (\Delta H^0_{298} = -173 \ kJ \ mol^{-1}) \tag{4.12}$$

The equilibrium gas composition is given in Fig. 4.58.

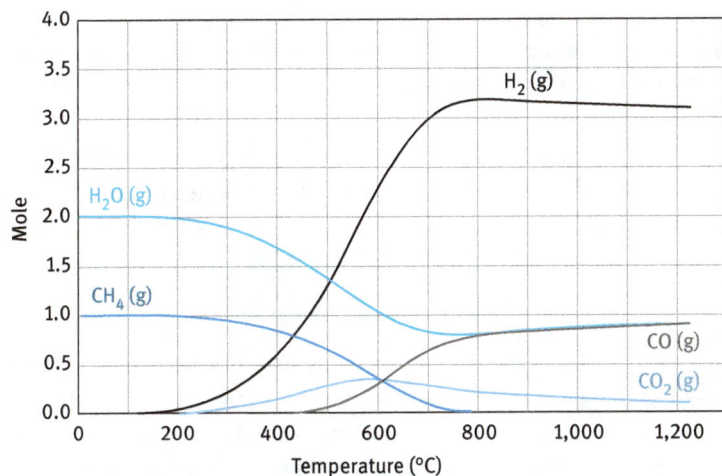

Fig. 4.58: Equilibrium gas composition at 1 bar as a function of temperature at the steam to methane molar ratio 2:1.

Obviously, the hydrogen and carbon monoxide content is increasing with temperature increase. At higher temperatures dry reforming of methane is also becoming important, thus the concentration of CO_2 passes through a maximum (Fig. 4.58):

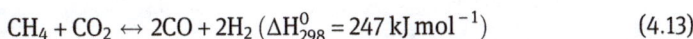

$$CH_4 + CO_2 \leftrightarrow 2CO + 2H_2 \ (\Delta H^0_{298} = 247 \ kJ \ mol^{-1}) \tag{4.13}$$

The conversion of methane to CO and hydrogen is favored at high T, low pressure and high steam to methane ratio [or steam to carbon (S/C)]. In practice, an increase of temperature is limited by thermal and mechanical stability of tubes, as the process requires the use of multitubular reactors. In particular, thick walls would worsen heat transfer, while tubes with thin walls might be subjected to severe damage at high temperatures, including bending, formation of cracks, etc.

> While working in industry I once had a discussion about steam reforming catalysts with a plant manager of an ammonia production plant, and I was suggesting that a novel catalyst will improve the performance by having a higher approach to equilibrium. The reply was that the company should first replace 500 tubes, as they look like "macaroni".

The operating steam to hydrocarbon ratio must be higher than the stoichiometric level to avoid carbon formation on the catalyst by cracking reactions, and to provide enough steam to operate the water-gas shift reaction later in the process. Thus, in steam reforming of natural gas the steam to carbon ratio is 1 to 3:3.8.

The presence of excess steam in the process gas to the reformer results in the formation of carbon dioxide by the water-gas shift reaction. Thus, the gas leaving the steam reformer also contains between 7 and 15% carbon dioxide.

In ammonia plants the methane reforming reaction from the tubular (primary) reformer is continued in the secondary reformer via the introduction of air to the reactor (Fig. 4.59). The combustion of the air produces temperatures around 1250 ° C, resulting in further reforming of methane.

In ammonia or hydrogen production units, additional hydrogen is obtained by a standalone production step where water-gas-shift reaction is done with the carbon monoxide produced upstream. This reaction, which is mildly exothermic, occurs at a lower temperature than steam reforming:

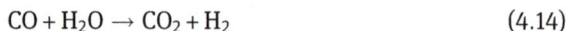

$$CO + H_2O \rightarrow CO_2 + H_2 \tag{4.14}$$

and is usually conducted in industry in two process steps with iron- and copper-based catalysts called the high- and low-temperature shift reactions, respectively. Thereafter carbon dioxide is removed by absorption using either KOH or amines (monoethanolamine or activated MDEA) scrubbing systems, and traces of residual carbon monoxide are converted to methane over a nickel-based methanation catalyst. The final step is ammonia synthesis, which will be discussed in detail below. The iron catalyst has not changed significantly since its first introduction by BASF in 1913, although there are some recent developments to apply Ru/C instead of iron. A variety of different converters has been developed and will be considered in the subsequent section devoted to ammonia synthesis.

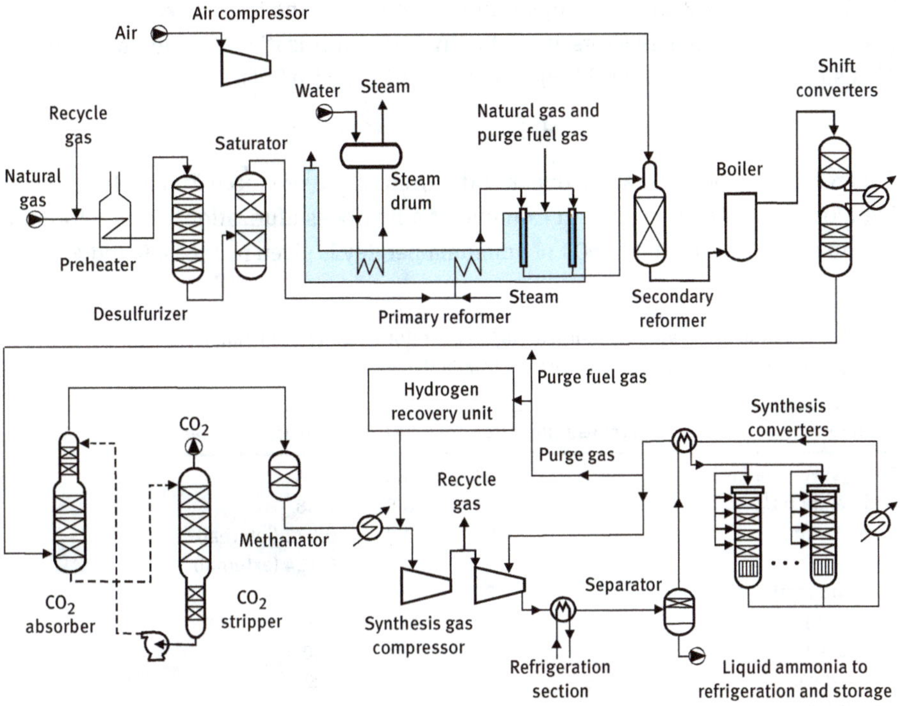

Fig. 4.59: Single train ammonia synthesis. (From [34]).

4.5.2 Process conditions for sulfur removal and primary reforming

Because the hydrocarbon feedstock for the production of hydrogen or synthesis gas contain catalytic poisons, which are detrimental for nickel catalysts, feed purification is performed in order to remove such poisons, namely sulfur and chlorine containing compounds. They are removed first by the hydrogenation of organic sulfur, nitrogen or chlorine compounds (for example, mercaptanes, etc.) to hydrogen sulfide using, for example, cobalt/molybdenum hydrodesulfurization catalysts (eq. 4.7). These reactions are exothermic and require hydrogen, which is generated in steam reforming. The cobalt/molybdate component is sulfided during commissioning and operates thereafter in a similar same way as in refinery hydrotreating.

At the end of its lifetime, the spent cobalt/molybdate catalyst in a typical ammonia plant can contain 0.5–3.0 wt% sulfur and 2–10 wt% carbon depending on the operation severity. As adsorbed hydrogen and carbon deposits could be present in hydrodesulfurization catalysts, making them pyrophoric, after use catalyst discharge is organized in such a way that prior to it the reactor must be flushed with an inert gas until the catalyst temperature has decreased to ~25–30 °C.

Because of very low levels of sulfur compounds in the feedstock, a temperature rise in the fixed-bed reactors used for hydrodesulfurization is seldom observed. This step is followed by scrubbing hydrogen sulfide with ZnO:

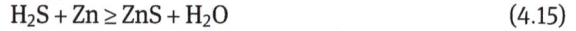

$$H_2S + Zn \geq ZnS + H_2O \qquad (4.15)$$

This reaction is not a catalytic one and the front of the produced ZnS is slowly moving along the reactor length. An example of a hydrodesulfurization/steam reforming unit in a plant producing 1,360 t of ammonia per day is given in Tabs 4.9 and 4.10.

Tab. 4.9: Conditions in hydrodesulfurization/steam reforming in an ammonia plant producing 1,360 t of ammonia per day (AM76 design).

Parameter	Hydrodesulfurization	ZnO	Primary reformer
Catalyst volume (m^3)	24	60	22.4
Bed height (m^3)	5	6.8	10.8
Diameter (m)	3.2	3.2	0.072 (internal)
			0.114 (external)
Pressure (bar)	35	34	31
T inlet (°C)	350	350	540
T outlet (°C)			780
CH$_4$, out (vol%)			8.6
Number of tubes			504
Lifetime (years)	5–25		4–8

Tab. 4.10: Conditions in hydrodesulfurization/steam reforming in an ammonia plant producing 1,360 t of ammonia per day (AM76 design).

Flow	Amount	T
Hydrocarbons (vol%): C1 = 98.7, C2 = 0.3, C3 = 0.03, C4 = 0.04, N$_2$ = 1, CO$_2$ = 0.02, H$_2$S <10 ppm	41,000 m^3 h^{-1}	471 °C
H$_2$ recycle (vol%): H$_2$ = 61.5, N$_2$ = 20.3, CH$_4$ = 0.2, Ar = 02	1,500 m^3 h^{-1}	45 °C
Steam to reformer	6,800 kg h^{-1}	367 °C

A new development in the area of hydrodesulfurization was made in 2010 by Johnson Matthey Catalysis in collaboration with Prof. Avelino Corma (Valencia). The new catalyst provides three-in-one functionality, combining full HDS conversion, H$_2$S absorption and ultrapurification. For a new plant design with an HDS vessel upstream of a ZnO based H$_2$S removal system, the HDS vessel can be completely removed from the design.

4.5.3 Kinetics and mechanism

In the classical work conducted by Temkin and his co-workers [35], reforming kinetics was studied on nickel foil at atmospheric pressure in a flow circulation system in the temperature range of 470–900 °C. Nickel foil was selected because it is difficult to avoid diffusion restrictions of the reaction, even for fine catalyst grains of commercially supported catalysts. It was concluded that the ratio of the partial pressures in the exit mixture was approximately close to the equilibrium constant of the water-gas shift reaction at 800 °C. A retardation of the reforming rate by hydrogen was observed at 470–530 °C. In a more extensive work by the same group [36], the following mechanism was proposed with two routes, the second one being the water-gas shift reaction.

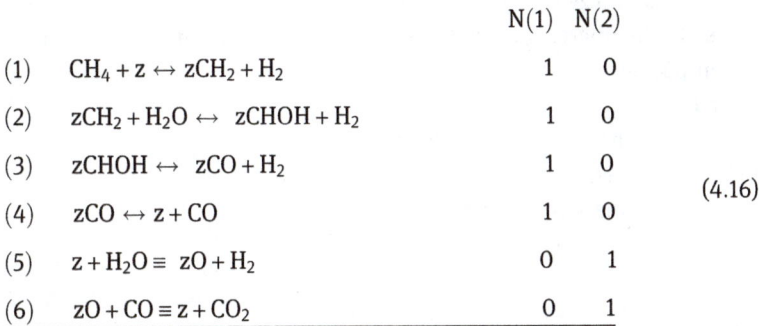

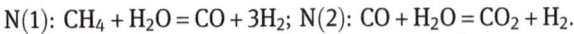

		N(1)	N(2)
(1)	$CH_4 + z \leftrightarrow zCH_2 + H_2$	1	0
(2)	$zCH_2 + H_2O \leftrightarrow zCHOH + H_2$	1	0
(3)	$zCHOH \leftrightarrow zCO + H_2$	1	0
(4)	$zCO \leftrightarrow z + CO$	1	0
(5)	$z + H_2O \equiv zO + H_2$	0	1
(6)	$zO + CO \equiv z + CO_2$	0	1

$$(4.16)$$

N(1): $CH_4 + H_2O = CO + 3H_2$; N(2): $CO + H_2O = CO_2 + H_2$.

In (4.16) z is the surface sites, steps 5 and 6 are quasi-equilibria, while other steps are reversible.

The kinetic equation for steam reforming [route $N^{(1)}$] has the form:

$$r = \frac{k_1 P_{CH_4}(1 - x_{(1)})}{\left(1 + I_1 P_{H_2O} + I_2\left(\frac{P_{H_2}^2}{P_{H_2O}}\right) + I_3\left(\frac{P_{H_2}^3}{P_{H_2O}}\right)\right)\left(1 + K_5\left(\frac{P_{H_2O}}{P_{H_2}}\right)\right)} \tag{4.17}$$

where k_1, k_2, k_3, k_4, k_{-1}, k_{-2}, k_{-3}, k_{-4} are the rate constants of corresponding elementary steps; $I_1 = k_{-1}/k_2$, $I_2 = k_{-1} \times k_{-2}/k_1 \times k_2$, $I_3 = k_{-1} \times k_{-2} \times k_{-3}/k_2 k_3 k_4$, and K_5 is the equilibrium constant of step 5.

At sufficiently low hydrogen pressure the rate of the reforming reaction can be expressed as:

$$r = k P_{CH_4}(1 - X_1) \tag{4.18}$$

where $X_{(1)} = (P_{CO} P_{H_2}^3 / K_1 P_{CH_4} P_{H_2O})$ reflecting the approach of reaction (1) in eq. 4.16 to equilibrium, K_1 is the equilibrium constant of reaction (1).

The first-order in methane, which follows for the equilibrium, corresponds to the dissociative adsorption of methane as the rate-determining step.

The simplified mechanism was later on extended to address some important kinetic (reaction orders at elevated pressures) and mechanistic aspects of steam reforming.

Thus the formation of carbon dioxide might proceed by the interaction between adsorbed oxygen atom and adsorbed CO rather than as a Eley-Riedal type of mechanism (step 6 in eq. 4.16). It is also probable that adsorbed H atoms participate in the reaction steps.

The model of Xu and Froment [37] was based on the experimental data obtained in a plug flow reactor between 500 and 550 °C and the total reaction mixture pressure from 3 to 15 bars. The ratio of partial pressures of hydrogen to methane and water to methane at the reactor inlet was 1.25, and 3 or 5, respectively. The catalyst had grains of size 0.18–0.25 or 0.3–0.4 mm. The rate of methane consumption decreased with increasing hydrogen partial pressure and increased with increasing steam partial pressure. The order of the methane consumption rate with respect to methane was significantly lower than unity.

The mechanism of steam reforming proposed by Xu and Froment [37] consists of three routes, each comprising a single rate-determining step, while the other steps are at quasi-equilibria.

		N(1)	N(2)	N(3)
(1)	$H_2O + z \equiv zO + H_2$	1	1	2
(2)	$CH_4 + z \equiv zCH_4$	1	0	1
(3)	$zCH_4 + z \equiv zCH_3 + zH$	1	0	1
(4)	$zCH_3 + z \equiv zCH_2 + zH$	1	0	1
(5)	$zCH_2 + zO \equiv zCH_2O + zH$	1	0	1
(6)	$zCH_2O + z \equiv zCHO + zH$	1	0	1
(7)	$zCHO + z \leftrightarrow zCO + zH$	1	0	0
(8)	$zCO + zO \leftrightarrow zCO_2 + z$	0	1	0
(9)	$zCHO + zO \leftrightarrow zCO_2 + zH$	0	0	1
(10)	$zCO \equiv CO + z$	0	1	0
(11)	$zCO_2 \equiv CO_2 + z$	0	1	1
(12)	$2zH \equiv zH_2 + z$	2	0	2
(13)	$zH_2 \equiv H_2 + z$	2	0	2

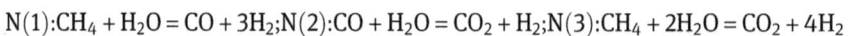

$$N(1):CH_4 + H_2O = CO + 3H_2; N(2):CO + H_2O = CO_2 + H_2; N(3):CH_4 + 2H_2O = CO_2 + 4H_2$$

$$(4.19)$$

The rate of steam reforming is then given by the sum of two routes (the first and the third):

$$r^{(1)} = \frac{k_1 P_{CH_4} P_{H_2O} [z]^2}{P_{H_2}^{\frac{5}{2}}} (1 - X_{(1)}) r^{(3)} = \frac{k_1 P_{CH_4} P^2_{H_2O} [z]^2}{P_{H_2}^{\frac{5}{2}}} (1 - X_{(3)}) \qquad (4.20)$$

where:

$$[z] = \frac{1}{1 + K_{H_2O} \left(\frac{P_{H_2O}}{P_{H_2}} \right) + K_{CO} P_{CO} + K_{H_2} P_{H_2} + K_{CH_4} P_{CH_4}} \qquad (4.21)$$

The kinetic model was able to describe experimental data, although there were still some mechanistic deficiencies. The experimental dependence on the reaction rate of the steam partial pressure cannot be explained using the theoretical value of pre-exponential factor for steam adsorption coefficient. The model of Xu and Froment [37] also suggests high coverage of non-dissociated methane, which is unlikely.

A microkinetic model from Aparicio [38] comprised 13 elementary steps, such as: adsorption and stepwise dissociation of methane to adsorbed CH_3, CH_2, CH, C and H species; adsorption and dissociation of steam to OH and H; formation of CHO by reaction of adsorbed carbon and hydroxyl; dissociation of CHO to adsorbed CO and H atoms; molecular adsorption of CO_2 with subsequent reaction with an adsorbed H atom to form formate (COOH) intermediate; desorption of CO and H (with recombination). The assumption of the involvement of surface intermediates CHO and COOH was based on low-temperature data, while these intermediates were not observed at high temperature. This indicates a danger of direct application of a microkinetic approach relying on experimental data, which correspond to conditions (temperature, pressure, catalyst) deviating substantially from conditions of real catalysis. In addition, the model is unnecessary complicated not allowing an explicit rate expression to be obtained.

Further mechanistic elaborations by Avetisov et al. [39] resulted in the mechanism that takes into account stepwise dissociation of methane, adsorption of water with dissociation and a step of the interaction between adsorbed carbon monoxide and adsorbed oxygen:

		N(1)	N(2)
(1)	$CH_4 + 2z \leftrightarrow zCH_3 + zH$	1	1
(2)	$zCH_3 + z \leftrightarrow zCH_2 + zH$	1	1
(3)	$zCH_2 + z \leftrightarrow zCH + zH$	1	1
(4)	$zCH + z \leftrightarrow zC + zH$	1	1
(5)	$zC + zOH \leftrightarrow zCOH + z$	1	1
(6)	$zCOH + z \leftrightarrow zCO + zH$	1	1
(7)	$zCO \equiv z + C$	1	0
(8)	$zCO + zO \leftrightarrow 2z + CO_2$	0	1
(9)	$H_2O + 2z \equiv zOH + zH$	1	2
(10)	$zOH + z \equiv zO + zH$	0	1
(11)	$2zH \equiv z + H_2$	3	4

(4.22)

$N(1):CH_4 + H_2O = CO + 3H_2; N(2):CH_4 + 2H_2O = CO_2 + 4H_2$

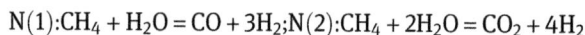

This mechanism leads to a very complicated rate equations (notation of the authors of [39]):

$$r_{CH_4} = k_1 P_{((CH)_4)} (1 - (1 / K_{((1))}((P^3_{(H_2)} P_{CO}) / P_{(H_2)} P_{((CH)_4)} (1 + (k_{(-8)}/k_{(-7)})(P_{((CO)_2)}/P_{CO})[z]/$$
$$(1 + (k_{(-8)}/k_{(-7)}) K_{((2))}(P_{(H_2O)}/P_{(H_2O)})[z])))$$

(4.23)

with:

$$[z] = \cfrac{1}{1 + K_{CH_3}\left(\cfrac{P_{CH_4}}{P^{\frac{1}{2}}_{H_2}}\right)\left(1 - \left(\frac{(1-X)}{DEN}\right)\right) + K_{CO}P_{CO}}$$

(4.24)

$$DEN = 1 + A_1 P^{\frac{1}{2}}_{H_2} + A_2 P_{H_2} + A_3 P^{\frac{3}{2}}_{H_2} + A_4 \frac{P^{\frac{5}{2}}_{H_2}}{P_{H_2}} + A_5 \frac{P^3_{H_2}[z]}{P_{H_2O}\left(1 + \left(\frac{k_{-8}}{k_{-7}}\right)K_{(2)}\left(\frac{P_{H_2O}}{P_{H_2}}\right)[z]\right)}$$

(4.25)

$$A_1 = \frac{k_{-1}}{k_2} K^{\frac{1}{2}}_{-11}, A_2 = \frac{k_{-1}k_{-2}}{k_2 k_3} K_{-11}, A_3 = \frac{k_{-1}k_{-2}k_{-3}}{k_2 k_3 k_4} K^{\frac{3}{2}}_{-11}$$

(4.26)

$$A_4 = \frac{k_{-1}k_{-2}k_{-3}k_{-4}}{k_2 k_3 k_4 k_9}\left(1 + \frac{k_{-5}}{k_6}\right)K^{\frac{5}{2}}_{-11}, A_5 = \frac{k_1}{K_{(1)}k_{-7}}$$

(4.27)

where $K_{CO} = k_{-7}/k_7$.

Simplifications, however, are possible, leading to eq. (4.18), which is recommended for practical purposes, especially at sufficiently large steam to hydrogen ratios.

4.5.4 Technology

In steam reforming for ammonia production, the conditions for most of plants are designed to afford certain methane content in the outlet of the primary reformer to be adequate for the downstream secondary reforming where the ratio between hydrogen and nitrogen of 3 mol/mol should be achieved. Some examples of operation conditions for steam reforming in ammonia synthesis plant are given in Tab. 4.11.

Tab. 4.11: Plant data for steam reforming in ammonia synthesis of 1,000 metric ton per day (mtpd) with 19 m^3 of steam reforming catalyst.

Lifetime (months)	9	15	24
Feed (Nm^3 h^{-1})	29,850	33,198	29,846
S/C (mole mole^{-1})	3.45	3.43	3.41
Tube inlet (°C)	489	470	478
Tube outlet (°C)	798	799	796
Pressure inlet (bar)	32.5	32.7	32.9
Pressure drop (bar)	2.4	2.3	2.8
CH$_4$ leakage (vol%)	9.6	9.3	9.1

Inlet methane 85–88%, ethane 5–6%.

The amounts of CO and CO_2 the conditions in Tab. 4.11 would be almost in the same range as methane (9–11%), with a somewhat higher concentration of CO_2 (11% CO_2 vs 9% CO).

Increased heat duty (outlet temperature) in a primary reformer leads to lower methane leakage, however, it increases the risk of excessive temperature rise in the combustion zone of a secondary reformer.

Conventional reformers for ammonia synthesis production were designed with a steam to carbon ratio of 3.3–3.6 mol/mol, while there is a trend nowadays to diminish this ratio as it reduces overall production costs.

The optimum conditions for hydrogen production, including the steam to carbon ratio, could be different from ammonia synthesis plants. They are dictated by, for example, such considerations as the feedstock. Hydrogen plants can operate based on natural gas or light hydrocarbons through to naphtha. As in refineries hydrogen production is integrated with other refinery units, feedstock to steam reformer could contain a mixture of off-gases from the catalytic reformer, catalytic cracker and other units.

In a steam reformer (Fig. 4.60) the gas passes up through vertical reactor tubes with a supported nickel catalyst, typically ca. 10 m long and ca. 10 cm in diameter. The long catalyst tubes are suspended vertically in rows within the furnace. Heat is supplied to the tubes using burners. For effective operation, heat must be transferred rapidly from the furnace itself to the surface of the catalyst, particularly at the top of the tubes.

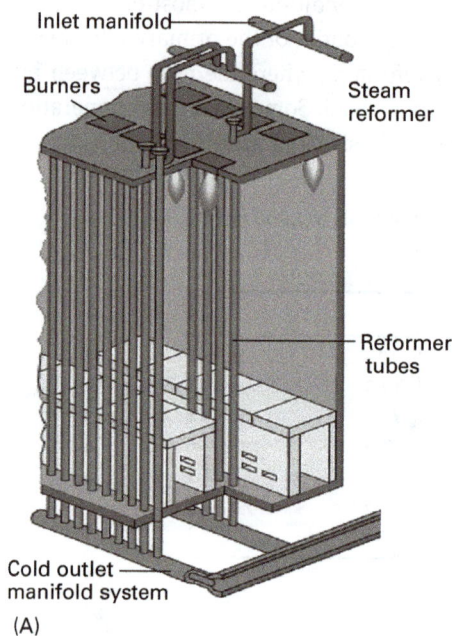

Fig. 4.60: View of a steam reformer.

The hydrocarbon should react rapidly with the steam, which helps to avoid the cracking reactions and thus catalyst deactivation. For inactive catalysts, overheating of the reactor tubes can occur. One of the ways to avoid overheating, besides introducing an active catalyst, is also shaping the catalyst in a way that allows rapid gas mixing, efficient heat transfer from the wall of the tube to the catalyst and uniform temperature gradient throughout the tube. Careful loading of the catalyst is also very important to achieve uniform gas flow and pressure drop through each tube. When these conditions are not reached, flow maldistribution occurs, leading to variable tube temperatures and hot spots.

A penalty of operating a reformer with hot tubes, for any reason, is that a temperature only 10 °C above the design level can reduce the tube life by up to 50%. The cost of tube failures resulting from the use of low activity catalysts or decreasing the steam ratio is very high.

Commercial tubular reformers operate at space velocities of 2,000–8,000 h^{-1} (STP) which are dictated by heat transfer and to lesser extent mass transfer. High flow rates and the need to minimize pressure drop require the application of rather large catalyst pellets. Therefore, a variety of shapes (rings or spheres with multiple holes, wagon wheels, clover leafs, etc.) have been proposed to minimize pressure drop on one hand and on another improve catalyst effectiveness factor, which is rather low (0.03–0.1).

Catalyst performance in primary reforming reactors is evaluated by several criteria, such as methane leakage (approach to equilibrium), pressure drop, tube wall temperature (hot spots), resistance against failures and catalyst lifetime.

The tube wall temperature is influenced by the catalyst (Ni content; surface area; shape; resistance to poisoning by S, Cl, As; degree of reduction) and process conditions (steam/carbon ratio; average heat flux; content of higher hydrocarbons; hydrogen content; pressure; inlet/outlet T; space velocity; operating failures).

While the process gas is fed downwards inside the tubes, the flue gas can go downwards in top-fired reformer furnaces or upwards in side wall, terrace wall or bottom-fired furnaces. The tube wall temperature is the highest on bottom-fired furnaces compared to other types of furnaces for the same process gas outlet temperature.

4.5.5 Catalysts

The catalyst in steam reforming of hydrocarbons generally operates at severe conditions, thus high thermal stability is important. In addition, as deactivation caused by the formation of carbon deposits and coke could be detrimental for the catalyst performance, the catalytic phase, promoters or supports should be carefully selected. Carbon formation is, for example, favored at a low steam to carbon ratio and acidic catalyst support, while the presence of hydrogen decreases the generation of carbon. There are peculiar features in the carbon or coke formation during the steam reforming of methane, making it different from coke in many other reactions related to hydrocarbon transformations. In particular, carbon forms needles, growing as filaments with the active catalytic metal (nickel) at the top. This can lead to structural damages of the catalysts and should be avoided as it leads to catalyst replacement.

Although noble metals display higher intrinsic activity than nickel and are more resistant to coke, in industrial formulations nickel is applied because of the availability and costs. The choice of support is also crucial, as the proper carrier material should ensure long lasting activity and mechanical stability. The surface area of supports is rather low (5–80 m^2 g^{-1}).

Among thermally stable supports with low acidity, α-alumina, magnesium aluminate, MgO and CaO should be mentioned. CaO, although possessing the highest thermal stability, has relatively low mechanical strength, therefore it is used as calcium aluminate. Calcium aluminate is very hard and has high initial crushing strength. Under the high pressure of carbon oxides calcium can, however, react with them somewhat, diminishing the crush strength. The silica content should be kept minimal (< 0.2 wt%), as it is volatile under normal operation conditions.

MgO should be used with a precaution, because it can hydrate at below 425 °C, giving magnesium hydroxide and dramatically weakening the catalyst. Reformer startups and shut-downs with catalysts including MgO in the formulations should therefore be done in a dry atmosphere (without steam) when temperature is below the critical hydration temperatures.

Potassium compounds are sometimes used as promoters as they are effective in lowering the acidity and allowing gasification of carbon with steam. The drawback of utilizing potassium is in its volatility at high temperature. For example, one commercial supplier offers the catalyst containing 25% NiO and 8.5% K_2O on calcium aluminate for primary reforming of naphtha, while for steam reforming of natural gas the recommended catalyst is a 10-holed ring (19 × 16 mm) containing 14% NiO on $CaAl_{12}O_{19}$ without potassium. For a mixed feed (natural gas/LPG) the catalyst formulation from this supplier (Süd Chemie, now Clariant) is 18% NiO, 1.6% K_2O on $CaK_2Al_{22}O_{34}$ in the form of a 10-holed ring (19 × 12 mm).

Another company (Johnson Matthey) manufactures several types of catalysts to be used depending on the feedstock type and plant conditions. These catalysts are nickel oxide on α alumina, calcium aluminate or a lightly alkalized nickel oxide catalyst on a calcium aluminate support.

Even if the specification above refers to NiO content, catalyst suppliers can ship primary reforming catalysts in the oxidized or pre-reduced, passivated form.

For example, Haldor Topsoe A/S uses magnesium aluminate as a support with the nickel content above 12% and manufactures both nickel oxide and pre-reduced catalyst. In order to initiate the reforming process immediately during the start-up, it is advised to charge the top 10–15% in the tubes with the catalyst in the pre-reduced form.

Nickel can be incorporated onto the support by multiple impregnations, which usually allows synthesizing stronger catalysts than by precipitation. After impregnation and calcinations, nickel oxide can be reduced, preferably by hydrogen than by steam and hydrogen. The extent of reduction depends on the support nature, the reduction temperature and time. Too high temperature and reduction time might lead to sintering. Reduction is done at a temperature close to the outlet temperature of the reformer, which at the end leads to a catalyst with low metal dispersion.

In the old formulations for methane steam reforming the amount of nickel could be rather high, reaching 25% with rather homogenous nickel distribution along catalyst grains. As, previously mentioned however, the reaction is heavily influenced by mass transfer with rather low catalyst effectiveness factor. In view of this, it is apparently clear that with such nickel distribution not all metal is involved in catalysis, thus nowadays nickel profiling is applied. For an average nickel concentration ~15% it is higher at the exterior of the catalyst and is diminishing along the grain depth.

The catalyst shape is extremely important for steam reforming catalysts, because it influences activity and the heat transfer coefficient. The shape should be designed to maximize the heat transfer rate, and subsequently the reaction rate, and minimize the tube wall temperature. An optimum shape also decreases pressure drop by increasing the voidage and size. In addition, by increasing geometrical surface area the activity could also be increased with a decrease in the tube wall temperature.

Although use of Raschig ring-shaped catalysts with different sizes was a common practice for several decades, modern steam reformers with increased heat transfer duty require more complex shapes (some of them are presented in Fig. 4.61). Such complexity is created by using multiple holes or adding flutes to the outside structure.

Fig. 4.61: Complex shapes for steam reforming catalysts.

It should be noted, however, that each hole diminishes mechanical strength, which decreases with an increase of the size of holes. At the same time, geometrical surface area is improved, pressure drop is lowered, affording also better heat transfer from the tube wall to the reacting gases and thus lower tube wall temperature (Fig. 4.62).

A comparison between the plant operation of a spherical catalyst with seven holes, which has been in operation in one of the ammonia synthesis plants since 2010, with another shaped catalyst is given in Tab. 4.12. In this table the temperature approach to equilibrium is defined as the difference between the actual and equilibrium temperature. Obviously, lower numbers indicate a close approach to equilibrium and thus better catalyst performance.

Careful loading is important in every steam reformer tube in order to have even distribution and to avoid side reactions, such as carbon formation. Such even loading is done using a special type of socks as discussed in Chapter 3. After filling each sock, vibration of the tube could be done to ensure uniform packing and minimization of the shrinkage during operation. Thereafter pressure drop is measured for all tubes. Such pressure drop measurements could be done after half filling and complete filling of the tubes. As already mentioned in Chapter 3, tubes displaying a pressure drop outside ±5% from the average should be refilled or additionally vibrated if the pressure drop is too low.

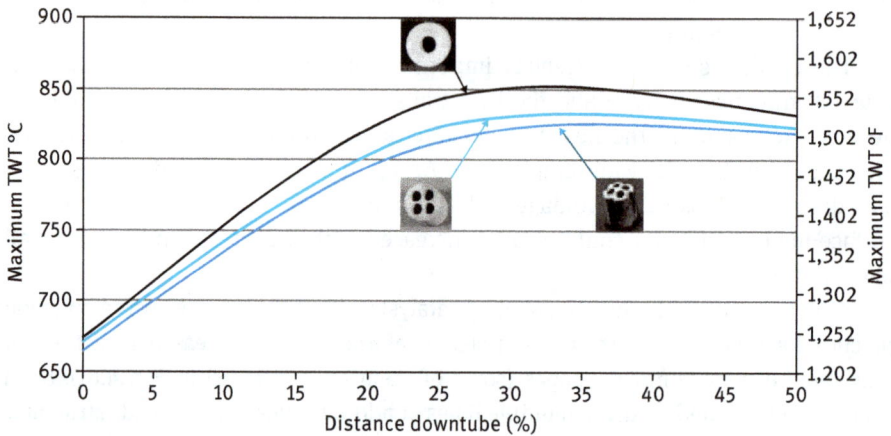

Fig. 4.62: Maximum wall temperature for different shapes. (From [40] copyright © Uhde GmbH 2005.).

Tab. 4.12: Comparison between spherical and cylindrical catalysts. Plant data. (From [41].).

Catalyst type	Run time	Volume (m³)	S/G	Inlet pressure (bar)	ΔP (bar)	T_{in}(°C)	T_{out}(°C)	Approach to equilibrium (°C)
	1 year	32	3.84	31	2.6	472	796	6
	1 year	29	3.6	29.6	1.6	481	771	1–2

4.6 Ammonia synthesis

4.6.1 General

Large-scale production of ammonia is one of the most important catalytic processes. The process was discovered in 1909 by Fritz Haber and his co-workers (Fig. 4.63), who demonstrated that it was possible to produce ammonia at a scale of 2 kg day^{-1} using osmium as a catalyst. The technology was further developed by BASF under the leadership of Carl Bosch. Such development required the construction of new high-pressure equipment and the application of a less exotic and expensive catalyst. Efforts of Alvin Mittash and co-workers, who tested over 2,500 formulations, resulted in Fe catalyst promoted with alumina, calcium oxide and potassium. The very same catalyst is essentially used in hundreds of industrial

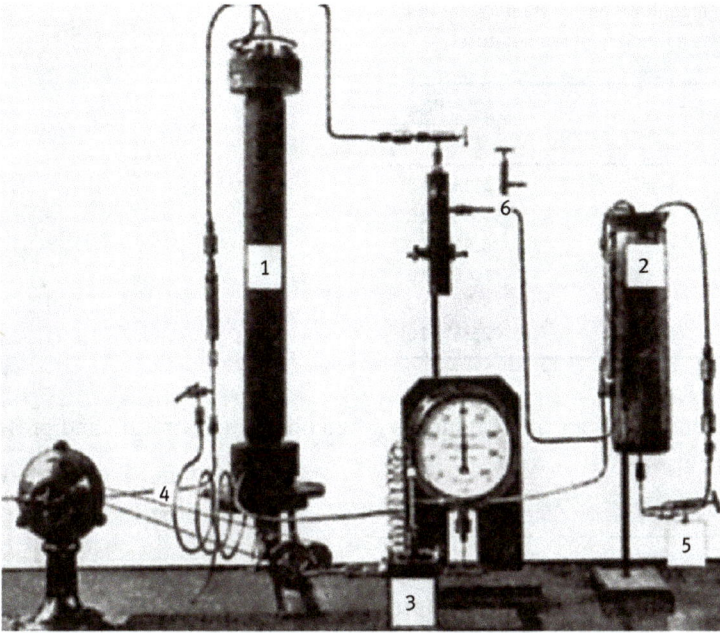

Fig. 4.63: The equipment of Haber and LeRossignol (1909).

installations with a capacity that grew from several tons a day in the original column erected in Oppau (now part of BASF site in Ludwigshafen) to single trains with 1,500–1,800 tons per day. Haber and Bosch were later awarded Nobel prizes, in 1918 and 1931, respectively, for their work.

Development of much more active Ru/C catalysts led to commercial application of Ru catalysts in few plants in the 1990s as will be discussed later.

Although the chemistry of ammonia synthesis is superficially very simple, this reaction, which operates at high pressures and temperatures, gave rise to the development of many fundamental concepts, such as surface heterogeneity, structure sensitivity, virtual pressure, rate-determining steps, microkinetics, etc. Reactor design is also very rich and sophisticated, with different types of reactors employed in industry.

4.6.2 Thermodynamics

The ammonia synthesis reaction:

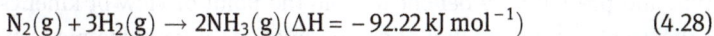

$$N_2(g) + 3H_2(g) \rightarrow 2NH_3(g)(\Delta H = -92.22\,\mathrm{kJ\,mol^{-1}}) \tag{4.28}$$

is exothermic, with a volume decrease ($\Delta V < 0$) and is limited by thermodynamics. Therefore, high P and low T should be preferred (Tab. 4.13).

Temperature (°C)	$K_{eq} = \dfrac{p_{NH_3}^2}{p_{N_2} \cdot p_{H_2}^3}$
300	4.34×10^{-3}
400	1.64×10^{-4}
450	4.51×10^{-5}
500	1.45×10^{-5}
550	5.38×10^{-6}
600	2.25×10^{-6}

Equilibrium concentrations of ammonia as a function of temperature and pressure are given in Fig. 4.64.

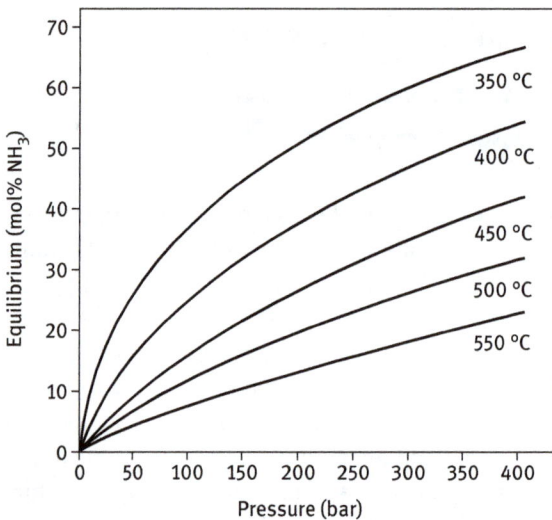

Fig. 4.64: Effect of pressure and temperature on equilibrium ammonia concentration at an inlet hydrogen to nitrogen ratio 3:1. (From [11]).

A catalyst of high activity is thus needed that should be able to operate at low temperature. Unfortunately, high activity could be achieved for rather inexpensive iron catalysts at 400–450 °C, leading to severe equilibrium limitations. The process is thus conducted industrially at 15–30 MPa between 300 and 550 °C. High temperature and pressure are beneficial from the point of view of kinetics, although at the expense of conversion limitations because of thermodynamic restrictions and costs, associated with high-pressure technology.

Ammonia synthesis reaction is exothermic, thus in order to operate efficiently equilibrium should be continuously shifted and heat should be removed. This is done by organizing the process in several stages. After one bed of catalyst, concentration of ammonia is becoming rather substantial (Fig. 4.65), coming closer to the equilibrium line. Thereafter either intermediate cooling is done by heat exchangers or ammonia is removed through quenching (adding cold reactants). In this case, the concentration line (dotted) is not going parallel to x-axis in Fig. 4.65. In should be noted that at the inlet of the catalyst bed, the reaction is far from equilibrium, while at the outlet because of high ammonia concentration the reverse reaction is becoming more prominent influencing in a negative way the forward reaction.

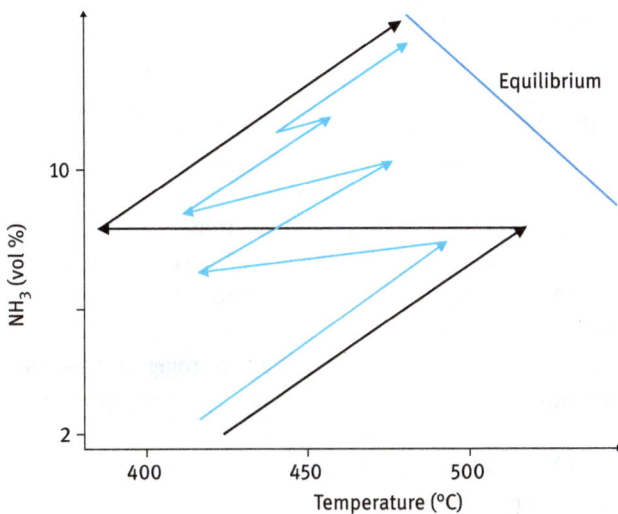

Fig. 4.65: Concentration – temperature diagram in ammonia synthesis.

The single pass through a reactor yield is then ~15%, calling for the reactants recycling. This is done by cooling gases in order to liquefy ammonia, removing it as liquid, and returning hydrogen and nitrogen in the reactor.

4.6.3 Kinetics

Kinetics of ammonia synthesis has received a lot of attention through the years. Schematically, in the simplest treatment, the ammonia synthesis mechanism near equilibrium was pictured in the following way:

1. $N_2 + * \Leftrightarrow *N_2$

2. $\underline{*N_2 + 3H_2 \equiv * + 2NH_3}$ (4.29)

    $N_2 + 3H_2 = 2NH_3$

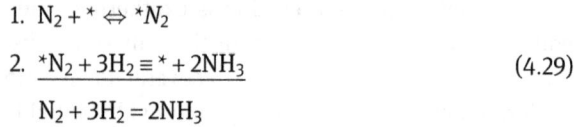

Here $*N_2$ is an adsorbed intermediate in the form of dinitrogen, step 1 is reversible and step 2 is in equilibrium.

Based on the experimental data, which covered the pressure range from below 1 atm up to 500 atm, it was proposed that near equilibrium the reaction rate is described by the following equation, which is often referred to in the literature as the Temkin-Pyzhev equation:

$$r = k'P_{N_2}\left(\frac{P_{H_2}^3}{P_{NH_3}^2}\right)^m - k''\left(\frac{P_{NH_3}^2}{P_{H_2}^3}\right)^{1-m}$$ (4.30)

where m is a constant ($0 < m < 1$). Under equilibrium, the reaction rate equals 0; therefore, $k_+/k_- = K$, where K is the equilibrium constant. Hence, only one of the constants, either k_+ or k, together with m should be determined from the experimental data. Equation (4.30) was based on the assumption that nitrogen chemisorption on an energetically non-uniform surface determines the rate of the overall reaction. The second step is considered to be an equilibrium involving the adsorbed nitrogen and the gas-phase hydrogen and ammonia.

Although in the original derivation it was supposed that nitrogen is adsorbed in the molecular form, an assumption on the dissociation nitrogen adsorption also leads to the same equation.

At high pressures, eq. (4.30) should be modified to include the deviations from the laws of ideal gases and to incorporate the effect of pressure on the reaction rate depending on the volume change of activation. Therefore, eq. (4.30) at high pressures contains not partial pressures, but fugacities.

A concept of fugacity was used also by Nielsen, Kjaer and Hansen [42] who proposed the following equation:

$$r = \frac{\dfrac{k_A p_{N_2} - k_B p_{NH_3}^2}{p_{H_2}^3}}{1 + K_o \dfrac{p_{NH_2}}{(p_{H_2}^{3/2})^{2m}}}$$ (4.31)

where p is fugacity (or effective pressure).

At low partial pressures of ammonia eq. (4.30) predicts infinite reaction rates, thus for such conditions a modified Temkin, Morozov and Shapatina equation [43] could be used:

$$r = \frac{k_{\pm P_{N_2}}^{1-m}\left(1 - \dfrac{P_{NH_3}^2}{KP_{H_2}^3 P_{N_2}^2}\right)}{\left(\dfrac{l}{P_{H_2}} + \dfrac{P_{NH_3}^2}{KP_{H_2}^3 P_{N_2}^2}\right)^m \left(\dfrac{l}{P_{H_2}} + 1\right)^{1-m}} \tag{4.32}$$

This equation, which provides a more statistically significant fit to the experimental data, generated by ICI, is based on the assumption that the reaction rate is determined by two slow irreversible steps. The first step is nitrogen chemisorption, and the second step is the addition of hydrogen to molecular adsorbed nitrogen:

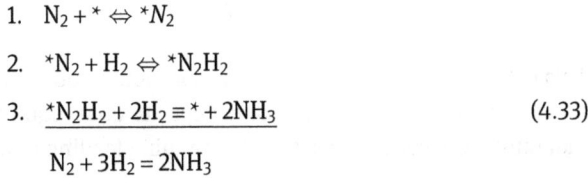

1. $N_2 + {}^* \Leftrightarrow {}^*N_2$
2. ${}^*N_2 + H_2 \Leftrightarrow {}^*N_2H_2$
3. ${}^*N_2H_2 + 2H_2 \equiv {}^* + 2NH_3$ \hfill (4.33)

$$N_2 + 3H_2 = 2NH_3$$

In eq. (4.32), which can also be derived with the supposition of dissociative nitrogen adsorption and which could be transformed to a Temkin-Pyzhev equation closer to equilibrium, $l = k_{-1}/k_{+2}$.

Equations of Ozaki-Taylor-Boudart:

$$r = \frac{\dfrac{k_A P_{N_2} - k_B (P_{NH_3})^2}{(P_{H_2})^3}}{\left[\dfrac{1 + K_c(P_{NH_3})}{(P_{H_2})^{3/2}}\right]^{2\alpha}} \tag{4.34}$$

and Nielsen:

$$r = kP_{H_2}^{\frac{1}{2}} P_{N_2}^{\frac{1}{2}} \tag{4.35}$$

were also proposed to be applicable at low ammonia concentrations [11]. Equation (4.34) could be also used at conditions far from equilibrium, when not only the first step in (4.29) is kinetically significant.

Recent microkinetic analysis by several groups (Dumesic, Stolze, Norskov) led to the following reaction mechanism, which is thought to operate on both Fe and Ru catalysts:

1. $N_2(g) + * \equiv N_2*$

2. $N_2* + \ * \Rightarrow 2N*$

3. $H_2(g) + 2* \equiv 2H*$

4. $N* + \ H* \equiv NH* + \ *$

5. $NH* + \ H* \equiv NH_2* + \ *$ (4.36)

6. $NH_2* + \ H* \equiv NH_3* + \ *$

7. $\underline{NH_3* \equiv NH_3(g) + *}$

$N_2 + 3H_2 = 2NH_3$

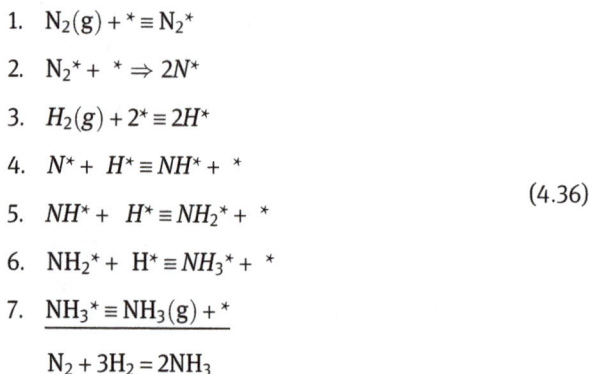

This mechanism is based on rapid adsorption of molecular nitrogen and slow dissociation to atomic nitrogen while other steps are quasi-equilibrated. This in essence means that only the second step is rate determining leading to the following rate equation:

$$r = k_{+2}\theta_{N_2}\theta_* - k_{-2}\theta_N^2 = k_{+2}K_1P_{N_2}\theta_*^2 - k_{-2}\left[\frac{P_{NH_3}}{K_3K_4K_5K_6\left(\sqrt{K_7P_{H_2}}\right)^3}\right]^2 \theta_*^2 \qquad (4.37)$$

where the fraction of vacant sites is expressed by:

$$\theta_* = \frac{1}{1+K_1P_{N_2} + \sqrt{K_7P_{H_2}} + \frac{1}{K_6}P_{NH_3} + \frac{P_{NH_3}}{K_5K_6\sqrt{K_7P_{H_2}}}\left(1 + \frac{1}{K_4\sqrt{K_7P_{H_2}}}\left(1 + \frac{1}{K_3\sqrt{K_7P_{H_2}}}\right)\right)}$$

$$(4.38)$$

Simulations of ammonia synthesis on potassium promoted iron for mechanism (4.36) using rate and equilibrium constants from surface science studies [44] demonstrated the ability of microkinetic analysis to rather accurately describe experimental data. The agreement between the models (not requiring any data fitting) and experiments obtained with a commercial catalyst is rather good, considering that modeling involves an extrapolation of nine or more orders of magnitude in pressure. Certain discrepancies could be associated with incorrect determination of the number of sites. Thus, the prediction of ammonia conversion can be improved considerably if the number of assumed sites is increased by a factor of only 3.

4.6.4 Catalysts

4.6.4.1 Iron
Typical commercial unreduced iron-based catalysts (Fig. 4.66) contain ~0.5–1% K_2O, 2–4% Al_2O_3, 2–4% CaO, while the rest is magnetite Fe_3O_4 (and some wustite FeO).

Fig. 4.66: Commercial iron catalyst for ammonia synthesis.

The formulations are essentially the same as developed by Alwin Mittash in 1909. Alumina functions as a structural promoter, enabling the production and preparation of an open and porous structure. During the initial catalyst activation, reduction starts on the outside of the granule, and dissolved alumina separates out of solid solution, forming in the pores between the iron crystallites. This minimizes further growth during completion of reduction and later operation, giving small 20–49 nm Fe crystallites.

Calcium oxide, and other basic promoters (for example, MgO, < 0.5%) react with silica impurities in the raw materials to form glassy silicates, which themselves can enhance the thermal stability of the reduced iron. The main benefit is to minimize any neutralization of the K_2O promoter, which would diminish its effectiveness. The role of K is more complicated. This component is considered to be an electronic promoter, functioning by considerably increasing the intrinsic activity of the high surface area iron particles produced on reduction. Such an enhancement of activity caused by formation of potassium ferrites is seen up to about 0.8% potash, while above this level the activity falls. The chemical action of potassium is associated with a decrease of the adsorption energy on iron, lowering concentration of adsorbed ammonia and thus increasing the number of sites available for nitrogen adsorption.

Natural magnetite with low impurity levels is nowadays used to produce ammonia synthesis catalysts.

Preparation of the catalyst by fusion of a mixture of magnetite and the promoters (added as carbonates, oxides or hydroxides, or as nitrate in the case of potassium) is described in Chapter 2. The resulting catalyst is a solid solution of wustite and alumina in the magnetite crystal lattice. The surface area of the final catalyst is rather small, in the range 1–2 m^2 g^{-1}.

The concentration of alumina should be less than the solubility of alumina in magnetite, which corresponds to a maximum content of about 3% alumina.

Ammonia synthesis catalysts are prepared in the form of magnetite, which must be activated by reduction to metallic iron before use. The reduction process is done when the unreduced catalyst is loaded in an ammonia converter prior to operation. On reduction, the catalyst becomes porous and develops a surface area around 20 m^2 g^{-1}.

> In plants using large volumes of catalyst, the reduction process can take several days. This is because it should be carried out slowly to avoid high levels of water, which then lead to the growth of the small crystallites and decrease catalyst activity. I once took part in reduction of ammonia synthesis and had to stay ~30 h in the control room without leaving it to ensure that no damage to the catalyst would occur.

Assistance in catalyst reduction is usually provided by catalyst suppliers and engineering companies, because plant operators rarely face this operation (the lifetime of ammonia synthesis catalysts can be 14–15 years and the catalyst is replaced when the vessel operating at a high pressure requires inspection because of regulations). In axial reactors with a large diameter the linear gas velocity is relatively low and the reduced catalyst can suffer some re-oxidation caused by back-mixing of the reducing gas. Gas flow should therefore be as high as possible. The same is applied in radial converters, where high flow rates are required to provide uniform reduction profiles. As the water content of the gas exiting the converter should be limited (max. 3,000 ppm) it is carefully controlled. During reduction more catalyst becomes reduced and the heat from exothermic ammonia synthesis supplements the heat from the start-up heater. This allows an increase in the gas rate while keeping temperature rise of a bed inlet temperature to about 5 °C h^{-1} until a maximum outlet temperature of 500 °C is reached. The gas rates can be increased as more beds are reduced.

In order to diminish reduction time ammonia synthesis catalysts are manufactured in the pre-reduced form. After such reduction done by a catalyst manufacture, the catalyst is carefully stabilized by a controlled flow of a mixture of nitrogen and air, resulting in less than 10% of iron re-oxidation to thin oxide film around iron particles. This procedure allows a starting reduction at a lower temperature (330 °C vs 400 °C for non-reduced catalysts) giving a shorter reduction time at a plant site. Such shorter procedure brings substantial savings as during catalyst reduction the ammonia synthesis train, including steam reforming of natural gas, is in full operation, consuming several dozens of thousands m^3 per hour of natural gas. It is thus desirable to reach the full production capacity as soon as possible, without compromising, however, long-term catalyst activity.

The pre-reduced catalysts are more expensive than the unreduced one, thus operating costs are balanced with the higher catalyst price. Typically, pre-reduced catalysts are used in only the top reactor sections, simplifying the beginning of the reduction process.

An example of a start-up procedure is given by Shepetovsky et al. [45]. The ammonia converter located in Berezniki, Russia with a capacity 1,360 t day^{-1} was revamped from axial flow to axial-radial flow by Ammonia Casale. It had three beds with axial-radial flow in the top bed (Fig. 4.67) and radial flow of gases through the second and third beds. For the first bed with axial-radial flow, 13.5 t of pre-reduced catalyst was used, while 22 and 90 t were used in the second and third beds respectively.

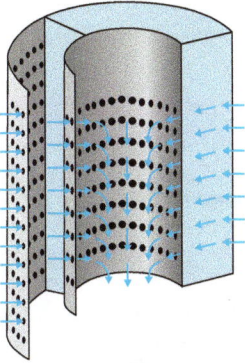

Fig. 4.67: Axial-radial flow distribution. Most of the gases pass in the radial direction. The balance is passed through the top in axial direction to eliminate the need for a top cover of the catalyst bed.

It is a typical procedure to load catalysts from a small hopper or a large bag fitted with a suitably flexible chute. Moving around the chute provides an even level of loading, keeping the desired catalyst bulk density in all parts of the catalyst bed. In the example above, during the catalyst loading after filling each 40 cm of bed height the layer was levelled using first shoveling and subsequently a vibrator.

The temperature, gas-flow rate and pressure profiles are presented in Fig. 4.68.

The following procedure was applied. Heating of catalyst in the first bed, initially to 300 °C and then to 450 °C, took 48 h. The amount of water released during this period was 0.15–0.22 vol%. Thereafter, for another 6 h the catalyst was reduced in the first bed and simultaneously reduction of catalyst in the second bed was initiated. The catalyst in the first bed was completely reduced at 500 °C and in the second bed the reduction continued at 450 °C for 6 h and then for another 6 h at 450–475 °C. Simultaneously reduction in the third bed started at 450 °C. During this 12 h-period the water content at the outlet was 0.1–0.9 vol%. Reduction of the catalyst in the bed 2 was finalized at 500 °C while reduction for the third bed continued for 14 h. Water content at the outlet was 0.08–0.7 vol%. At the end of this period ammonia was produced with 99.82% purity. The total duration of the process was 91 h. Plant performance data a few weeks after the catalyst reduction are given in Tab. 4.14.

After another 2 weeks the ammonia output was 60 t h^{-1} (1,450 t day^{-1}) with an ammonia content at the reactor outlet of 16–18 vol%.

Ammonia synthesis is one of several catalytic processes where catalyst manufacturers guarantee catalysts performance. Comparison between guaranteed parameters, mean values during warrant tests and performance at nominal capacity is given in Tab. 4.15.

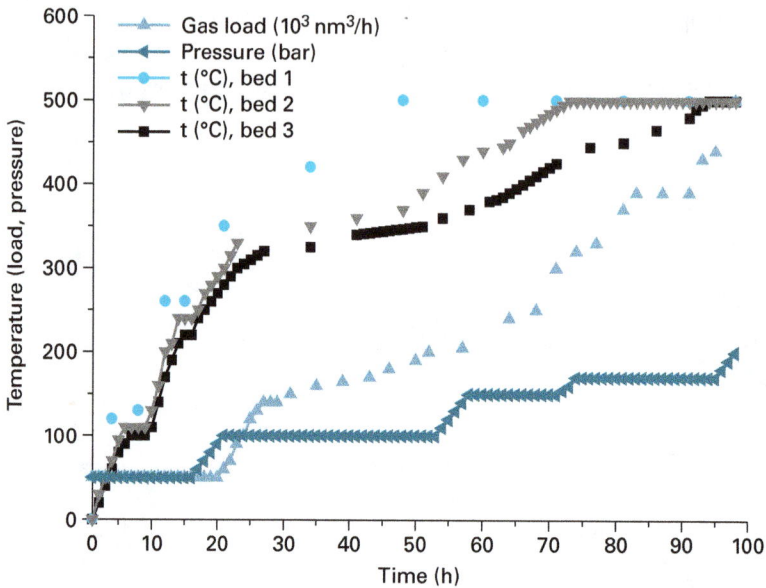

Fig. 4.68: An example for ammonia synthesis catalyst reduction procedure. (Redrawn from [45]).

Tab. 4.14: Plant performance data few weeks after start-up. (Modified from [45]).

Date	Pressure (MPa)	Gas flow 10^3 (m³ h⁻¹)	Inlet/outlet (°C)			Outlet/inlet		ΔP, (MPa)	NH₃ output (t h⁻¹)
			Bed I	Bed II	Bed III	Reactor (°C)	% NH₃		
11.09.2004	22.8	834	440	470	402	348	11.7	0.40	51
			520	520	470	146	2.9		
12.09.2004	22.1	786	434	469	414	349	11.8	0.37	51.3
			521	520	477	146	2.3		
13.09.2004	22.3	822	421	442	424	350	11.8	0.40	53.8
			509	500	479	151	2.3		
14.09.2004	22.8	775	425	445	445	348	12.0	0.38	53.7
			510	505	478	148	2.1		

4.6.4.2 Ruthenium

Part of the industrial production now takes place with a more expensive ruthenium-based catalyst (the KAAP process) rather than with cheaper iron because this more active catalyst allows reduced operating pressures. Little information is

Tab. 4.15: Warranty testing data. (Modified from [45]).

Parameter	Units	Guarantee	Mean values	Values at nominal capacity
Inlet gas flow	10^3 m³ h⁻¹	790	932	790
Inlet reactor T, not more	°C	175	172	172
Outlet reactor T, not more	°C	329	327	329
ΔP	MPa	0.56	0.601	0.42
Outlet pressure	MPa	21.0	22.1	20.0
Ammonia output	t day⁻¹	1,375	1412.7	1,375
Inerts	%	12–14	15	14

available, however, on the commercial production of the ruthenium-based ammonia synthesis catalyst.

Ruthenium, promoted by alkali metals and barium and then supported on a cheap carbon support, has provided a significant increase in activity compared with iron catalysts. K and Cs apparently facilitate nitrogen adsorption, allowing electron charge transfer to Ru, similar to iron-based catalysts. A magnesium oxide support has been claimed to give a longer life. Ru/C demonstrated much higher activity than iron at moderate hydrogen to ammonia ratios and ammonia contents, while the rate of ammonia production is strongly inhibited by ammonia. Operation at lower temperatures and pressures is thus favorable over Ru, and this catalyst should not be used at higher ammonia concentrations.

The history of Ru catalyst development and scale-up at BP is documented in the literature [46].

In 1992, low-pressure ammonia synthesis was implemented using a KBR Advanced Ammonia Process (KAAP) technology at the Methanex (formerly Pacific Ammonia) plant in Kitimat, British Columbia, Canada. This was the result of a retrofit of an existing plant with a new radial flow ammonia converter filled with Ru on a stable graphite carbon support. Ru catalyst is 20 times as active as the iron catalyst.

During the scaling up of the process, a process demonstration unit was erected capable of producing about 2 tons of ammonia per day. The process demonstration unit ran in various tests for over 8 years. A major problem during the start-up was deactivation caused by lower levels of contaminants in the process gas compared to the cylinder gas used in the laboratory reactors. Such low levels enhanced reduction of the promoters, thereby deteriorating catalyst performance at the demonstration plant. Diminishing the promoter loadings mitigated catalyst deactivation.

A wood-based carbon was selected as a catalyst support. The challenge is the carbon support scale-up involved in the development of furnace facilities for continuous treatment of the active carbon precursor at temperatures up to 2,500 K and controlled air oxidation.

In the first KAAP plant the effluent gas contained about 20% ammonia. The catalyst was reduced at 300 °C, which took less than one day, with the evolution of only small amounts of water. The catalyst was operated at lower pressures and temperatures than the iron catalyst, and contrary to iron could also tolerate the presence of 4,000 ppm of carbon monoxide. Because of high costs of the natural gas this plant was first closed temporary in 2000–2002 and then finally in 2006.

In 1998 two grassroots KAAP plants producing ammonia at 1,850 t day^{-1} began operating in Trinidad and Tobago using one bed of iron catalyst and three beds of ruthenium catalyst. A third KAAP plant in Trinidad started up in 2002 and a fourth began operating in 2004.

Start-up of other KAAP plants in Trinidad, Egypt and Venezuela was in 2008–2010. In new designs, lower pressure (90 bar) and temperature, less gas compression and less high-pressure equipment lead to lower operating and capital costs, less downtime while still being capable of maintaining the required ammonia yield per pass.

4.6.5 Reactors and process design

The ammonia synthesis reaction occurs in a loop where hydrogen and nitrogen synthesis gas is circulated continuously through a converter containing the catalyst (Fig. 4.69).

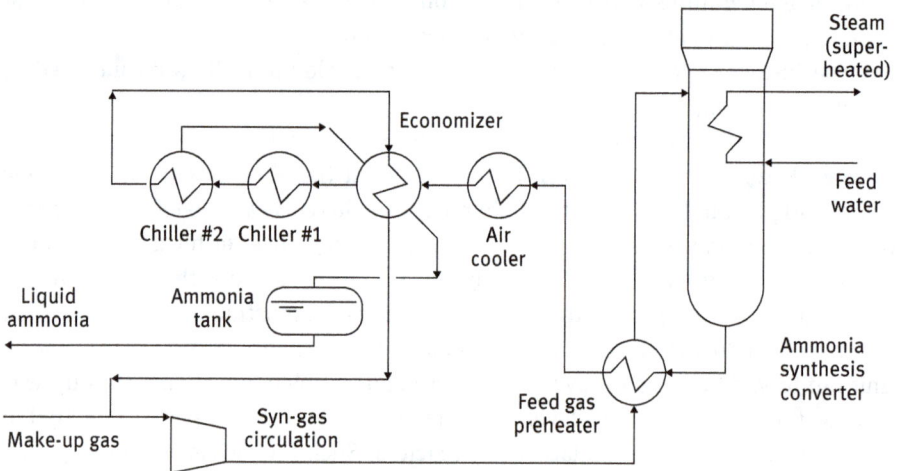

Fig. 4.69: Ammonia synthesis loop. (From [47]). Reproduced with permission.

Ammonia is condensed (condensation temperature in industry is typically −6 °C) and removed from the loop. In order to avoid accumulation of inert gases in the loop, such as methane and argon, they are continuously removed in a purge gas stream.

In the past, axial flow reactors were used, operating at high pressure (300 bar). Larger catalyst particles (6–10 mm) were required in order to limit the pressure drop through the catalyst. As the catalyst effectiveness factor (at large values of Thiele modulus) is inversely proportional to particle size, strong diffusional limitations were present. By designing converters in such a way that gas flows radially through the catalyst bed, it is possible to decrease the overall pressure drop, using smaller catalyst particles (1.5–3 mm) exhibiting higher activity per unit volume. Revamp of axial flow plants to radial flow operation mode does not lead, however, to higher ammonia production. Instead, pressure in the converter is decreased to 20–22 MPa, lowering production costs.

As explained above, because of equilibrium limitations, reactors typically include either interstage cooling or the addition of cold synthesis gas, an operation known as quenching.

Several reactor options are installed in commercial ammonia plants. In the Haldor Topsoe S-200 radial converter, two catalysts beds are applied with an intermediate cooling (Fig. 4.70A). The S-300 reactor from the same company has three beds and two interbed exchangers. This converter enables a higher conversion for the same catalyst volume or smaller catalysts volumes and lower investment costs for new plants. Similar designs (with three beds and radial flow from outside to inside) are available from other companies (for example, Uhde GmbH, Fig. 4.70B). Kellogg axial converters (Fig. 4.70C) have several beds with quenching.

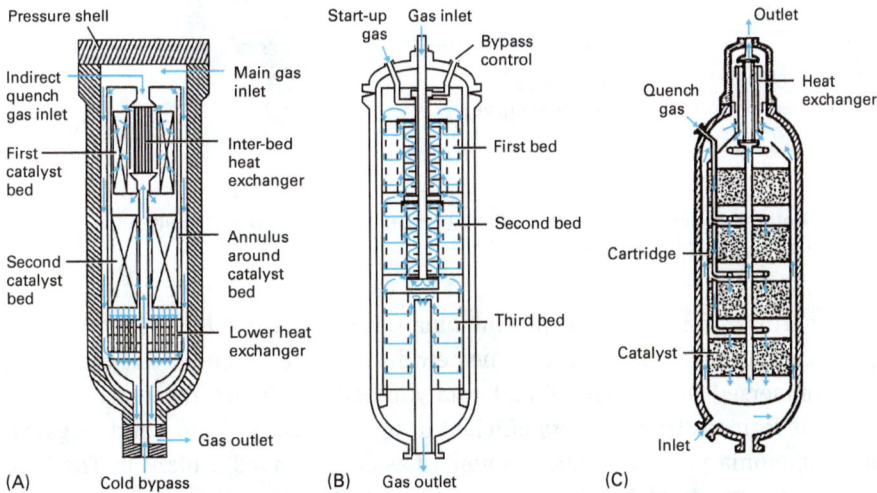

Fig. 4.70: Ammonia synthesis converters with: (A) reactor with two beds, from [48] copyright UNIDO and IFDC, (B) reactor with 3 beds (from [49]), (C) reactor with 3 beds and quenching.

In Fig. 4.71 a horizontal ammonia synthesis reactor is shown that allows removal of the catalyst basket from the converter without aid of heavy cranes usually needed to lift the catalyst basket out of vertical converters.

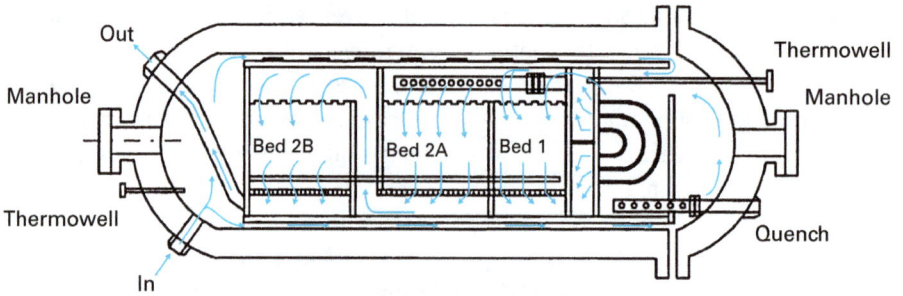

Fig. 4.71: KBR horizontal ammonia synthesis reactor. (From [48]).

In the reactors discussed above, the catalysts were distributed in several beds operated adiabatically in an attempt to follow the optimal equilibrium curve (Fig. 4.72).

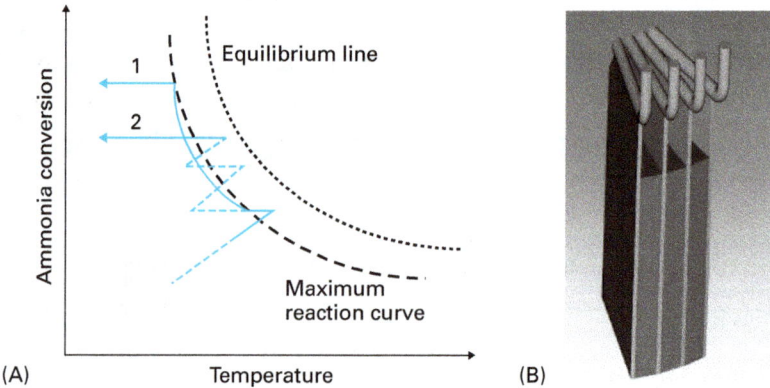

Fig. 4.72: Isothermal ammonia converter with (A) temperature profiles with (1) multitubular reactor, (2) isothermal design; (B) cooling plates.

In a departure from this concept Ammonia Casale developed a pseudo-isothermal ammonia synthesis converter. The new design is based on the use of cooling plates immersed in the axial-radial catalytic bed to remove the reaction heat while it is formed. This is a very efficient design and can help to build mega capacity ammonia plants. So far, Ammonia Casale revamped a plant in Thailand running at capacity of 753 mtpd to boost it to 953 mtpd with energy savings of 0.96 Gcal MT^{-1}. Significant reductions in the circulation rates and specific duties of heat exchangers were obtained. Lately, reactor internals were installed in another plant in Canada [50].

In ammonia dual-pressure process by Uhde (Fig. 4.73), the synthesis section comprises a once-through section followed by a synthesis loop. In this way, large capacities can be achieved (4,000–5,000 metric tons per day).

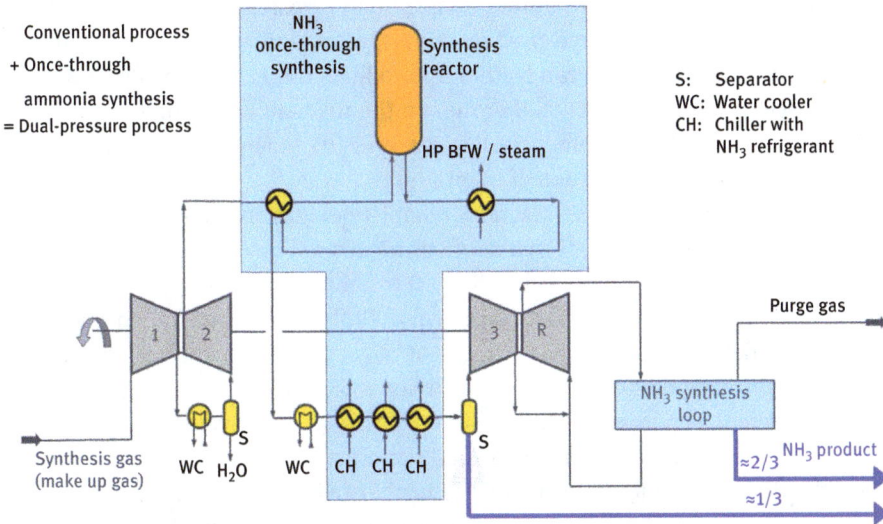

Fig. 4.73: Ammonia dual-pressure process by Uhde [51].

The ammonia synthesis loop consists of two stages. An indirectly cooled once-through converter operates at approximately 110 bar. The major part of ammonia produced in this converter after cooling is separated from the gas products. Overall, the once-through converter generates one-third of the total ammonia. The ammonia synthesis loop per se operates at approximately 210 bar.

Although many ammonia plants were revamped to improve performance efficiency there are still some barriers to their implementation of new technologies. Among these barriers, limited knowledge of process technologies outside the plant should be mentioned, as small operators are not aware of all the best technologies. Plant operators also want more data on new technologies to prove its reliability and are not eager to be the first to make risky investments. In additional, key barriers such as low energy costs, lack of corporate R&D, capital availability and corporate priorities could hinder more widespread implementation of new technologies.

4.7 Oxidation

4.7.1 General

Oxidation reactions are very important for production of intermediates, different chemicals and monomers for the polymer industry (Tab. 1.3). A key issue in many catalytic oxidation reactions is selectivity, as even a superficially minor increase in selectivity brings substantial economic returns because of large production capacity. The challenges in improving selectivity are related to (a) potentially higher reactivity of the desired product to further oxidation, (b) intermediates in complex oxidation reactions can lead to by-products rather than the desired ones, (c) presence of different functional groups in the reactant also prone to oxidation in addition to the desired functionalization and (d) homogeneous radical reactions can contribute to formation of by-products.

Typically, during oxidation reactions complete oxidation can also take place in addition to partial oxidation, which is often the desired one. Selectivity control by variation of reactant concentration is difficult to achieve, thus temperature remains as the main parameter to regulate selectivity. Typically, the activation energy of complete oxidation is much higher (by 20–40 kJ/mol) than for the selective oxidation, thus with temperature increase selectivity is diminished. This points out on a need of efficient temperature control. Overheating in such exothermal processes should be avoided as it worsens the overall performance. For consecutive reactions when the desired intermediate product is further oxidized, selectivity toward this intermediate product is decreasing with conversion. Consequently, an optimum conversion should be typically achieved, which is done by regulating the contact time (in continuous processes) and applying an excess of the organic compound, while oxygen is in deficiency.

The latter operation strategy is also related to safety aspects, which are extremely important, as many mixtures of organic compounds are explosive. Thus, safe processing requires operation below or above explosion limits. Low concentration of the organic substrate implies low rates and need to recycle substantial amounts of the oxidant. The reaction mixture can be also diluted with steam. Introduction of oxygen at different points in the reactor is another operation strategy.

From the engineering viewpoint, an important characteristic of catalytic oxidation reactions, besides explosivity, is their high exothermicity. Apart from few exceptions (e.g. oxidation of SO_2 to SO_3) adiabatic reactors are not suitable, as overheating of the catalyst bed leads to deactivation or thermal runaway. In order to facilitate the heat removal catalytic oxidation is thus often conducted in multitubular or fluidized bed reactors. Multitubular reactors do not have a uniform temperature profile, and hot spots exist moving along the reactor length as the catalyst deactivates with time-on-stream. From the point of view of heat removal, fluidized bed reactors can be considered as isothermal. On the other hand, such arrangement results in back-mixing diminishing selectivity. This could be partially avoided by having several segments (Fig. 4.74a) or using a riser (similar to FCC), where the catalyst is separated in a cyclone and send back to the reactor. In Fig. 4.74a, a possibility to use cooling agents at different stages of the multistage fluidized bed column is illustrated. Temperature regulation can be also done by quenching with the cold reactant (Fig. 4.74b).

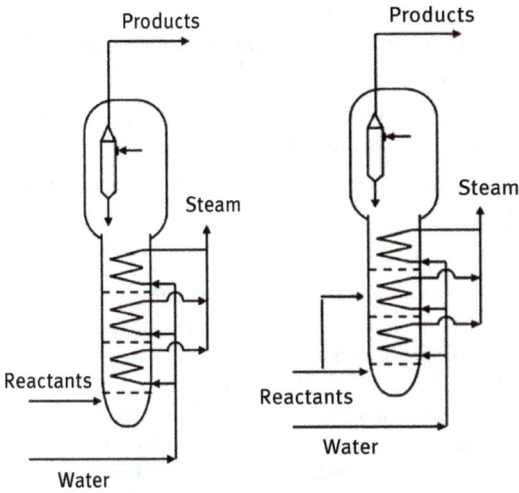

Fig. 4.74: Partial segmentation in the fluidized bed reactor (a) without and (b) with quenching.

Selectivity could be also controlled by a proper catalyst design and addition of inhibitors, which diminish the overall catalyst activity, improving selectivity. For example, some known catalyst poisons (halogens, selenium) are added to silver-based catalyst for oxidation of ethylene to ethylene oxide. In the same process, di-chloroethane is introduced to the feed diminishing many-fold activity, increasing, however, selectivity toward the desired product of partial oxidation. Selectivity could be also regulated by selecting a proper type of the support, its porosity or size of the catalyst grains.

The catalyst composition and morphology play an important role. Several groups of catalytic materials are used for oxidation, such as metals (Cu, Ag, noble metals), oxides of transition metals and mixed oxides of, for example, V, Bi, Mo, W.

The mechanisms of oxidation using heterogeneous catalysts were discussed in Section 1.2.4.1.

4.7.2 Epoxidation of olefins

4.7.2.1 Ethylene oxide

One of the most versatile organic intermediates is ethylene oxide with an epoxide ring, which can be readily opened with, for example, water giving ethylene glycol. Opening of the ring in ethylene oxide by reacting with alcohols, fatty alcohols and ammonia (amines) give, respectively, polyols, surfactants and ethanolamines. EO is also an important monomer in production of polyesters and polyurethanes. Annual worldwide production of EO is expected to exceed 32 Mt by 2023 [52].

Oxidation of ethylene is done at elevated pressures (between 1 and, respectively, 2.2 or 3 MPa for oxygen or air-based processes) using silver catalysts with metal loading 7–20 wt% on an inert support with no acid centers to avoid side reactions. The surface area of the support is rather low (<2 m^2/g) and only high-temperature stable supports (such as α-alumina or SiC) are used. One of the reasons to use a low surface area support is to avoid oxidation of ethylene oxide in the pores when there are internal mass transfer limitations.

> This is an interesting example how knowledge in reaction kinetics and catalytic reaction engineering, namely dependence of selectivity to the intermediate product in consecutive reactions on Thiele modulus (Section 3.5.6), can guide design of catalytic materials. Many textbooks on chemical reaction engineering do not emphasize clearly such links between theory and practical chemical applications. As a result, sometimes "classical" chemists do not appreciate the impact that kinetic analysis and reaction engineering can bring to their everyday work on for example organic reactions or materials development. I remember a colleague, an organic chemist, who attended a conference on chemical engineering and was complaining that "this *dc/dt* is not chemistry at all". Apparently, the driving force for attending that conference was its location (marvelous city of Florence in Italy).

Catalyst manufacturers keep information about the composition of catalysts confidential; therefore, details of formulations are not readily available. Several promoters, for example, salts or other compounds of alkali (Li, K, Na) and alkaline earth metals (Ba, Ca) are added to the catalyst significantly improving selectivity. Application of Rb or Cs oxide in combination with Re and its copromoter (e.g., S, P, Mo, Mn or Cr) was also reported. Light alkali (e.g., LiCl) can be used in combination with the heavy ones (e.g., CsCl). Moreover, chlorine compounds (e.g., 1,2-dichloroethane) are introduced to the reaction gases, having a negative impact on activity, improving, however, selectivity by suppressing total oxidation of ethylene to carbon dioxide and water.

These measures allow to reach 80–90% selectivity with oxygen as an oxidation agent. Selectivity is declining with time-on-stream due to several reasons (poisoning, sintering, dust formation, etc.). The range of temperatures is 220–250 °C. Too high temperature obviously leads to overoxidation and formation of carbon dioxide, while catalysts might not be active enough at too low T. Oxidation of ethylene to epoxide is mildly exothermic ($\Delta H = -105$ kJ/mol) while complete oxidation of ethylene and ethylene oxide to carbon dioxide and H$_2$O is much more exothermic ($\Delta H = -1324$ and -1220 kJ/mol, respectively).

Besides CO$_2$, water and ethylene oxide traces of acetaldehyde and formaldehyde are formed and must be removed or separated from the recycle gas stream. Catalyst life is typically 2–5 years. Thereafter, silver is recovered from the spent catalyst.

Either air or oxygen is used for oxidation of ethylene oxide synthesis in multitubular reactors (Fig. 4.75) containing several (up to 20) thousands of 6–13 m long tubes with internal diameter of 20–50 mm filled with the catalyst in the form of

Fig. 4.75: Reactor for ethylene oxide synthesis: (a) outside view and (b) schematics [53].

spheres or rings with a diameter of 3–10 mm. The contact time is approximately 1 s at 230–290 °C and 10–30 bar.

As in other selective oxidation processes, a special care should be taken not only about conversion to avoid unwanted complete oxidation, but also about concentration range of ethylene in the mixtures with oxygen, since such mixtures are explosive. As a result, ethylene conversion per pass is limited being 20–65% in the air-based compared with 7–15% for the oxygen-based process. The catalyst productivity is approximately twofold higher in the oxygen-based process, which also displays better selectivity to ethylene oxide. The oxygen-based process is organized in the concentration domain above the explosion limit. In the oxygen-based process, methane and argon are applied (Tab. 4.16) to improve the overall safety.

Incomplete conversion calls for utilization of gas recycling which in the case of air-based process means, on one hand, a recycle of a large amount of nitrogen in the recycle gas and, on the other hand, much larger amounts of purge gas to prevent build-up of nitrogen concentration in the recycle stream. Subsequently, the off-gas leaving the primary reactor still contains substantial amount of ethylene which should be processed before venting into the atmosphere. Figure 4.76 illustrates the flow scheme of an oxygen-based process.

When oxygen being more expensive than air is used, selectivity is higher, losses of ethylene are lower as well as the equipment size is smaller. Pressure is typically 1–2 MPa, which does not influence selectivity in the oxidation reaction but improves downstream absorption. Ethylene is taken in excess compared to oxygen. Due to formation of CO_2 it should be efficiently removed. Recycled gas along with fresh oxygen and ethylene (and inhibitor, not shown) is fed into the reactor 1. Steam is generated in steam generator 3. Outlet gases are cooled in the heat exchanger 2, cooler 4 and thereafter send to absorber 5, where ethylene oxide is dissolved in water due to its

Tab. 4.16: Process parameters in ethylene oxide production processes.

Parameter	Air based	Oxygen based
Ethylene concentration, vol%	2–10	15–40
Oxygen concentration, vol%	4–8	5–9
CO_2 concentration, vol%	5–10	5–15
Ethane concentration, vol%	0–1	0–2
Argon concentration, vol%		5–15
Methane concentration, vol%		1–60
Temperature, °C	220–277	220–235
Pressure, MPa	1–3	2–3
Ethylene conversion, %	20–65	8–12
Overall ethylene oxide yield	63–75	75–82

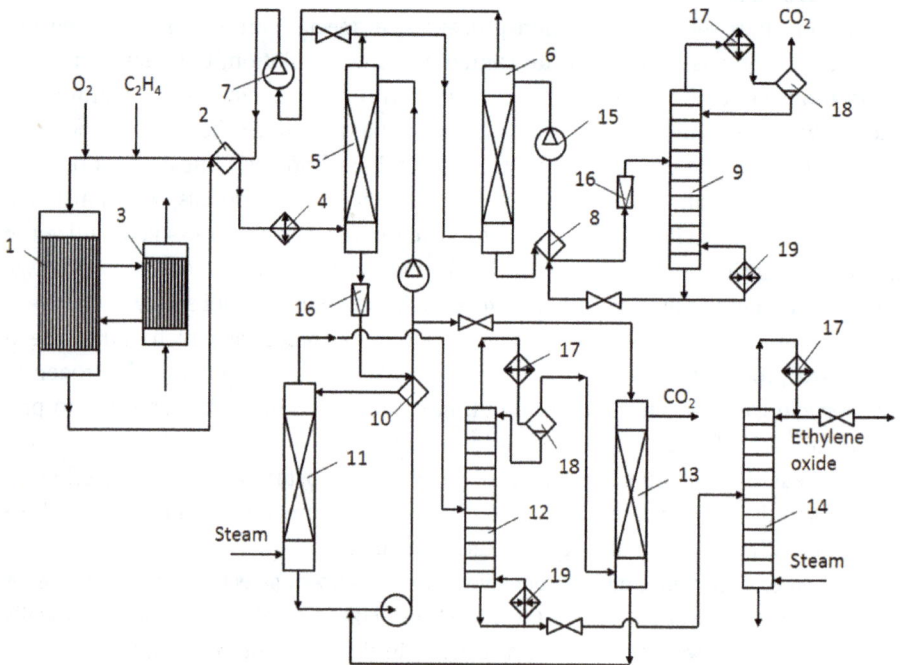

Fig. 4.76: Oxygen-based synthesis of ethylene oxide. 1, reactor; 2, 8, 10, heat exchanges, 3, steam generator; 4, cooler, 5, 6, 13, absorbers; 7, compressor; 9, stripper (desorber); 11, column; 12, 14, distillation columns; 15, pump; 16, valve; 17, condenser; 18, separator; 19, reboiler. Modified from [54].

miscibility. In addition, a part of CO_2 is also adsorbed. A part of the gas (unreacted ethylene and CO_2, along with some impurities) is recycled while the rest is routed to absorber 6, where CO_2 is removed using, for example, hot pot scrubbing or water solution of alkanol amines. Subsequently, in a stripper 9 higher temperatures than for absorption are applied and CO_2 is released. The lean solvent is sent back to the absorber. After a pressure release to approximately 0.5 MPa (expander 16), water solution of ethylene oxide and CO_2 from the bottom of absorber 5 passes through a heat exchanger 10 to column 11 where ethylene oxide and CO_2 is distilled from water. The latter after passing a heat exchanger 10 is sent back to absorbers 5 and 13. After distillation (column 12) ethylene oxide (bottom) is sent to the final distillation (column 14), while a part of ethylene oxide (top of column 12) is captured in the absorber 13 with water and rerouted to column 11.

"Ethylene oxide is one of the most dangerous large-scale chemicals present in petrochemical industry and an ability of its safe production is a benchmark, reflecting a level of technological advancement and maturity of chemical industry in a particular country" told a seemingly old lecturer to a group of students, including the author, when back in 1984 we took the course on chemistry and technology of petrochemicals and basic organic chemicals at Mendeleev University. In fact, the lecturer, probably at that time younger than the author nowadays, did not elaborate too much on the technological aspects of this technology.

The author lately was involved in several research project on oxidation of ethylene to ethylene oxide, resulted in few scientific publications, and have visited few years ago production facilities in Dzerzhinsk, where 4 multitubular reactors with capacity 50,000 t/a each were put on stream in yearly 1980s. It is instructive to review the historical development at that particular company [55]. The original selectivity of a silver containing catalyst supported on corundum was 70–72%, which was quite high in those days. The only by-product was carbon dioxide apart from almost negligible amounts of aldehydes. The reactors had 13,500 tubes with internal diameter of 21 mm and the catalyst bed length of 7 m. An organic compound was used as a heat carrier. The operation pressure was and still is 2 MPa. The inlet contained (vol%) %: 15–20 C_2H_4, 6.5–6.8 O_2, 7–9 CO_2, and nitrogen as the rest. After the revamp in 2005 nitrogen was switched to methane. The reactor capacity was improved by 20%. The plant operates with a more modern catalyst affording 84% selectivity upon introduction of less than 5 ppm of a chlorocontaining compound. A more selective catalyst allows to decrease the generated heat as less products of total oxidation are formed and to diminish the consumption of ethylene and oxygen per ton of product. With not that selective catalyst the company experienced a lot of problems in the initial phase of running the plant. Just during the first year there were 333 unexpected shutdowns, sometimes up to 10 per day. Such frequent stops of the process induced by sudden, without visual causes, sharp jumps of temperature and pressure in the outside-catalyst space of the reactor, continued for the next few years leading to a considerable decrease in capacity.

A large heat of the main reaction, substantially increasing when the selectivity is lost and complete oxidation prevails, operation close to an explosion limit and high sensitivity of the reactor to operational parameters and temperature fluctuations were obvious reasons for such frequent stops.

In practice after a sudden temperature increase in the bed there was an increase in the pressure drop and a subsequent decrease in the overall gas flow through the reactor, increasing obviously the contact time. Then after 15–30 s the exit gas temperature immediately reached 400–600 °C.

Thermal runaways happened when as a method to mitigate catalyst aging and to maintain production capacity, changes of the operation parameters (e.g. elevation of temperature and promoter concentration) were carried out. Thereafter, such increases in temperature occurred at certain time points.

Analysis of the reactor performance demonstrated that there was a distribution of catalyst activity in different tubes. The same holds for heat exchange efficiency. Tubes with anomalously high activity and low heat exchange efficiency had increased parametric sensitivity, leading to thermal runaway in those tubes in the case of small disturbances of process parameters. Ignition of the reaction mixture happened downstream the tubes when the volume was expanded. Somewhat similar behavior will be discussed in Section 4.7.3.2 in connection to thermal runaways in o-xylene oxidation plants.

There were several other reasons for thermal runaways such as mechanical damage of shaped catalysts, attrition and formation of dust. Corrosion products coming from pipelines could also influence thermal runaways. Finally, catalyst exposure to the liquid sorbent led to leaching of promoters across the reactor diameter and the reactor length, especially in the tubes at the reactor periphery. Subsequently selectivity was decreased and more heat was released.

The first generation of high-selectivity catalysts introduced in the late 1980s allowed an increase of initial selectivity values by more than 6 percentage points (e.g., from 80%) to give start-of-cycle selectivity values of 86% or larger bringing enormous saving in ethylene feedstock costs. Activity and stability of these catalysts were lower compared with the traditional ones, thus selectivity declined more rapidly and the lifetime of high-selectivity catalysts was 1.5–2 years being approximately half of the lifetime of traditional EO catalysts. More recently developed so-called high-performance catalysts [56] by CRI Catalyst Company (Fig. 4.77) allowed approximately the same initial selectivity and a significantly slower performance decline. According to real plant data, the average selectivity over the lifetime of a catalyst is approximately 87%.

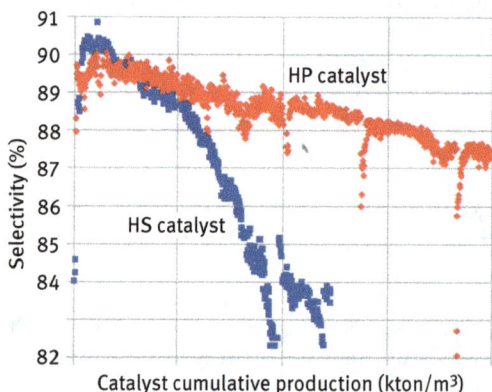

Fig. 4.77: Selectivity curves comparing high selectivity (HS) and high-performance (HP) catalysts over their live time at the same operating conditions [56].

4.7.2.2 Propylene oxide

A similar process for propylene oxide has not been developed, since there are no catalysts available up to now affording selective oxidation of propylene. Thus, other routes have been practiced along the years, including the chlorohydrin routes resulting in large amount of wastes (2.2 tons of salt per ton of product)

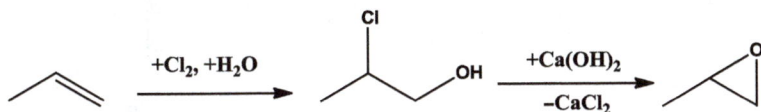

An alternative route is to use oxidation of ethylbenzene giving a peroxy compound, which subsequently reacts with propylene resulting in propylene oxide and a secondary alcohol.

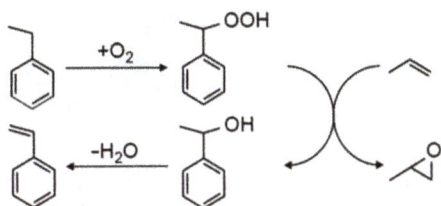

The later after dehydration gives styrene (2.5 tons per ton of propylene oxide).

Recently, a more direct method for propylene oxide synthesis was developed based on hydrogen peroxide giving water as the only by-product.

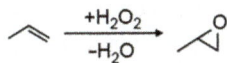

The process flow scheme is presented in Fig. 4.78. Oxidation is done in methanol solution with titanium silicate TS-1 as the catalyst. Water is formed as a product from H_2O_2 along with glycols, which are generated in small amounts as by-products.

4.7.3 Oxidation of alkanes to anhydrides

4.7.3.1 Maleic anhydride

Maleic anhydride has numerous industrial uses and is of significant commercial interest worldwide. The primary use of maleic anhydride is in the manufacture of polyester and alkyd resins. Benzene was used as the dominating feedstock

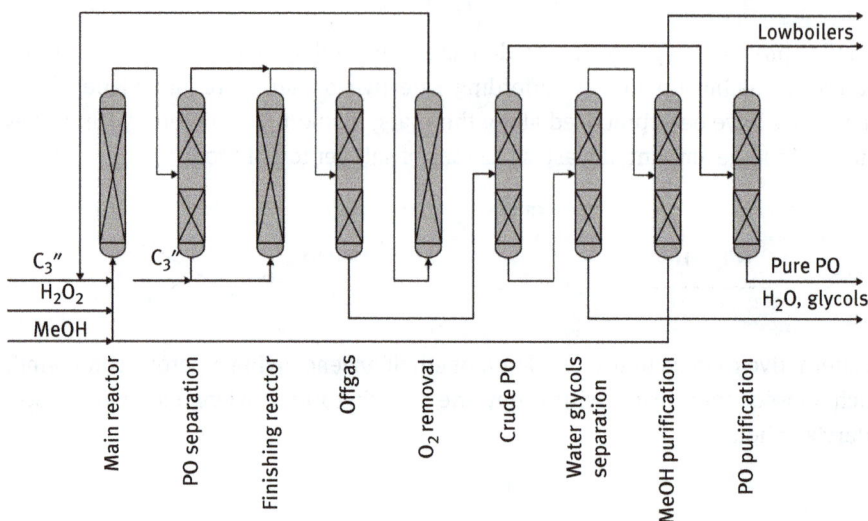

C$_3$"
C$_3$"
H$_2$O$_2$
MeOH

Pure PO
H$_2$O, glycols

Main reactor | PO separation | Finishing reactor | Offgas | O$_2$ removal | Crude PO | Water glycols separation | MeOH purification | PO purification

Fig. 4.78: Oxidation of propylene to propylene epoxide with H$_2$O$_2$ [57].

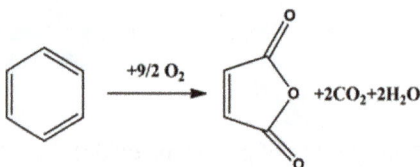

being in the recent years continuously replaced by oxidation of n-butane or n-butane–n-butene mixtures with a high paraffin content (Fig. 4.20).

The traditional process begins by mixing benzene with an excess of air to give concentrations from 1 to 1.4 mol%. A low benzene concentration must be utilized in order not to exceed the flammability limit of the mixture. The reaction gas mixture then passes over the catalyst in a multitubular (diameter ca. 25 mm) fixed-bed reactor at an optimum pressure range of 0.15–0.25 MPa. The desired reaction is highly exothermic, causing hot spots of 340–500 °C. As a rule, commercial catalysts for benzene oxidation are supported by an alumina or silica carrier and have a surface area of 1–2 m^2/g. A typical catalyst is comprised of V$_2$O$_5$, MoO$_3$ and some other promoters such as phosphorus, alkali and alkaline earth metals, tin, boron and silver. Benzene is passed through the catalyst at the concentration of 60–130 g benzene per liter of catalyst per hour. Conversion 97–98% is achieved with an initial selectivity over 74%. The lifetime of the catalyst can be as long as 4 years depending mainly on the reactor operating temperature and the purity of the starting materials.

Recovery of the significant part of the product (40–60%) is done by cooling the reactor outlet stream to 55 °C. Condensed maleic anhydride must be removed as soon as possible to avoid prolonged contact with water in the reaction gas, otherwise resulting in formation of maleic acid. The rest of maleic anhydride that cannot be recovered is washed out with water as maleic acid. Water scrubbing and subsequent dehydration of maleic acid at approximately 130 °C is required to purify and reform the remaining maleic anhydride. A side reaction is isomerization of maleic acid to fumaric acid.

The processes developed or in the developmental stage for the oxidation of C4 hydrocarbons to maleic anhydride imply fixed-bed (Fig. 4.79), fluidized bed and transport bed reactors.

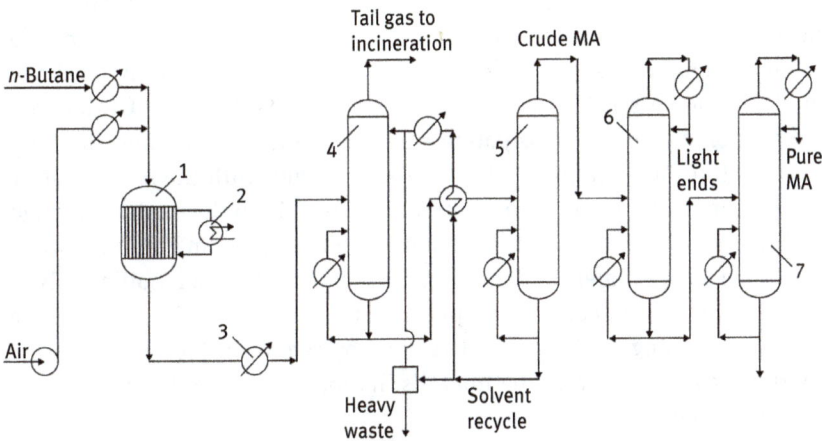

Fig. 4.79: Synthesis of maleic anhydride by butane oxidation in a fixed-bed reactor.

The reaction is the complex one, requiring extraction of eight hydrogen atoms and insertion of three oxygen atoms to butane along with a ring closure. High concentration of butane is advantageous, but dangerous since butane and oxygen could be within explosion limits, thus the concentration of butane in air is approximately 1.8 mol%. A major advantage of C4-based process compared to benzene oxidation is that no carbon is lost in the reaction to form maleic anhydride. Moreover, the feedstock is cheaper, not carcinogenic contrary to benzene, and the flammability limits for C4 hydrocarbons are lower. Oxidation of C4 hydrocarbons at 80% conversion and 70% selectivity gives comparable weight yields as for benzene oxidation. In the fixed-bed process (Fig. 4.79) oxidation is done in tubular reactors similar to benzene oxidation with vanadium and phosphorus oxide (V–P–O) unsupported catalysts. In the catalyst formulations having phosphorus–vanadium ratio of 1.2, several promoters (Li, Zn, Mo, or Mg, Ca, and Ba) are applied improving conversion and selectivity compared to unpromoted catalysts. Higher phosphorus–vanadium ratios increase conversion but lead to shorter catalyst lifetime. An oxidation reactor operates at 400–480 °C and 0.3–0.4 MPa. Formation of water is higher in butane oxidation processes compared to benzene oxidation, therefore only a small amount of maleic anhydride can be condensed from the reactor effluent. The remaining product is washed out as maleic acid in a scrubber and then dehydrated similar to the case of benzene oxidation. Tail gases after the absorption column are incinerated, as recovery of unreacted C4 hydrocarbons is substantially more difficult than in the case of benzene. An alternative recovery method utilizes absorption of the product from the reactor outlet by organic solvents and will be explained below for the case of the fluidized bed reactor.

Application of such fluidized bed reactors allows increasing the concentration of butane in the feedstock compared to fixed-bed operation by 50–100%. The fluidized bed process has the advantage of a particularly uniform temperature profile without hot spots improving the process selectivity. A particular problem of this process is mechanical stress on the catalyst (V–P–O), its abrasion, and erosion at the heat-dissipating surfaces. On the other hand, the fluidized bed constitutes an extremely effective flame barrier, with the result that the process can operate at higher C4 concentrations (i.e., within the explosion range) than the fixed-bed process. The process flow scheme is given in Fig. 4.80. In the recovery system, crude maleic anhydride is distilled from the high boiling solvent in a fractional distillation column (stripper in Fig. 4.80) while the solvent is recycled. Further refining of the product is done by removing the light ends and recycling the heavier ones from the final distillation column to the stripper. The tail gases are incinerated.

Butane oxidation can be also done utilizing a transport reactor (Fig. 4.21), where oxidation of the reactants by the lattice oxygen and subsequent reduction of V^{+5} to V^{+4} is separated in space from oxidation of V^{+4} to V^{+5}. Comparison between oxidation of butane to maleic anhydride in the fluidized bed and a transport reactor-regenerator tandem is given in Tab. 4.3. Despite clear advantages of the latter process with higher throughput and inherent safety, its industrial implementation was not very successful, and the production was stopped after some time in operation.

4.7.3.2 Phthalic anhydride synthesis

4.7.3.2.1 Feedstock
Commercialization of phthalic anhydride (PA) production goes back to 1872 when BASF developed the naphthalene oxidation process (Fig. 4.81). Details of physical and chemical properties of phthalic anhydride are presented in Lorz et al. [59].

The most important derivatives of phthalic anhydride, which is produced globally with a capacity ~4.5 million tons per year, are phthalic esters which can be

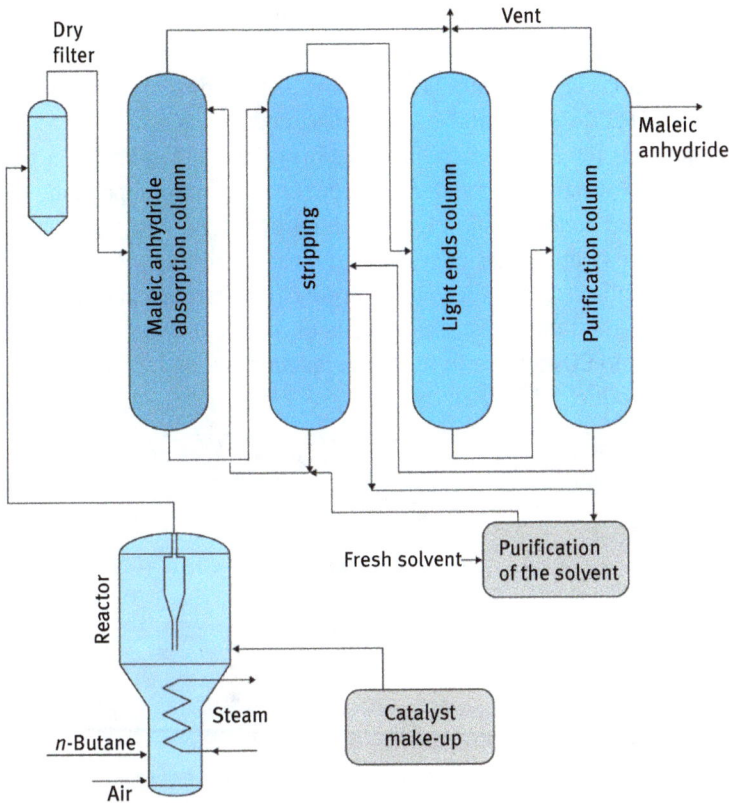

Fig. 4.80: Fluidized bed process for oxidation of butane in maleic anhydride with solvent recovery (modified after 58).

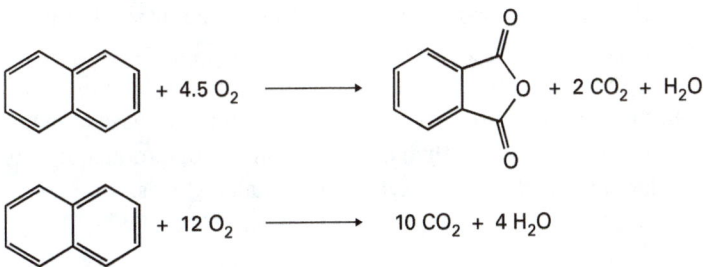

Fig. 4.81: Oxidation of naphthalene to PA.

used as plasticizers, polyester resins and dyes. In particular the diester, dioctyl phthalate (DOP) has a widespread use, improving flexibility of polyvinyl chloride, which otherwise is a rather rigid material. Until the early 1960s coal tar naphthalene was the dominant feedstock, while later on gas-phase oxidation of o-xylene over vanadium-based catalysts gained industrial importance and almost completely replaced oxidation of naphthalene.

In industry, oxidation of o-xylene is done in fixed-bed tubular reactors (15,000–30,000 tubes) at 300–400 °C and 0.5–1.8 vol% of xylene in air.

One of the reasons for the feedstock switch was o-xylene availability from petroleum, while quantities of naphthalene derived from coal tar were dependent on the production of coke and were not sufficient for the growing PA and plasticizers markets. The other reason is atom efficiency. As can be seen from Fig. 4.81, naphthalene oxidation inevitably results in the consumption of two carbons and the generation of two moles of carbon dioxide per each mole of product formed. Oxidation of o-xylene does not lead to CO_2 release in the main reaction, while it could be certainly formed as a side product (Fig. 4.82).

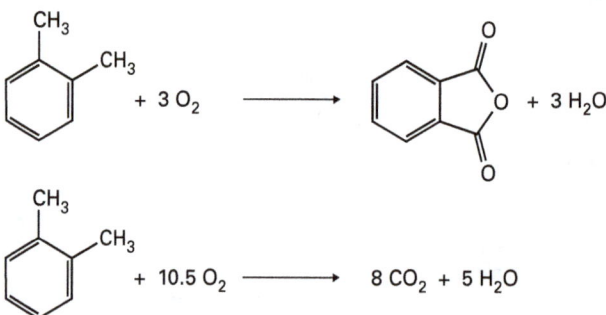

Fig. 4.82: Oxidation of o-xylene to phthalic anhydride.

In addition to feedstock availability and unnecessary CO_2 emissions, in some areas and in particular in cold climates, handling of naphthalene, which is solid at room temperature, was challenging.

Although more than 90% of PA is nowadays produced from o-xylene there are still a few plants operating with naphthalene that are mainly located close to major steel mills. Naphthalene is much cheaper than o-xylene as it is a by-product of coke production. Few installations utilize mixed o-xylene-naphthalene feeds.

4.7.3.2.2 Thermodynamics

The oxidation reactions are not limited by equilibrium and are highly exothermic. High activation energy means that even with a small temperature increase, heat production increases substantially, leading to prominent hot spots in fixed-bed reactors. The reaction enthalpy of the o-xylene oxidation is $-1{,}108.7$ kJ mol^{-1} at 298 °C while it is $-4{,}380$ kJ mol^{-1} for complete oxidation. As overall selectivity in o-xylene oxidation is ~80% at almost complete conversion, heat generation in this reaction is significant, reaching 1,300–1,800 kJ per mole of o-xylene. Heat removal is thus essential. As mentioned above, oxidation is performed in fixed-bed tubular reactors. Another option would be to use fluidized bed reactors, which are very efficient in controlling temperature, maintaining a uniform temperature throughout the bed. Such option was in fact implemented in only few places for naphthalene oxidation, where V_2O_5–K_2SO_4 on silica was applied, allowing good fluidization properties. As such support does not allow high yields of phthalic anhydride in o-xylene oxidation there are apparently no industrial o-xylene- based fluidized bed reactors currently in operation, although development work in companies and academia might still continue.

It should be noted that the feedstock as well the product can be explosive in mixtures with air. Thus, a special care should be taken to ensure that the feedstock and product are not within explosion limits with air. It can be seen from Fig. 4.83 illustrating general principles of reactions which can lead to explosions that at a certain temperature, say 500 °C, at low pressures the whole system is below the first explosion limit. When the concentration of hydrocarbons increases (partial pressure increases) explosions can occur, while with a further partial pressure increase safe operation is once again possible.

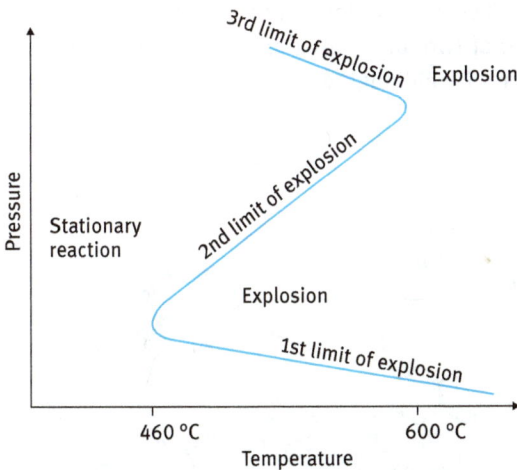

Fig. 4.83: Examples of reactions with explosions.

At one of the PA plants that I visited very often while working in industry, there were ignitions observed with the reaction mixture when the catalyst was not able to convert o-xylene efficiently during reactor start-up. At the outlet of the reactor there is a gas cooler, with surface-to-volume ratio different from that of the pipe connecting the reactor with the gas cooler. The length of the gas cooler is ~6 m and I could see from the outside that the color of the gas cooler at least for 1m from the inlet was somewhat reddish because of ignition of the unconverted reactant. These observations strongly indicated the presence of branched chain reactions. Once the catalyst was back to conventional operation, such ignition terminated.

In another PA plant with naphthalene as a feedstock I was shown a production site with five reactors, where at some point there were six reactors. The reactor had completely disintegrated and the reactor upper hood was never found as the reaction mixture ignited, most probably caused by heat transfer salts entering the reactor as the reactor tubes were broken.

The lower explosion limit for o-xylene or naphthalene in air is 47 g m^{-3} (STP). Compression and preheating of air represent additional costs, thus the increase of o-xylene loadings above this limit was addressed throughout the years. Many plants operate now at 60, 80 or even 100 g m^{-3}, as it was shown that an explosion starting in the catalyst bed is quenched by the catalyst heat capacity. Extra thickness in the upper part of the reactor combined with adequate rupture disks is implemented to avoid explosions in the free space above the catalyst bed. The probability of phthalic anhydride auto-ignition is minimal because the PA auto-ignition temperature (580 °C) is well above typical operation temperatures (salt bath temperature 345–350 °C).

4.7.3.2.3 Mechanism
Fig. 4.84 summarizes the important by-products formed in the oxidation of o-xylene. The reaction requires breaking of two and formation of 12 bonds, thus a detailed analysis of all reaction pathways is challenging.

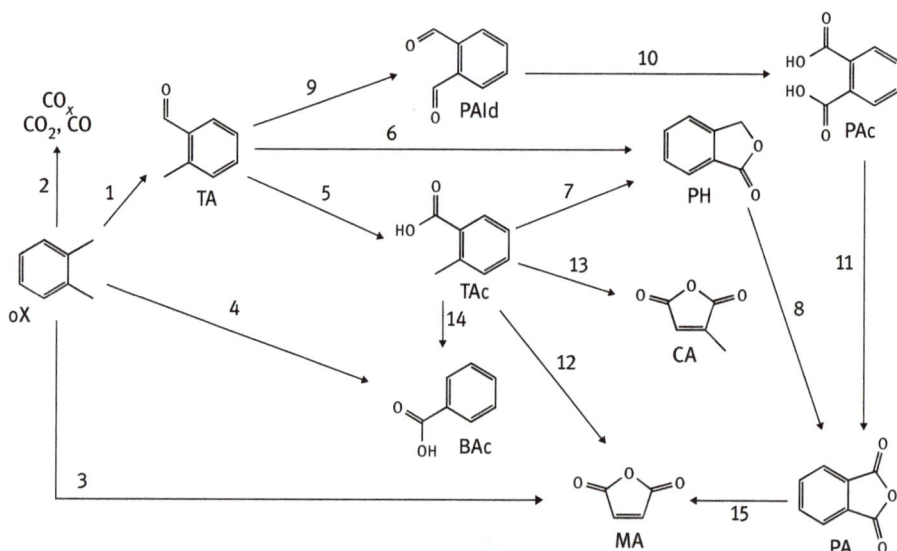

Fig. 4.84: Oxidation of o-xylene. (From [60], reproduced with permission).

The major intermediates in the selective oxidation path are tolualdehyde (TA) and phthalide (PH), while such intermediates as toluic (TAc), phthalic (PA) acids, and phthalaldehyde (PAld) were also proposed. A non-selective route, i.e., a route that does not result in PA, includes the formation of benzoic (BA) acid, citraconic aldehyde and maleic anhydride (MA) and COx (CO and CO_2).

The composition of the product gas is presented in Tab. 4.17.

Tab. 4.17: Main by-products in the reaction gas for the latest generation catalysts (PA yield 94.5–95.6%).

MA	BA	CA	PH	TA
3.3–4.0 wt%	0.4–0.8 wt%	0.35–0.5 wt%	0.02–0.07 wt%	0.005–0.02 wt%

The by-product formation is crucial not only in terms of product selectivity. In some instances excessive presence of by-products deteriorates the product quality, in particular the Hazen number, which reflects the PA color, should be rather low in some applications, between 5 and 10.

In addition to the main by-products, given in Tab. 4.17, other carbon-containing by-products are carbon monoxide and carbon dioxide. In general, the reactor yields do not exceed 112–115 kg PA/100 kg o-xylene, which corresponds to ~82% of stoichiometric yield. Taking into account losses during condensation and distillation, the yields of pure PA are about 112 kg PA/100 kg o-xylene.

For the sake of comparison a few key numbers are also given here for oxidation of naphthalene. The heat of reaction is 1,788 kJ mol^{-1}, while in complete combustion 5,050 kJ mol^{-1} is evolved. The main by-products are maleic anhydride and naphthoquinone (0.05–1.0 wt%) and the PA yield at the reactor outlet for the latest generation catalysts reaches 105–106 kg kg^{-1} naphthalene.

4.7.3.2.4 Catalyst

Oxidation reactions with either of the feedstock (o-xylene or naphthalene) are carried out in tubular reactors cooled by a molten salt. Ring types of catalysts are applied, which replaced spherical catalysts used in the old plants.

For naphthalene partial oxidation V_2O_5 promoted by K_2SO_4 was applied on a moderately porous silica carrier. For o-xylene oxidation porous supports were found to be detrimental and largely non-porous carriers, such as silicon carbide, quartz, steatite ceramics or porcelain with dimensions such as 8 × 6 × 5 mm or 7 × 7 × 4 mm and filling density 0.88–0.93 kg l^{-1} are used. Low area of inert ceramic supports and coating of them with vanadia/titania in the form of a thin layer ("eggshell" catalyst) with a thickness of up to 100 μm is thus needed to avoid internal mass transfer limitations, which would otherwise diminish catalyst activity and selectivity to phthalic anhydride.

Finely divided titanium dioxide (anatase) and vanadium pentoxide oxide layer together with promoters (antimony oxide, rubidium oxide, cesium oxide, niobium and phosphorous) are introduced on the support by, for example, spray-drying. In the monolayer formed on the surface there are chemical bonds between V ion and Ti surface ions via oxo bridges. The layer is a two-dimensional sheet of the formula $VO_{2.5}$, which is believed to contain both strongly bound tetravalent and pentavalent vanadium. Both of the valence states take part in the oxidation of o-xylene by a typical redox mechanism. Titania is considered to inhibit desorption of the many intermediates involved in the overall reaction.

Formation of bulk (crystalline or amorphous) vanadia, if it is introduced in excess, might have a negative impact on catalysts, thus up to 8–10% of vanadium pentoxide in combination with titanium dioxide is used. Details of the catalysts properties and synthesis were discussed by Cavani et al. [61].

In the past some of the catalysts, such as a silica gel supported naphthalene oxidation catalyst containing V_2O_5 and K_2SO_4, required the addition of SO_2 for activation and longer service life, an option that is not required with the modern catalysts.

In industry for low o-xylene loadings, two-zone catalysts (i.e., combination of catalysts with low and high activity) were used, while higher o-xylene loadings require three- and four-zone catalysts (Fig. 4.85).

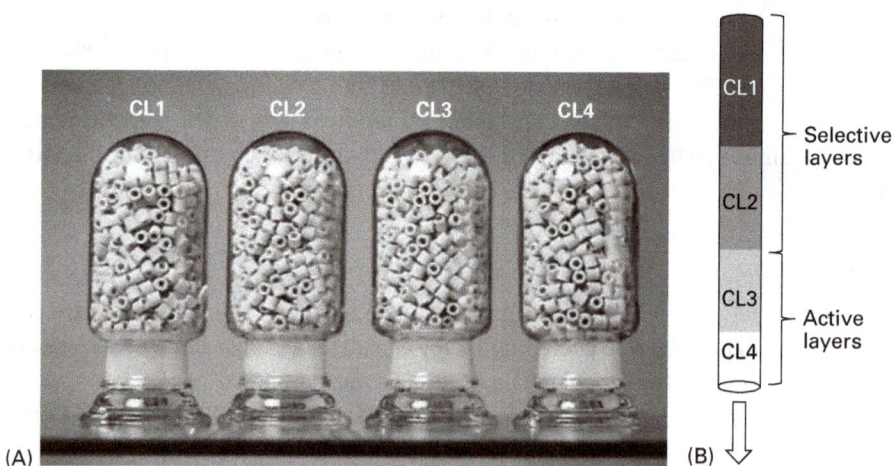

Fig. 4.85: Four-zone BASF O4-68 catalyst for o-xylene load of 80 g Nm^{-3}. (From [62].

The optimal amount of the active phase varies depending in which catalyst zone of the multitubular reactor the catalyst is located, as different reaction stages require different catalyst compositions. It could be easily anticipated that because of high reaction exothermicity, the top catalyst layer should be able to minimize the temperature increase in the hot spot and at the same time ensure fast start-up of cold reaction gas. In the second layer, most of o-xylene conversion takes place and the catalyst should have high selectivity with the lowest activity. Two final layers in the four-zone catalyst system should be more active and less selective ensuring good conversion of remaining underoxidation products to phthalic anhydride and complete combustion of overoxidation products to CO_2. The amount of cesium in different layers is decreasing from top to bottom. Besides helping to avoid excessive hot spots, dopants such as Cs are also used to minimize formation of maleic anhydride and carbon oxides.

Such operation with several layers allows for minimizing the negative influence of the hot spot, which is typically moving along the reactor length with time-on-stream (Fig. 4.86).

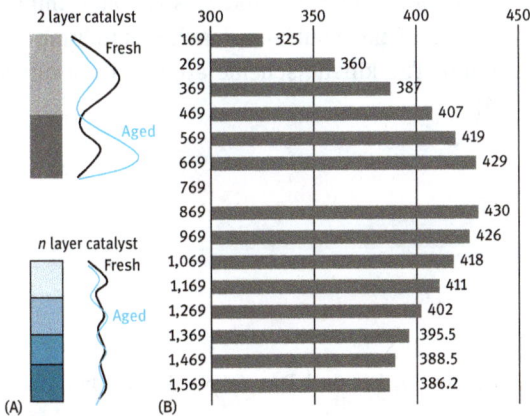

Fig. 4.86: Temperature profile along the reactor length: (A) typical behavior, (B) experimentally observed in the first zone of a two-zone catalyst.

BASF, one of the commercial catalyst suppliers, introduced in 2011 a five-layer catalyst O4-88 (Fig. 4.87) with first three selective layers affording phthalic anhydride yield up to 116.5 wt%. The hot spot is located in the CLO layer. Details on the composition are not available, while the patent literature [63] disclosed, for example, the pre-catalyst layer of Ag/V(O) bronze (mixed oxides of silver and vanadium with atomic ratio Ag/V < 1) dispersed on Mg silicate rings. This formulation should not be, however, necessarily the one used commercially.

OxyMax® phthalic anhydride catalysts from Clariant for selective oxidation of o-xylene, naphthalene and mixed feedstocks, produced by fluid-bed coating process, also consist of up to five catalyst layers.

Fig. 4.87: Five -zone BASF O4-88 catalyst for o-xylene load up to 100 g Nm^{-3}. (From [62]).

4.7.3.2.5 Kinetics

The complexity of the reaction prevents detailed analysis, and mostly some simplified schemes were discussed in the literature. One of such approaches is to lump all intermediates into one pseudo-compound. The kinetic scheme applied most often was suggested by Calderbank et al. [64]:

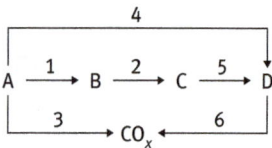

where A, B, C, D are o-xylene, o-tolualdehyde, PH and PA, respectively. Oxidation of PA (step 6) is considered insignificant.

The following rate equations were derived:

$$r_1 = k_1 a P_A, \quad r_2 = k_2 a P_B, \quad r_3 = k_3 a P_A, \quad r_4 = k_4 a P_A, \quad r_5 = k_5 a P_C, \quad r_6 = k_6 a P_D \tag{4.39}$$

where

$$\alpha = \frac{k_c P_{O_2}}{k_c P_{O_2} + P_A(m_1 k_1 + m_3 k_3 + m_4 k_4) + P_B m_2 k_2 + P_C m_5 k_5 + P_D m_6 k_6} \tag{4.40}$$

and m is the moles of oxygen that react per mole of xylene consumed. The values of m are 1, 6.5, 3, 1, 1 and 3.5 for steps 1–6, respectively.

4.7.3.2.6 Technology, reaction and separation

Oxidation of o-xylene is done industrially in multitubular reactors containing up to 25,000 tubes with ~2.5 cm internal diameter. In the initial design, reactors contained a salt bath circulation pump and a fitted internal cooler. This allowed temperature control by a molten salt, which is eutectic of potassium and sodium nitrates, with a melting point of 141 °C. At a later date external cooling was introduced. The molten salts circulate through a waste heat boiler, which is used to generate saturated steam. In case of industrially high o-xylene loadings used nowadays, reactors use radial salt flow (contrary to axial flow in previous designs) for improved heat transfer (Fig. 4.88).

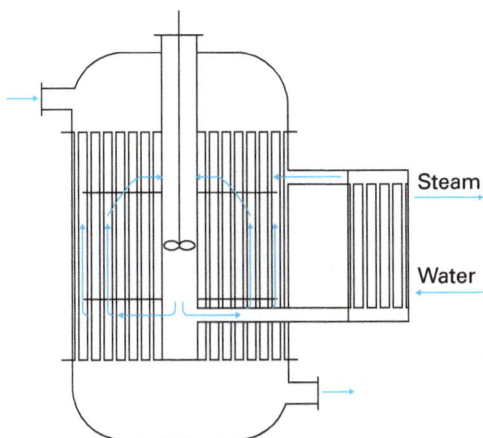

Fig. 4.88: Reactor with radial salt flow (DWE type). (From [65], copyright Elsevier.).

The oxidation air is pre-heated to about 180 °C and charged with feedstock in concentrations up to 100 g Nm^{-3}. The air rate is typically 4 Nm^3 per reactor tube and per hour.

Sometimes cleaning of air can be very important. Once while working in industry I visited a plant, which was build using a licensed technology and was a carbon copy of another plant also located in Europe. Unfortunately, after the plant start-up, the values of selectivity guaranteed by the technology licenser could not be reached and the catalyst deactivated. It turned out that troublesome plant was located at a riverbank (literary) and the air was contaminated with aerosol containing potassium. The plant manager proudly showed an installation with high-efficient air purification from aerosols using fibrous filters developed by Petryanov in 1939 and used shortly after the Second World War to protect personnel of uranium mining facilities from exposure to radioactivity. The self-made addition to the existing technology of phthalic anhydride production mitigated the catalyst contamination problem completely. After learning that a decade before this visit I was working at the same institution as Prof. Petryanov, who even attended my PhD defense, the plant manager was happy to hear more details about the history behind discovery of the aerosol filter.

As could be seen from Fig. 4.89 no recycle is needed as high conversion levels are achieved. Operating salt temperatures are in the range 325–425 °C. Initial start-up temperature is ~375 °C and it takes approximately 2 weeks to stabilize the catalyst

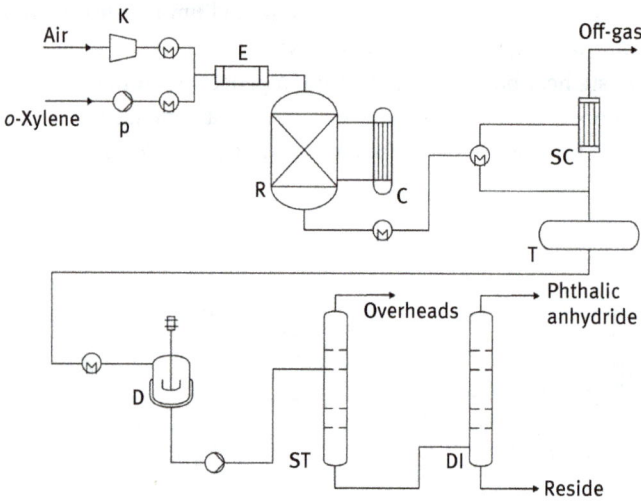

Fig. 4.89: PA production technology [66]. K, air compressor; P, o-xylene pump; E, evaporator; R, reactor; C, salt bath cooler; SC, switch condensers; T, crude phthalic anhydride tank; D, predecomposer; ST, stripper column; DI, distillation column.

activity and operate at 350–355 °C. When the reaction takes off a hot spot is generated in the reactor, which exceeds salt bath temperature by 50–60 °C as it is in the range 410–430 °C. The temperature declines along the reactor length as the generated heat is related to reaction rate, which obviously declines with increased conversion. With operation time the catalyst deactivates and the hot spot moves downwards along the reactor. In order to compensate for the activity loss the salt bath temperature is increased during operation to 360–365 °C at the middle of the catalyst lifetime and to 370–380 °C at the end. Higher operation temperature results in lower selectivity, higher phthalide yields, worsening of the product quality, which is seen as increase of the color number, and finally lower phthalic anhydride yields. Thus after 4–5 years of operation the catalysts have to be replaced.

Fig. 4.90 illustrates a typical behavior of o-xylene oxidation catalysts with time-on-stream, while Fig. 4.91 displays operation with mixed o-xylene-naphthalene feed with an equal ratio between reactants.

Loading of catalysts in industrial multitubular reactors is a time-consuming procedure, as it should guarantee uniform flow distribution. With such multizone catalysts loading is arranged by filling first one zone, measuring pressure drop in all tubes, levelling off loading and proceeding with another zone.

The reactor effluent gas leaving the reactor is cooled down first in the gas cooler (Fig. 4.89). Downstream cooler switch condensers equipped with U-type finned tubes are applied. Such tubes are alternatively heated or cooled with a hot or cold heat-transfer medium. By cooling the gas the crude PA desublimates on the fin tube surface.

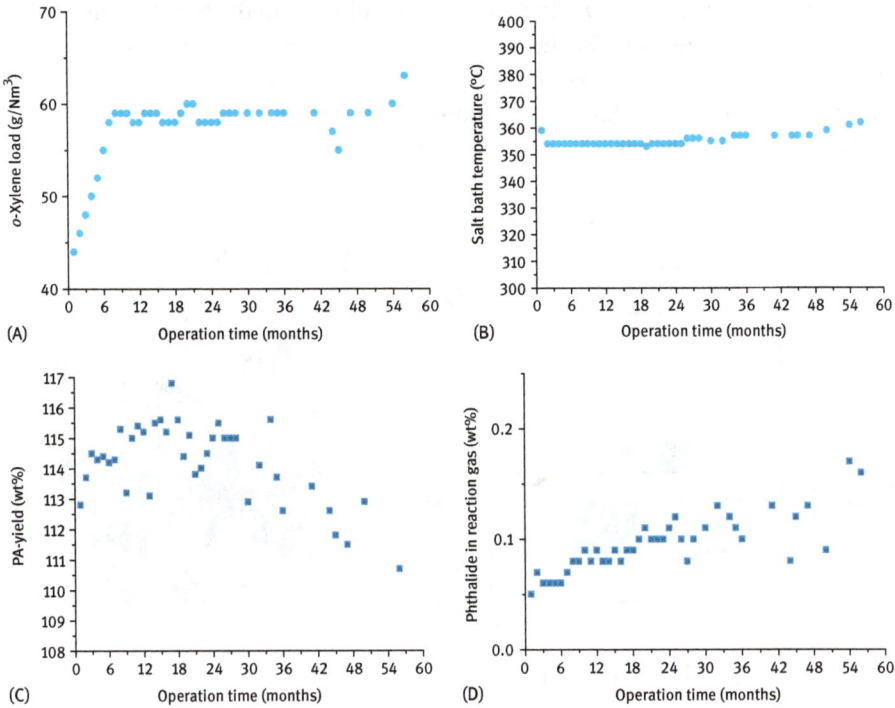

Fig. 4.90: Experience in a commercial plant with loading of o-xylene 60 g Nm^{-3}.

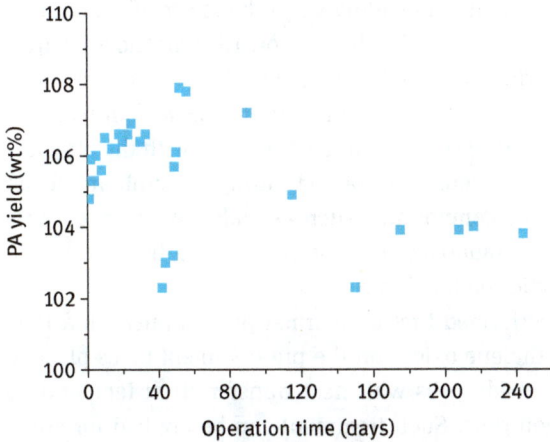

Fig. 4.91: Experience in a commercial plant with mixed o-xylene-naphthalene feed.

Each switch condenser can be loaded with a certain quantity of crude phthalic anhydride, which is defined by the surface area of switch condenser. After full loading a switch condenser is isolated from the gas stream and heated up to melt the crude phthalic anhydride from the fin tubes and direct it to the purification section, first through a crude PA tank. The heat removed from the reactor process gas during the loading phase is absorbed by a heat transfer oil circulating through the fin tubes. The same oil at a higher temperature is used for unloading. A scheme of a Rollechim switch condenser is presented in Fig. 4.92.

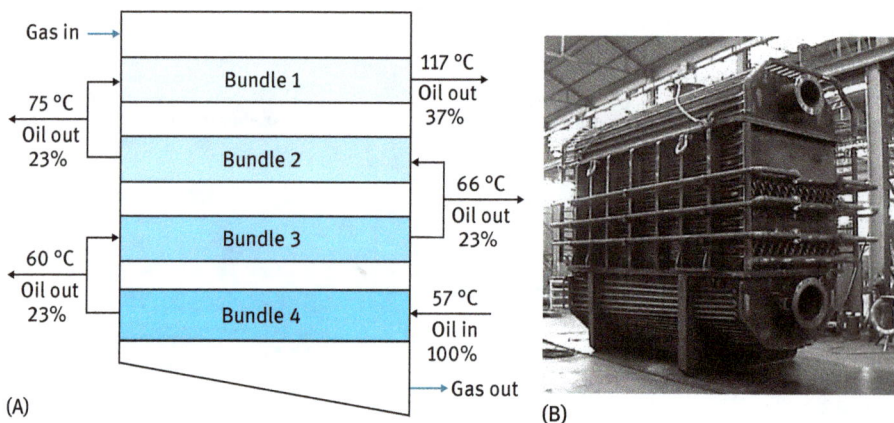

Fig. 4.92: Switch condensers. (Figures from [67]).

As visible from this Fig. 4.92 the downflow mode of operation is applied, as the reaction gas enters at the top. Such operation allows deposition of most of the solid product on the upper most bundle. During the unloading stage the lower tube bundle is washed by molten PA, and in this way removing highly corrosive maleic anhydride and phthalic acid from the tubes, which was challenging in up-flow operations.

High efficiency of the switch condenser affords more than 99.5% of phthalic anhydride recovery and thus less than 0.5% of the crude PA escapes with the off-gas.

The off-gas leaving the switch condensers is passed through a scrubbing tower where the major part of the organic compounds, such as maleic anhydride, is removed by water scrubbing. Carbon monoxide cannot be removed by such scrubbers, thus either thermal or catalytic combustion is used.

The product purification is performed first by thermal pre-treatment and then by vacuum distillation. For naphthalene oxidation the pre-treatment takes place in a cascade of vessels heated by outside coils with heat transfer oil at temperature 230–300 °C with 10–24 h retention time. Such treatment can be applied for crude products with low contents of impurities. Chemical treatment was also used to destroy the by-product naphthoquinone.

The purification in o-xylene oxidation is intended to destroy phthalic acid by dehydration to PA, and removal of water and low-boiling compounds such as maleic anhydride, o-tolualdehyde and benzoic acid.

After the predecomposer (where phthalic acid is dehydrated) phthalic anhydride is introduced into the first distillation column (Fig. 4.89). It is separated in this column from the low-boiling components, which are taken from the column top. In the subsequent column purified phthalic anhydride is removed from the top and is delivered to flaking, liquid storage or directly to downstream facilities, while the residue is discharged from the bottom.

Both distillation columns operate under vacuum, which is generated by ejectors generally driven by steam.

4.7.4 From alcohols to aldehydes: oxidative dehydrogenation of methanol to formaldehyde

Formaldehyde is produced industrially from methanol by partial oxidation and dehydrogenation with air

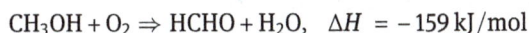

$$CH_3OH \Rightarrow HCHO + H_2, \quad \Delta H = +84 \, kJ/mol$$

$$CH_3OH + O_2 \Rightarrow HCHO + H_2O, \quad \Delta H = -159 \, kJ/mol$$

mainly in the presence of silver catalysts, steam and excess methanol at either 680–720 °C with methanol conversion approximately 99% (so-called complete conversion) or at 550–650 °C giving incomplete conversion of methanol 70–80%. In the latter case, the product is distilled and the unreacted methanol is recycled. The advantage of lower temperature and incomplete conversion is in less prominent side reactions. The reaction temperature depends on the excess of methanol in the methanol–air mixture. The composition of the mixture must be outside the explosion limits. Addition of steam is needed to diminish coking on silver catalysts.

Oxidation can be done also without steam only with excess air in the presence of a modified iron–molybdenum–vanadium oxide catalyst at 250–400 °C with methanol conversion of approximately 98 – 99%.

Among side reactions, oxidation of hydrogen to water, decomposition of formaldehyde to CO and hydrogen, oxidation of methanol and formaldehyde to CO_2 and water should be mentioned.

Another important factor affecting the yield of formaldehyde and methanol conversion, besides the catalyst and temperature, is the addition of inert compounds to the reactants. Water is added to the spent methanol–water-evaporated feed mixtures, and nitrogen is added to air and air–off-gas mixtures, which are recycled to dilute the methanol–oxygen reaction mixture.

There are several options for the process based on silver catalysts. One variant used for many decades included incomplete conversion with a single-stage methanol separation and recycle (Fig. 4.93).

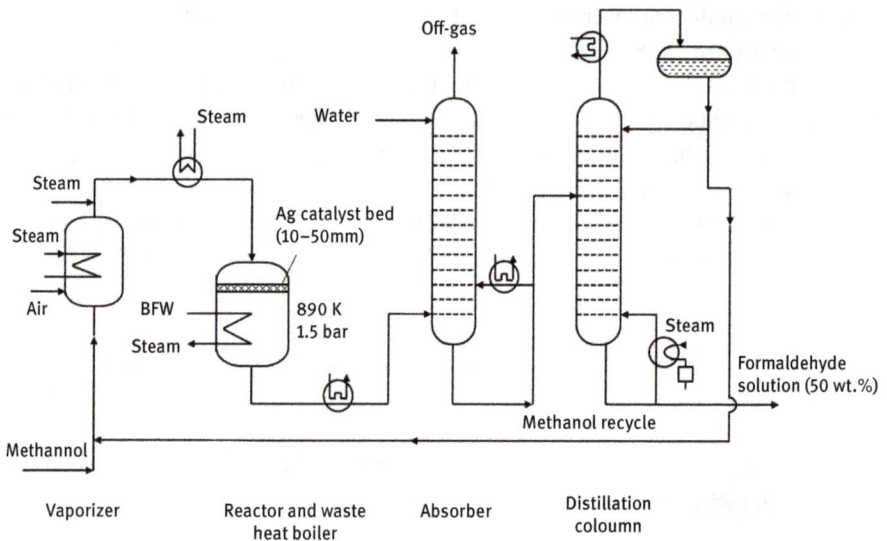

Fig. 4.93: Silver-based methanol oxidation to formaldehyde [20]. Reproduced with permission.

First, a mixture of methanol and water is evaporated and sent to the adiabatic reactor containing a shallow bed (25–30 mm) of a silver catalyst. There is a separate flow to the vaporizer of the fresh process air. The content of the "inert" gases (nitrogen, water and CO_2) should be selected to prevent formation of explosive methanol–oxygen mixtures. A typical methanol-rich feed (40–45% methanol, 20–25% air and the rest steam) enables safe operation outside the flammability limits. The average lifetime of a catalyst bed depends on impurities such as inorganic materials in the air and methanol feed. The catalyst bed is located immediately above a cooler (water boiler). Temperature of the outlet gases diminishes to 150 °C and simultaneously superheated steam is produced. Such rapid cooling is needed to avoid oxidation of formaldehyde to CO_2. In the absorption tower the process gas is flowing countercurrently with water. Methanol is recovered at the top of the distillation column and is recycled to the bottom of the evaporator. The product containing up to 55 wt% formaldehyde and less than 1 wt% methanol along with approximately 0.01 wt% formic acid is taken from the bottom of the distillation column. A special care should be taken on preventing corrosion as formaldehyde solutions are corrosive.

Capital and operating costs for distillation and recycling equipment needed for recovery of unconverted methanol can be reduced by approximately 12% and 20% in the case of a complete conversion process (Fig. 4.94), which also has several variants. The fixed-bed adiabatic reactor is, however, essentially the same as in incomplete conversion with a 15–30 mm deep, 2–3 m diameter bed of Ag catalyst (Fig. 4.95). Several mesh fractions are typically used with the finest on the top

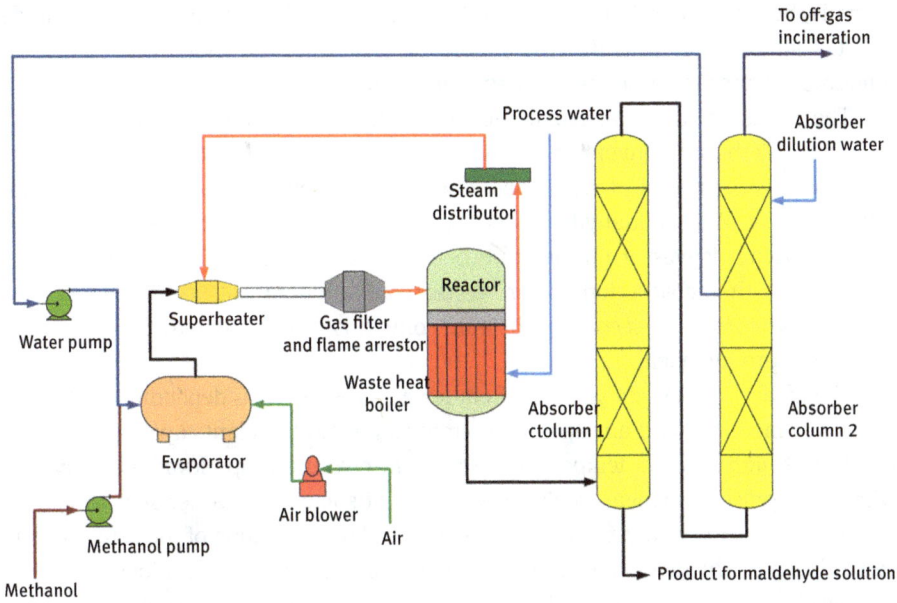

Fig. 4.94: Silver-based complete methanol oxidation to formaldehyde process [68]. Reproduced with permission.

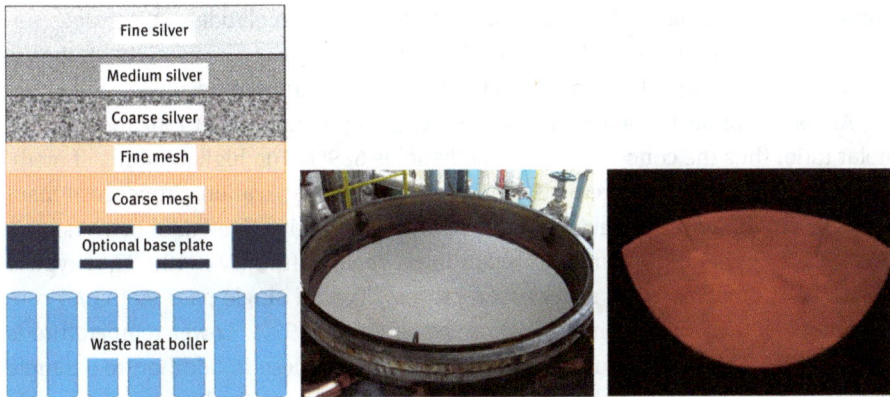

Fig. 4.95: Silver catalyst (left) bed configuration, center: fresh bed, right: in operation [68]. Reproduced with permission.

and the largest granules on the bottom of the bed. The bed is arranged to prevent the finest silver grains from slipping through the structure.

The feed comprising methanol, air, steam and tail-gas recycle is introduced to the reactor at 340–350 °C and approximately 1–1.5 bar. Safety considerations are important in methanol oxidation process. Addition of steam helps to increase selectivity and mitigates catalyst deactivation by lowering coke formation and sintering.

There is, however, a limit on the amount of steam, because the target formaldehyde product strength of typically 55% should be achieved, and moreover, additional amounts of water increase the downstream treatment and recycling costs.

The residence time is approximately 0.01 s to minimize decomposition of formaldehyde. Methanol conversion is 98.5–99% with selectivity to formaldehyde of 90–92% with the rest of products being CO, CO_2 and small amounts of formic acid. Rapid cooling of formaldehyde gas is done using a waste heat boiler system. The exit gas stream leaves at temperature between 120 and 200 °C, and then passes through the first absorption tower. Water containing recaptured methanol and formaldehyde from the second absorption column is sent to the reactor, enhancing the overall product yield.

The lifetime of the catalyst life is from few months to a year depending on the impurities in methanol and air, operation temperature and the plant capacity. Sintering of silver at high reaction temperatures results in pressure drop elevation. As typical with many processes where catalyst deactivates at some point, a decision should be taken to replace the catalyst. Such decision is based on evaluation of the catalyst costs versus lower productivity, higher operations costs and safety considerations.

Besides silver-based technology an alternative process employs a metal oxide catalyst.

The catalyst composition in the latter so-called Formox process is rather complex; for example, it can contain iron and molybdenum oxides $Fe_2(MoO_4)_3$ with an excess of MoO_3 giving Mo:Fe atomic ratio of 1.5–2.0. Iron molybdate structure is stabilized by such additives as V_2O_5, CuO, Cr_2O_3, CoO and P_2O_5. The catalyst lifetime is typically 18–24 months because the catalyst is more tolerant to poisons.

An excess of air is used in this process with approximately 13:1 air to methanol molar ratio, thus the concentration of methanol is 6–9%. Too high amounts of methanol would lead to catalyst reduction and subsequent deactivation. Proper heat management avoiding overheating is needed to prevent sintering. Methanol conversion is done at atmospheric pressure and much lower temperature (270–400 °C, a typical value 340 °C), compared to silver-based process. Typical single-pass methanol conversion is 97–99% with formaldehyde selectivity of 92–95%. Lower temperature is beneficial from the viewpoint of higher selectivity. All these in addition to a simple method of steam generation contribute to the advantages of the Formox process. At the same time, the volume of gas is 3–3.5 higher than in the silver-based technology resulting in more expensive larger capacity equipment.

Figure 4.96 displays the flow scheme for the Formox process. Methanol is fed through a steam-heated evaporator and after mixing with the fresh process air and the recycled off-gas passes through a heat exchanger heated by the product stream.

The catalyst pellets or rings (Fig. 4.97) of the size 3–4 mm are placed in multitubular reactors of a diameter approximately 2.5 m containing thousands of tubes (diameter 2–3 cm and 1.0–1.5 m length). The reaction heat is removed by a

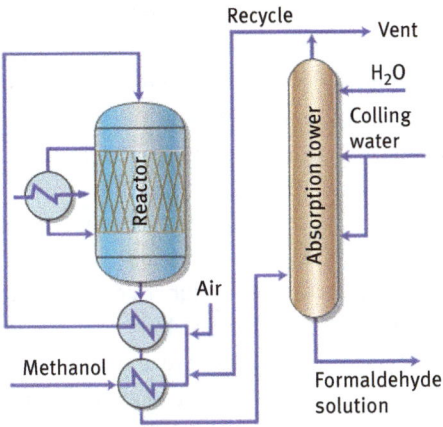

Fig. 4.96: Synthesis of formaldehyde by oxidation on Fe–Mo catalyst [69].

Fig. 4.97: Fe–Mo catalysts of different shapes (a) [70]; (b) [71] for synthesis of formaldehyde by methanol oxidation.

high-boiling heat-transfer fluid (oil, Dowtherm, molten salt) held at approximately 270–290 °C. This fluid is circulating outside of the tubes generating steam in a boiler.

A typical feature of highly exothermal reactions in multitubular reactors are hot spots, which in the case of methanol oxidative dehydrogenation on iron molybdate reaches 340–350 °C.

The outlet gases are cooled to 110 °C to avoid formaldehyde and sent to the bottom of an absorber column to which the process water is added at the top. The product containing up to 55 wt% formaldehyde and 0.5–1.5 wt% methanol is further purified by ion exchange to diminish the formic acid content. The overall plant yield is 88–91 mol%. The absorber off gas (N_2, O_2, CO_2 with few percent of dimethyl ether, carbon monoxide, methanol and formaldehyde) is noncombustible as such and requires catalytic incineration at 450–550 °C or alternatively combustion with additional fuel at 700–900°C.

Both types of plans using either silver or iron catalysts despite operation at very different conditions give similar values of methanol conversion and formaldehyde yields. According to available comparative analysis of both processes [72] if cheap methanol is available, silver can be considered a better option, especially for smaller operations. For larger plants with capacity of up to 100,000 tons per year the metal oxide catalysts can be more suitable compensating a higher capital expenditure related to fivefold higher air flow and subsequently higher compression costs. Plants using more productive silver catalysts may present lower operating costs, lower catalyst costs as silver can be regenerated contrary to only Mo from iron molybdate and lower steam production. The tail gas in the silver process contains about 20% hydrogen and can be burned directly eliminating carbon monoxide and other environmentally harmful organic compounds. In an alternative iron oxide process, burning of the tail gas with too low amounts of flammable compounds requires a catalytic incinerator or addition of extra fuel, which adds to the production costs. Metal oxide catalyst is more resistant to poisons and because of low temperature is less susceptible to thermal degradation giving a better quality formaldehyde product with less impurities, such as formic acid, heavy metals and unreacted methanol. A further advantage of the process using iron oxide with a substantially longer catalyst life are less downtime problems.

4.7.5 Ammoxidation

4.7.5.1 Acrylic acid

Acrylic acid is produced by air-based heterogeneous catalytic oxidation of propene in the gas phase using multicomponent mixed metal oxide catalysts. The reaction is occurring through two exothermic steps with first formation of acrolein and water

$$CH_2 = CH - CH_3 + O_2 \rightarrow CH_2 = CH - CH = O + H_2O, \ \Delta H = -340.8 \, kJ/mol$$

and subsequent oxidation of acrolein to acrylic acid

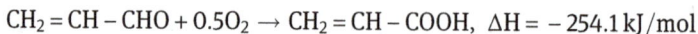

$$CH_2 = CH - CHO + 0.5O_2 \rightarrow CH_2 = CH - COOH, \ \Delta H = -254.1 \, kJ/mol$$

These two steps require different reaction conditions and catalysts to produce optimum conversion and selectivity in each step. For the first step, supported multicomponent systems containing Bi-Mo-Fe-Co-Sb-K mixed oxides are used. Acrolein yields depend not only on the chemical compositions of these catalysts, but also on their physical properties, such as shape, porosity, pore-size distribution and specific surface area, as well as on the reaction conditions and construction of the reactor.

There was a significant improvement in the yield of the product because of the catalyst development. In the 1950s, the yield with Cu_2O catalyst was 10–20%, which was elevated to 45–50% in 1960s on Bi-Mo-P containing catalyst. Bi-Mo-Fe-Co mixed oxides allowed a further increase of the yield to 80–85%. A further improvement was obtained with Bi-Mo-Fe-Co-Sb-K mixed oxides.

The catalysts for oxidation of propene to acrylic acid are essentially the same as applied in propylene ammoxidation (Section 4.7.5.2) when oxidation is done in the presence of ammonia. In the latter case, the overall yield of acrylonitrile is lower than the combined yield of acrolein and acrylic acid. Both reactions share the same mechanistic features. Oxidation of propylene to acrolein (the first step in formation of acrylic acid) follows Mars-van-Krevelen mechanism with a number of steps, each requiring their own active sites. Therefore, multicomponent catalysts are applied and catalytic performance strongly depends on the catalyst structure, which is discussed in more detail in Section 4.7.5.2.

The process flow diagram of the oxidation section of acrylic acid production is shown in Fig. 4.98.

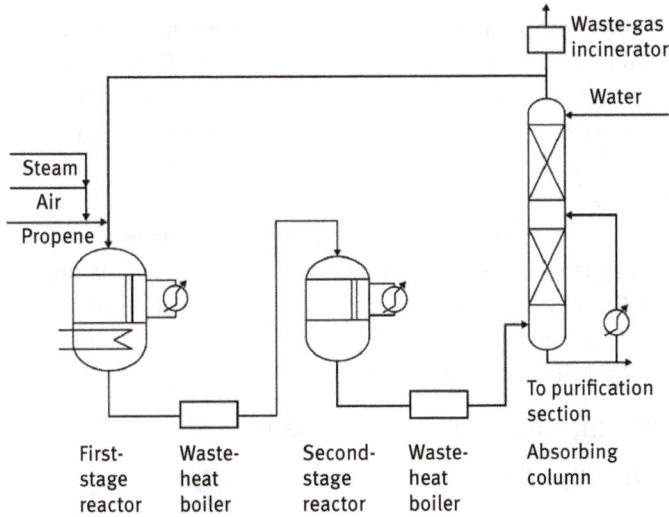

Fig. 4.98: Schematic diagram of the oxidation section in acrylic acid production.

Air, propylene and steam are fed to reactors consisting of catalyst-filled tubes with molten salt circulating on the shell side for removal of the reaction heat. The salt passes through a steam boiler to produce steam for heat recovery. The gas composition is 5–10% propylene, 30–40% steam with air being the rest. An excess of air is used to keep the catalyst in the oxidized state. Steam can be replaced by inert gas. Other reaction parameters: temperature 300–400 °C, contact time 1.5–3.5 s, and inlet pressure: 150–250 kPa. With time the yield of the product is decreasing and/or pressure drop is increasing to a point when a catalyst charge is replaced. Such replacement can happen after a relatively long catalyst lifetime (up to 10 years)

At high propene conversion (up to 98%), acrolein is formed in the yields of 83–90%. The main by-product is acrylic acid (5–10% yields) with acetaldehyde and acetic acid being other by-products. The reactor effluent is rapidly quenched at the exit to prevent subsequent reactions of acrolein. The catalysts used in the second step (multicomponent metal-oxide catalysts containing Mo, V, W and some other elements) require reaction temperatures from 200 to 300 °C and contact times from 1 to 3 s. They give almost 100% conversion of acrolein and yields of acrylic acid larger than 90%. The effluent gas from the second-stage multitube reactor in Fig. 4.98 is cooled to about 200 °C and then fed to the absorbing column to be scrubbed with water. Because the effluent gas contains a large amount of steam, acrylic acid is usually obtained as an aqueous solution of 20–70 wt%. The waste gas (nitrogen, excess oxygen, CO_2 and propylene) along with some residual organic compounds should be incinerated.

4.7.5.2 Acrylonitrile

The worldwide capacity for acrylonitrile ($CH_2 = CH-CN$), an important monomer for the manufacture of useful plastics, is currently approximately 7 million metric tons per year. The average capacity of acrylonitrile plants is around 170,000 tons/year. The main application of acrylonitrile is in production of acrylic fibers, acrylonitrile-butadiene-styrene, adiponitrile, nitrile–butadiene copolymers as well as acrylamide for water treatment polymers.

Prior to 1960s the most popular process of acrylonitrile production was addition of hydrogen cyanide to acetylene. Currently manufacture of acrylonitrile mainly follows the Standard Oil of Ohio (SOHIO) process, based on ammoxidation of propylene in fluidized bed reactors using propylene, ammonia and air as reactants and giving as by-products acetonitrile and hydrogen cyanide

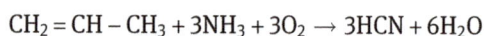

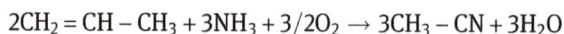

$$CH_2 = CH - CH_3 + 3NH_3 + 3O_2 \rightarrow 3HCN + 6H_2O$$

$$2CH_2 = CH - CH_3 + 3NH_3 + 3/2O_2 \rightarrow 3CH_3 - CN + 3H_2O$$

in commercially attractive quantities (ca. 25% total yield). The side reactions are favored at higher temperature and pressure, and longer residence time. The amounts of HCN are larger than that of acetonitrile. Besides these products carbon dioxide is also formed. The complex reaction network in the synthesis of acrylonitrile is shown in Fig. 4.99.

The overall process has, nevertheless, high selectivity to acrylonitrile not requiring further recycling steps. The costs of acrylonitrile production are closely linked to the market price of reagents, in particular propylene, which contributed substantially (> 70%) to the total cost.

A schematic view on the reaction mechanism is shown in Fig. 4.100.

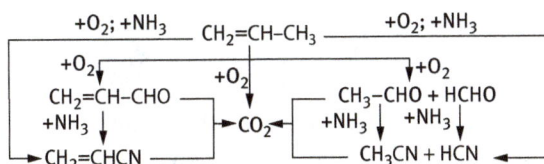

Fig. 4.99: Reaction network in acrylonitrile synthesis.

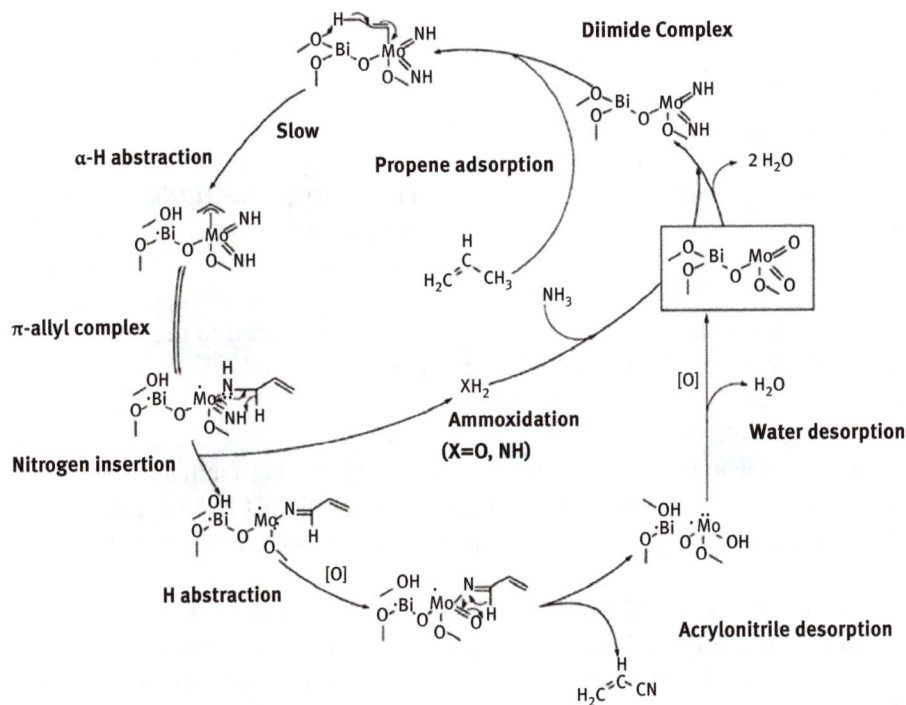

Fig. 4.100: Ammoxidation mechanism [73]. Reproduced with permission.

Ammonia interacts with the bifunctional active centers, generating an extended ammoxidation site containing ammonia as = NH. On this site propylene is inserted by α-hydrogen abstraction forming an allylic complex. Further nitrogen insertion and proton abstraction gives acrylonitrile, which then desorbs from the surface.

Regeneration of the reduced surface takes place by oxygen (O^{2-}) from the lattice oxygen, which is filled with oxygen from the gas phase. An ammoxidation catalyst thus possesses multifunctional and redox properties. As mentioned earlier, ammoxidation catalysts can be used also for synthesis of acrolein, even if in the presence of ammonia the yields of acrolein are small. An explanation for this is that surface species (e.g., π-allyl complex in Fig. 4.100) which are precursors for acrolein react in a fast way to acrylonitrile. Thus, the rate of acrylonitrile formation is higher than that of acrolein.

The bismuth phosphomolybdate catalysts supported on silica are the dominant catalysts used commercially with such promoters as K, Cs, Mg, Mn, Co, Ni, Fe, W and Re. Historically to the mixed metal oxide Fe-Bi-Mo-O, first divalent cobalt and nickel molybdates were added improving activity and selectivity, followed by introduction of alkali metal promoters [74].

The catalyst should fulfil several functions, such as α-H abstraction (provided by Bi^{3+}, Sb^{3+} or Te^{4+} in the composition), olefin chemisorption and nitrogen insertion (Mo^{6+}, Sb^{5+}), while a redox couple (Fe^{2+}/Fe^{3+} or Ce^{3+}/Ce^{4+}) enhances the transfer of lattice oxygen between the bulk of the catalyst and its surface.

In the chemical composition of catalysts, multimetal molybdates are used comprising bismuth molybdate as α-$Bi_2Mo_3O_{12}$ phase, divalent metal (Ni, Co, Mg, Fe) molybdates $M^{2+}MoO_4$ of α and β structure and trivalent $Fe_2Mo_3O_{12}$ dispersed in silica (50% w/w). The catalysts also contain antimonate with the rutile structure, made up of four metal antimonate cations and a redox couple of Fe, Ce, U, and Cr. As mentioned earlier, alkali metals (e.g., Rb, K, Cs) are added to the active phase along with other promoters (P, B, Mn). An example of a complex ammoxidation catalyst chemical composition of the active phase is given below: $(K, Cs)_{0.1}(Ni, Mg, Mn)_{7.5}(Fe,Cr)_{2.3}Bi_{0.5}Mo_{12}O_x$.

A coprecipitation procedure with aqueous solutions of bismuth nitrate and ammonium molybdate is used in catalyst preparation, followed by heat treatment of the dried particles at approximately 500 °C to crystallize the bismuth molybdate phase.

Because of high reaction exothermicity ($\Delta H_o = -515$ kJ/mol), ammoxidation reaction is conducted in a fluidized-bed reactor needed to remove the excess of heat. Such operation imposes several restrictions on the catalyst size which should be in the range 40–100 μm (Fig. 4.101) and mechanical properties, which should account

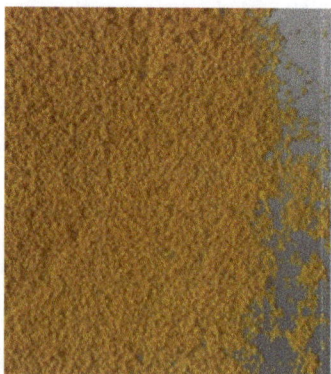

Fig. 4.101: Ammoxidation catalyst [71].

for the abrasive environment in a fluidized bed reactor. The catalyst should be hard enough and resistant to attrition. To meet these requirements, bismuth molybdate catalyst is supported on silica.

The steady-state kinetics of propene oxidation is first order in propene and zero order in oxygen when oxygen concentration is above stoichiometry. The apparent activation energy for acrylonitrile formation is the same as for acrolein formation (~83 kJ/mol); moreover, kinetic regularities are similar (first order in propene and zero order in oxygen and ammonia) pointing out on a similar mechanism.

The SOHIO process (currently INEOS Technologies ammoxidation technology) operates with near stoichiometric amounts of reactants in temperature range of 400–510 °C and at 50–200 kPa (0.5 to 2 bar) giving propylene conversion above 95%, and the yield of acrylonitrile approximately 80%. In the SOHIO process (Fig. 4.102), propylene, ammonia after evaporation and air are passed through a fluidized bed reactor containing the catalyst.

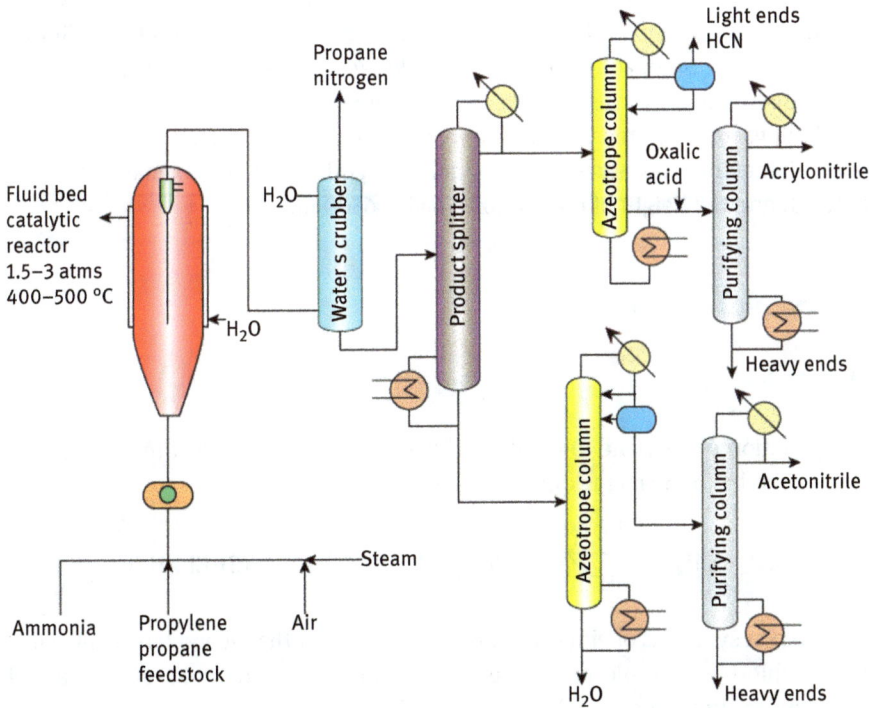

Fig. 4.102: Flow scheme of acrylonitrile synthesis.

The ratio between reactants is 1: (0.9 ÷1.1): (1.8 ÷ 2.4) for propylene, ammonia and oxygen, respectively. An excess of oxygen is required to keep the catalyst in a proper state needed for maintaining high activity and selectivity. The fluidized bed reactor is used to control temperature during this exothermal reaction (enthalpy −670 kJ/mol). As a general comment, it can be stated that typically a reaction heat above 500 kJ/mol should not be handled in fixed-bed multitubular reactors. Usually the heat recovered is used to produce high-pressure steam.

Catalyst deactivation is a serious problem; however, by stabilization of the Mo phase, addition of excess of MoO_3 phase during preparation and periodic addition of the catalyst during operation it is possible to extend the catalyst lifetime to 10 years. Fine particles (Fig. 4.101) are required for proper fluidization and the optimal product yield. The contact time is 6–10 s.

The internal heat transfer coils in the reactor are used as baffles to improve fluidization, minimizing back-mixing. This finally leads to higher selectivity. An expansion chamber at the reactor top is used for separation of smaller and larger catalyst particles, the former ones are returned through the cyclones to the reactor inlet. The reactants pass through the reactor only once. A heat exchanger located downstream the reactor is used for generation of medium pressure steam. Thereafter, the reactants are quenched, and unreacted ammonia is neutralized in water to remove remaining ammonia. The product mixture containing water and organic compounds is split into two phases in the product splitter. Acrylonitrile, acetonitrile and hydrogen cyanide are separated by multiple distillations. The column where HCN is removed operates under slight vacuum to prevent release of toxic HCN to atmosphere.

The diameter of the reactor for acrylonitrile synthesis is approximately 9–10 m (Fig. 4.103) and the height of the fluidized bed is 6500 mm.

4.8 Oxychlorination

4.8.1 Overview

Oxychlorination of ethylene to dichloroethylene over $CuCl_2$ catalyst ($\Delta H = -295$ kJ/mol) followed by thermal cracking of the latter

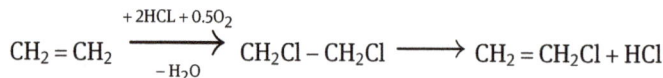

$$CH_2 = CH_2 \xrightarrow[-H_2O]{+2HCL+0.5O_2} CH_2Cl - CH_2Cl \longrightarrow CH_2 = CH_2Cl + HCl$$

is a key step in synthesis of vinyl chloride, a monomer for the second largest plastic – polyvinyl chloride. Oxychlorination reaction is strongly exothermic, and a careful temperature control is essential both to ensure high selectivity and prevent rapid catalyst deactivation.

In the so-called balanced process (Fig. 4.104), ethylene oxychlorination and dichloroethylene pyrolysis are coupled along with direct chlorination of ethylene in a single process to increase the throughput of vinyl chloride monomer (often abbreviated as VCM). In this way there is no net consumption or production of HCl. Vinyl

Fig. 4.103: Fluidized bed reactor for ammoxidation.

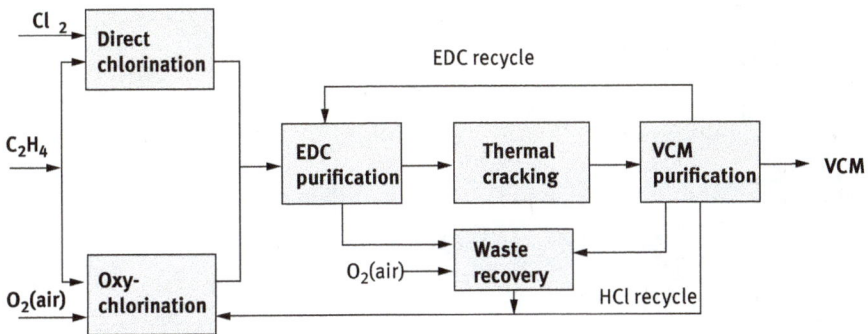

Fig. 4.104: Balanced process of VCM manufacturing.

chloride is made only by thermal cracking of 1,2-dichloroethane, formed by either direct chlorination or oxychlorination of ethylene.

The first exposure of the author to industrial chlorination was almost 35 years ago as a trainee at a chemical plant in a remote location where a range of pesticides were produced using various chlorination reactions. Apart from personal memories, related to daytime temperature of −30 °C to −35 °C for a whole month of traineeship and accommodation in a factory dormitory with a very noisy life 24/7 as neighbors were working in shifts, the author remembers that wearing a gas mask was obligatory in the chlorination section because of green clouds of chlorine. This was certainly not that sort of green chemistry, which one would like to use industrially.

4.8.2 Catalysts and reactors

In industry the reaction is carried out in the vapor phase in either fluidized bed reactors (Fig. 4.104a) or fixed-bed reactors (Fig. 4.105b) with oxygen being supplied as pure gas (oxygen-based process) or as conventional air (air-based process). Fixed-bed oxychlorination generally operates at 230–300 °C and 150–1,400 kPa, while lower temperatures can be used in fluidized bed reactors (220–245 °C) operating at 150–500 kPa. More efficient heat removal can be achieved in fluidized bed reactors.

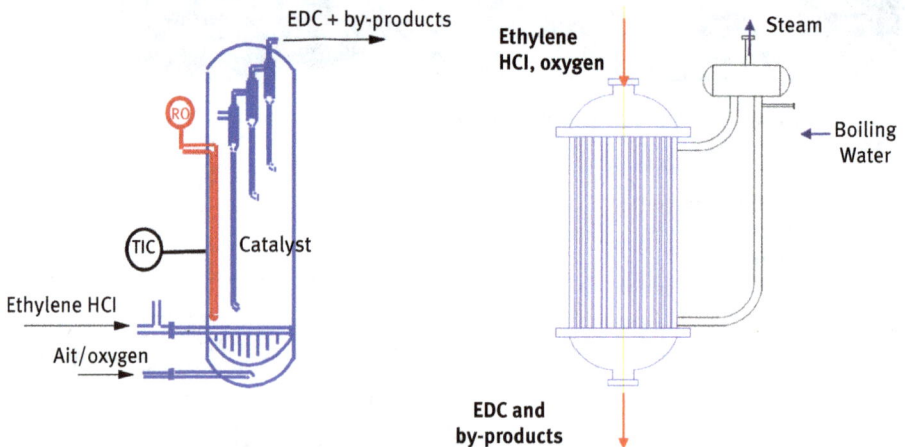

Fig. 4.105: Reactors for ethylene oxychlorination [75].

As catalysts, copper chloride as an active phase with promoters on porous supports such as alumina (e.g. 8–16 wt% $CuCl_2/\gamma Al_2O_3$) is applied. It is almost generally accepted that the mechanism of oxychlorination involves a three-step redox process: (i) reduction of $CuCl_2$ to $CuCl$ with simultaneous chlorination of ethylene

$$CH_2 = CH_2 + 2CuCl_2 \rightarrow ClCH_2 - CH_2Cl + Cu_2Cl_2$$

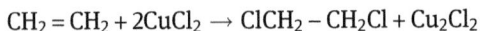

(ii) oxidation of CuCl to give an oxychloride and (iii) closure of the catalytic circle by rechlorination with HCl, restoring the original CuCl$_2$. The last two steps give the overall reaction

$$Cu_2Cl_2 + 2HCl + 0.5O_2 \rightarrow 2CuCl_2 + H_2O$$

The active site probably involves an isolated Cu$_x$Cl$_y$ complex, which is anchored to the high-surface-area γ-Al$_2$O$_3$ support.

A particular challenge with copper chloride is its volatility at reaction temperature, which can be decreased by adding potassium chloride. The latter results, however, in lower activity, improving at the same time selectivity by decreasing formation of ethyl chloride. The control over copper chloride content is needed as an excess would lead to catalyst caking. KCuCl$_3$ and K$_2$CuCl$_4$ have low melting points and the eutectic mixture with copper chloride melts at 150 °C. Other alkali metals and rare earth chlorides (e.g., MgCl$_2$ or LaCl$_3$) can be also added to the catalyst formulation as promoters or to inhibit by-product formation. Their role is also in maintaining a high Cu^{2+} concentration, as it is critical for high activity, selectivity and stability.

Catalysts are usually prepared on a large scale by impregnating a suitable alumina support with aqueous solutions of cupric and potassium chlorides. In the classical method the solution containing cupric chloride and other salts is added to the support followed by holding for 30 min and drying of the catalyst at approximately 110 °C for 16–20 h in dry air. Even if good distribution of the active phase is achieved during impregnation, drying results in nonuniform crystallization and formation of the catalytically active phase with poor dispersion. This in turn negatively influences catalytic performance. Alternative methods to achieve better dispersion include impregnation during fluidization or by support immersing in a melt of cupric-potassium chlorides, application of nonaqueous solutions, dry impregnation with copper chloride or with a salt of copper, which can be decomposed followed by annealing with HCl and oxygen.

In fluidized bed reactors microspheroidal catalyst powder with a particle size of 40–60 μm is used (Fig. 4.106a,b), while in fixed-bed reactors loaded with catalyst pellets a much larger size of catalyst particles in the mm range should be utilized (Fig. 4.106c) to avoid too high pressure drop.

During operation because of attrition the average size is decreasing which is compensated by adding the fresh catalyst. As an example, a vinyl chloride plant with a capacity 300,000 tons per year with 50 tons of catalyst hold-up requires 15 tons per of catalyst to compensate for the losses. For adequate fluidization the fraction of the particles below 40 μm should be 25–35%. The copper content does not practically change.

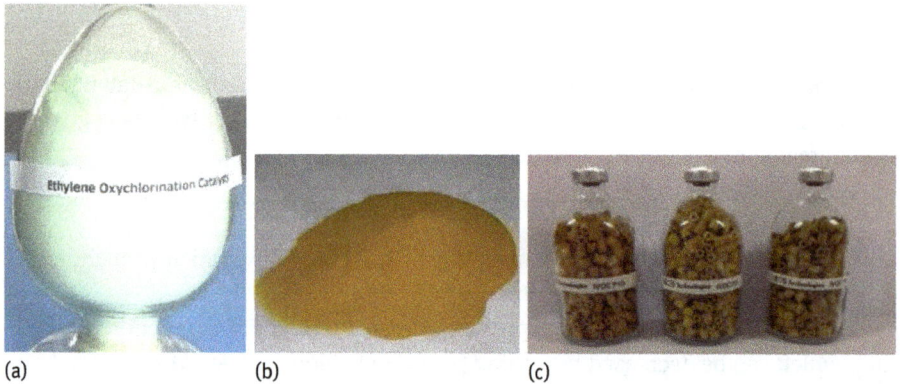

Fig. 4.106: Ethylene oxychlorination catalysts for fluidized bed reactors from different suppliers (a) [76] and (b) [75] and (c) for fixed bed applications [75].

One of the main process problems with oxychlorination is a possible tendency for the catalyst particles to *cake* or to stick to the cooling bundles within the reactor. This results in poor temperature control, poor fluidization of agglomerated particles, local overheating and a loss in selectivity. Erosion in the industrial reactor walls leads to a release of iron, whose penetration into the catalyst worsens selectivity by increasing the yield of the total oxidation products.

Temperature control for a very exothermic reaction is easier to organize in fluidized beds, compared to fixed-bed tubular reactors. Higher temperature besides promoting dehydrochlorination of ethylene dichloride to vinyl chloride and formation of other chlorinated products also leads to increased oxidation of ethylene, elevated formation of coke and more losses of copper by sublimation.

In fluidized bed reactors (Fig. 4.105a) the heat is controlled by adding cold reactants and having an internal heat exchanger generating steam. Due to high gas flow rates, industrial fluidized bed reactors often operate in the turbulent fluidization regime. These reactors despite periodical makeup of the catalyst to compensate for losses in cyclones and mechanical degradation allow effective reaction heat removal, near isothermal operations, negligible impact of external mass- and heat-transfer limitations, as well as minimal internal mass transport limitations. The latter is important because, as discussed in Chapter 3, operation in the internal mass transfer regime negatively influences selectivity to the intermediate product.

The reaction conditions for oxychlorination, which are optimum from the viewpoint of the ratio between reactants, require operation within the explosive limits. An appropriate design of fluidized bed reactors with certain measures against radical chain propagation is thus needed. Special devices for mixing reactants to exclude formation of large volumes of explosive mixtures outside the fluidized bed, allow safe operation and low by-products selectivity making strict feed control less critical. Besides adequate mixing of reactants, conditions for fluidization should be properly ensured, which impose limitations not only on the flow rate and the catalyst particles size, but also on the reactor diameter and thus production capacity.

Clear advantages of fluidized bed reactors are their isothermal operation and a possibility to compensate the catalyst losses. This allows to maintain constant catalyst activity, and therefore a stable reactor performance, product quality and feed consumption. The drawbacks include corrosion of the reactor internals needed for heat removal and lower conversion compared to the tubular reactor.

Tubular reactors (Fig. 4.105b) allow higher conversion, especially HCl exceeding 99%. Heat is removed by evaporation of steam condensate; thus, a uniform temperature profile is difficult to achieve. Formation of hot spots with too high temperature can be mitigated by diluting the catalyst with inert diluents of different types and materials (Fig. 4.107) and/or by varying the copper chloride content in the catalyst.

(a)   (b)

Fig. 4.107: Inert materials for dilution of catalyst fixed beds in ethylene oxychlorination (a) standard and high performance diluents from INEOS [75], (b) graphite pellet 5 × 5 mm and alumina ring 6 × 6 × 2.5 mm diluents from BASF [77].

Several options of catalyst loading in tubular reactors can be used. For instance, in a single reactor up to four different layers of a catalyst containing the same amount of CuCl$_2$ (8.5 wt%) can be loaded with different degree of dilution. The first layer contains 7% of the catalyst, followed by 15%, 40% and finally 100% of the catalyst without any dilution.

Mixing of the catalyst and the inert material should be properly done to avoid any dust and local overheating, minimize nonuniform pressure drop along the reactor and ensure good performance in terms of catalysts life and selectivity. This was in particular important when silica particles with irregular shape were used as dilutants. Mixing can be done by a catalyst manufacturer, which delivers catalyst mixtures for direct loading in the reactor.

A better control with a diluted catalyst can be obtained in a series of several separate reactors, which could be further improved by having different catalyst compositions with varying amounts of CuCl$_2$. Spherical catalysts supplied previously by Stauffer Chemical Company did not require dilution with an inert. Dilution of catalysts containing 6%, 10%, 18% CuCl$_2$, and promoted with 3%, 3%, and 2%

KCl with inerts was also reported in the first two beds, while the last bed contained only the more concentrated catalyst [3].

Conditions in different reactors are different. Obviously, they are more severe in the first reactor, thus the catalyst life is shorter (one year) and can be increased to 18 months and 3 years, respectively, in the second and third reactors. When several types of catalysts are applied hot spots are present close to the position in tubes where the catalyst type changes.

In tubular reactors, which can operate with air or oxygen, control of the oxygen content is required to ensure operation outside of explosive conditions. Another option is to split air between several beds. Half of the overall amount of the air can be added to the first reactor, while the rest is split between the second and third reactors to minimize hot spots. Replacement of air with oxygen reduces the amount of vented gas.

4.8.3 Reaction network and kinetics

The main reactions shown in Fig. 4.108 follow the route with oxygen involvement (A) or without it (B).

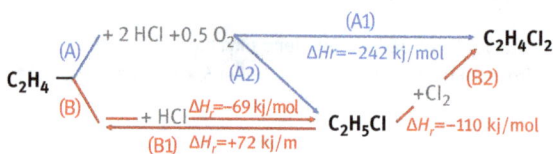

Fig. 4.108: Reaction network in ethylene oxychlorination [78].

All chloroorganic by-products except ethyl chloride result from secondary 1,2-di-chloroethane transformations, with dichloroethane dehydrochlorination and partial oxidation being the primary reactions in such transformations. Vinyl chloride and chloroethanol are converted further to 1,1,2-trichloroethane, trichloroacetaldehyde, trichloromethane, tetrachloromethane, dichloroethene, CO_2 and CO. Ethyl chloride is formed by ethene hydrochlorination.

Typically, the reaction kinetics follows first order dependence on ethylene and oxygen partial pressures, while HCl does not influence the reaction rate, affecting, however, selectivity in total oxidation and also selectivity to dichloroethane, which decreases upon increasing partial pressure. The detailed kinetic analysis of oxychlorination was performed in [79] where a following mechanism was proposed:

1. $CuCl_2{}^* + 2HCl \longrightarrow CuH_2Cl_4{}^*$ 2

2. $CuH_2Cl_4{}^* + C_2H_4 \rightarrow CuH_2Cl_4 \cdot C_2H_4{}^*$ 2

3. $CuH_2Cl_4 \cdot C_2H_4{}^* \rightarrow CuH_2Cl_3 \cdot C_2H_4Cl^*$ 2

4. $CuH_2Cl_4{}^* + CuH_2Cl_3 \cdot C_2H_4Cl^* \rightarrow 2CuH_2Cl_3{}^* + C_2H_4Cl_2 \,(\text{fast})$ 2   (4.41)

5. $CuH_2Cl_3{}^* + O_2 \rightarrow CuH_2Cl_3O_2{}^*$ 1

6. $CuH_2Cl_3O_2{}^* + 3CuH_2Cl_3{}^* \rightarrow 2CuH_2Cl_4{}^* + 2CuCl_2{}^* + 2H_2O \,(\text{fast})$ 1

$$4HCl + 2C_2H_4 + O_2 \rightarrow 2C_2H_4Cl_2 + 2H_2O$$

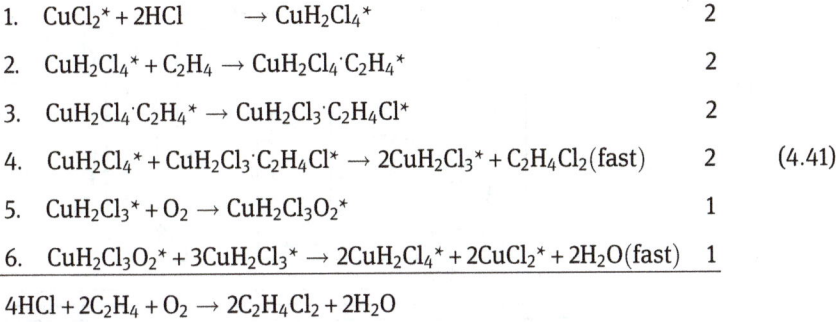

On the right-hand side, the stoichiometric (Horiuti) numbers of a particular step are given. The following kinetic equation corresponds to this scheme

$$r_{C_2H_4Cl_2} = \frac{K_1 K_2 k_3 P_{C_2H_4} P_{HCl}^2}{1 + \dfrac{k_3}{k_{-2}}\left(1 + \dfrac{k_2}{k_{-1}} P_{C_2H_4}\right) + K_1 P_{HCl}^2 \left[1 + \dfrac{k_3}{k_{-2}} + K_2 P_{C_2H_4}\left(1 + \dfrac{k_3}{2k_5 P_{O_2}}\right)\right]} \tag{4.42}$$

Equation (4.42) can be simplified to

$$r_{C_2H_4Cl_2} = \frac{K_2 k_3 P_{C_2H_4} 2k_5 P_{O_2}}{K_2 k_3 P_{C_2H_4} + 2k_5 P_{O_2}} \tag{4.43}$$

4.8.4 Technology

Oxychlorination can be done either in air or oxygen atmosphere. The flow scheme of an air-based process is given in Fig. 4.109. In this process, ethylene and air are fed in slight excess of stoichiometry enabling high conversion of HCl. As the vent system is large because of the presence of nitrogen with the highest molar flow rate such operation is aimed at minimizing the loss of excess ethylene in the vent stream. In the oxygen-based process only a small portion of the vent gas is purged, being significantly lower than in the air-based process. Typical feedstock conversion is approximately 94–99% based on ethylene and 98.0–99.5% for HCl. Selectivity to dichloroethane is 94–97%.

Direct chlorination is done in reactor 1 where there is a constant level of liquid containing the catalyst ($FeCl_3$). Heat of the reaction is removed by evaporation of 1,2-dichloroethane, which is partly condensed and partly goes to distillation. Oxychlorination is done, for example, in a fixed-bed reactor 5 operating under 0.5 MPa at 260–280 °C. Ethylene, recycle gas and HCl are preliminary mixed, thereafter oxygen (as air) is added to the reactor. The mixing sequence and composition are selected to avoid expositions. The heat of reaction is removed by generation of steam. The outlet gases (unconverted ethylene, oxygen and HCl, as well

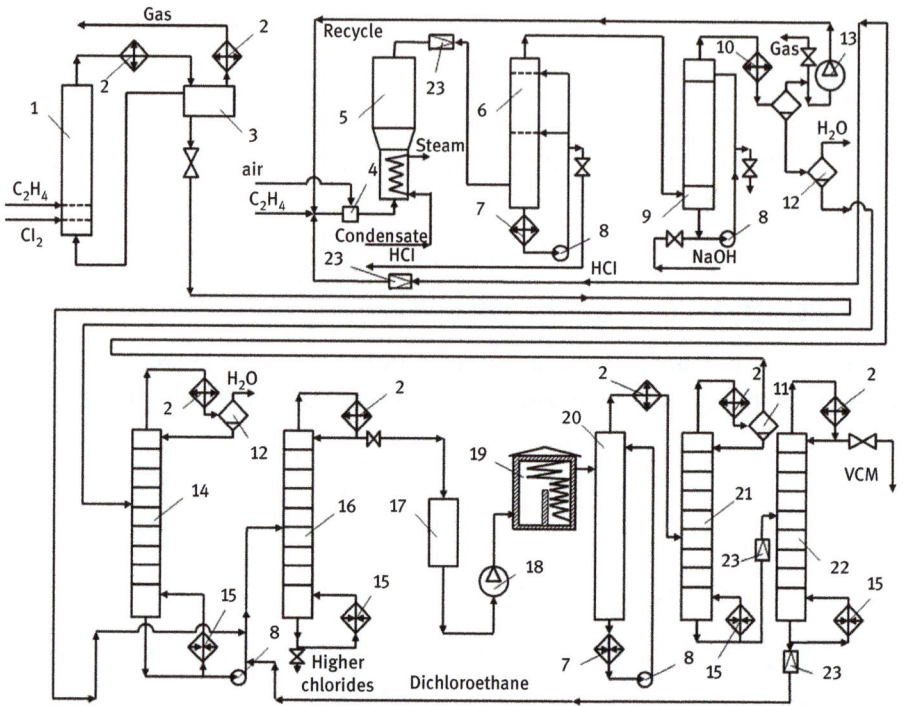

Fig. 4.109: Synthesis of vinyl chloride: 1, chlorator; 2, condenser; 3, collecting vessel; 4, mixer; 5, reactor; 6, 20, direct heat exchanger; 7, 10, heat exchangers; 8, circulation pump; 9, scrubber; 11, 12, separators; 13, compressor; 14, drying column; 15, boiler; 16, 21, 22, distillation columns; 17, vessel; 18, pump; 19, furnace; 23, expansion valve [54].

as 1,2 dichloroethane) are cooled in a heat exchanger 6 with a mixture of water and 1,2 dichloroethane. Partially cooled mixture is treated in a scrubber 9 with alkali removing HCl and CO_2. After cooling in a heat exchanger *10*, the condensate is separated (pos.11), while the recycled gases (ethylene, oxygen and inerts) after compression are returned to oxychlorination. Part of this stream is purged. Condensate from 11 flows to separator 12, where heavier 1,2-dichloroethane is separated from water. It still contains some water, thus drying is done in column 14. Two streams of dichloroethane (one from direct chlorination and another unreacted dichloroethane after pyrolysis) are then combined and distilled (pos. 16). Pyrolysis of 1,2-dichloroethane into vinylchloride and HCl is done in a furnace 19 at 1.5 – 2.0 MPa and 500 °C. The gases are cooled in a heat exchanger 20 by a recycle of 1,2-dichloroethane and thereafter cooled with water. Distillation in column 21 allows to separate HCl, which is then send to oxychlorination, and the bottom part containing vinyl chloride and unreacted 1,2-dichloroethane. The latter mixture is separated in column 22 operating under pressure. The bottom part – unreacted 1,2 dichloroethane – is sent back to distillation (pos. 16).

Utilization of oxygen allows a significant reduction in the volume of the vent system and operation at lower temperatures resulting in improved operating efficiency and product yield. Larger than stoichiometric excess for ethylene is used in oxygen-based process compared to air based. Downstream processing includes cooling of the reaction effluent, its purification from the traces of unconverted HCl, separation of dichloroethane and water also by condensation as in the air-based process, recompression, reheating and recycling. Higher selectivity to dichloroethane is achieved by lowering per pass conversion of ethylene. A purge stream of approximately 2–5% is needed to avoid accumulation of such impurities as, for example, carbon oxides, argon or light hydrocarbons, either introduced with the feed or formed during the reaction.

In fluidized bed reactors, the preheated gas mixture enters the base of the vessel and temperature is controlled either by cooling coils or a bundle of cooling tubes in the fluid bed. The oxychlorination reaction takes place at a relatively low temperature, around 220–225°C and at a low pressure of about 2 bar.

At higher temperature, more by-products are formed as both ethylene oxidation and dichloroethane cracking become more prominent. The latter gives by-products with higher levels of chlorine substitution. Moreover, elevation of temperature enhances sublimation of $CuCl_2$ and thus catalyst deactivation.

By-products in oxychlorination are chloral CCl_3CHO, 1,1,2-trichloroethane, chloroform, 1,2-dichloroethylene and ethyl chloride. Removal of chloral is especially important as it can polymerize in the presence of strong acids. Caustic washing is done for this purpose preventing thereby polymerization and formation of solids, which can otherwise clog processing equipment downstream.

References

[1] https://condorchem.com/files/catalogos/Air%20Treatment%20-%20ONLINE.pdf
[2] Beskov, V.S., Safronov, V.S. (1999) General Chemical Technology and the Fundamentals of Industrial Ecology. Moscow, Russia: Khimia.
[3] Lomic, G, Kis, E., Grabovac, M., Marinkovic-Neducin, R. (2002), Investigation of unusual inactivity of industrial platinum-rhodium catalyst gauze, Scanning 24: 140–143.
[4] Chernyshev, V. I., Zjuzin, S. V. (2001) Improved start-up for the ammonia oxidation reaction, Platin. Met. Rev. 45: 3440.
[5] http://www.catalysis.ru/resources/institute/Publishing/buclet/BIC-Booklet-2014_en.pdf
[6] Lloyd, L. (2011) Handbook of Industrial Catalysts. Springer.
[7] Maurer, R., Bartsch, U. (2001) Krupp Uhde Nitric Acid Technology, Enhanced plant design for the production of azeotropic nitric acid. Heraeus Nitric Acid Conference, Johannesburg, 2001. http://www.cheresources.com/azeotrop.pdf.

[8] Hömmerich, U., Rautenbach, R. (1998) Design and optimization of combined pervaporation/ distillation processes for the production of MTBE. J. Membr. Sci. 146:53.

[9] Noskov, A.S. (2006) Industrial catalytic reactors and their peculiarities, Industrial Catalysis in Lectures. Moscow, Russia: Kalvis.

[10] http://www.uop.com/reforming-ccr-platforming/.

[11] Bartolomew, C.H., Farrauto, R.J. (2006) Fundamentals of Industrial Catalytic Processes. Wiley.

[12] Krishna, R., Sie, S.T. (1994) Strategies for multiphase reactor selection. Chem. Eng. Sci. 49:4029.

[13] Dudukovic, M.P. (2009) Frontiers in reactor engineering. Science 325:698.

[14] Müller, H. Sulfuric Acid and Sulfur Trioxide, Ullmann's Encyclopedia of Industrial Chemistry. DOI: 10.1002/14356007.a25_635.

[15] Peters, T.A., van der Tuin, J., Houssin, C., Vorstman, M.A.G., Benes, N.E., Vroon, Z.A.E.P., Holmen, A., Keurentjes, J.T.F. (2005) Preparation of zeolite-coated pervaporation membranes for the integration of reaction and separation. Catal. Today 104:288.

[16] Koch Modular Process Systems, LLC. Pilot Plant Services Group, http://www.pilot-plant.com/ reactions.htm.

[17] www.cdtech.com/techProfilesPDF/CDTECHEB.pdf.

[18] https://inside.mines.edu/~jjechura/Refining/07_Catalytic_Cracking.pdf. Copyright © 2017 John Jechura

[19] http://en.wikipedia.org/wiki/File:FCC_Chemistry.png

[20] Moulijn, J.A., Makkee, M., Van Diepen, A. (2013) Chemical Process Technology, 2nd Edition. Chichester, Wiley.

[21] Zeolites and Catalysis: Synthesis, Reactions and Applications, Cejka, J., Corma, A., Zones, S. (Ed.) Wiley, Weinheim, 2010.

[22] BASF (2011) Multistage Reaction Catalysts (MSRC): A Breakthrough Innovation in Fluid Catalytic Cracking (FCC) Technology. http://www.catalysts.basf.com/refining.

[23] https://refiningcommunity.com/wp-content/uploads/2017/11/Boron-Based-Technology-an-Innovative-Solution-for-Resid-FCC-Unit-Performance-Improvement-Clark-BASF-FCCU-Galveston-2018.pdf

[24] http://www.uop.com/fcc

[25] Vogt, E.T.C., Weckhuysen, B.M. (2015) Fluid catalytic cracking: recent developments on the grand old lady of zeolite catalysis, Chem. Soc. Rev. 44: 7342–7370.

[26] Fernandes, J.L., Verstraete, J.J., Pinheiro, C.I.C., Oliveira, N.M.C., Ribero, F.R. (2007) Dynamic modeling of an industrial R2R FCC unit. Chem. Eng. Sci. 62:1184.

[27] http://www.treccani.it/portale/opencms/handle404?exporturi=/export/sites/default/ Portale/sito/altre_aree/Tecnologia_e_Scienze_applicate/enciclopedia/italiano_vol_2/273-298ITA3.pdf&%5D

[28] Zecevic, J., Vanbutsele, G., de Jong, K.P., Martens, J.A. (2015) Nanoscale intimacy in bifunctional catalysts for selective conversion of hydrocarbons, Nature, 528: 245–248.

[29] Kolesnikov, I.M., Catalysis and production of catalysts, Moscow, Tekhnika, 2004, 400p.

[30] http://www.treccani.it/export/sites/default/Portale/sito/altre_aree/Tecnologia_e_Scienze_ applicate/enciclopedia/inglese/inglese_vol_2/273-298_ING3.pdf

[31] http://www.treccani.it/portale/opencms/handle404?exporturi=/export/sites/default/ Portale/sito/altre_aree/Tecnologia_e_Scienze_applicate/enciclopedia/inglese/inglese_vol_ 2/309-324_ING3.pdf&%5D

[32] Rana, M.S., Samano, V., Ancheyta, J., Diaz, J.A.I. (2007), A review of recent advances on process technologies for upgrading of heavy oils and residua, Fuel 86: 1216–1231.

[33] https://www.eni.com/en_IT/attachments/innovazione-tecnologia/technological-answers/scheda-est-eng.pdf

[34] http://csd.newcastle.edu.au/chapters/Fig1_2.png

[35] Bodrov, I.M., Apelbaum, L.O., Temkin, M.I. (1967) Kinetics of the reaction between methane and water vapor catalyzed by nickel on a porous support. Kinetics and Catalysis 8: 821–829.

[36] Khomenko, A.A., Apel'baum, L.O., Shub, F.S., Snagovsky, Yu.S., Temkin, M.I. (1971) Kinetics of the reaction of methane with water vapor and a reversible reaction of carbon monoxide hydrogenation on nickel. Kinetika i Kataliz 12: 423–431.

[37] Xu., J, Froment, G.F. (1989) Methane steam reforming, methanation and water-gas shift: I. Intrinsic kinetics. AIChE J. 35: 88–96.

[38] Aparicio, L.M. (1997) Transient isotopic studies and microkinetic modeling of methane reforming over nickel catalysts. J. Catal. 165: 262–274.

[39] Avetisov, A.K., Rostrup-Nielsen, J.R., Kuchaev, V.L., Bak Hansen, J.-H., Zyskin, A.G., Shapatina, E.N. (2010) Steady-state kinetics and mechanism of methane reforming with steam and carbon dioxide over Ni catalyst. J. Mol. Catal. A. 315: 155–162.

[40] Beyer, F., Brightling, J., Farnell, P., Foster, C. (2005) Steam reforming – 50 years of development and challenges for the next 50 years. AIChE 50th Annual Safety in Ammonia Plants and Related Facilities Symposium, Toronto.

[41] Dul'nev, A.V., Obysov, A.V. (2012) Experience from the industrial operation of natural gas reforming supported Ni catalysts and ways to improve them. Catal. Ind. 4: 19.

[42] Nielsen, A., Kjaer, J., Hansen, B. (1964) Rate equation and mechanism of ammonia synthesis at industrial conditions. J. Catal., 3: 68–79.

[43] Temkin, M.I. (1979) The kinetics of some industrial heterogeneous catalytic reactions. Adv. Catal. 28: 173–281.

[44] Aparicio, L.M., Dumesic, J.A. (1994) Ammonia synthesis kinetics: Surface chemistry, rate expressions, and kinetic analysis. Top. Catal. 1: 233.

[45] Shepetovsky, D.B., Brodskaya, I.G., Ozolin, P.A. (2007) Loading, reduction and examination of catalysts "CA-KJ" in industrial column of ammonia synthesis with productivity NH3 1360 t/d on "OAO AZOT" (Berezniki). Catal. Ind. N3, 42–47.

[46] Brown, D.E., Edmonds, T., Joyner, R. W., McCarroll, J.J., Tennison, S.R. (2014) The genesis and development of the commercial BP doubly promoted catalyst for ammonia synthesis, Catal. Lett. 144: 545–552.

[47] Luzzi, A., Lovegrove, K., Filippi, E., Fricker, H., Schmitz-Goeb, M., Chandapillai, M., Kaneff, S. (1999) Technoeconomic analysis of a 10 MWe solar thermal power plant using ammonia-based thermochemical energy transfer. Solar Energy 66:91.

[48] Fertilizer Manual. Kluwer, 1998.

[49] http://www.thyssenkrupp-uhde.de/de/kompetenzen/technologien/fertiliser/ammonia kharnstoff/75/94/ammonia-conversion.html

[50] http://www.iffcokandla.in/data/polopoly_fs/1.2465622.1437673182!/fileserver/file/507742/filename/020.pdf

[51] http://www.iffcokandla.in/data/polopoly_fs/1.3253588.1472738687!/fileserver/file/681710/filename/4F.pdf

[52] https://www.prnewswire.com/news-releases/global-ethylene-oxide-market-report-2018—forecast-to-2023-the-growing-demand-for-pet-bottles-from-the-packaging-industry-300755179.html

[53] Makhlin, V.A. (2009) Development and analysis of heterogeneous catalytic processes and reactors, Theor. Found. Chem. Eng., 43: 261–275.

[54] Lebedev, N.N. (1988) Chemistry and Technology of basic organic and petrochemical synthesis, Chimia.

[55] Krupnov, P.V., Deryugin, A.V., Chesnokov, B.B. (2007) Experience and technical revamp of Dzerzhinsk ethylene oxide and glycols plant of OAO Sibur Neftekhim, Catal. Ind., 2:43–45.

[56] https://www.digitalrefining.com/article/1001406,Enhancements_in_Ethylene_Oxide_ Ethylene_Glycol_manufacturing_technology.html#.XUG1RW8zb4c

[57] http://www.metrohm-applikon.com/Downloads/Process_Application_Note_AN-PAN-1007- HPPO-propylene-oxide-production.pdf

[58] http://www.treccani.it/portale/opencms/handle404?exporturi=/export/sites/default/ Portale/sito/altre_aree/Tecnologia_e_Scienze_applicate/enciclopedia/inglese/inglese_vol_ 2/615-686_ING3.pdf

[59] Lorz, P.M., Towae, F.K., Enke, W., Jäckh, R., Bhargava, N., Hillesheim, W. Phthalic Acid and Derivatives. In: Ullmanns Encyclopedia, vol. 27, page 135, DOI: 10.1002/14356007.

[60] Marx, R., Wölk, H.-J., Mestl, G., Turek, T. (2011) Reaction scheme of o-xylene oxidation on vanadia catalyst. Appl. Catal. A.Gen. 398:37.

[61] Cavani, F., Caldarelli, A., Luciani, S., Cortelli, C., Cruzzolin, F. (2012) Selective oxidation of o-xylene to phthalic anhydride: from conventional catalysts and technologies toward innovative approaches. Catalysis (RSC) 24:204–222.

[62] https://catalysts.basf.com/public/files/literature-library/32012BF-9796-7152_BR_PUE_ Catalysts_Ansicht.pdf

[63] US Patent 8153825 (2012) Preparation of phthalic anhydride by gas phase oxidation of o-xylene, BASF.

[64] Calderbank, P.H., Chandrasekharan, K., Fumagalli, C. (1977) The prediction of the performance of packed-bed catalytic reactors in the air-oxidation of o-xylene. Chem. Eng. Sci. 32:1435.

[65] Anastasov, A.I. (2003) Deactivation of an industrial V_2O_5–TiO_2 catalyst for oxidation of o-xylene into phthalic anhydride. Chem. Eng. Proc. 42:449.

[66] http://d-nb.info/1030112800/34

[67] http://www.rollechim.com/switch.htm

[68] Miller, G.J., Collins, M. (2017) Industrial production of formaldehyde using polycrystalline silver catalyst, Ind. Eng. Chem. Res. 56: 9247–9265.

[69] http://www.treccani.it/portale/opencms/handle404?exporturi=/export/sites/default/ Portale/sito/altre_aree/Tecnologia_e_Scienze_applicate/enciclopedia/inglese/inglese_vol_ 2/615-686_ING3.pdf&%5D

[70] https://www.clariant.com/en/Solutions/Products/2019/04/15/07/20/FAMAX-200-Series

[71] http://www.glp.com.au/wp-content/uploads/2014/02/General-SC-Overview-BCT.pdf

[72] https://www.phxequip.com/resource-detail.40/comparing-the-different-formaldehyde-pro duction-processes.aspx

[73] Burrington, J. D., Kartisek, C. T., Grasselli, R. K. (1984) Surface intermediates in selective propylene oxidation and ammoxidation over heterogeneous molybdate and antimonate catalysts, J. Catal. 87: 363–380.

[74] Brazdil, J. F. (2017) A critical perspective on the design and development of metal oxide catalysts for selective propylene ammoxidation and oxidation, Applied Catal. A., 543: 255– 233.

[75] https://www.ineos.com/globalassets/ineos-group/businesses/ineos-technologies/presenta tions/edc-oxychlorination-catalysts.pdf

[76] http://www.honorshinechemc.com/photo/pc3635343-catalyst_for_ethylene_oxychlorina tion_honor_1c_high_copper_content_oc_catalyst_nc_200.jpg

[77]   https://catalysts.basf.com/public/files/literature-library/1120115938_BR_Oxicloration-FBC_
       USLetter.pdf
[78]   Vajglová, Z., Kumar, N., Eränen, K., Peurla, M., Murzin, D. Yu., Salmi, T. (2018) Ethylene
       oxychlorination over $CuCl_2/\gamma$-Al_2O_3 catalyst in micro- and millistructured reactors, J. Catal.,
       364: 334–344.
[79]   Bakshi, Yu.M., Gel'bstein, A.I., Gel´perin, E.I., Dmitrieva, M.P., Zyskin, A.G., Snagovkii, Yu.S.
       (1991) Investigation of mechanism and kinetics of additive oxychlorination of olefins. V.
       Kinetic model of the reaction, Kinet. Catal., 32: 740–748.

Acknowledgments

I would like to express my gratitude to very many colleagues who shared their experience on various aspects of catalysis, in particular to the late Prof. M.I. Temkin and Dr. N.V. Kul'kova, as well as Dr. A.K. Avetisov (all from Karpov Institute of Physical Chemistry), Dr. R. Touroude (University of Strasbourg), Dr. E. Schwab (BASF) and Dr. I.L. Simakova (Boreskov Institute of Catalysis). The everyday interactions with colleagues from the Laboratory of Industrial Chemistry and Reaction Engineering at Åbo Akademi University (Dr. K. Eränen, Dr. N. Kumar, Prof. J.-P. Mikkola, Dr. P. Mäki-Arvela and Prof. J. Wärnå) and collaboration with numerous colleagues and PhD students from Åbo Akademi University and other institutions are gratefully acknowledged. Please, forgive me for not mentioning you all. Very special thanks go to my colleague, friend and brother-in-science Prof. T. Salmi, who has influenced me through the decades in very many ways.

The moral support of my family was very important in writing the first and the second edition of this book. I am especially grateful to E. Murzina for her technical and spiritual contribution and especially endless patience and tolerance.

https://doi.org/10.1515/9783110614435-005

Recommended reading

Anderson, B.G., de Jong, A.M., van Santen, R.A. (2002) In situ measurement of heterogeneous catalytic reactor phenomena using positron emission. In: Haw, J.F., editor. In-situ Spectroscopy in Heterogeneous Catalysis. Weinheim, Germany: Wiley-VCH, pp. 195–237.

Atherton, J., Houson, I., Talford, M. (2011) Understanding the reaction. In: Houson, I., editor. Process Understanding for Scale-Up and Manufacture of Active Ingredients. Chapter 4, Weinheim, Germany: Wiley-VCH, pp. 87–125.

Aris, R. (1975) The Mathematical Theory of Diffusion and Reaction in Permeable Catalysts. Vol. 1. Oxford: Clarendon Press.

Augustine, R.L. (1996) Heterogeneous Catalysis for the Synthetic Chemist. New York: Marcel Dekker, Inc.

Bartolomew, C.H., Farrauto, R.J. (2006) Fundamentals of Industrial Catalytic Processes. Wiley. Hoboken, NJ.

Beller, M., Renken, A., Rutger A. van Santen (2012) Catalysis. From Principles to Applications. Weinheim, Germany: Wiley-VCH.

Berger, R.J., Stitt, E.H., Marin, G.B., Kapteijn, F., Moulijn, J.A. (2000) Chemical reaction kinetics in practice. CATTECH 4: 2.

Boudart, M., Djega-Mariadassou, G. (1984) Kinetics of Heterogeneous Catalytic Reactions. Princeton, NJ: Princeton University Press.

Boudart, M. (1995) Turnover rates in heterogeneous catalysis. Chem. Rev. 95: 661.

Butt, J.B. (2000) Reaction Kinetics and Reactor Design. New York: Marcel Dekker.

Campanati, M., Fornasari, G., Vaccari, A. (2003) Fundamentals in the preparation of heterogeneous catalysts. Catal. Today 77: 299.

Chorkendorff, I., Niemanstverdriet, J.W. (2003) Concepts of Modern Catalysis and Kinetics. Weinheim, Germany: Wiley-VCH.

Clark, A. (1970) The Theory of Adsorption and Catalysis. New York: Academic Press.

De Jong, K.P. (2009), Synthesis of Solid Catalysts, Weinheim, Germany: Wiley-VCH

Dittmeyer, R., Emig, G. (2008) Simultaneous heat and mass transfer and chemical reaction. In: Ertl, G., Knözinger, H., Schüth, F., Weitkamp, J., editors. Handbook of Heterogeneous Catalysis. Weinheim, Germany: Wiley-VCH, pp. 1727–1784.

Doraiswamy, L.K., Sharma, M.M. (1984) Heterogeneous reactions: Analysis, examples and reactor design. Vol. 1, Gas-Solid and Solid-Solid Reactions, Vol. 2, Fluid-Fluid and Fluid-Fluid-Solid Reactions. New York: John Wiley.

Dumesic, J.A., Rudd, D.F., Aparicio, L. (1993) The Microkinetics of Heterogeneous Catalysis. American Chemical Society, Washington, DC.

Dutta, S., Gualy, R. (2000) Build robust reactor models. Chem. Eng. Prog. 37–51.

Eigenberger, G. (1992) Fixed Bed Reactors, Ullmann's Encyclopedia of Industrial Chemistry. Vol. B4, 199.

Euzen, J.P., Trambouze, P., Wauquier, J.P. (1993) Scale-up Methodology for Chemical Processes. Paris: Technip.

Fogler, H.S. (1998) Elements of Chemical Reaction Engineering (3rd edn). Englewood Cliffs, NJ: Prentice Hall.

Geus, J.W., van Dillen, A.J. (2008) Preparation of supported catalysts by deposition-precipitation. In: Knözinger, H., Schueth, F., Weitkamp, J., editors. Handbook of Heterogeneous Catalysis, 2.4.1. Weinheim, Germany: Wiley-VCH, pp. 428–467.

Gladden, L.F., Mantle, M.D., Sederman, A.J. (2006) Magnetic resonance imaging of catalysis and catalytic processes. Adv. Catal. 50: 1.

https://doi.org/10.1515/9783110614435-006

Frank-Kamenetskii, D.A. (1955) Diffusion and Heat-transfer in Chemical Kinetics. Princeton, NJ: Princeton University Press.

Froment, G., Bischoff, K. (1990) Chemical Reactor Analysis and Design (2nd edn). New York: Wiley.

Haber, J., Block, J.H., Delmon, B. (1995) Methods and procedures for catalyst characterization. Pure Appl. Chem. 67: 1257–1306.

Hagen, J. (2006) Industrial Catalysis: A Practical Guide. Weinheim, Germany: Wiley-VCH.

Hammer, B., Norskov, J.K. (2000) Theoretical surface science and catalysis – calculations and concepts. Adv. Catal. 45: 71.

Ertl, G., Knözinger, H., Weitkamp, J., editors. (1997) Handbook of Heterogeneous Catalysis. 5 Volume Set. Weinheim, Germany: Wiley-VCH.

Hanefeld, U., Lefferts, L., editors (2018). Catalysis. An Integrated Textbook for Students. Weinheim, Germany: Wiley-VCH.

Hessel, V., Löwe, H., Mueller, A., Kolb, G. (2005) Chemical Micro Process Engineering. Weinheim, Germany: Wiley-VCH.

Horiuti, J., Nakamura, T. (1967) Theory of heterogeneous catalysis. Adv. Catal. 17: 1.

Horiuti, J. (1973) Theory of reaction rates as based on the stoichiometric number concept. Ann. NY Acad. Sci. 213: 5.

Ioffe, I.I., Reshetov, V.A., Dobrotvorskii, A.M. (1985) Heterogeneous Catalysis. Leningrad: Khimia.

Kiperman, S.L. (1979) Foundations of Chemical Kinetics in Heterogeneous Catalysis. Moscow: Khimia.

Kolasinski, K.W. (2002) Surface Science. Foundations of Catalysis and Nanoscience. Chichester: Wiley.

Kolesnikov, I.M. (2004) Catalysis and Production of Catalysts. Moscow: Teknika.

Laidler, K.J. (1987) Chemical Kinetics. New York: Harper and Row.

Levenspiel, O. (1998) Chemical Reaction Engineering (3rd edn). Weinheim, Germany: Wiley.

Lloyd, L. (2011) Handbook of Industrial Catalysts. Berlin: Springer.

Logan, S.R. (1996) Fundamentals of Chemical Kinetics. Edinburgh Gate: Longman.

Marceau, E. (2006) Characterizing a supported catalyst: A methodological approach. In: Murzin, Yu. D., editor. Nanocatalysis. ISBN: 81-308-0056-X. Kerala, India: Research Signpost. pp. 127–175.

Marceau, E., Carrier, X., Che, M., Clause, O., Marcilly, C. (2008) Ion exchange and impregnation. In: Knözinger, H., Schueth, F., Weitkamp, J., editors. Handbook of Heterogeneous Catalysis, 2.4.2.. Weinheim, Germany: Wiley-VCH. DOI: 10.1002/9783527610044.hetcat0022, pp. 467–484.

Marceau, E., Carrier, X., Che, M. (2009) Impregnation and drying. In: de Jong, K.P., editor. Synthesis of Solid Catalysts. Weinheim, Germany: Wiley-VCH, pp. 59–82.

Masel, R.I. (2001) Chemical Kinetics and Catalysis. New York: Wiley-Interscience.

Missen, R.W., Mims, C.A., Saville, B.A. (1999) Chemical Reaction Engineering and Kinetics. New York: John Wiley & Sons.

Deustchmann, O., editor. (2012) Modelling and Simulations of Heterogeneous Catalytic Reactions: From the Molecular Process to the Technical System. Weinheim, Germany: Wiley-VCH.

Moulijn, J.A., Makkee, M., van Diepen, A.E. (2013) Chemical Process Technology. Second Edition. Chichester: John Wiley and Sons.

Murzin, D., Salmi, T. (2016) Catalytic Kinetics. Science and Engineering. Amsterdam: Elsevier.

Murzin, D. Yu. (2015) Catalytic Reaction Technology. Berlin: Walter de Gruyter.

Ostrovskii, N.M. (2001) Kinetics of Catalyst Deactivation. Moscow: Nauka.

Petrov, L.A. (1992) Application of graph theory to study of the kinetics of heterogeneous catalytic reactions. Math. Chem. 2: 1.

Petrov, L.A., Alhamed, Y., Zahrani, A.Al., Daous, M. (2011) Role of chemical kinetics in the heterogeneous catalytic studies. Chinese J. Catl. 32: 1085.

Pilling, M.J., Seakins, P.W. (1995) Reaction Kinetics. Oxford: Oxford University Press.

Poling, B.E., Prausnitz, J.M., O'Connell, J.P. (2001) The Properties of Gases and Liquids. McGraw-Hill. New York.

Post, T. (2010) Understanding the real world of mixing. Chem. Eng. Prog. 25.

Ross J.R.H. (2019), Contemporary Catalysis. Fundamentals and Current Applications. Amsterdam: Elsevier.

Rothenberg, G. (2008). Catalysis. Concepts and Green Applications. Weinheim, Germany: Wiley-VCH.

Salmi, T.O., Mikkola, J.-P., Wärnå, J.P. (2010) Chemical Reaction Engineering and Reactor Technology. CRC Press, Boca Raton, Fl.

Satterfield, C.N. (1970) Mass Transfer in Heterogeneous Catalysis. Cambridge, MA: MIT press.

Schmidt, F. (2001) New catalyst preparation technologies observed from an industrial viewpoint. App. Catal. A. Gen. 221: 15.

Schmidt, L.D. (1998) The Engineering of Chemical Reactions. Oxford: Oxford University Press.

Smith, J.M. (1981) Chemical Engineering Kinetics. McGraw-Hill. New York.

Schüth, F., Hesse, M. (2008) Catalyst forming. In: Ertl, G., Knözinger, H., Schüth, F., Weitkamp, J., editors. Handbook of Heterogeneous Catalysis. Weinheim, Germany: Wiley-VCH, pp. 676–699.

Snagovskii, Yu.S., Ostrovskii, G.M. (1976) Modelling of Kinetics of Heterogeneous Catalytic Reactions. Moscow: Khimia.

Somorjai, G.A. (1994) Introduction to Surface Chemistry and Catalysis. New York: Wiley-Interscience.

Stolze, P. (2000) Microkinetic simulation of catalytic reactions. Prog. Surf. Sci. 65: 65.

Stitt, E.H., Simmons, M.J.H. (2011) Scale-up of chemical reactions. In: Houson, I., editor. Process Understanding for Scale-up and Manufacture of Active Ingredients. Weinheim, Germany: Wiley-VCH, pp. 155–198.

Thomas, J.M., Thomas, W.J. (1997) Principles and Practice of Heterogeneous Catalysis. Weinheim, Germany: Wiley-VCH.

Temkin, M.I. (1979) The kinetics of some industrial heterogeneous catalytic reactions. Adv. Catal. 28: 173.

Twygg, M. (1997) Handbook of Industrial Catalysis. Manson Publishing, London.

van Santen, R.A., van Leeuwen, P.W.N.M., Moulijn, A.A., Averill, B.A. (2002) Catalysis: An Integrated Approach. Elsevier, Amsterdam.

van Santen, R.A., Neurock, M. (2017) Modern Heterogeneous Catalysis. An Introduction. Weinheim, Germany: Wiley-VCH.

van Santen, R.A. (2006) Molecular Heterogeneous Catalysis. Weinheim, Germany: Wiley-VCH.

Önskan, Z.P, Avci, A.K. (2016) Multiphase Catalytic Reactors, Theory, Design, Manufacturing and Application. Weinheim, Germany: Wiley-VCH.

Index